Property of the National Institutes of Health

POLYPEPTIDES AND PROTEIN STRUCTURE

POLYPEPTIDES AND PROTEIN STRUCTURE

Alan G. Walton
Case Western Reserve University
Cleveland, Ohio

ELSEVIER • NEW YORK
New York • Oxford

Elsevier North Holland, Inc.
52 Vanderbilt Avenue, New York, New York 10017

Sole distributors outside the USA and Canada:

Elsevier Science Publishers B. V.
P. O. Box 211, 1000 AE Amsterdam, The Netherlands

© 1981 by Elsevier North Holland, Inc.

Library of Congress Cataloging in Publication Data

Walton, Alan G.
   Polypeptides and protein structure.

   Bibliography: p.
   Includes index.
   1. Proteins. I. Title.
QD431.W34   574.19'245   80-19982
ISBN 0-444-00407-6

Desk Editor   Robert Glasgow
Design   Edmée Froment
Design Editor   Glen Burris
Cover Design by Paul Agule Design
Production Manager   Joanne Jay
Compositor   General Graphics Services
Printer   Halliday Lithograph

Manufactured in the United States of America

Contents

Preface ix

SECTION I UNITS AND RULES

Chapter 1 Primary Structure 3
1.1 Configuration and Nomenclature of Amino Acids 4
1.2 Polymerization of Amino Acids 8
1.3 Determination of Sequence 9
1.4 Peptides—Role of Sequence 13
1.5 Role of Homology 15

Chapter 2 Secondary Structure 17
2.1 Bond Rotation 17
2.2 Rotational Constraints 18
2.3 Helices 30
2.4 Relation of Helical Parameters 32
2.5 Known Helical Structures 33
2.6 Summary of Helical Structures 42
2.7 The β-Bend 43
2.8 Random Structures 45
2.9 π_{LD} or β-Helices 46

Chapter 3 Prediction of Conformation 50
3.1 Conformation-Directing Properties of Amino Acids 50
3.2 Application of Amino Acid Conformational Properties 52
3.3 The Chou–Fasman Approach 54
3.4 Other Predictive Methods and Comparison of Data 59
3.5 Basis for Qualitative Conformation Prediction 59

3.6	Energetics of Chain Folding	61
3.7	Prediction of Tertiary Structure	61

SECTION II EXPERIMENTAL METHODS

Chapter 4 X-Ray Diffraction — 71

4.1	Bragg's Law	71
4.2	Two-Dimensional Unit Cells and Miller Indices	72
4.3	Three Dimensions and Crystal Systems	73
4.4	Diffraction by Powders	75
4.5	Obtaining Unit-Cell Parameters from Powder-Diffraction Patterns	77
4.6	Crystalline Perfection	80
4.7	Preferentially Oriented Microcrystalline Specimens	81
4.8	Intensity of Diffraction	82
4.9	Diffraction by a Helical Chain	84
4.10	Calculation of Diffraction from a Discontinuous Helix	87
4.11	Additional Complexities of Helical Arrays	89
4.12	Single Crystals of Globular Proteins	94
4.13	Classification of Protein Domain Structure Based on X-Ray Diffraction Analysis	96
4.14	Active Sites	103

Chapter 5 Electron Diffraction and Microscopy — 105

5.1	Electron Diffraction	105
5.2	Application of Electron Microscopy to Fibrous Proteins	113
5.3	Electron Imaging	116

Chapter 6 Infrared Spectroscopy — 119

6.1	Absorption of Radiation in a Vibrating System	119
6.2	Amide Bands	121
6.3	Polarization of Radiation and Dichroism	123
6.4	Vibrational Analysis	124
6.5	Dichroism and Conformation	130
6.6	Amide A and B Bands	132
6.7	Side-Chain Vibrations	134
6.8	Infrared Spectroscopy of Fibrous Proteins	134
6.9	Spectroscopy of Globular Proteins	136

Chapter 7 Raman Spectroscopy — 140

7.1	Polarization of Raman Lines	142
7.2	Vibrational Analysis	142
7.3	Amide Lines and Conformation	143
7.4	Deuteration	144
7.5	Conformational Analysis of Proteins	145
7.6	Side-Chain Vibrations	149
7.7	Comparison of Proteins in Solid and Solution States	150
7.8	Resonance Raman Spectroscopy and Structure/Mechanism Studies	151

Contents

Chapter 8 Electronic Spectroscopy — 154
    8.1 Types of Spectroscopy — 154
    8.2 Polarized Light — 154
    8.3 Absorption Phenomena — 156
    8.4 Ultraviolet Absorption and Conformation — 168
    8.5 Linear Dichroism — 170
    8.6 Topographical Mapping by Ultraviolet Spectroscopy — 171

Chapter 9 Optical Rotatory Dispersion (ORD) and Circular Dichroism (CD) Spectroscopy — 172
    9.1 Optical Rotation — 172
    9.2 Dispersive Effects and ORD — 174
    9.3 Circular Dichroism Spectroscopy — 179
    9.4 Utilization of CD Spectra for Determining Structure — 180
    9.5 Circular Dichroism and the Structure of Proteins — 183
    9.6 Circular Dichroism of the Polyproline Helix — 185
    9.7 Effects of Side Chains — 186

Chapter 10 Electronic Emission: Fluorescence Spectroscopy — 189
    10.1 Internal Conversion Processes and Intersystem Crossing — 189
    10.2 Lifetime of States and Excitation Coefficients — 191
    10.3 Origins of Intrinsic Fluorescence in Proteins and Polypeptides — 193
    10.4 Uses of Fluorescence Spectroscopy — 195
    10.5 Fluorescence Depolarization — 198
    10.6 Fluorescent Lifetime — 201
    10.7 Energy-Transfer Processes — 202
    10.8 Complexation Processes — 205
    10.9 Extrinsic Fluorescence — 211

Chapter 11 Nuclear Magnetic Resonance — 213
    11.1 The Resonance Phenomenon — 213
    11.2 The NMR Experiment — 216
    11.3 Relaxation Processes — 217
    11.4 Chemical Shift — 219
    11.5 Spin-Spin splitting — 220
    11.6 Free-Induction Decay Spectra — 221
    11.7 Multiple Resonance Effects — 223
    11.8 Paramagnetic Effects — 224
    11.9 Structural Information for Proteins and Peptides — 225
    11.10 Summary — 233

SECTION III THERMODYNAMIC AND HYDRODYNAMIC PROPERTIES OF PROTEINS

Chapter 12 Proteins in Solution — 237
Alex M. Jamieson
    12.1 Osmotic Pressure — 238
    12.2 Light Scattering — 243

12.3	Intrinsic Viscosity, Sedimentation Coefficient, and Diffusion Coefficients	247
12.4	Sedimentation Velocity	250
12.5	Dynamic Light Scattering	254
12.6	Measurement of Partial Specific Volume	257
12.7	Gel-Permeation Chromatography and Gel Electrophoresis	259
12.8	Sedimentation Equilibrium	265
12.9	Small-Angle X-Ray Scattering	268
12.10	Comparison of Techniques for Molecular-Weight Determination	271
12.11	Structural Characterization: Radius of Gyration and Related Information	272
12.12	Structural Characterization: Hydrodynamic Volume and Related Information	282
12.13	Flow Dichroism	286
12.14	Transient Electric Birefringence	288
12.15	Dielectric Relaxation	288
12.16	Concluding Remarks on Structural Characterization	295

SECTION IV PROTEIN SYSTEMS

Chapter 13 Collagen and Connective Tissue — 299
    13.1 Tropocollagen — 299
    13.2 Fibrillar Structure — 309

Chapter 14 Molecular Structure and Function of Muscle — 316
    14.1 Muscle Structure — 316
    14.2 Thick Filaments — 317
    14.3 Thin Filaments — 324
    14.4 Muscle Contraction — 337

Chapter 15 Proteins of the Blood — 341
    15.1 Hemoglobin — 341
    15.2 Immunoglobulins — 347
    15.3 Serum Albumin — 354
    15.4 Blood-Clotting Proteins — 355
    15.5 Summary — 364

Appendix A Fourier-Transform Analysis — 365

Appendix B Symmetry and Interpretation of Structure — 369

Index — 379

Preface

This book addresses the principles of protein structure determination and aspects relating structure to function. The theme underlying the approach is that the polypeptide chain contains information that codes for chain shape, chain assembly, molecular assembly, and molecular function. To some extent, the simpler aspects of the code may be unraveled by comparison with synthetic polypeptides. The more complex aspects of the code may only be understood by observing the biologic structure using physical methodology, and by comparing this structure with other biologic structures that have similar but somewhat different structure and code. Thus the book attempts to approach structure elucidation on a systematic basis, relating various levels of structure to the code (amino acid sequence).

This book is directed to the student (undergraduate/graduate) who has some background in physical methodology but is facing a first exposure to protein structure determination. In most chapters mathematical development has been kept to the minimum necessary to utilize the technique or concept in question. A background in calculus (in rare cases, vector analysis) and in physical chemistry would be useful, although many of the chapters may be read and, hopefully, understood without extensive background in either. The book is therefore meant for chemists, biochemists, biophysicists, and macromolecular science students, for all of whom a course in protein structure, using an "ab initio" approach, is appropriate.

The text is divided into four sections relating to fundamental parameters, physical methods, hydrodynamics, and protein systems. Obviously a book of this limited scope excludes many techniques and methods; rather, relatively tried and true methods are emphasized. On the other hand, some very new methods are included because of their apparent significance. Some of these new methods may very well not prove particularly effective in the long run, since this area is changing very quickly. No apology is made here for such an approach, since the author deems it important to acquaint the student with the current status of the field.

Rather than tackle the function of isolated proteins, the fourth section of the book relates to protein systems. Rarely do proteins function independently of other macromolecules in their environment, and consequently a perspective of systems is sought in which proteins function in tandem with other entities. The systems chosen—tissue, muscle, and blood—are of increasing complexity and necessarily much remains unknown about these systems. Consequently, the student is exposed to the problems, as well as some answers, in the protein structure field.

Finally, several of the figures in the text are the results of original and previously unpublished data. Much helpful comment on assembling the material has been provided by my colleagues; Chapter 11 is based on written material provided by Dr. Edgar Jaynes, and Chapter 12 (Section Three) is entirely the work of my colleague, Professor Alex Jamieson. Extensive help with, and comments on, the text have been received from Doctors Freifelder, Geil, and Blackwell, as well as from many students who have made use of the material in class. The author is particularly grateful to Dr. Jane Richardson for providing unpublished material on protein classification and to the many authors who have given permission for reproduction of figures, chief among these being Dr. Bruce Fraser. Finally, the production could not have been completed without the artistic help of Mark Soderquist, Don Solomon, and Randy Sparer, or the help of Peggy Buccieri and Helen Bircher.

Alan G. Walton

POLYPEPTIDES AND PROTEIN STRUCTURE

I

UNITS AND RULES

1

Primary Structure

Nature has chosen, by the process of chemical evolution, to build protein molecules, with their myriad structures and functions, from 21 major (and a few minor) amino acids. These amino acids are bound together in linear chains. Because the chain usually folds, amino acids that are not adjacent in the chain interact with one another; thus, all of the structural and functional aspects of proteins result from the properties of the amino acids and their interactions. When the chain contains relatively few amino acids (2 to 50) it is called a peptide; the terms polypeptide and protein are usually reserved for a molecule containing between fifty and several thousand residues in the chain. In nature each protein molecule of a particular type has the same number and same arrangement of amino acids in the chain sequence and consequently has the same molecular weight.

There are many potential combinations of amino acid residues in a peptide chain of a given size. For example, a peptide having 25 residues joined together could have $(21)^{25}$ different sequences if there are 21 possible amino acids. Since there are probably no more than a few hundred biologically active peptides existing in nature containing 25 residues, clearly only a few of the possible combinations have been selected by the evolutionary process.

The general formula for all but one of the amino acids commonly found in proteins is shown in Figure 1.1. Except for proline, which will be discussed later, each is an α-amino acid, meaning that both the carboxyl and the amino group are attached to the same α-carbon atom. The carbon atoms in the side chain are designated β, γ, δ, and so on, as shown in Figure 1.2. If the side chain R group is not simply hydrogen, there are four different substituents on the α-carbon and thus the amino acid is optically active.

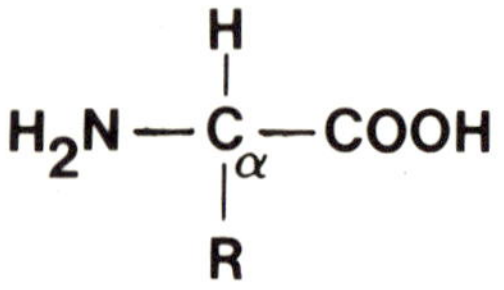

H₂N—C$_\alpha$—COOH
         |
         H
         |
         R

Figure 1.1 General formula for all α-amino acids except proline.

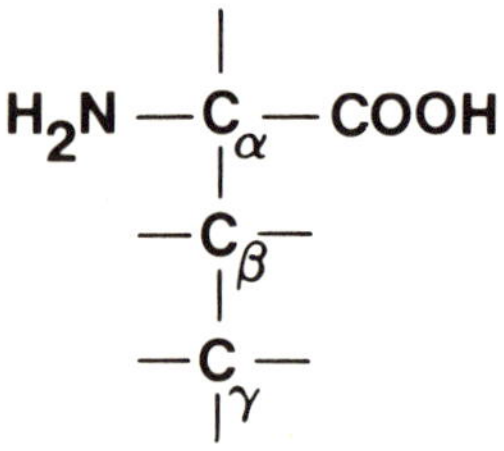

H₂N—C$_\alpha$—COOH
         |
       —C$_\beta$—
         |
       —C$_\gamma$—

Figure 1.2 Side-chain nomenclature.

1.1 Configuration and Nomenclature of Amino Acids

In mammalian proteins the amino group is attached to the α-carbon atom and the optical configuration of that carbon is L-type. One says that the biologic amino acids are all L-amino acids (except glycine, which is not optically active). However, in discussing amino acids it is not necessary to state that it is of L-type because by international convention an amino acid is presumed to be L unless otherwise specified. The commonly occurring amino acids and their three-letter and one-letter abbreviations are shown in Tables 1.1 and 1.2; it is usual to use the three-letter form. (See Figure 1.3.)

The amino acid proline (and its derivatives) differs from all other common amino acids since, owing to the pyrrolidine ring structure, it is a secondary amine. Nature uses this unique structure to place special constraints on the manner in which a polypeptide chain can fold.

Figure 1.3 Isomeric configuration of amino acids. **(a)** L isomer; **(b)** D isomer.

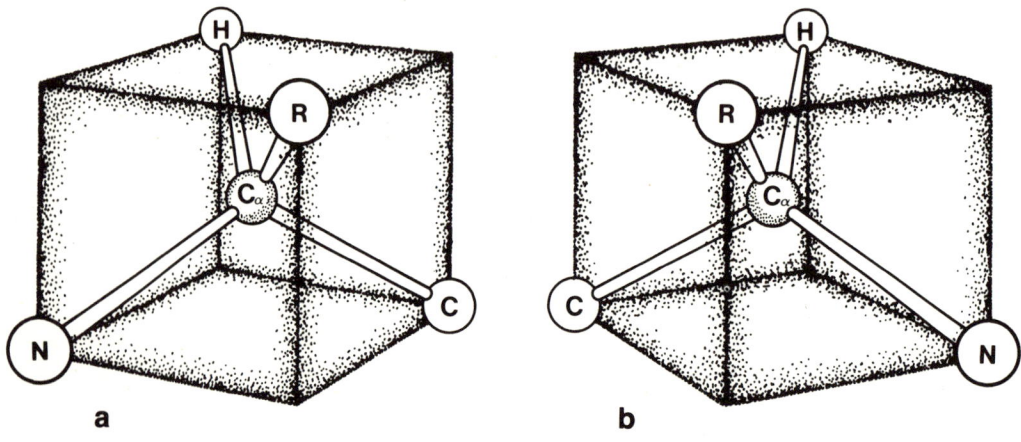

Table 1.1 The Commonly Occurring Amino Acids

Zero Side-Chain Functionality

$$H_2N-\underset{\underset{H}{|}}{\overset{\overset{H}{|}}{C}}{}^{\alpha}-COOH \qquad H_2N-\underset{\underset{CH_3}{|}}{\overset{\overset{H}{|}}{C}}-COOH$$

　　　Glycine　　　　　　　Alanine
　　　Gly　　　　　　　　　Ala

These amino acids have nonpolar hydrocarbon side chains that undergo weak physical interaction with similar residues. The group in the side chain is sufficiently hydrophobic that these amino acids are often found away from a hydrating environment; that is, they are buried deep within the folded protein.

$$H_2N-\overset{\overset{H}{|}}{\underset{\underset{H_3CCH_3}{\diagup\diagdown}}{\underset{CH}{|}}}C-COOH \qquad H_2N-\overset{\overset{H}{|}}{\underset{\underset{\underset{H_3CCH_3}{\diagup\diagdown}}{\underset{CH}{|}}}{\underset{CH_2}{|}}}C-COOH$$

　　　Valine　　　　　　　Leucine
　　　Val　　　　　　　　Leu

$$H_2N-\underset{\underset{\underset{\underset{\delta\ CH_3}{|}}{\gamma\ CH_2}}{\beta\ CH-CH_3}}{\overset{\overset{H}{|}}{C}}{}^{\alpha}-COOH$$

　　　Isoleucine
　　　Ile

Monofunctional Side Chains

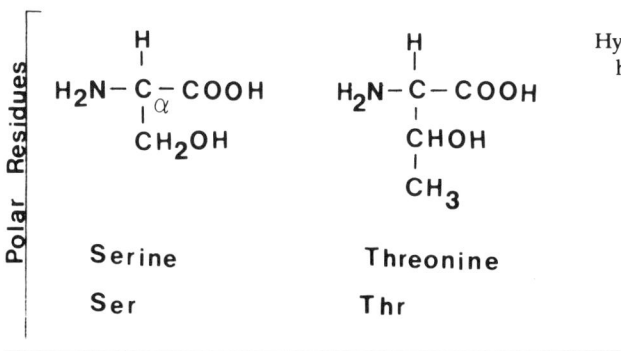

Hydroxylic side chains capable of hydrogen bonding.

　　　Serine　　　　　　　Threonine
　　　Ser　　　　　　　　Thr

(continued)

Table 1.1 The Commonly Occurring Amino Acids (*continued*)

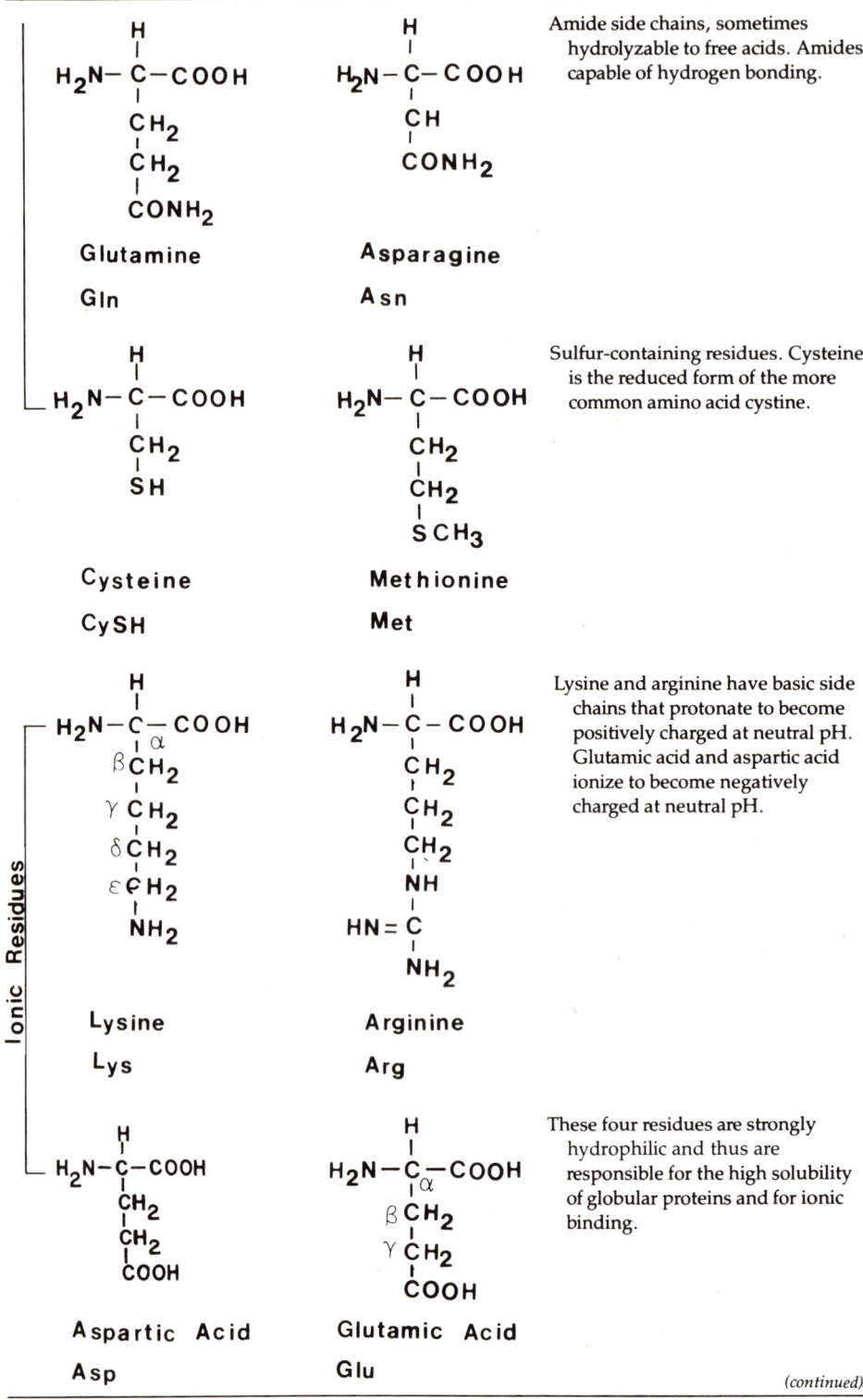

(*continued*)

Table 1.1 *(continued)*

Imino Acid

Proline

Pro

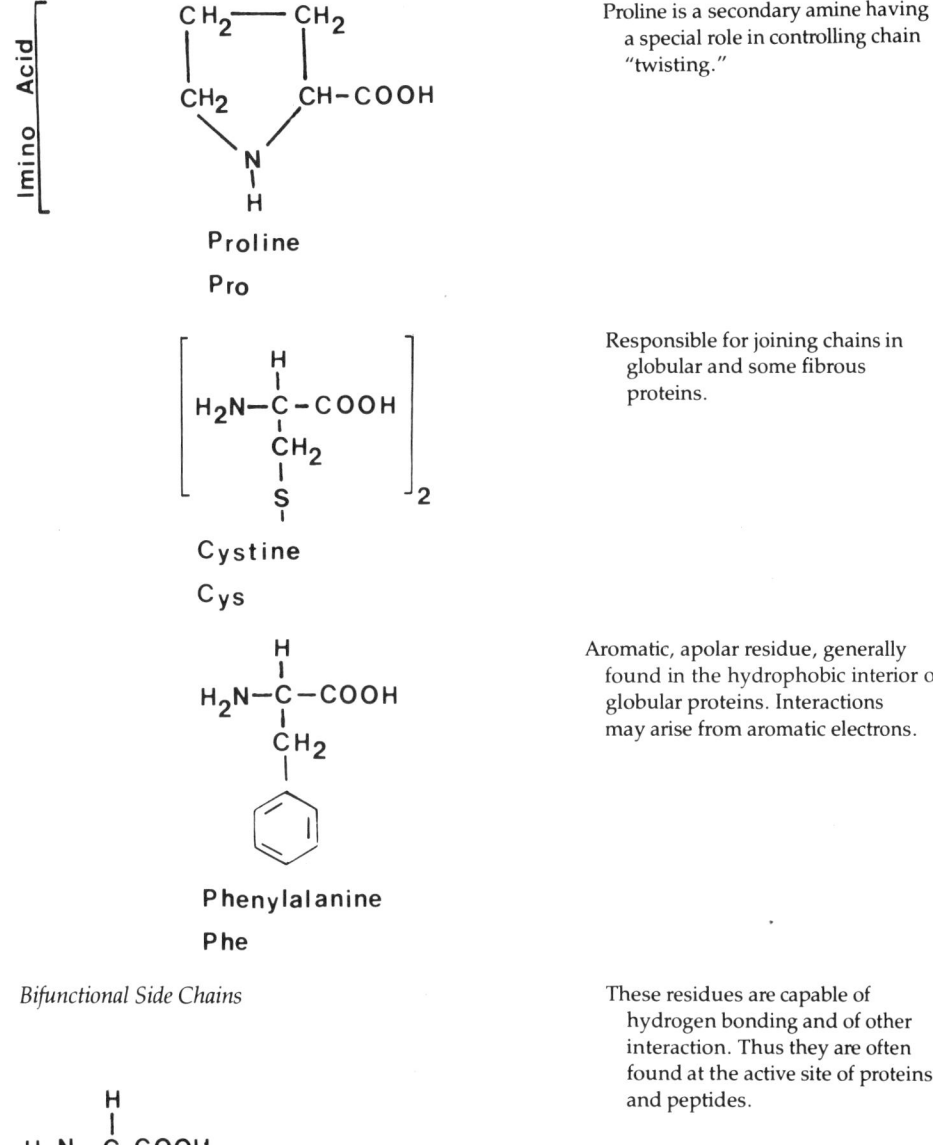

Proline is a secondary amine having a special role in controlling chain "twisting."

Cystine

Cys

Responsible for joining chains in globular and some fibrous proteins.

Phenylalanine

Phe

Aromatic, apolar residue, generally found in the hydrophobic interior of globular proteins. Interactions may arise from aromatic electrons.

Bifunctional Side Chains

These residues are capable of hydrogen bonding and of other interaction. Thus they are often found at the active site of proteins and peptides.

Tryptophan

Trp

Tyrosine

Tyr

Histidine

His

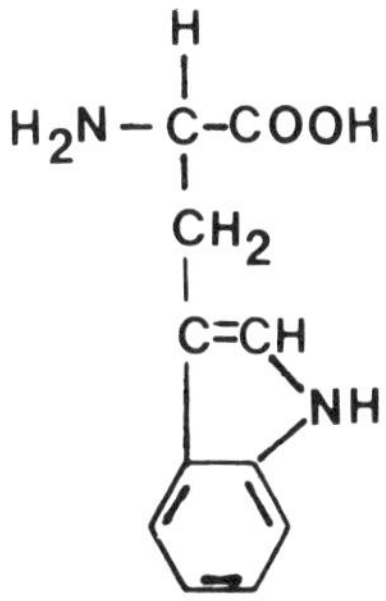

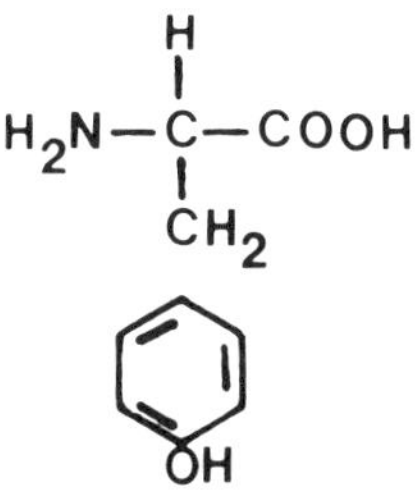

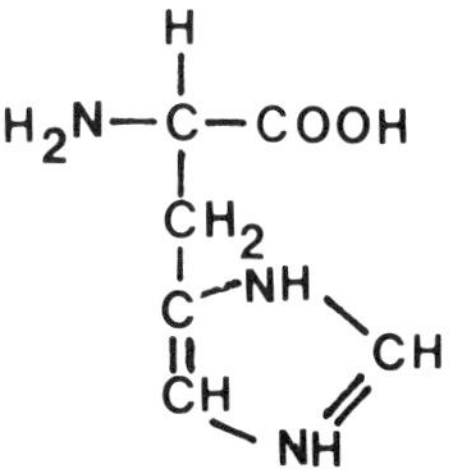

Table 1.2 Abbreviations for the Common Amino Acids

Amino acid	3-Letter symbol	1-Letter symbol	Amino acid	3-Letter symbol	1-Letter symbol
Alanine	Ala	A	Lysine	Lys	K
Arginine	Arg	R	Methionine	Met	M
Asparagine	Asn	N ⎫B*	Phenylalanine	Phe	F
Aspartic Acid	Asp	D ⎭	Proline	Pro	P
Cysteine	Cys	C	Serine	Ser	S
Glutamine	Gln	Q ⎫Z†	Threonine	Thr	T
Glutamic Acid	Glu	E ⎭	Tryptophan	Trp	W
Glycine	Gly	G	Tyrosine	Tyr	Y
Histidine	His	H	Valine	Val	V
Isoleucine	Ile	I	Unknown or Other	—	X
Leucine	Leu	L			

*B is used when the amino acid in question could be Asn or Asp.
†Z is used when the amino acid in question could be Gln or Glu.

The amino acids are commonly divided into subgroups based upon the properties of the side chain R group, since to a large extent this dictates the properties (e.g., binding, chain shape) of the region of the protein occupied by the residue. Table 1.1 also lists some of the important properties of the side chains.

1.2 Polymerization of Amino Acids

Amino acids are joined by the reaction shown in Figure 1.4. The CO — NH bond formed is the peptide bond. Except for proline, the amino acids are joined only by peptide bonds, producing a backbone that is a repeat sequence of

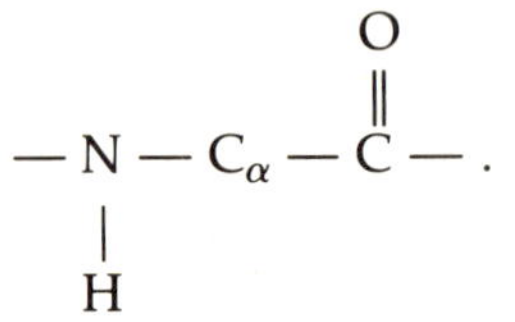

$$-\text{N}-\text{C}_\alpha-\overset{\overset{\text{O}}{\|}}{\text{C}}-\,.$$
$$\,\,\,\,|$$
$$\,\,\,\,\text{H}$$

The bond formed is always (except when proline is involved) a CO — NH peptide bond. Similarly, the process of hydrolyzing proteins proceeds only by cleavage of the peptide bond. The formation of a peptide bond imposes particular constraints on the ability of the polypeptide chain to fold into various shapes because the bond is planar; this will be considered in the next chapter.

Figure 1.4 Condensation of amino acids to give linear-chain polypeptide.

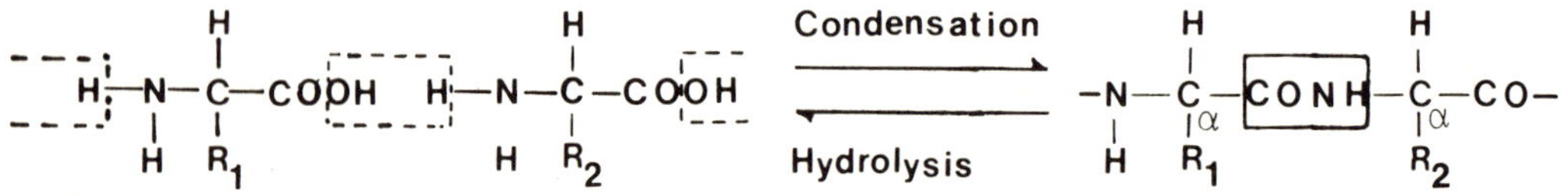

A particular protein has a unique sequence of amino acids that is known as the *primary structure* of the polypeptide chain.

It is possible in the laboratory to synthesize polypeptides containing only one amino acid. If this were alanine, for example, the primary structure would be — ala — ala — ala, which is more commonly abbreviated (ala)$_n$ or described precisely as poly-α-L-alanine. Such a molecule is called a polyamino acid. Most simple polyamino acids have been synthesized and studied, mainly as a means of determining the rules that govern the shape of a polypeptide. The shape is called the conformation or the *secondary structure*.

Polypeptides containing two or more amino acids having a random sequence can be synthesized. The notation copoly (A^{20}:B^{80}) means that residues A and B are in the ratio 20% to 80% or 1:4. Because of the nature of the polymerization process there exists a variety of distributions along the chain. A particular sequence might be, for example,

–B–B–A–A–B–B–B–B–A–B–B–A–B–B–B–B–B–B–B–B . . .

Although there are methods available for calculating the distribution of amino acids from reaction kinetics and/or nuclear magnetic resonance spectra, the required experimental data are usually difficult to obtain. It should be noted that, unlike naturally occurring proteins and polypeptides, each molecule of a random copolypeptide may have a different sequence and molecular weight.

It is also possible to synthesize a polypeptide having a repeating sequence, for example,

–A–B–B–B–B–A–B–B–B–B–A–B–B–B–B . . .

which would be denoted poly (AB$_4$) or (AB$_4$)$_n$. The overall composition of the two polypeptides, random and sequential, is the same; however, the primary sequence differs. The importance of the primary sequence is that the properties of these polypeptides are likely to be quite different.

These differences are greater when the polypeptide contains three or more amino acids. Two immediately obvious effects of primary sequence are those relating to the solubility of the polypeptide and its shape (conformation).

Some naturally occurring proteins (e.g., globular proteins) have no detectable repeating sequences. On the other hand, fibrous proteins have a repeat sequence in which either the same residue repeats in a regular array or residues (amino acids) of similar properties repeat in a periodic manner. Examples of such proteins and sequences (shown in parenthesis) are silks (AB), collagen (ABC), and tropomyosin (AB$_2$CD$_3$).

1.3 Determination of Sequence

The primary structure or sequence of a polypeptide chain is determined by several specific procedures using a standard battery of enzymes. The following outline is provided only as a general strategy, although a number of problems may be encountered that require special enzymes or methodology.

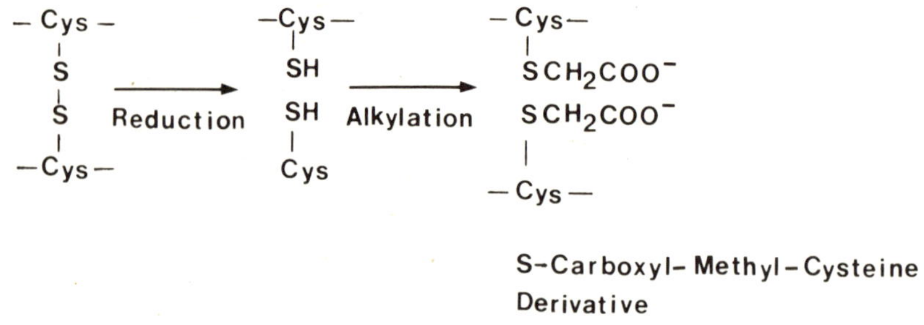

```
- Cys -              -Cys-           - Cys-
    |                   |               |
    S                   SH              SCH₂COO⁻
    |       Reduction                   Alkylation
    S                   SH              SCH₂COO⁻
    |                   |               |
  -Cys-                 Cys            - Cys -
```

S-Carboxyl-Methyl-Cysteine
Derivative

Figure 1.5 Reaction scheme for breaking of disulfide bonds.

Step 1. Breaking of all disulfide bonds

Disulfide bonds may be broken by either reduction or oxidation. A common procedure is to reduce the disulfide bond by reaction with β-mercaptoethanol, and then to alkylate the product with iodoacetic acid in order to prevent — S — S — bond reformation. Thus the product is not cysteine but S-carboxymethylcysteine. (See Figure 1.5.)

Step 2. Hydrolysis of the sample to determine amino acid composition

On hydrolysis of the polypeptide chain all peptide bonds are cleaved. This releases the amino acids, whose concentrations are then determined, usually by an automated amino acid analyzer. The condition of hydrolysis is typically 6N HCl at 110° C for 24 hours or more, under an inert atmosphere; this procedure destroys Trp and hydrolyzes Gln and Asn to Glu and Asp, respectively. In addition, Ser and Thr are partially destroyed. These last two amino acids may be estimated accurately by performing 24-, 36-, and 43-hour analyses and extrapolating back to zero time. Trp is usually analyzed separately by a spectrophotometric titration of the N-bromo succinimide. It is not possible to determine the concentration of Gln and Asn by amino acid analysis; they are, however, identified by the sequential analysis procedure explained in the following section.

Step 3. Sequential Analysis

The fundamental chemical procedure of sequence determination is the reaction of the chain with phenylisothiocyanate (called PITC or the Edman reagent); this cleaves the N-terminal amino acid forming a phenylthiohydantoin derivative, which may be identified (Figure 1.6).

The procedure is then repeated although this is limited by complications arising from by-products. Routine analysis of polypeptides with up to about 30 components may be carried out using a commercially available instrument, the automatic sequenator. If the polypeptide chain contains more than 30 residues, it is necessary to break the chain into smaller fragments.

Where possible, the cyanogen bromide cleavage technique is used; otherwise enzymes are used to cleave the polypeptide chain.

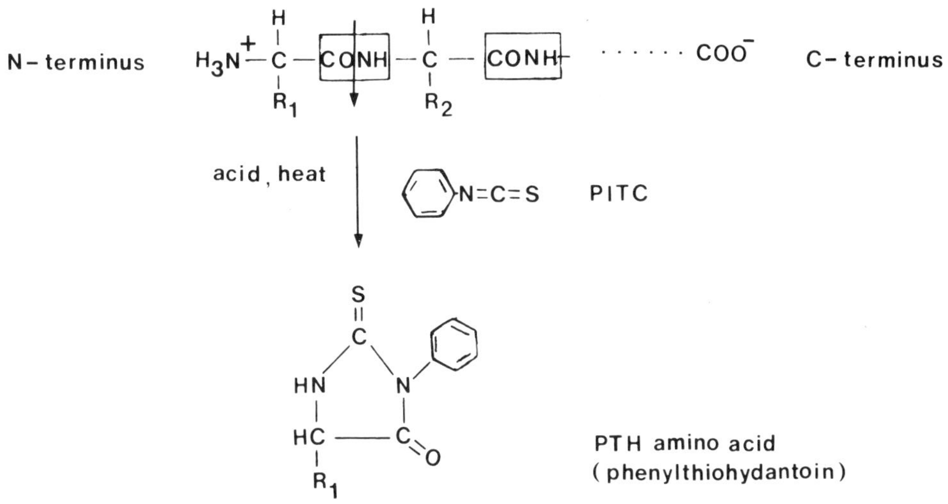

1. Primary Structure

Figure 1.6 Reaction scheme for the cleavage of N-terminal peptide residues. Polypeptide chains are written, by convention, with the N-terminus [NH$_2$ (protonated)] to the left.

Step 4. The Cyanogen Bromide Cleavage Technique

Cyanogen bromide cleaves polypeptide chains specifically at the C-terminal side of methionine, converting the residue to homoserine, as shown in Figure 1.7.

Since many proteins contain a relatively small number of methionine residues, the cyanogen bromide cleavage method produces a small number of fragments, many of which are suitable for sequential analysis without further fragmentation.

Enzymatic cleavage. Enzymatic cleavage of most proteins is somewhat less specific than cyanogen bromide cleavage, but is a very important method

Figure 1.7 Reaction scheme for polypeptide-chain cleavage (by cyanogen bromide) adjacent to a methionine residue.

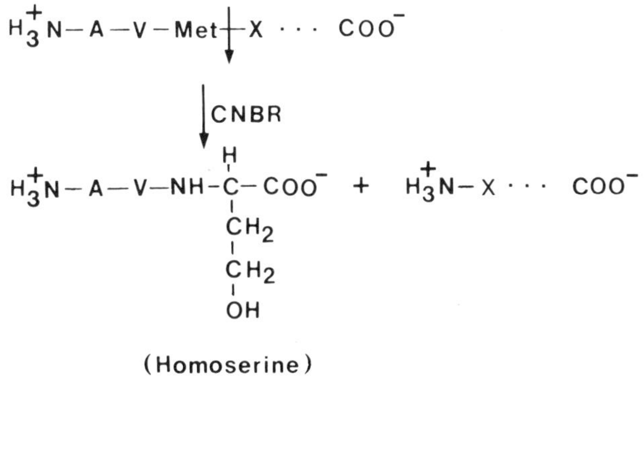

for breaking long polypeptide chains into manageable portions and for aiding in the identification of the sites of cyanogen bromide cleavage. Two types of enzymes are generally used; the *endopeptidases,* which cleave the protein chain sequentially from the C-terminal end of the polypeptide chain; and *hydrolytic enzymes,* which cause internal chain scission. Typical endopeptidases are carboxypeptidase A and B. Carboxypeptidase A will cleave successive residues from the C-terminus until it comes to Arg, Lys, or Pro, or a residue adjacent to Pro. Carboxypeptidase B, on the other hand, is specific to Arg and Lys but again will not cleave residues adjacent to Pro. A commonly used hydrolytic enzyme is *trypsin,* which cleaves only on the carboxyl side of Arg and Lys except when these residues are followed on the C-side by Pro. Another hydrolytic enzyme is chymotrypsin, which cleaves at the C-side of Phe, Trp, Tyr, and Leu, and occasionally Asn and Met. Pro blocks cleavage if it is C-linked to the normally cleavable residues.

After the initial cleavage into large peptide fragments, the arrangement of these fragments in the polypeptide chain is determined. Two different strategies are usually used. If complete cyanogen bromide cleavage produces fragments of reasonable size, a partial cleavage is carried out; this results in a polypeptide in which two or more fragments are combined (i.e., uncleaved). These large fragments are separated by column chromatography and the amino acid composition and molecular weight is determined. A comparison with the amino acid sequence, composition, and molecular weights of the fragments in the total digest often allows the ordering to be accomplished. In the second method, fragments are produced that overlap the normal cleavage positions. In this procedure a combination of two cleavage procedures (e.g., cyanogen bromide and trypsin) provides overlapping fragments that allow the ordering procedure to be carried out.

Combining the fragment order and sequence of the amino acids within the fragments yields the primary structure. However, if there are cystine residues present, it is important to know where they form internal — S — S — crosslinks.

Step 5. Location of linked cysteines (cystines)

If we consider a hypothetical sequence in which there were originally two cystine residues (four Cys/2), then on reduction, these give cysteine residues where the side chain radical is — CH_2SH. Suppose the sequence is

$$^+H_3N \ A \ Cys \ A \ Met \ A \ Cys \ B \ Met \ B \ Cys \ C - Met \ C \ Cys - DDD \ COO^-$$

with cleavage arrows ① ↓ ② ↓ ③ ↓ ④ above the Cys positions.

when A, B, C, and D are known amino acid residues. Initially we do not know whether Cys ① is sulfur linked to Cys ②, ③, or ④ in the original intact peptide. However, if instead of first reducing the cystines to cysteines, they are left in their cystine form (— S — S —), then cyanogen bromide cleavage, indicated by arrows, will give a different product for the various possible linkages. If Cys ① is linked to Cys ② and Cys ③ to Cys ④, then the

cyanogen bromide products will be

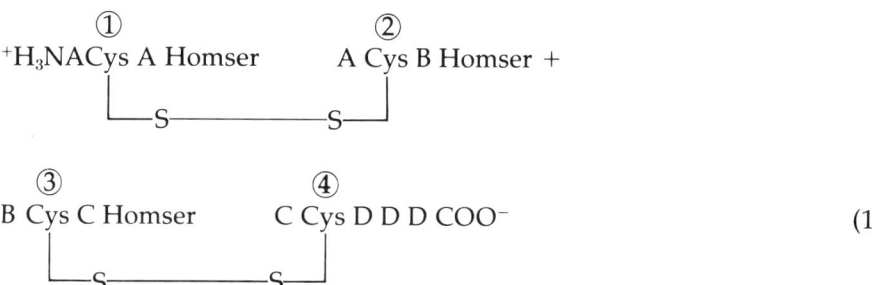

 ①   ②
⁺H₃NACys A Homser A Cys B Homser +
   └─S────────S─┘

 ③   ④
B Cys C Homser C Cys D D D COO⁻   (1)
  └─S────────S─┘

Alternatively, if Cys ① is linked to Cys ③ and Cys ② to Cys ④, the products will be

 ①   ③
⁺H₃NACys A Homser B Cys C Homser +
   └─S────────S─┘

 ②   ④
A Cys B Homser C Cys D D D COO⁻   (2)
  └─S────────S─┘

A third different combination would be achieved if ① is linked to ④, and so on. Amino acid analysis of the two fragments produced in reaction 1 would show residues A associated only with B; and B, C, and D in the other fragment. Reaction 2 would show A, B, and C associated in one fragment and A, B, C, and D in the other.

1.4. Peptides—Role of Sequence

The simplest primary sequence is found in short polypeptide chains—specifically peptides—and a great deal of effort has been expended in attempting to assess the role of sequence in determining the activity of peptides. Peptides consist of two or more amino acids; they have relatively short chains, suggesting very little three-dimensional architecture in the sense of globular proteins, and they exert a strong chemical stimulus. In a sense, the peptides represent the simplest functionality of globular proteins. They are generally conceived to act chemically by undergoing cooperative binding to a substrate, which may be protein, carbohydrate, nucleic acid, lipid, or combinations of these possibilities. This binding process, often at a cell surface, modifies the function of the substrate and turns on or off physiologic processes. Many hormones are peptides. After secretion from endocrine glands into the bloodstream, the hormones either interact with a specific receptor in the cell membrane or interact directly with DNA, possibly by removal of repressor molecules.

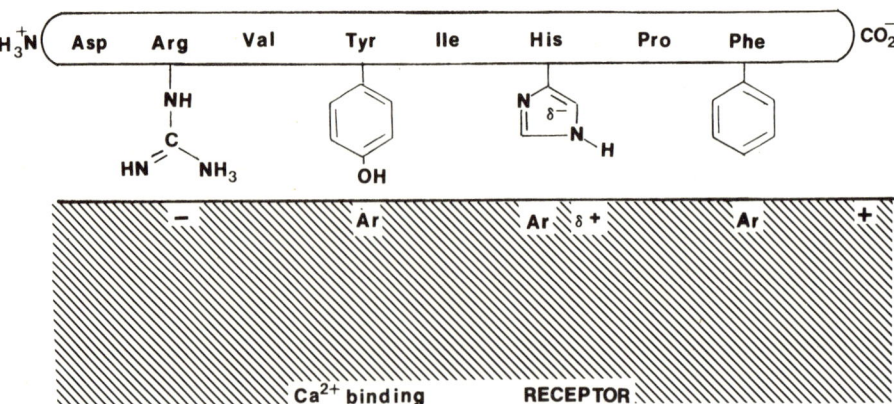

Figure 1.8 Schematic representation of a molecule of the peptide hormone angiotensin interacting with a substrate receptor site. The model is hypothetical in that the nature and arrangement of substrate residues are not known.

As an example of the role of sequence and of side chain interactions, we shall consider only one of many peptide hormones—angiotensin. Angiotensin is a hormone that interacts via the cell receptor mechanism, causing increase in blood pressure and stimulation of steroid secretion. It is an octapeptide Asp Arg Val Tyr Ile His Pro Phe. The questions we address here are: what amino acid residues are vital to the function of this peptide and how do side chains modify its activity?

Without any particular knowledge of the binding site for angiotensin, we might conjecture that it could have the schematic form shown in Figure 1.8.

This hypothesis is based upon the known attractive forces between molecules (i.e., ionic, van der Waals, H-bonds, hydrophobic bonds, etc.). For example, the positively charged arginine residue is paired with a negative group in the receptor; tyrosine is shown to interact with an aromatic group (or hydrogen bonding site or both) in the substrate. Although not shown in the diagram, it is also possible that valine and isoleucine interact with the substrate by hydrophobic forces and aspartic acid by ionic binding, and so on.

The exact mode of binding of hormones is not yet known, nor is the sequence of the receptor site.

Peptide chemists have attacked this problem by modifying the amino acid sequence or just the side chains of peptides, and performing physiologic studies to ascertain whether biologic activity is maintained. For angiotensin, literally hundreds of different combinations and replacements have been studied. The two major questions that are asked are the following:

1. Is the specific residue necessary or will a similar residue undergo a similar interaction with the receptor?
2. Does the substitution of a different residue or side chain modification change the overall shape (conformation) of the molecule simultaneously?

From studies of more than 100 synthetic analogues of angiotensin II, a composite picture of the role of the various peptide residues may be estab-

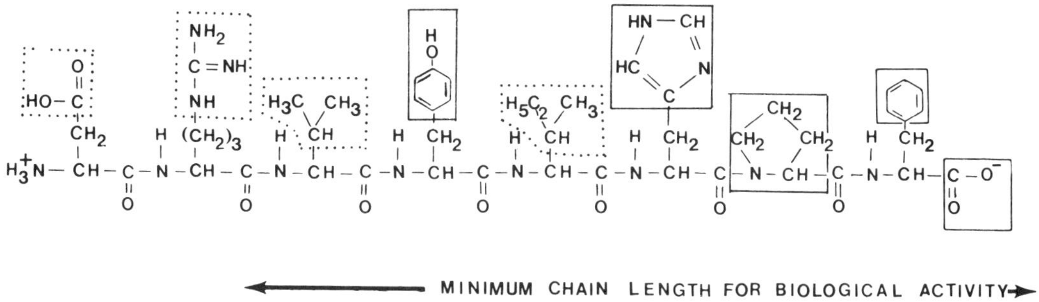

Figure 1.9 Minimum structural requirements of angiotensin II for biologic activity;— groups necessary for activity; .... groups unnecessary for activity.

lished. Tyr⁴, structures resembling the imidazole ring, and properties of His⁶, Pro⁷, Phe⁸, and the C-terminal carboxyl group are all necessary for full biologic activity. The N-terminal amine, Asp¹, the carboxyl group, the Arg² side chain, and the Ile⁵ side chain are not necessary, that is, other substituents give rise to active compounds. Val³ may be replaced with proline to give better than 50% activity so we may also class the Val³ side chain as not necessary.

We conclude, from the preceding observations, that:

1. Several residues are necessary to the biologic activity of angiotensin (true also for other peptide hormones), and that consequently the binding of these residues to the receptor site is cooperative; that is, the arrangement of substrate binding residues is such that a concerted effect is exerted by the configuration and arrangement of several residues.
2. Different types of interaction forces are involved in the binding process, including ionic attraction (C-terminal carboxyl), aromatic π-electron interaction (Tyr⁴, Phe⁸), probably hydrogen bonding (Tyr⁴ hydroxyl), and nonpolar van der Waals forces (Ile⁵, Val⁵, etc).
3. The chain shape is probably important because of the constraint applied by the pyrrolidine ring in Pro⁷.

A summary of these features is shown in diagrammatic form in Figure 1.9.

1.5 Role of Homology

Another method of assessing the role of the primary sequence of peptides is to examine sequence *homology*, that is the common elements occurring in different species. Considerable effort has gone into tracing the molecular evolution of peptides and proteins based on the sequence of equivalent molecules passing from the simplest organisms to complex mammals. It is not, of course, always easy to find or isolate the appropriate material.

We shall examine here changes in primary sequence of a fairly close class of creatures for one class of peptides, eledoisin and related peptides. These peptides rapidly lower blood pressure, cause contraction of extravascular smooth muscle, and stimulate digestive gland secretions.

Table 1.3 Homology of Eledoisin-Related Peptides

Origin	\multicolumn{14}{c}{Residue no.}
	1	2	3	4	5	6	7	8	9	10	11	12	13	14
Eledoisin (Octopus)			Glx	Pro	Ser	Lys	Asp	Ala	Phe	Ile	Gly	—	Leu	Met
Substance P (Bovine)			Arg	Pro	Lys	Pro	Gln	Gln	Phe	Phe	Gly	—	Leu	Met
Physalaemin Frog			Glx	Ala	Asp	Pro	Asn	Lys	Phe	Tyr	Gly	—	Leu	Met
Phyllomedusin Frog			Glx	Asn	—	Pro	Asn	Arg	Phe	Ile	Gly	—	Leu	Met
Ranatensin Frog			Glx	Val	—	Pro	Gln	Trp	Ala	Val	Gly	—	Leu	Met
Bombesin Toads	Glx*	Gln	Arg	Leu	Gly	Asn	Gln	Trp	Ala	Val	Gly	His	Phe	Met
Alytesin Toad	Glx	Gly	Arg	Leu	Gly	Thr	Gln	Trp	Ala	Val	Gly	His	Leu	Met

*The amino terminal Glx is pyrrolidone carboxylic acid.

Very large differences in sequence are observed for this set of peptides (see Table 1.3) in contrast to others of similar size. In this case only two residues are common to the sequences, Gly^{11} and Met^{14}. However, the properties of several nonhomologous residues are similar, for example, Leu^{13} and Phe^{13} are hydrophobic, and all residues (Phe and Ala) in position 9 are also hydrophobic. Position 7 involves hydrogen bonding, predominantly amide groups, and there is usually a common N-terminal residue (Glx).

We shall return to the theme of homology when examining the effect of sequence on conformation and tertiary structure.

Further Reading

Blackburn, S. *Protein Sequence Determination.* Dekker, New York, 1970.

Dayhoff, M.O. *Atlas of Protein Sequence and Structure.* National Biomedical Research Foundation, Washington, D.C., 1972. Supplements 1, 1973; 2, 1976; 3, 1978.

Hirs, C.H.W. and S.N. Timasheff, eds. *Methods in Enzymology,* Volume XLVII. Enzyme Structure, Part E. Academic Press, New York, 1977.

Needleman, S.B. *Protein Sequence Determination.* Springer-Verlag, New York, 1970.

Needleman, S.B. *Advanced Methods in Protein Sequence Determination.* Springer-Verlag, New York, 1977.

2

Secondary Structure

2.1 Bond Rotation

It is evident that the conformation or secondary structure of a polypeptide chain will be determined by the rotation of bonds in the polypeptide backbone.

In the two-dimensional representation given in Figure 2.1 we find three sets of rotation angles ϕ_x, ψ_x and ω_x. These three rotation angles correspond to rotations about the N—C_α bond (ϕ), the C_α—C bond (ψ) and the peptide C—N bond (ω). Conformational analysis seeks to determine the factors and forces that control these rotation angles, and thus attempts to understand and predict the conformation or spatial structure of the polypeptide chain.

At first sight it would appear that, even for a simple peptide, there are an infinite number of possibilities for the combination of rotation angles. There are, however, many constraints on the rotation angles that limit them to discrete angular regions. These constraints fall into the following categories:

1. Electron delocalization of the peptide bond, which creates double-bond character and produces "stiff," virtually "untwistable" bonds.
2. Peptide residue sequence symmetry.
3. Rotation regions are limited by steric "contact" between atoms (i.e., no two atoms can occupy the same space). This space is defined by London–van der Waals forces between atoms.
4. Polar and/or ionic forces dictate a conformation that maximizes favorable interactions (within the framework of the other limitations).
5. Hydrogen bonding can generally occur if the chain backbone folds back on itself (NH···O=C$<$) and many conformations are optimized by the hydrogen bonding pattern.
6. Torsional potentials, arising from potential overlap of electronic orbitals, place constraints on the favored bond geometry.

18   I. Units and Rules

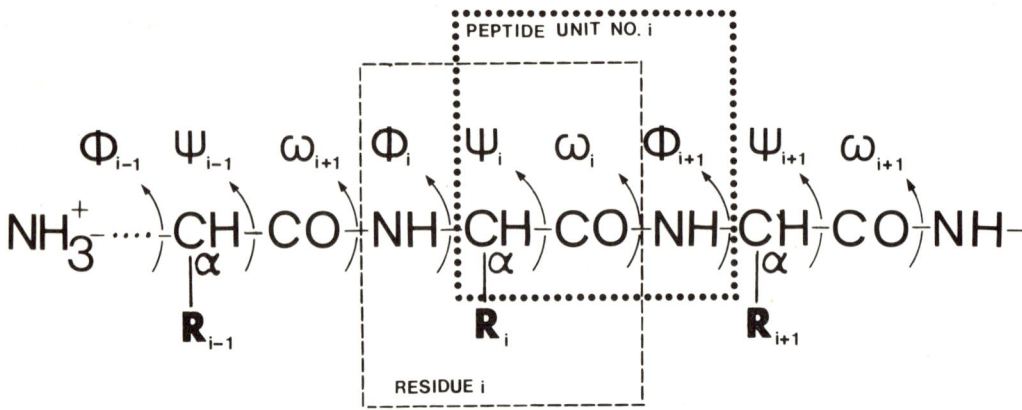

Figure 2.1 Possible bond rotations in the backbone of a polypeptide chain.

According to the latest nomenclature* a perspective drawing of a section of polypeptide chain representing two peptide units is presented in Figure 2.2.

2.2 Rotational Constraints

Electron Delocalization

X-ray crystallographic studies of simple dipeptides have shown that, whereas the N—C$_\alpha$ bond length is 1.47 Å, as expected for a normal N—C covalent bond, the C—N peptide bond length is 1.32 Å. This short bond compares with the value of 1.25 Å for the average C=N bond length in model compounds and is taken as strong evidence for double-bond character in the peptide bond. If one takes a direct proportionality between bond length and the number of electrons in the bond, then there would be ~70% double bond character or on the average 1.7 electrons in the N—C bond.

The origin of the electron delocalization can be seen if the electronic orbitals of the constituent O, C, and N (and H) atoms are studied.

First, the nitrogen atom (N), with $2s^22p^3$, electronic configuration, is normally tricovalent, as it is in the peptide bond. For purposes of considering the double-bond character of the peptide bond, the electronic structure of the nitrogen atom is best represented in terms of hybridization, which gives: three electrons in the three trigonal planar sp^2 orbitals; and two, the lone pair, in a nonhybridized p$_x$ orbital, perpendicular to the plane of the sp^2 orbitals. (See Figure 2.3.)

The nitrogen atom may then enter into single covalent (σ) bond formation with an electron in the sp^2 orbital of carbon. Carbon, with C$\cdots 2s^22p^2$, normally hybridizes sp^3, but in the formation of double bonds is represented by three electrons in sp^2 orbitals and a single electron in a nonhybridized p$_x$ orbital. (See Figure 2.4.) Carbon is in turn bonded (σ) to an oxygen atom O$\cdots 2s^22p^4$. There

---

*IUPAC Information Bulletin on Nomenclature, Symbols, Units and Standards No. 10, February 1971. Since many papers in the literature conform to the old notation, which is 180° out of phase for each rotation, care should be exercised when interpreting rotation angles.

2. Secondary Structure

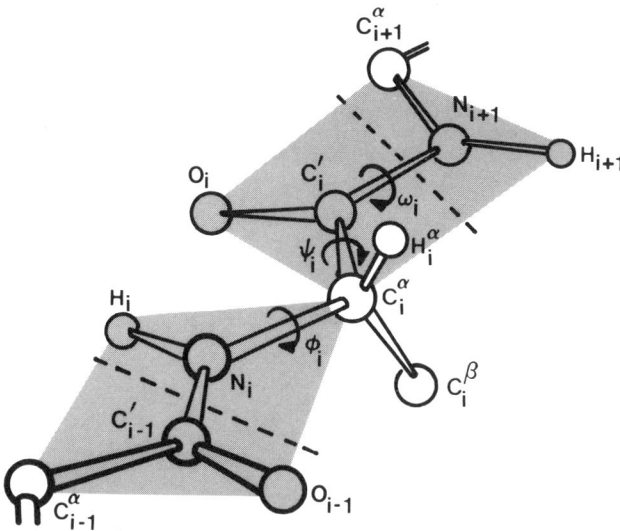

Figure 2.2 Perspective drawing of a section of polypeptide chain representing two peptide units. The limits of a *residue* are indicated by dashed lines, and recommended notations for atoms and torsion angles are indicated. The chain is shown in a fully extended conformation ($\phi_i = \psi_i = \omega_i = +180°$), and the residue illustrated is in the L-configuration (see Figure 1.3).

are at least two possible electronic configurations for oxygen that can account for the peptide bond (Figure 2.5). The first would present oxygen as a nonhybridized atom with three orthogonal 2p orbitals, p_z containing one electron, combining with a carbon sp^2 to form a σ-bond; one electron in the π-bonding p_x orbital and two in the nonbonding orthogonal p_y orbital. Alterna-

Figure 2.3 Spatial configuration of electron orbitals in the nitrogen atom.

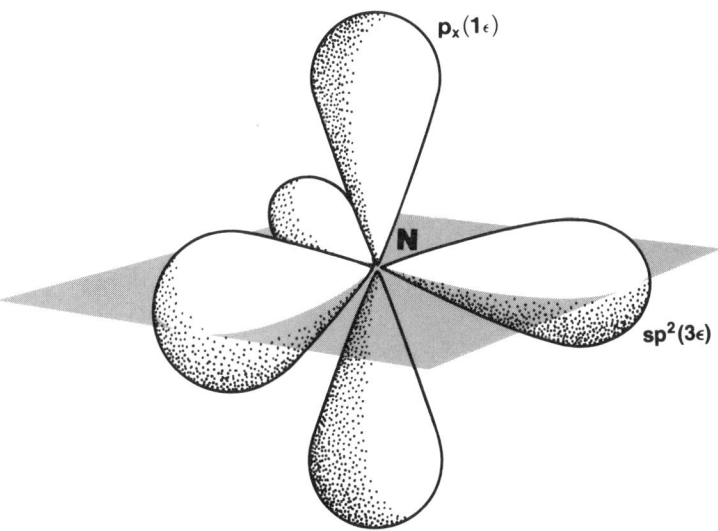

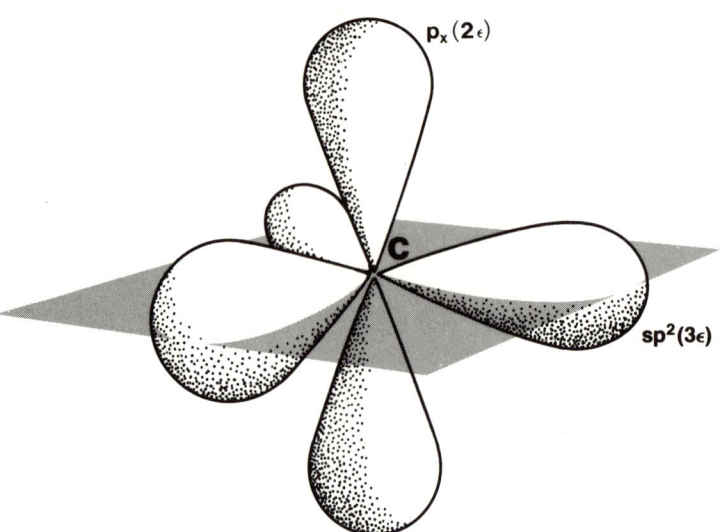

Figure 2.4 Spatial configuration of electron orbitals in the carbon atom.

tively, the orbitals could be represented by two sp orbitals, one containing two electrons, the other σ-bonding to the carbon sp²; and two p orbitals p_x with one electron (for π-bonding) and the p_y orbital with two nonbonding electrons. These two situations are spatially equivalent.

When these three atoms (N, C, O) are bonded together then, one O(sp_z or p_z) orbital and one C(sp²) orbital overlap to form a σ-bond, as do C(sp²) and N(sp²). Maximum interaction between the p_x orbitals is achieved if they are parallel and form a conjugated π-bonding system through which the four p_x electrons [O(1), C(1), N(2)] traverse. Consequently, we may expect the three atoms O, C, N to be coplanar; and since the σ-bonding of the hydrogen atom to nitrogen involves an sp²/s overlap in the same plane as the other σ bonds, the hydrogen atom is also expected to be coplanar. Furthermore, the α-carbon atom is also bonded through sp² orbitals of N or C, rendering two more atoms coplanar with the peptide bond. Thus, in all, six atoms are expected to lie in the same plane as the peptide bond (see Figure 2.6).

The C_α atom is not expected to donate delocalized electrons since it is tetracovalent and is presumably sp³ tetrahedrally hybridized. Comparison of the orbital angles, with bond angles from x-ray diffraction shows that the angle $\angle C_\alpha CN$ of 116° is between the 109° sp³ tetrahedral angle and the 120° of the sp² trigonal arrangement (Figure 2.7). The other two angles $\angle CNC_\alpha = 122°$ and $\angle NC_\alpha C$ at 111° are close to the trigonal and tetrahedral angles. (The hybridization of electrons in the peptide bond may be tackled by more quantitative methods including the Hückel molecular orbital approach. We shall examine quantum mechanical methods in more detail when considering electronic spectroscopy.)

The implications of the electron delocalization in the peptide bond are far-reaching. In brief they are as follows:

1. Double-bond character in the peptide bond prevents rotation and fixes ω at or near 0° or 180°.

2. Secondary Structure

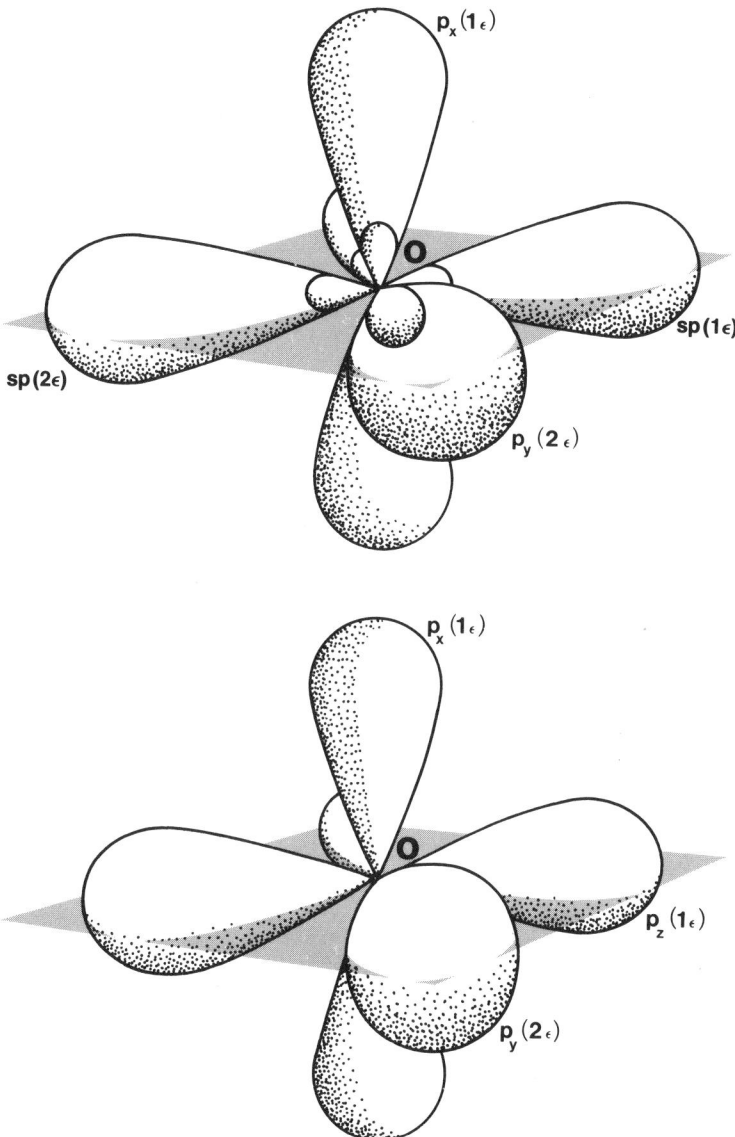

Figure 2.5 Possible spatial configurations of electron orbitals surrounding oxygen.

2. The double-bond character allows only a *cis* or *trans* configuration. The *trans* configuration ($\omega = 180°$) is energetically preferred in virtually all peptides and proteins.
3. Delocalization of electrons involves three atoms and is the basis of most forms of electronic spectroscopy of polypeptides.
4. The steric effect of the delocalization is to render six atoms $2C_\alpha$, C, N, O, and H(—N) coplanar.

We now see that instead of three rotatable bond angles, each peptide unit now possesses essentially two (ψ, ϕ).

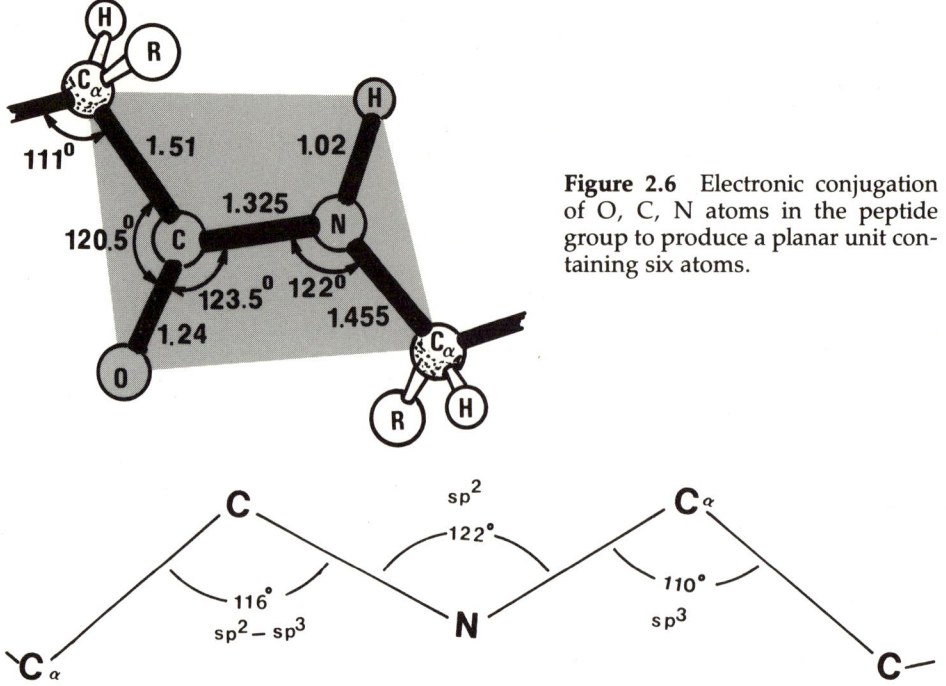

Figure 2.6 Electronic conjugation of O, C, N atoms in the peptide group to produce a planar unit containing six atoms.

Figure 2.7 Bond angles and expected electron hybridization of bonds and atoms in the polypeptide backbone.

Peptide Residue Sequence Symmetry

It would seem that if we have a polypeptide with n residues, it is necessary to consider 3n rotational variables simultaneously to characterize the shape or conformation of the chain (i.e., ψ, ϕ, and ω for each residue). However, as we have seen, the ω angles are held fixed at approximately $\omega = 180°$ because of electron delocalization. If there is sequence symmetry in the polypeptide chain it is often possible to reduce the number of rotational variables to a trivial number. For example, if we are dealing with a poly(amino acid), the simplest polypeptide sequence, it is evident that the optimal interaction between residues occurs if *all residues have the same environment*—that is, the forces imposed on residue $i - 1$ will be the same as i and $i + 1$. Under these circumstances (and provided there are no external perturbations in the form of thermal fluctuations) we may enunciate the Equivalence Condition.

Equivalence Condition

If all residues in the chain are the same,

$$\phi_1 = \phi_2 = \phi_3 = \cdots \phi_{i-1} = \phi_i = \phi_{i+1} \cdots = \phi_n$$

$$\psi_1 = \psi_2 = \qquad\qquad = \psi_i \qquad\qquad = \psi_n$$

$$(\omega_1 = \omega_2 = \qquad\qquad = \omega_i \qquad\qquad = \omega_n).$$

2. Secondary Structure

In other words, we have reduced the 3n rotational parameters to only two, ψ_i and ϕ_i, which are representative of the whole chain. We shall see later that for the above condition, or any sequence of repeating ϕ and ψ, the chain describes a helix or structure that possesses a linear axis.

If we consider a repeat polypeptide sequence of ABABAB, there are now four unique rotation parameters ψ_A, ϕ_A, ψ_B, ϕ_B. Similarly, for a polytripeptide ABC there are six variables, and so on. In practice we usually find $\psi_A = \psi_B = \psi_C$ and so on, but this is not necessarily so, either in theory or practice. Nevertheless a repeat peptide sequence leads to a repeat ψ, ϕ sequence and a chain with a linear axis. *Thus by applying the converse concept we must expect fibrous proteins that have linear axes to have peptide repeat sequences of some kind.*

On the other hand, in structures that have no linear axis (i.e., globular proteins) we would expect no peptide repeat and little or no bond rotational symmetry. Thus for globular proteins, conformational analysis is extremely complicated since it is necessary to consider many simultaneous rotational variables. Many attempts to circumvent these difficulties have been made; in particular, consideration is made of each successive pair of residues from one end of the polypeptide sequence to the other and then the process is repeated (iterated) with the hope that such a stepwise procedure optimizes the total conformation. For the present we shall consider only the implications of adjacent residue interactions, but will later look at more sophisticated approaches to the multivariable problem.

Steric "Contact" Nonbonded Interactions

Given that the angle ω is essentially fixed in polypeptides, we now turn our attention to limitations on the rotation of ψ and ϕ. To a first approximation, atoms may be treated as spherical entities that exclude the penetration of other atoms. These "van der Waals radii" are listed in Table 2.1. In parentheses are absolute minimum values found by Ramachandran in small (model) compounds.

The use of van der Waals radii for atoms allows us to demonstrate restrictions on the values of ψ and ϕ imposed by potential overlap of atoms. If we consider only the polypeptide backbone there are three rotational regions

Table 2.1 van der Waals Radii and Closest Distance of Approach*

	C	N	O	H
C	3.20 (3.00)	2.90 (2.80)	2.80 (2.70)	2.40 (2.30)
N		2.70 (2.60)	2.70 (2.60)	2.40 (2.20)
O			2.70 (2.60)	2.40 (2.20)
H				2.00 (1.90)

*In actual fact most atoms are not precisely representable by spheres since their orbital configuration extends further in some directions than others.

that are clearly not allowed. These are: (i) with $\psi = -180°$, $\phi = 0°$ the carboxyl oxygens of the peptide group overlap; (ii) with $\psi = 0°$, $\phi = -180°$ there is a potential overlap of hydrogen atoms; and (iii) with $\psi = 0°$, $\phi = 0°$ there is an overlap of O and H atoms. These situations are shown in Figure 2.8.

A steric map showing the allowable regions for ψ and ϕ based on the exclusion volumes of atoms (Ramachandran plot) is presented in Figure 2.9. It

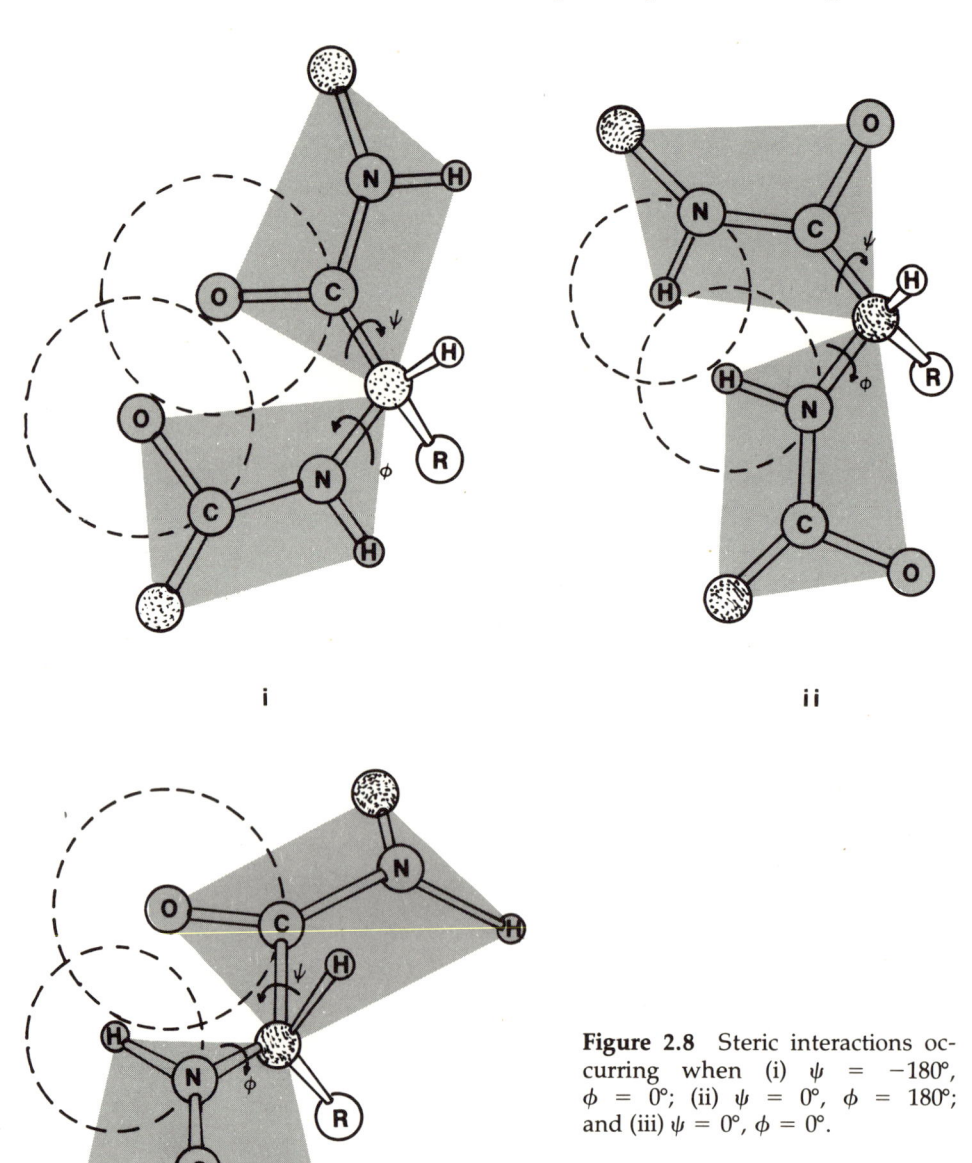

Figure 2.8 Steric interactions occurring when (i) $\psi = -180°$, $\phi = 0°$; (ii) $\psi = 0°$, $\phi = 180°$; and (iii) $\psi = 0°$, $\phi = 0°$.

2. Secondary Structure

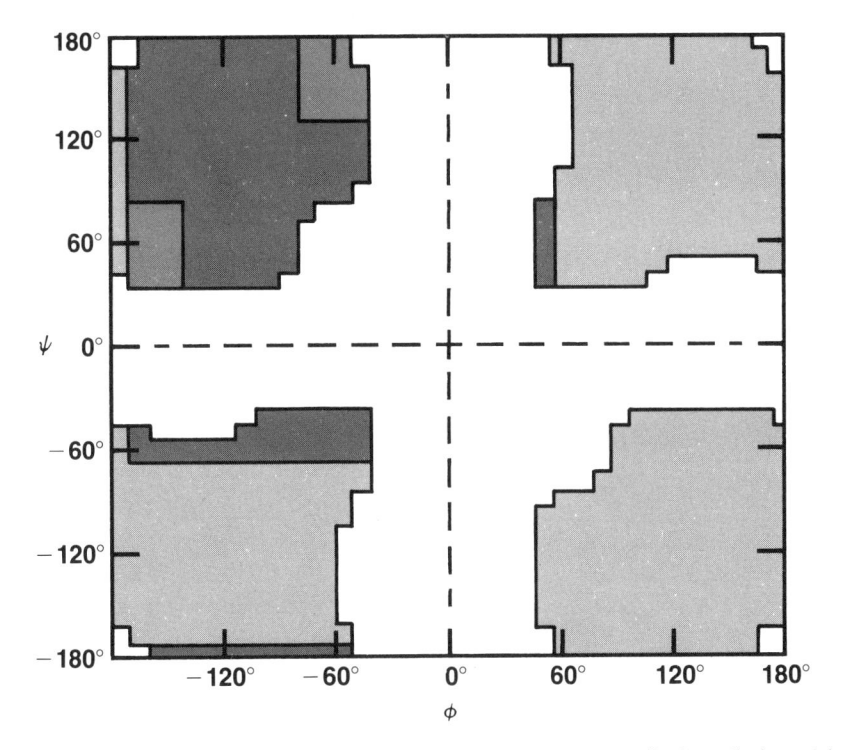

Figure 2.9 Allowed areas of the steric map for glycine (light color) and leucine (dark color).

can be seen that the residue with the smallest side chain—namely, glycine—restricts the ψ, ϕ rotation least (i.e., there is a large allowed rotation region in the Figure, light color). Furthermore, the steric map is symmetrical about lines joining $\psi = \mp 180°$, $\phi = \mp 180°$ and $\psi = \pm 180°$, $\phi = \pm 180°$ for glycine. This feature reflects the fact that there is a symmetrical distribution of atoms on the C_α, that is, hydrogens
$\begin{smallmatrix} H & H \\ \diagdown & \diagup \\ & C_\alpha \\ \diagup & \diagdown \end{smallmatrix}$
. As larger side-chain residues are substituted for hydrogen, the map loses its symmetry and the allowable regions shrink. D-residues give a similar Ramachandran map with a 180° transposition in ψ and ϕ.

If we wish to refine the maps in order to predict which values of ψ and ϕ are most likely, it is necessary to study the energetics of interaction between various atoms. This may be achieved by a variety of methods, including quantum mechanical and "classical" approaches. In the latter category two commonly used potential functions are the Buckingham potential

$$V = -A/r_{ij}^6 + B\exp(-\mu r_{ij}) \qquad (2.1)$$

and the Lennard-Jones potential

$$V = -A/r_{ij}^6 + B/r_{ij}^{12} \qquad (2.2)$$

where V is the potential energy due to nonbonded interactions of two atoms (i and j) separated by a distance r_{ij}. A, B, and μ are constants. The construction of a potential energy map of ψ, ϕ takes into account the nonbonded interactions and other interactions (described below) at discrete values of ψ and ϕ, which then allows the establishment of energy contours.

Other Interactions

Polar or ionic interactions

When amino acid residues in a chain carry a net charge (e.g., Glu or Lys residues) it is necessary to include an ionic interaction term of the form

$$V_I = \frac{e_i e_j}{\epsilon r_{ij}} \tag{2.3}$$

The e's are the charges for atoms i and j (in appropriate units), r_{ij} is the separation distance, and ϵ is the dielectric constant. Since ionic forces act over a much larger range than the London–van der Waals forces considered in the previous section, it is clear that an accurate expression of ionic or polar forces is necessary. The main practical difficulty has been in choosing an appropriate value for the dielectric constant ϵ. If the medium between atoms were truly a vacuum, and the charges were separated by sufficient distance, a value of $\epsilon = 1$ would be appropriate. If, on the other hand, a continuum of water separated the ions, then ϵ would equal 82. In fact, there is neither a definable continuum nor sufficient ionic separation for these extreme values to be used and values of $3 < \epsilon < 10$ have been used. Because of the uncertainty in the appropriate dielectric constant, calculated conformational energies should be compared only on a relative basis.

Apart from the discrete ionic charge, all sets of nonidentical atoms are polarized because of the relative electronegativity of the atoms causing dipolar charge separation. Whereas this has been a subject of considerable quantum mechanical investigation, a method proposed by Del Re has been used frequently to determine the magnitude of charge separation. The method is applied only to σ-bonds and originated in a semiempirical quantum mechanical approach. Although, again, the absolute magnitude of calculated charge separation may be questioned, the method is simple to use and provides a useful insight into the dipolar nature of peptides (or any organic molecule).

The charge readjustment between atoms is related to the ability of the atoms to influence electronic distribution (i.e., the average time that electrons spend in the neighborhood of atoms). The coefficient of induction δ (indicating the extent of inductive shift) is given by

$$\delta_A = \delta_A^0 + \Sigma \nu_{A\lambda} \delta_\lambda, \tag{2.4}$$

where λ refers to adjacent atoms, δ_A^0 is the standard inductive coefficient of atom A, and the coefficient $\nu_{A\lambda}$ represents the influence on A of adjacent atoms.

To understand how this relation works let us apply it to a molecule of methane (Figure 2.10a)

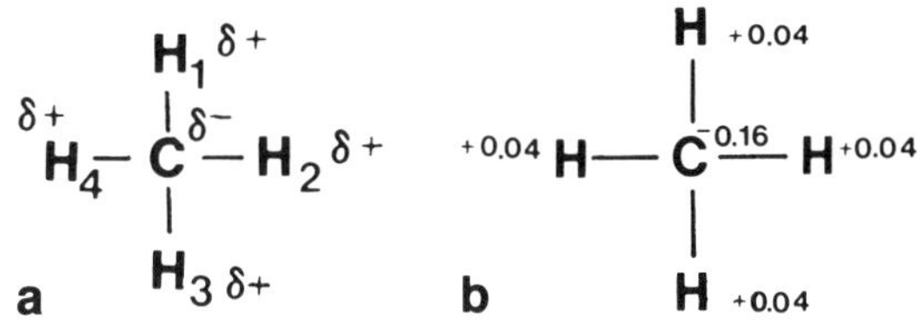

2. Secondary Structure

Figure 2.10 (a) Inductive shifts of electrons cause the creation of dipolar charge distribution for methane. (b) Charge distribution on methane using Del Re parameters.

$$\delta_{H_1} = \delta_H^0 + \nu_{HC}\delta_C \quad (= \delta_{H_2} = \delta_{H_3} = \delta_{H_4}) \tag{2.5}$$

$$\text{and } \delta_C = \delta_C^0 + \nu_{CH}(\delta_{H_1} + \delta_{H_2} + \delta_{H_3} + \delta_{H_4}) \tag{2.6a}$$

$$= \delta_C^0 + 4\nu_{CH}\delta_{H_1}. \tag{2.6b}$$

To solve these equations we need only a table of standard induction coefficients (δ_A^0) and "influence" coefficients (ν_{AB}). (See Table 2.2.) From the table and equations (2.5) and (2.6b),

$$\delta_H = 0 + 0.4\delta_C \tag{2.7}$$

$$\text{and } \delta_C = 0.07 + 4 \times 0.3 \times \delta_H. \tag{2.8}$$

Consequently, $\delta_C = 0.134$, $\delta_H = 0.054$.

In order to convert these induction coefficients to fractions of electron charge, the Mullikan equation is used.

$$Q_{AB} = (\delta_B - \delta_A)/2k_{AB} \tag{2.9}$$

Q_{AB} is the total charge on atoms A and B, and k_{AB} is another characteristic constant listed in Table 2.2.

$$Q_{HC} = (0.08)/2 = 0.04$$

$$Q_{CH} = -4Q_{HC} = -0.16$$

Therefore, the methane molecule may be represented as in Figure 2.10b, where the numbers represent fractional electron charge. A negative sign indicates a net excess of negative charge; a plus sign, a net excess of positive charge.

Numbers obtained from this type of calculation may be used in conjunction with equation (2.3) to estimate the dipolar interactions in peptide chains.

Table 2.2 Del Re Parameters for Calculating Charge Distribution

Bond	C–H	C–C	C–N	C–O	C–O⁻	N–H	N⁺–H	O–H	C–N⁺	C–F	C–Cl	C–S	S–S	S–H
k_{AB}	1.00	1.00	1.00	0.95	0.80	0.45	0.60	0.45	1.33	0.85	0.65	0.75	0.60	0.70
$\nu_{A(B)}$	0.30	0.10	0.10	0.10	0.10	0.30	0.30	0.30	0.10	0.10	0.20	0.20	0.10	0.30
$\nu_{B(A)}$	0.40	0.10	0.10	0.10	0.10	0.40	0.40	0.40	0.10	0.10	0.40	0.40	0.10	0.40
δ_A^0	0.07	0.07	0.07	0.07	0.07	0.24	0.31	0.40	0.07	0.07	0.07	0.07	0.07	0.07
δ_B^0	0.00	0.07	0.24	0.40	0.40	0.00	0.00	0.00	0.31	0.57	0.35	0.07	0.07	0.00

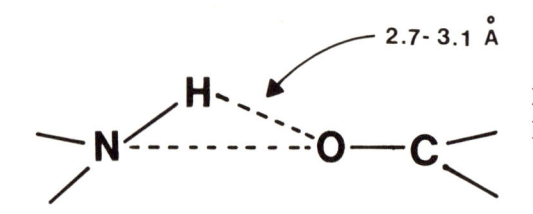

Figure 2.11 Schematic diagram of a peptide hydrogen bond.

Hydrogen bonding

In historical perspective, the concept that polypeptides form conformations that tend to optimize (maximize) hydrogen bonding represented a major step forward in structural analysis. Hydrogen bonds form between the hydrogen atom attached to the peptide nitrogen atom, and the carboxyl oxygen (Figure 2.11). The hydrogen and oxygen atoms may belong to peptide groups in the same chain, in which case we have *intra*chain hydrogen bonding; or in adjacent chains, giving rise to *inter*chain hydrogen bonding. Such bonds only form within a limited framework of atomic positions, the hydrogen bond length being between 2.7 and 3.1 Å in length. Originally it was postulated that hydrogen bonds N — H $\cdots$ O = C should be linear, but we know that deviations of up to 90° are possible.

From the important role hydrogen bonding has played in polypeptide chemistry it might be assumed that: (a) the hydrogen bond contributes the main driving force to develop a specific conformation; and (b) the details of hydrogen bond interactions are well understood. Neither (a) nor (b) is true. Rather, the type of hydrogen bonding arrangement seems to tip the balance between other interactions. Suitable interaction potentials for hydrogen bonds are still being developed.

Again two approaches, classical and quantum mechanical, have been put forward. Most of the classical approaches treat the hydrogen bond in terms of a combination of dipolar and London–van der Waals forces with some angular dependent term.

Torsional potentials

Restrictions to internal rotation of a bond arise from orbital interactions of substituents and bonding electrons. Such potentials generally contribute to only a small degree and are often represented by a simple angular function

$$V_T = \frac{V_o}{2}[1 \pm \cos\alpha\theta], \tag{2.10}$$

where θ is the bond rotation angle and α a constant appropriate to the geometry of the system.

The inclusion of the various interactions into the computation of a dipeptide ψ,ϕ map makes it clear that certain regions are preferred.

In Figures 2.12a and b it can be seen that both the glycine and L-alanine peptide maps bear a close resemblance to the steric map (Figure 2.9), but contours representing differences in conformational energy of one kilocalorie

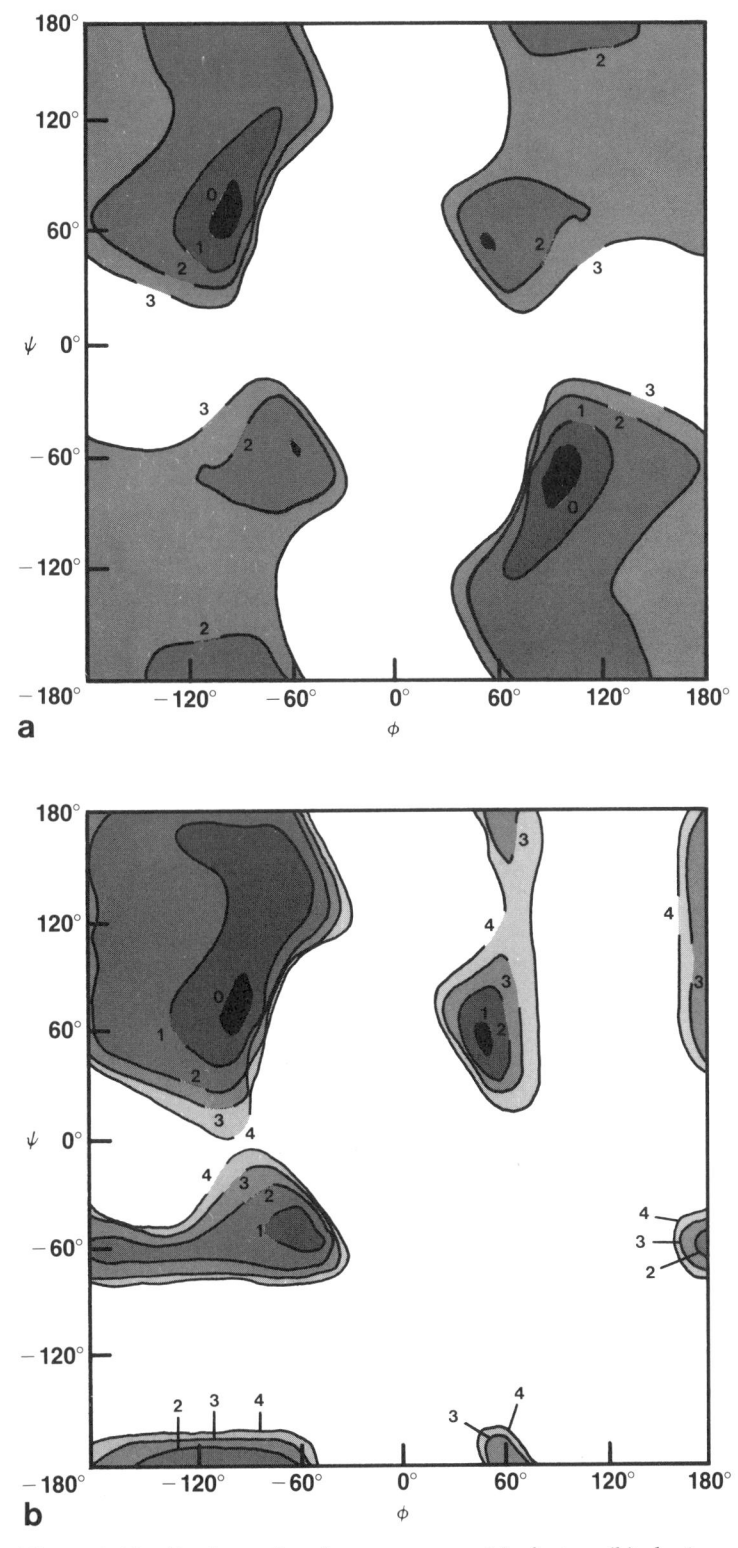

Figure 2.12 Conformational energy maps: **(a)** glycine; **(b)** alanine.

are drawn, such that the region surrounded by O is the most energetically favored.

Of course it is possible to extend this type of approach to various residues having different side-chain properties. In such cases the calculations are limited by the size of the side chain, possible side-chain rotations (χ rotations) and side-chain complexity. Nevertheless the steric energy maps show similar features but with the detailed energy minima lying at ψ,ϕ angles characteristic of the specific side chain. The implication of this observation is that, provided no external influences are brought to bear, *peptide pairs prefer a unique conformation*. That is to say, any one pair has one set of ψ,ϕ angles, which represents maximum stability. (A different pair may prefer the same ψ,ϕ angles also.)

We can now see that the regions of rotation that ψ and ϕ may occupy are strictly limited and, as we will see, correspond to conformations that are observed both in synthetic polypeptides and proteins. Our approach so far has ignored certain refinements that would need to be taken into account if a more precise predictive approach were to be required. These refinements include the following:

Calculations studied so far in this text involve only nearest neighbor peptide groups in the same chain; interactions not involving the nearest neighbor are often important. These interactions may be intrachain and/or interchain.

It is the *free* energy, not the potential energy, that determines the stable chain conformation.

The calculations are carried out for a vacuum conformation and solvent interactions remain to be examined.

Rather than present these refinements, we will now examine how the predicted favorable regions of steric energy maps (e.g. Figures 2.12a and b) correspond to definable shapes.

2.3 Helices

All polypeptide chains that follow the equivalence condition as spelled out in the previous section—that is, those that can be represented by a limited (repetitive) set of ψ,ϕ angles—are helices or degenerate helices. There is therefore a mathematical relationship between ψ,ϕ and helical parameters.

First let us examine and define the properties of polypeptide helices.

In Figure 2.13a and b polypeptide helices are shown with n = three peptide residues per turn. These peptides have been arbitrarily listed as A, B, and C but could be the same or different residues. The helix has a pitch of P; that means the residues fall in equivalent positions at a distance P apart. The translation along the helix axis for each peptide is h. Thus, for the example in Figure 2.13, h = P/3. Additional features of the helix shown in Figure 2.13a are that it is left-handed; that is, the helix chain rotates to the left as viewed from either end of the cylinder (left-handed helices denoted by minus before n), and also the radius of the helix is r and the pitch angle β. In polymer chemistry, the helix is defined by N_t; that is, by the number of residues N found in t complete turns of the helix. Figure 2.13 would therefore be a 3_1 (rational) helix. The word rational is sometimes used to indicate that slight distortions from this helix,

2. Secondary Structure

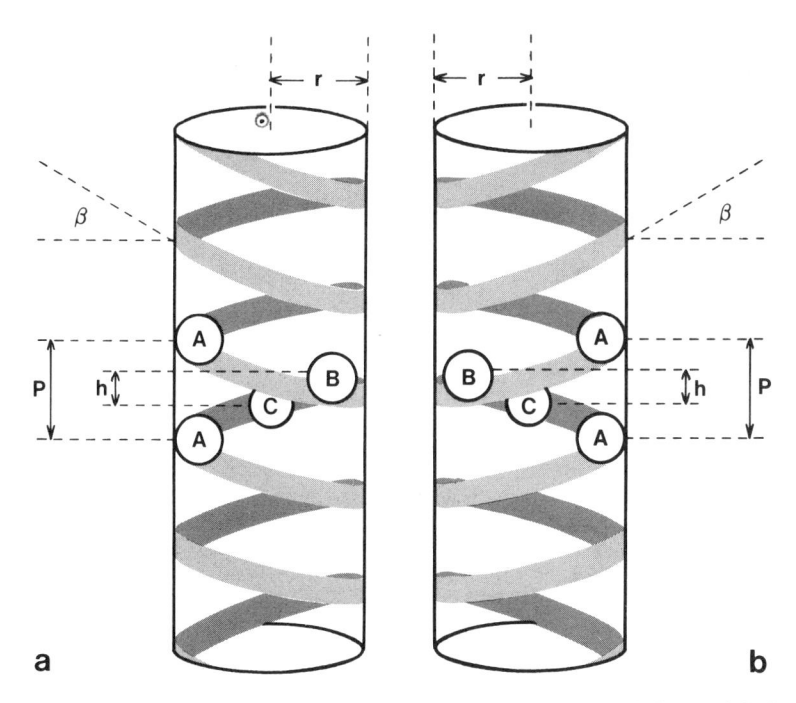

Figure 2.13 Representations of helical molecules: **(a)** left-hand helix; **(b)** right-hand helix.

which might more accurately be represented by, say, 301_{100}, would still be thought of as a 3_1 helix to a first approximation. If residues A, B, and C are the same, then the axis of the helix is a threefold screw axis, that is, rotations of 120° (and a translation of h) bring the helix to an equivalent position.

The minimum number of peptides possible in a repeat structure is two, as shown in Figure 2.14a. This structure is no longer strictly helical (n = ±2) but is an extended ribbon and is equivalent to a "squashed" or degenerate 2_1 helix. Again the residues are arbitrarily designated A and B to make the repeat more apparent. In this case h (peptide repeat) = P/2.

Figure 2.14b shows a nonintegral helix with 3.5 residues per turn. It is not, of course, necessary that residues repeat in adjacent turns of the helix. In this case, this (hypothetical) helix repeats every second turn and is thus a 7_2 helix. The most common helix in proteins, the so-called α-helix, is a nonintegral helix with 18 residues in five turns, that is, it is a rational 18_5 helix. In Figure 2.14b the residues have been arbitrarily labeled A through G to clarify the repeat pattern.

As we have seen previously, the nature of the peptide sequence affects the backbone rotation angles and hence the nature of the helix. Because of the geometric difficulties of depicting residues on three-dimensional helices, the method of two-dimensional representation has been developed. The right-hand 3_1 and 7_2 helices shown in Figure 2.13 and Figure 2.14 are repeated in the two-dimensional format in Figure 2.15. Left-hand helices can be represented with the sequence lines inclined to the left instead of the right. This nomenclature will prove useful when we are studying supercoiling of fibrous proteins.

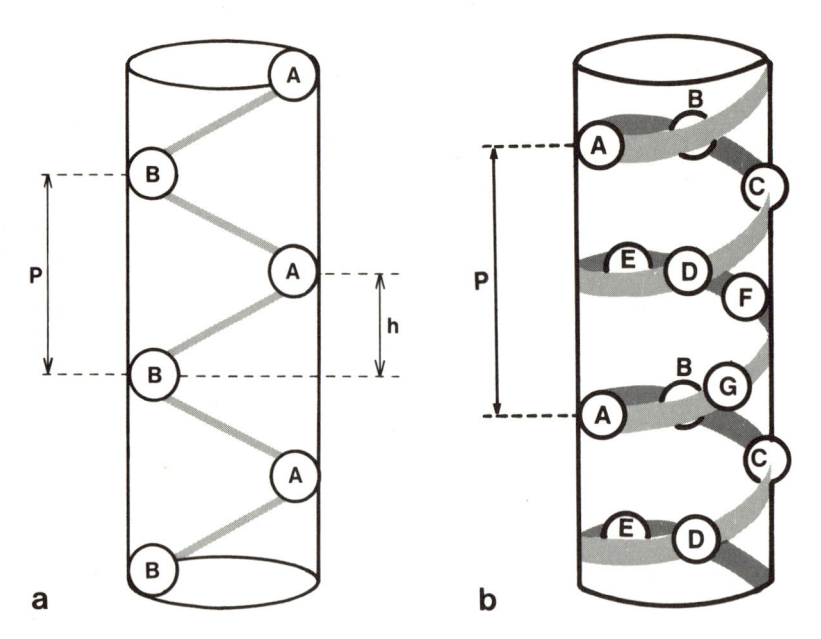

Figure 2.14 Representations of **(a)** a 2_1 helix (a ribbon); and **(b)** a nonintegral (7_2) helix.

2.4 Relation of Helical Parameters

We have now seen that various parameters—ψ, ϕ, h, N_t, and n—are used to characterize helices. Of these ψ and ϕ may be said to be theoretical parameters; h, N_t, and n, experimental (extracted from x-ray diffraction data, as in Chapter 4). The relationship between these two sets is as follows:

$$\cos\left(\frac{\theta}{2}\right) = +0.817 \sin\left(\frac{\phi+\psi}{2}\right) + 0.045 \sin\left(\frac{\phi-\psi}{2}\right) \tag{2.11}$$

Figure 2.15 **(a)** Representation of 3_1 helix of ABC residue sequence. **(b)** Representation of 7_2 helix of sequence ABCDEFG.

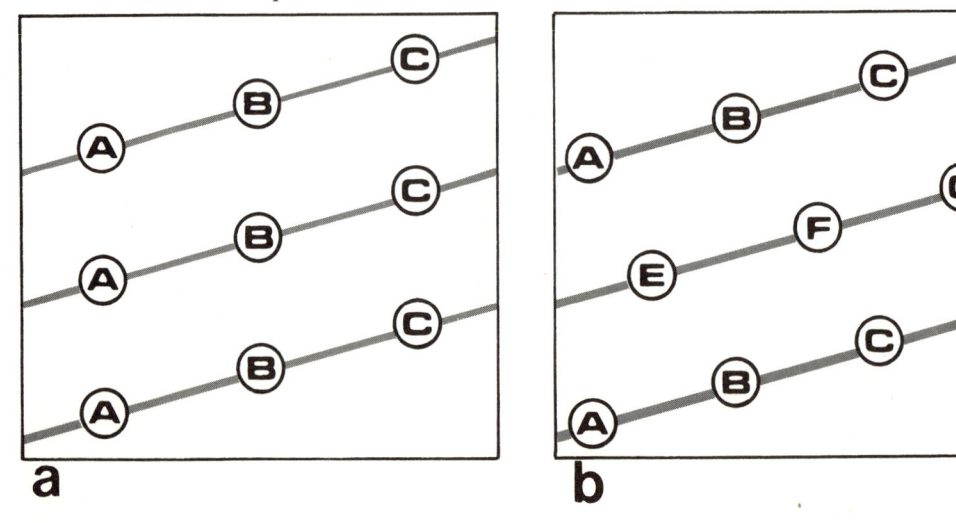

2. Secondary Structure

Table 2.3 Most Favorable Helical Parameters as Derived from Glycine and Alanine Steric Energy Maps

ψ	ϕ	h	n	N_t	Helix designation
−60°	−60°	1.50Å	+3.6	18₅	α-helix
+140°	−145°	3.5	+2	2₁	β-sheet (antiparallel)
+120°	−115°	3.25	+2	2₁	β-sheet (parallel)
+150°	−80°	3.1	+3	3₁	polyglycine II
+55°	+65°	1.33	+4	4₁	ω-helix

$$h \sin\left(\frac{\theta}{2}\right) = -2.967 \cos\left(\frac{\phi+\psi}{2}\right) - 0.664 \cos\left(\frac{\phi-\psi}{2}\right) \tag{2.12}$$

$$\theta = 360/n \tag{2.13}$$

$$n = N/t \tag{2.14}$$

The parameter θ is known as the unit "twist" and corresponds to the number of degrees of turn of the helix occupied by each residue.

From the above equations, values of h, n, N_t may be extracted by inserting values of ψ and ϕ such as those found optimal in Figures 2.12a and b. In Table 2.3 favorable values of ψ and ϕ have been placed in the two left-hand columns and corresponding values of h, n, and N_t on the right.

If we add to this list the situation in which there is a nonrepeating residue sequence and no repeating ψ and ϕ (i.e., the "irregular" structure), we find that our rather unrefined dipeptide approach accurately predicts the existence of each of the four major protein conformations: α-helix, β-sheet, polyglycine II (pGII), and irregular. Such a result is somewhat surprising when we remember that the equivalence rule of identical ψ and ϕ values for polypeptide chains, which led to the above predictions, is only strictly valid for poly(amino acids) in vacuum at 0° K!

2.5 Known Helical Structures

The α-Helix

The α-helix is a nonintegral 18₅ helix found commonly in native proteins and synthetic polypeptides. Synthetic poly(amino acids) have generally provided most information on the details of the α-helix conformation because of their ability to crystallize in a regular structure accessible to x-ray diffraction analysis. The characteristics of the helix are that it has 3.6 residues per turn and thus each residue imparts a 360/3.6 = 100° twist to the helix and the peptide repeat is 1.50 Å. It can be seen in Figure 2.16 that the hydrogen-bonding array is such that N — H ··· O = C groups are essentially in linear array and *nearly parallel to the axis of the α-helix*. This observation will be seen to be important when we study characterization methods, particularly infrared spectroscopy. All α-helices found in native proteins so far are right-hand (RH) helices (i.e., n = +3.6. This feature is probably due to the slight steric preference, induced by London–van der Waals forces, for RH α-helices (see Figure 2.12b) originating in side-chain interactions. An example of a synthetic polypeptide having a left-

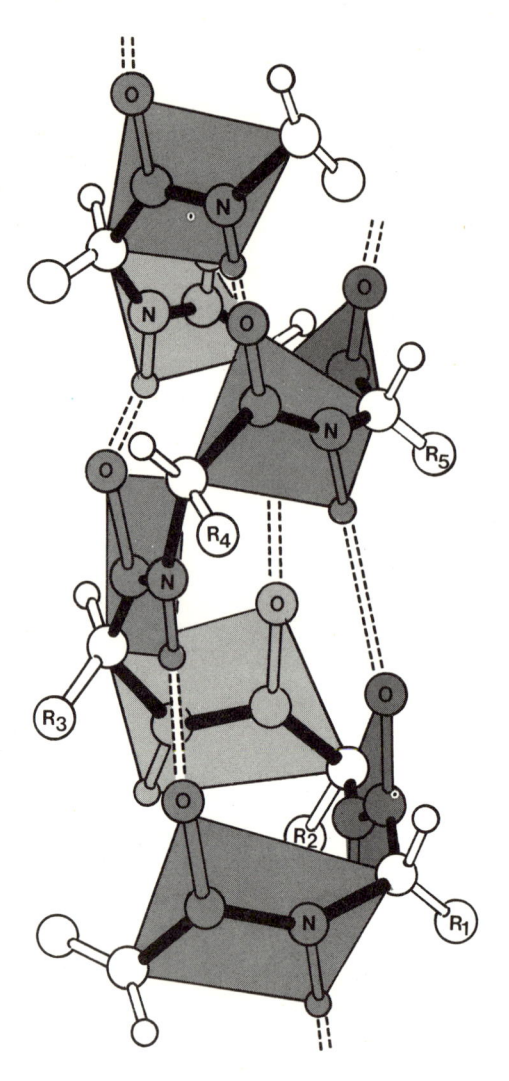

Figure 2.16 Arrangement of atoms in a right-hand α-helix. Note hydrogen bonds between first and fourth peptide, second and fifth, and so on.

hand (LH) α-helix is poly(β-benzyl-L-aspartate), but this helix is easily denatured and is therefore not very stable. Poly(D-amino acids) do, of course, possess stable LH α-helices, but are not found in nature.

The β-Sheet

The β-sheet is a "degenerate" 2_1 helix, better thought of as a ribbon structure. If the chains were fully extended (i.e., with $\phi_i = \psi_i = \omega_i = 180°$) then two adjacent antiparallel chains would form a structure such as that shown in Figure 2.17.

Antiparallel β-sheet (extended)

The terminology "antiparallel" arises from the fact that polypeptide chains have "direction"; that is, the sequence $\overrightarrow{NCC_\alpha}$ occurs in the opposite direction in the two chains. Further chains would be arranged in a similar manner above

2. Secondary Structure

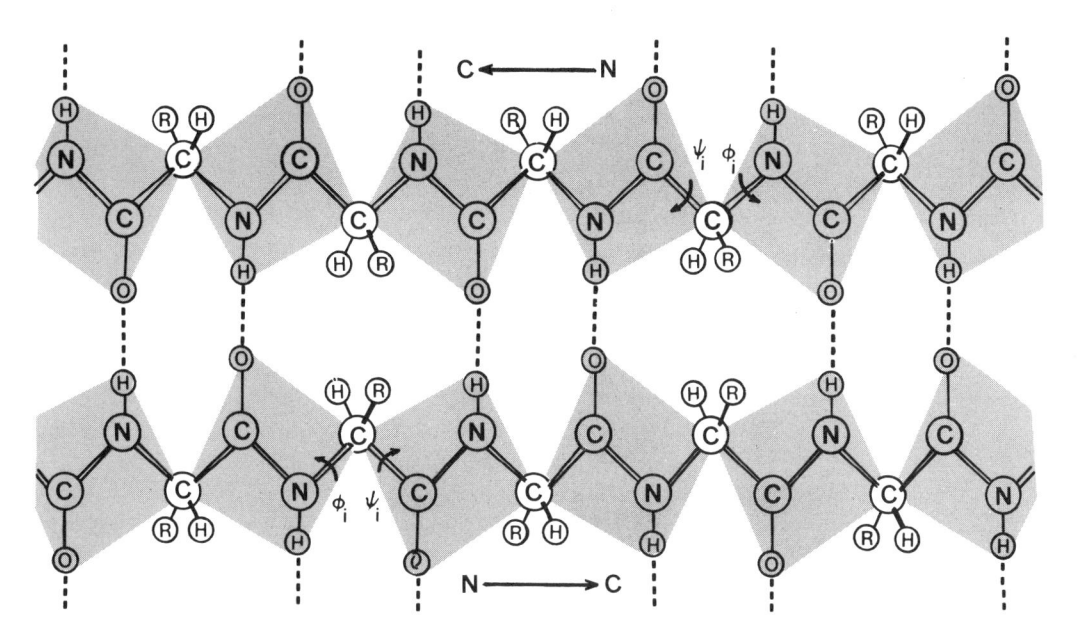

Figure 2.17 Hydrogen-bonding scheme for two antiparallel "β-sheet" chains.

and below the ones indicated, such that a "sheet" of antiparallel chains could be drawn in the plane of the page. At maximum extension the peptide repeat is ~ 3.5 to 3.6 Å. One notes, however, that if the chains are fully extended and are hydrogen bonded in the manner shown, the substituents R are oriented below the plane of the page but towards the similar substituent in the adjacent chain. Such an effect can, and probably does, cause some steric interference, which may be relieved by moving the C_α atoms out of the plane of the paper by rotating ψ and ϕ. In this manner the Rs are rotated away from each other and the sheet becomes "pleated" while maintaining the same hydrogen bonding scheme. A side view schematic of the pleated sheet is then as in Figure 2.18.

Figure 2.18 "Side-on" view of the pleated β-sheet. Dashed lines represent hydrogen bonds.

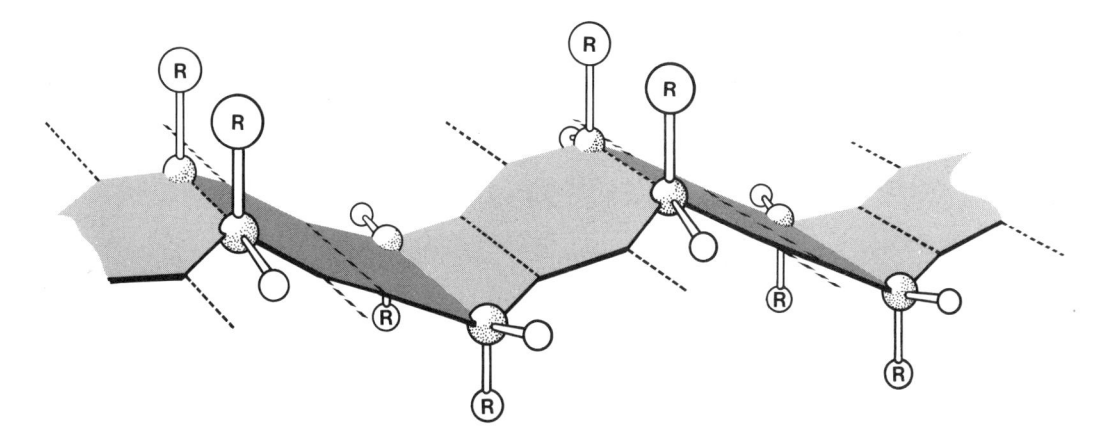

The effect of the chain puckering is to reduce the peptide repeat distance to between 3.4 and 3.5 Å. In silk fibroins the antiparallel pleated sheet has a peptide repeat of 3.475 Å.

The arrangement of hydrogen bonds in the β-sheet is such that they are *perpendicular* to the polypeptide chain axis in contradistinction from the α-helix.

Parallel β-sheet

It is possible to arrange the polypeptide chains with a suitable hydrogen bonding sequence such that the chains are parallel (Figure 2.19).

As can be seen, the hydrogen bonds are no longer linear and energy calculations show that the parallel β-sheet conformation (in vacuo) is not as stable as its antiparallel cousin. It seems likely that only isolated chain segments of globular proteins can comfortably take on such a structure and as yet no known synthetic polypeptide has been produced with a parallel (∥) β-sheet structure.

Most of the subtleties in β-sheet conformations arise from the relative staggering of chains and the subsequent arrangement of sheets. Figure 2.20a shows two possible arrangements of a polydipeptide in the plane of the sheet. Figure 2.20b shows two possible arrangements of the sheets (in the plane of the chains and perpendicular to the plane of the sheets) for a polydipeptide. It can be seen that if the bulk of the side-chain residues is different for the two residues then the sheet separation can be quite different, because all of one type of side chain lie on one side and the other on the opposite side. Such is actually the case in some silk fibroins where the simplest chain-repeat representation is (GlyX)$_n$. More complicated sequences clearly have more potential steric variations.

Figure 2.19 Hydrogen-bonding arrangement of two parallel chains in the β-sheet conformation.

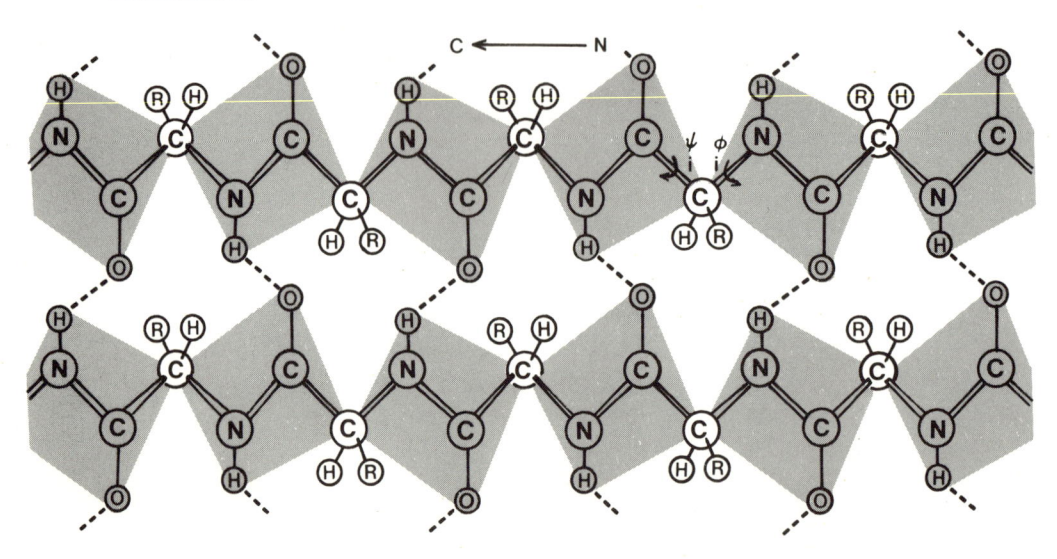

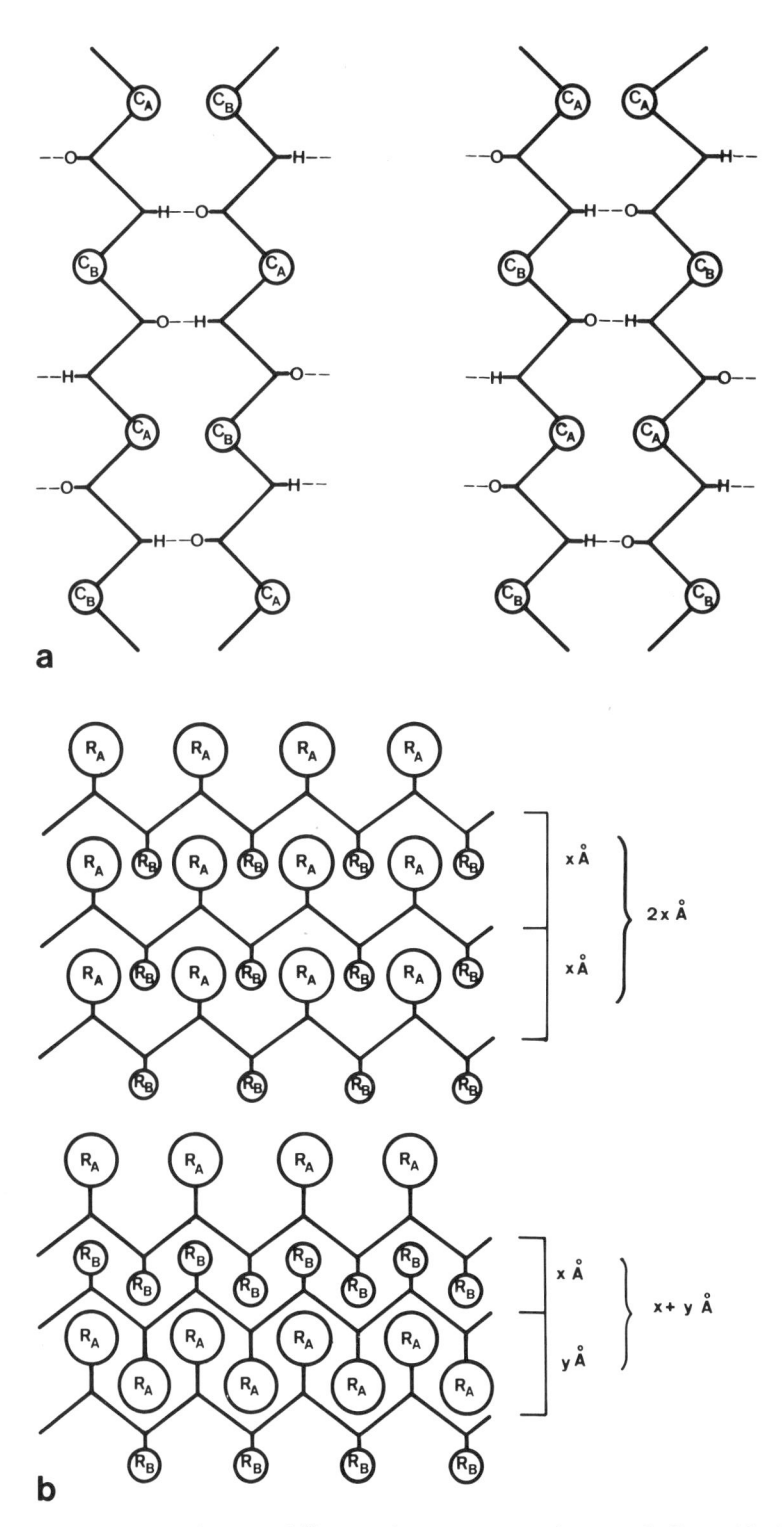

Figure 2.20 (a) Two different chain staggers for a polydipeptide in an antiparallel β-sheet arrangement. (b) Two different sheet separations for a polydipeptide in an antiparallel β-sheet arrangement.

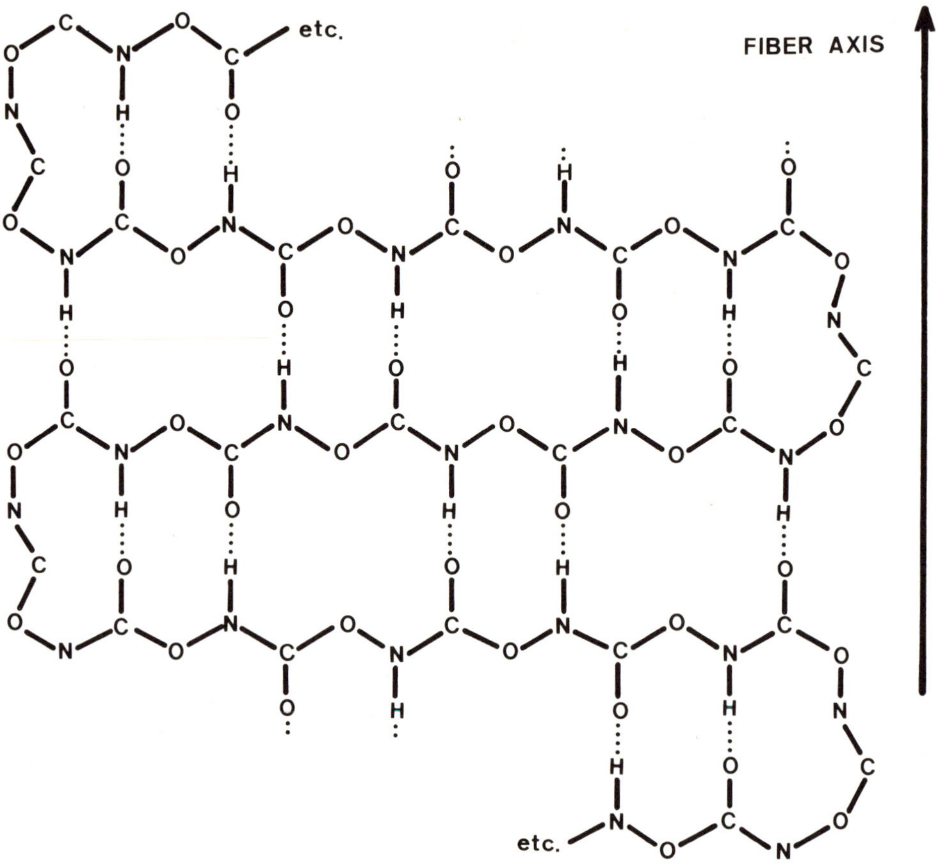

Figure 2.21 Cross-β sheet.

Cross-β structure

The cross β-sheet structure found in fibrous proteins is an antiparallel β-sheet structure in which the chains run perpendicular to the fiber direction and fold (presumably in a β-bend, see section 2.7) backwards and forwards across the fiber width (Figure 2.21). The hydrogen-bond direction is still perpendicular to the chain axis but the chain axis is perpendicular to the fiber axis. These effects are important when the physical methodology, particularly x-ray diffraction and infrared dichroism (bond orientation), are discussed.

Polyglycine II and Polyproline II (pGII and pPII) Structures

The pGII and pPII are both 3_1 helices, which are stabilized by different types of forces.

Polyglycine II (pGII) helix

Polyglycine in its helical form is unique in that the probability or energetics of forming LH or RH helices are equal and thus pGII in solution consists of an equal mixture. In the solid state the helices of pGII pack hexagonally and each

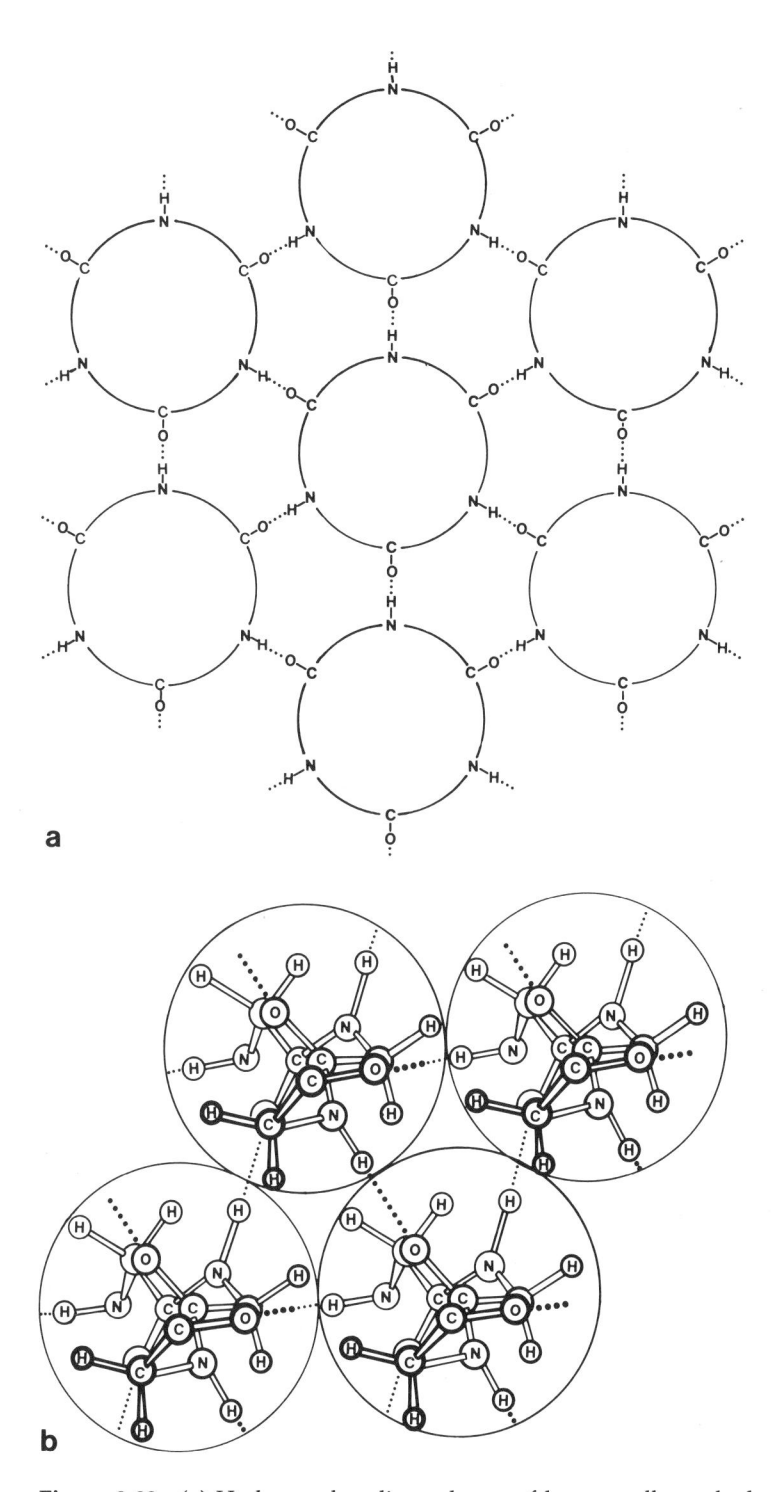

Figure 2.22 (a) Hydrogen bonding scheme of hexagonally packed polyglycine helices. (b) Atomic arrangement in 3_1 polyglycine II helices in a polyglycine crystal.

peptide group forms two hydrogen bonds NH · · · and C = O · · · to adjacent helices. Thus with three peptides per turn there are six hydrogen bonds, one to each of the adjacent chains. (See Figure 2.22)

The peptide repeat is 3.1 Å and the helix repeat = 3 × 3.1 = 9.3 Å. Introduction of other residues into the polyglycine helix generally causes disruption of the structure because the inter-helix displacement is increased by the bulk of side-chain residues and the hydrogen-bond length cannot be accommodated. Thus the presence of the pGII structure in nature appears to be limited to a form of silk fibroin that possess a long run of glycyl residues.

It is natural to ask, if the pGII structure is a 3_1 helix, then what is the pGI helix? The answer is that polyglycine from many solvents crystallizes in a β-sheet form (rippled sheet) and that the helix is the second form. Polyproline also has two conformations, one of which—the polyproline II helix (pPII)—is very similar to the polyglycine II (pGII) helix.

Polyproline II (pPII) helix

The proline residue, as we have seen previously, is unusual in that the pyrrolidine ring prevents rotation of the ϕ-angle. The ϕ-rotation is fixed at $-80°$ and the residue can only be compatible with helices having $\phi = -80°$. Neither α-helices nor β-sheets meet this requirement (see Table 2.3) and therefore the proline residue cannot be accommodated in such helices. However the pGII helix has $\phi = \pm 80°$ and therefore the prolyl residue will fit neatly into such a 3_1 helix (Figure 2.23). This constraint requires that the pPII helix be *only left-handed*. Furthermore, it is interesting to note that although pPII also crystallizes into a hexagonal array of helices in the solid state, since there is no hydrogen atom available in the peptide group, there is no intra-or interchain hydrogen bonding.

Figure 2.24 shows a schematic diagram of the pPII helix. The pPII/pGII helices are the basis of the collagen structure. Silk fibroins are also known with pGII and collagenlike structure.

The previous three conformations, α-helix, β-sheet, and pGII/pPII are the main regular conformations found in proteins. The other conformation suggested as energetically favorable in Table 2.3 is the ω-helix. Although not known in nature, the ω-helix is known for a synthetic polypeptide.

The ω-helix

The only other known conformation for polypeptides containing only L-amino acids is the unusual ω-helix. If a film of LH α-helical poly(β-benzyl-L-aspartate) cast from carbon tetrachloride is heated to 160° in vacuo, it undergoes a

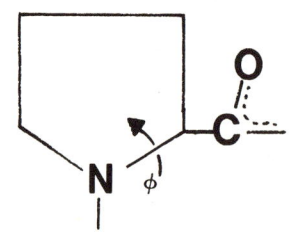

Figure 2.23 Restriction of the N—C_α bond rotation in the prolyl residue.

2. Secondary Structure   41

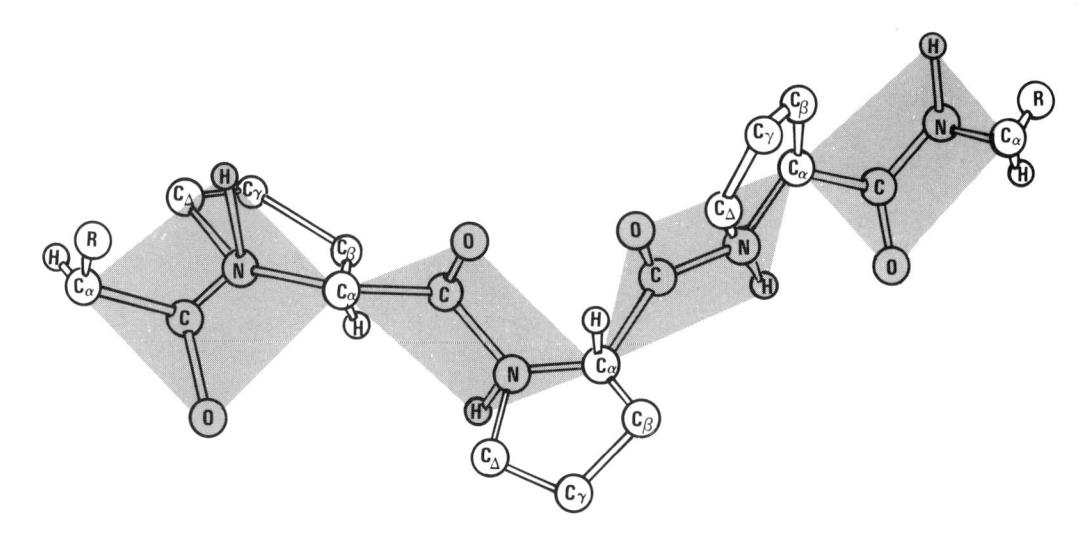

Figure 2.24 The 3_1 polyproline helix.

solid-state transition to a helix with fourfold symmetry (i.e., 4_1). The hydrogen-bonding scheme is similar to that of the α-helix but the symmetry and peptide repeat (1.325 Å) are quite different. Nevertheless the separation of helix turns $3.6 \times 1.5 = 5.4$ Å for the α-helix and $4 \times 1.325 = 5.3$ Å for the ω-helix are very similar. The end-on geometry of the ω-helix is shown in Figure

Figure 2.25 End-on view of the 4_1 ω-helix.

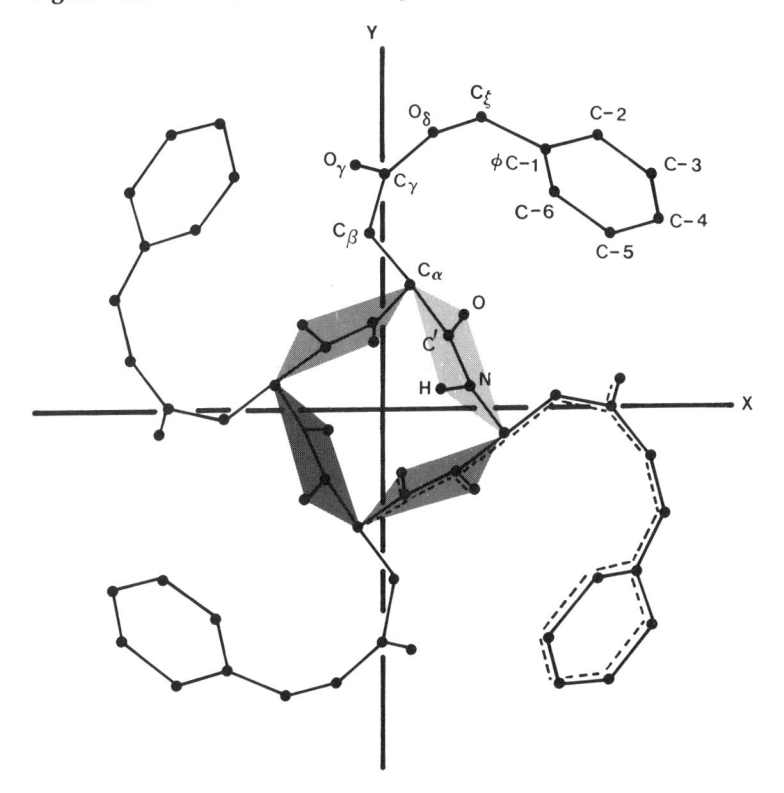

2.25. It is inferred that since it arose from a LH α-helix and has similar hydrogen-bond structure that it also is LH.

2.6 Summary of Helical Structures

We have seen that helical structures are generated by uniform residue interactions that manifest themselves through the polypeptide backbone rotation angles ϕ,ψ (and ω). The helices are characterized by (measurable) variables h, the peptide repeat; n, the number of residues per turn of the helix; and θ, the "twist" per residue* (= 360°/n). Also the symmetry of the helix is generally specified in terms of rational numbers (e.g., 18_5 helix).

The symmetry of the helix and the peptide repeat are obtained experimentally by x-ray diffraction (Chapter 4) and should bear some straightforward relation to ϕ and ψ. The sets of variables are in fact related by equations 2.11 and 2.12.

Table 2.4 Summary of Poly (-L-peptide) Helices

Structure	Helix Nom.	n	ϕ (deg)	ψ (deg)	ω (deg)	Å	θ (deg)	Comment
α-helix (RH)	18_5	3.6	− 57	− 47	+180	1.50	100	Commonly found in globular proteins and coiled in fibrous proteins.
α-helix (LH)	18_5	3.6	+ 57	+ 47	+180	1.50	100	Example, poly (-β-benzyl-L-aspartate).
β-sheet (parallel)	2_1	2	−119	+113	+180	3.25	180	Appears occasionally in adjacent chain segments of globular proteins.
β-sheet (anti-II)	2_1	2	−139	+135	+180	3.3–3.5*	180	Commonly found in proteins and synthetic polypeptides.
Polyglycine I								Similar to anti-II β, only a "rippled" β structure.
(pGII) (LH)	3_1	+3	− 80	+150	+180	3.1	120	Synthetic polyglycine may be LH or RH, LH pGII in some native silk.
Polyglycine II(RH)	3_1	−3	+ 80	−150	+180	3.1	120	
Polyproline I (pPI)	10_3	3.3(3)	− 83	+158	0	1.9	108	Synthetic poly-L-proline, only known: consistently *cis* peptide bond.
Polyproline II (pPII)	3_1	−3	− 80	+150	+180	3.1	120	Same conf. as LH pGII, found in supercoil form in collagen.
ω-helix	4_1	−4	+ 64	+ 55	+180	1.325	90	Found only in synthetic polypeptides, e.g., poly-β-benzyl-L-aspartate.

*The peptide repeat for anti-II β-structure depends upon the degree of "pucker" in the sheets.
RH = right-handed; LH = left-handed.
t (twist angle) = 360/n.

---

*The latest international convention uses "t" for the twist.

2. Secondary Structure   43

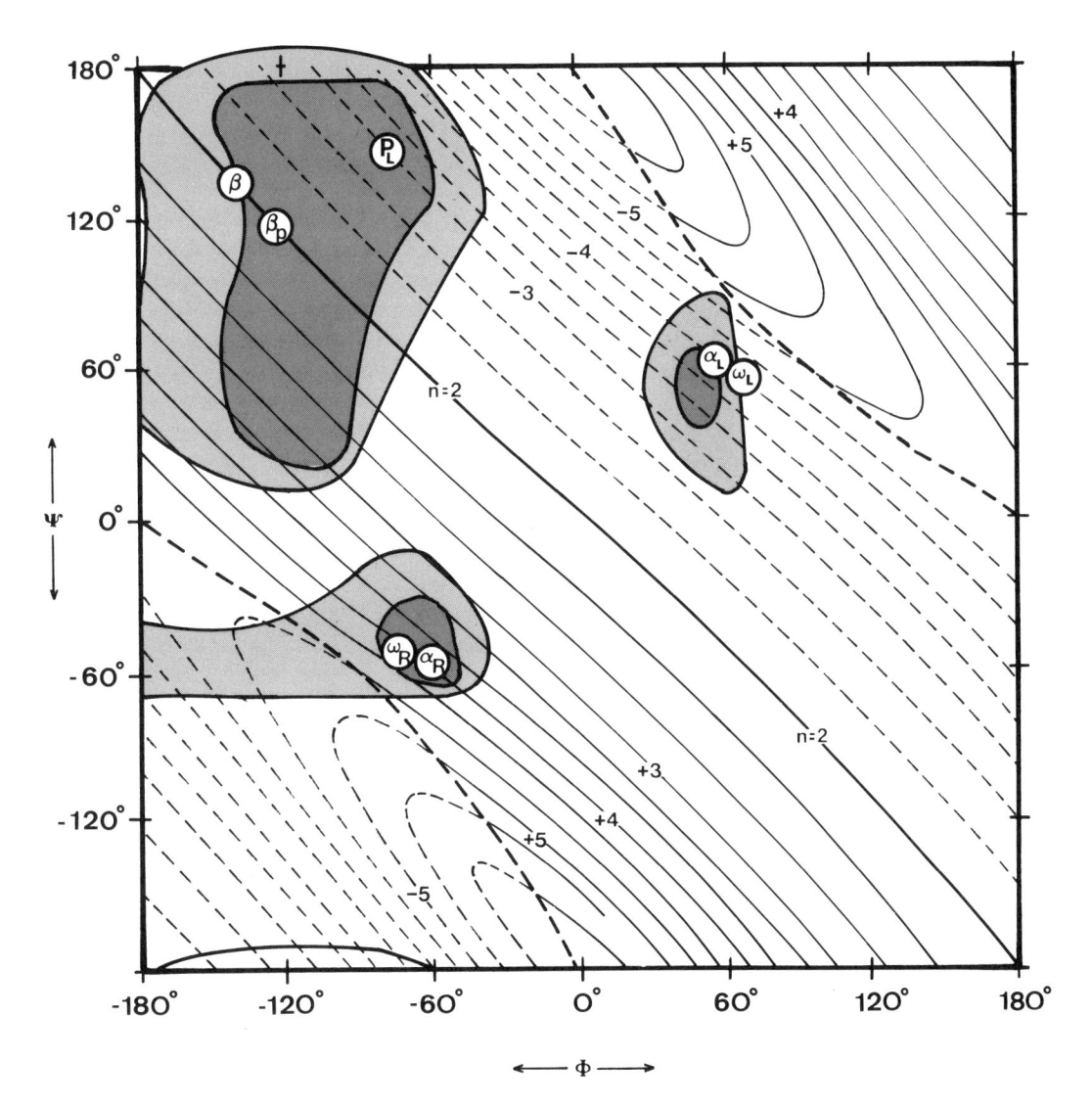

Figure 2.26 A steric energy plot for alanine showing the positions of the major conformations: α = α-helix (RH), β = antiparallel β-sheet. P_L is polyproline II = polyglycine II (LH). P_R is polyglycine RH. Contours are marked corresponding to the number of peptides per helical turn = n. Positive values are RH helices; negative values LH.

A summary of the parameters described in the preceding text is shown in Table 2.4

A ψ, ϕ plot for glycine/alanine with the positions of the various conformations shown is indicated in Figure 2.26.

2.7 The β-Bend

A conformational feature of many globular proteins that does not fit under the heading of helices is the β-bend. When polypeptide chains fold into globular structures, the chain changes direction abruptly, often through 180°. Although

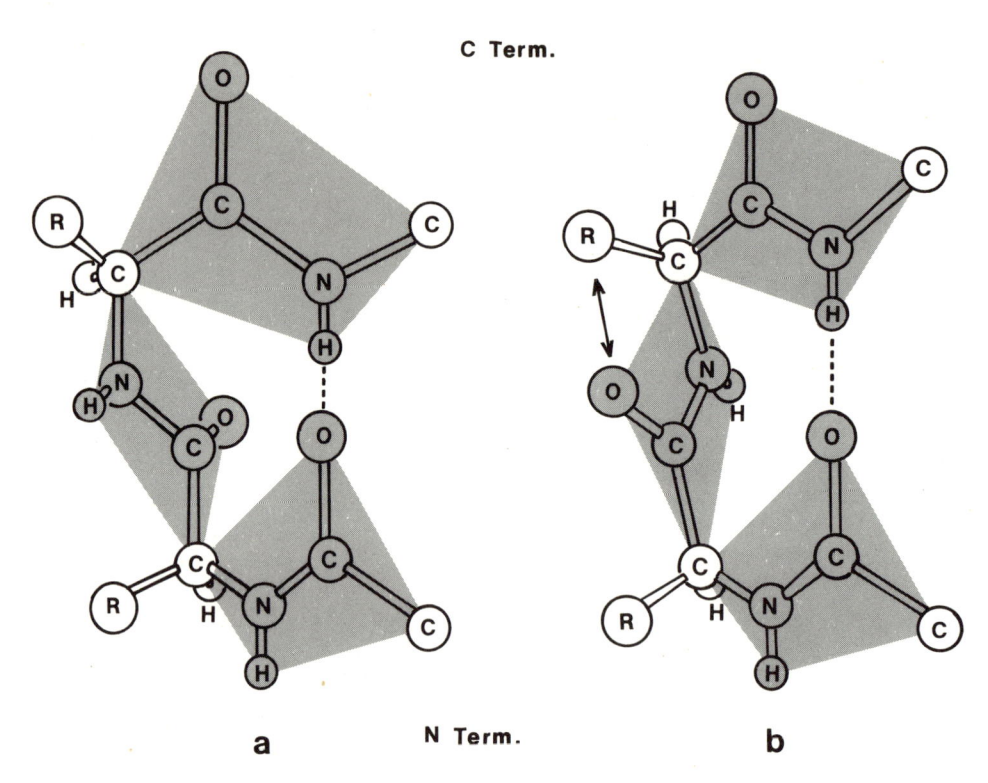

Figure 2.27 Schematic diagram of the two simplest forms of β-turn. Type b is stable only if a glycine residue is incorporated, because of short contact between the radical R on the C_α-carbon and the carbonyl oxygen, as shown.

we shall consider the reasons for such a change in the following chapter, it is pertinent to examine one of the most common features of polypeptide "hairpin" corners: the β-bend. Venkatachalam and others have pointed out that an energetically economical and space-saving way of turning a corner in a polypeptide chain involves four amino acid residues, which are often joined to chains in the β-sheet conformation. Two main possibilities are shown in Figure 2.27. In these figures three peptide residues (four C_α atoms) are shown. In terms of amino acid residues, the third from the N-terminus (3rd C_α) can only be glycine in the second representation because of the potential steric contact between the side chain R and the adjacent carboxyl. Nevertheless it can be shown that if this short contact can be accommodated, configuration (b) is energetically favored. Thus we would expect that glycine would be a common residue in the fold for two reasons: the steric fit for configuration (b) and also the large degree of rotational freedom that glycine possesses.

The tetrapeptide sequences found in β-bends of lysozyme are shown in Table 2.5. On a random basis, it would have been expected that only one glycyl residue would occur in the third position from the N-terminus, but as can be seen, the residue is present in four different β-folds, suggesting that its presence is helpful to the formation of such folds. This type of statistical approach to the role of residues has been utilized in the attempted prediction of peptide and protein conformation, as will be seen later.

2. Secondary Structure

Table 2.5 Bends Found in Egg White Lysozyme

Residue Number	Sequence
20–23	Tyr Arg *Gly* Tyr
36–39	Ser Asn Phe Asn
39–42	Asn Thr Gln Ala
47–50	Thr Asp *Gly* Ser
54–57	Gly Ile Leu Glu
60–63	Ser Arg Trp Trp
66–69	Asp Gly Arg Thr
69–72	Thr Pro *Gly* Ser
74–77	Asn Leu Cys Asn
85–88	Ser Ser Asp Ile
100–103	Ser Asp *Gly* Asp
103–106	Asp Gly Met Asn

2.8 Random Structures

We may define two classes of random structures: random in space but fixed in time; and random in space and time.

Irregular Structure

Globular proteins possess a primary structure in which there is no definable repeat pattern of residues and thus no repeat pattern of ϕ, ψ angles in significant stretches of the polypeptide chain. The residues find themselves in essentially different but constant environments at temperatures below the thermal denaturation point. Furthermore, each polypeptide chain of a specific type of globular protein will possess the same irregular conformation. It is true, of course, that there are minor fluctuations in configuration and perhaps conformation caused by thermal vibration and by substrate binding, but for the purposes of this book we shall refer to polypeptide chain conformation in native globular protein structure that is not one of the standard conformations spelled out above as "irregular." It is notable that the irregular structure of each protein is unique.

Random-Denatured Protein

If a protein is thermally denatured so that its normal three-dimensional structure is destroyed, the unfolded chains undergo freely fluctuating motion such that in any instant of time, the polypeptide chains from any given protein are *different* in conformation. The ϕ,ψ angles are not fixed for any peptide pair and rotate within the energetic constraints placed on the residue by surrounding residues. Although the time-average conformation of polypeptide chains for different proteins will, to some extent, be different depending on sequence, molecular weight, and residue content, the correlation (or more specifically lack of correlation) of position of distant residues may be similar for different proteins and expressible mathematically. This situation we will consider as "true random" with the proviso that the *trans*-configuration of the peptide bond is usually maintained.

Coil Structures

In an attempt to model denatured protein structure, a great deal of attention has been paid to the poly(amino acids) poly(glutamic acid) and polylysine. As noted previously, glutamate and lysyl residues are ionized at neutral pH. At high pH (> 10.5) the lysyl residue becomes deprotonated and the polypeptide takes on an α-helical conformation (Figure 2.28). Similarly poly(glutamic acid) is ionized (deprotonated) at neutral pH but becomes α-helical at pH < 4.5 (Figure 2.29).

Interest has centered around the charged forms, which have variously been called "random coil," "extended coil," "charged coil," and so on. The main point of contention is whether there is any short-range correlation of (ϕ,ψ) angles indicative of nonrandom structure. Mutual repulsion between the charges in the side chains suggests that a relatively extended chain might be formed. Indeed conformational energy predictions for ionized poly(glutamate acid) indicate a left-handed extended helix of approximately 2.4 residues per turn and a rise per residue of 3.2 Å. ($\phi = -96°$, $\psi = 141°$, i.e., somewhat similar to pPII.) However, such calculations do not include the effect of thermal fluctuations. Chain statistical approaches suggest that, whereas the ion repulsion does in fact prevent the polypeptide chain from collapsing into a denatured protein-type structure, there is no need to assume nonrandom structure. Circular dichroism measurements (which are not necessarily unambiguous) support the concept that there is indeed a short-range correlation of residue conformation, and for this reason we shall adhere to the terminology "coil" for these ionized structures rather than "random coil." In any case there is little doubt that the "coil" is distinctly different in its properties from the irregular and true random conformations.

2.9 Π_{LD} or β-Helices

Native Peptides

So far we have examined conformations that involve polypeptides having peptide units in the all L- (or all D-) configuration. A new class of polypeptide structures that has emerged in the past few years is that based on the alternation of L- and D-residues. In the native state, D-residues are not found in mammalian proteins but a large number of bacterial peptides and peptide antibiotics contain D-residues. Such is the case for the pentadecapeptide antibiotic gramicidin A, whose primary sequence is

   N-formyl-L-Val-Gly-L-Ala-D-Leu-L-Ala-D-Val-L-Val-D-Val-L-Trp-D-Leu-L-Trp-D-Leu-L-Trp-D-Leu-L-Trp-HN-CH$_2$-CH$_2$-OH

This antibiotic peptide is believed to form ion-conducting channels through the lipid bilayer of cell membranes. The channel is apparently formed by the peptide helix such that a hole of appropriate size for ion transport lies down the axis of the helix.

Although the helix nomenclature is not yet standardized, the Π-nomenclature arose because of a structure proposed for poly-L-amino acids (but never

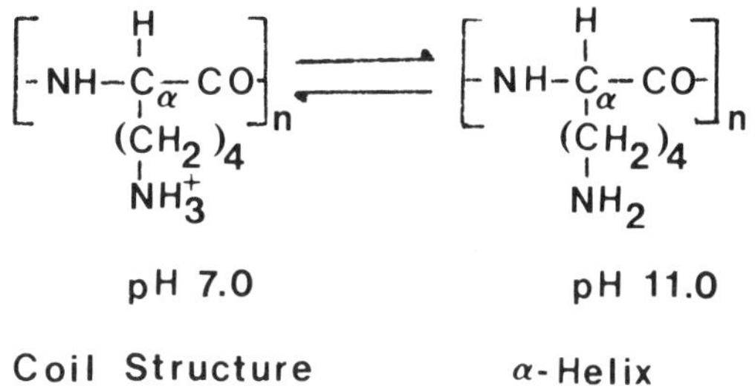

2. Secondary Structure

Figure 2.28 The correlation between conformation and ionization state for polylysine.

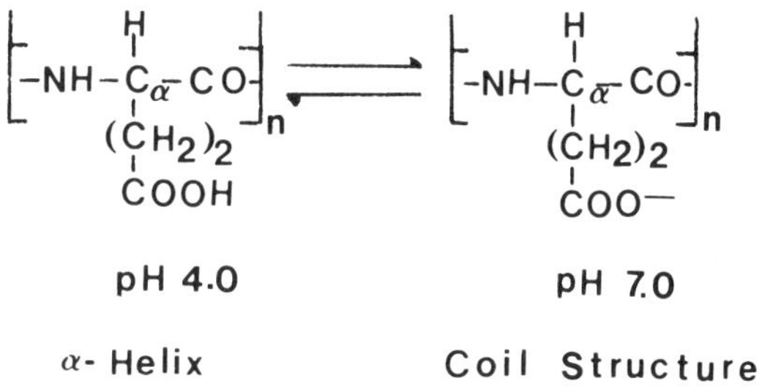

$$\left[-NH-\underset{\underset{\underset{NH_3^+}{|}}{(CH_2)_4}}{\overset{\overset{H}{|}}{C_\alpha}}-CO-\right]_n \rightleftharpoons \left[-NH-\underset{\underset{\underset{NH_2}{|}}{(CH_2)_4}}{\overset{\overset{H}{|}}{C_\alpha}}-CO-\right]_n$$

pH 7.0   pH 11.0

Coil Structure   α-Helix

Figure 2.29 The correlation between conformation and ionization state for poly(glutamic acid).

$$\left[-NH-\underset{\underset{\underset{COOH}{|}}{(CH_2)_2}}{\overset{\overset{H}{|}}{C_\alpha}}-CO-\right]_n \rightleftharpoons \left[-NH-\underset{\underset{\underset{COO^-}{|}}{(CH_2)_2}}{\overset{\overset{H}{|}}{C_\alpha}}-CO-\right]_n$$

pH 4.0   pH 7.0

α-Helix   Coil Structure

found), which is the "tightest" of the possible Π-helices, having 4.4 residues per turn. However, the β-helix nomenclature is particularly sensible because the hydrogen-bonding scheme is identical with the parallel β-sheet structure. Since the prevalent nomenclature in the literature is that of the Π helix we shall use it for the present purposes. Furthermore, we shall use the $\Pi_{LD}{}^6$, $\Pi_{LD}{}^8$, etc. nomenclature to indicate 6, 8, and so on, residues per turn (of which half will be L- and half D-). Figure 2.30 shows the side and end view of a $\Pi_{LD}{}^6$ helix. It is notable that in the gramicidin A sequence, Gly plays the role of a D-residue, allowing the formation of the Π-helix.

Most efforts, outside of those directed at small peptide structure, have centered recently on defining the range of possible Π-helices using synthetic sequential poly(amino acids). For example, poly(D-Ala-L-Ala), poly(D-Glu(OBzl)-L-Glu(OBzl), and poly(D-Val-L-Val) have been synthesized and reported.

Synthetic poly(L,D-peptides)

Due to improving methods over the past few years it has been possible to study various synthetic poly(amino acids) and their derivatives, which have strictly repeating L- and D-residues. The most thoroughly studied of these has been poly(-γ-benzyl-D-L-glutamate). Several structures are known for this polypeptide, including an α-helix (in which the rise per residue h = 1.47 Å, i.e. slightly smaller than the more usual 1.49–1.50 Å), at least two forms of Π_{LD} helices, and perhaps a ω-helix.

The conformational energetics of poly(L,D peptides) have been studied and indicate some interesting possibilities. It is of course possible to synthesize polypeptides of the type (L-A, D-A)$_n$ or (L-A, D-B), where A and B are α-helix or β-sheet favoring residues. Since these are polypeptides we have the possibility of four unique ϕ,ψ angles (two pairs), in which each

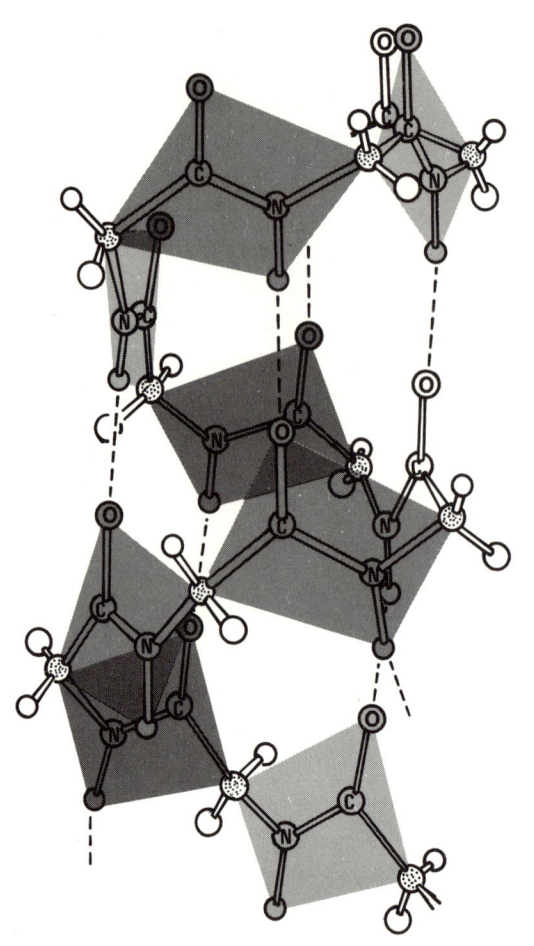

Figure 2.30 Representation of the $\Pi_{LD}{}^6$-helix.

member of the pair can lie in the appropriate energy region for α-helix or β-sheet. Most work, as previously mentioned, has been performed with the homopolydipeptide of γ-benzyl glutamate.

For the Π_{LD}-helix there are at least three helical possibilities based on ϕ,ψ angles in the parallel (and antiparallel) β-sheet rotation region of the conforma-

Table 2.6 Conformation Parameters for π-Helices

	$\pi_{LD}{}^4$	$\pi_{LD}{}^6$	$\pi_{LD}{}^8$
ϕ_L	−100 to −80	−112 to −102	−128 to −120
ψ_L	100 to 115	112 to 124	144 to 152
ϕ_D	110 to 130	140 to 152	132 to 142
ψ_D	−65 to −85	−124 to −134	−142 to −148
∠ (∠NC$_\alpha$C)	106°	110°	110°
n (dipeptides/turn)	2.25 to 2.32	3.12 to 3.18	4.09 to 4.14
h (Å) per dipeptide	2.38 to 2.44	1.54 to 1.58	1.16 to 1.20
E (Kcals)	−19.4	−17.9	−14.2
r (Å)	0.6	1.7	2.8

tion map. In Table 2.6 it can be seen that the three simplest Π_{LD}-helices that can be built from a parallel β-sheet bonding structure have approximately four, six, or eight peptides in one turn and we shall prefer to refer to these helices as $\Pi_{LD}{}^4$, $\Pi_{LD}{}^6$, $\Pi_{LD}{}^8$ respectively. The $\Pi_{LD}{}^4$-helix has been calculated to be the most stable structure and is believed to be one of the structures that poly(D-Glu-OBzl)-L-Glu(OBzl) assumes. It has a hole down the middle of the helix of radius 0.6 Å. As the number of residues per turn increases, the hole increases in size and the rise per residue decreases.

Apart from the single-stranded Π_{DL}-helices, poly(D–L peptides) also exhibit double-stranded polymorphism. In this case the (Π, Π_{DL})-helices belong to the same family of structures but the second strand is antiparallel to the first, giving an antiparallel β-type hydrogen-bonding scheme. It is noticeable that Π_{LD}-helices of left- and right-handedness are isoenergetic for infinite chain length. Values of ϕ and ψ corresponding to only the RH helices are quoted in Table 2.4.

Another structure found for low-molecular-weight poly(γ-benzyl-D-L-glutamate) is the sheet structure based on $\alpha+$ and $\alpha-$ conformations, in which all peptide carbonyls are pointing on the same side of the plane defined by the C_α atoms and the NHs are all on the other side. This structure has been called the β_{DL} or α-pleated sheet form. The latter is preferred here.

Further Reading

Dickerson, R.E. and I. Geis. *The Structure and Action of Proteins.* Benjamin, Menlo Park, California, Chapters 1 and 2, 1969.

Jaenicke, R. *Protein Folding.* Elsevier-North Holland, New York, 1980.

Ramachandran, G.N. Conformation of Polypeptides and Proteins. *Advances in Protein Chemistry* 23:284–438, 1968.

Schulz, G.E. and R.H. Schirmer. *Principles of Protein Structure.* Springer-Verlag, New York, Chapter 5, 1979.

3

Prediction of Conformation

One of the major aims of the molecular biophysicist for several decades has been the prediction of secondary, tertiary, and higher orders of structure from the sequence of individual amino acid residues without recourse to experimental methodology. Although this goal has not been fully realized, a number of methods have been developed that give us insight into the required features and that have been applied with some success.

3.1 Conformation-Directing Properties of Amino Acids

One method of approaching protein structure has been the study of simple model compounds. The simplest of these model compounds is the poly (amino acid). The concept is that if a polymer, $(A)_n$, is normally in an α-helical conformation in the solid state, then this is the preferred conformation for residue A and thus in some sequence ABC . . . , A would be an α-helix directing residue. If B and C were also α-helix directing residues, then the polypeptide $(ABC)_n$ would be expected to be α-helical. If B and/or C were not α-helix directing residues, then, depending upon the ability of A to "force" the conformation, an α-helix or some other conformation would result.

The conformation of most of the poly(α-L-amino acids) is now well established from solid-state structures and the data for the most favored and additional conformations are listed in Table 3.1.

Further information on the conformation-directing role of residues may be obtained by studying synthetic sequential polypeptides such as (Gly$_2$-Ala)$_n$, in which the alanyl residue appears compatible with the pGII structure. Often, however, the insertion of β-directing residues such as Val, Ser, and so on, into a sequence containing α-helix directing residues produces a random conformation.

3. Prediction of Conformation

Table 3.1 Conformational Properties of Poly(Amino Acids)

Polymer residue	Conformation 1	Conformation 2	Notes
Gly	β-sheet	pGII	(Rippled sheet)
Ala	α-helix	β-sheet	Oligomers are β-sheet
Val	β-sheet		
Leu	α-helix		
Ile	β-sheet		
Ser	β-sheet		
Thr	β-sheet		
Asp*	α-helix		
Glu*	α-helix	(β-sheet)	Salts Ca, Sr, Ba are β-sheet.
Lys†	α-helix	β-sheet	Salts, HPO$_4$, are β-sheet; also product from high pH, high temperature is β-sheet.
His	α-helix		
Phe	α-helix		
Tyr	β-sheet		
Trp	α-helix		
Cys	β-sheet		
Met	α-helix		
Pro	pPII	pPI	
Arg	α-helix		

All β-sheets are antiparallel.

*Polyaspartic acid and polyglutamic acid in the solid state are in the neutral carboxyl form. The conformational influence in solution at neutral pH where the side chain is ionized is one of α-helix disruption.

†Polylysine in the solid state is normally in the form of the HBr salt. The charged residue in solution at neutral pH would have an α-helix disruptive influence.

Nevertheless the general rule may be formulated that:

1. Amino acid residues that are not branched at the carbon and do not have an active group attached to the C_β tend to form α-helices or appear in α-helical regions in proteins (Figure 3.1).
2. Amino acid residues with branched C_β configurations form β-sheets. This observation is in accord with the deductions based on conformational energy calculations in Chapter 2. The branched C_β side chain

Figure 3.1 Branching at C_β causes conformational effects based on steric volume. Unbranched side chains with carboxyl or amino groups also fall into the α-directing category.

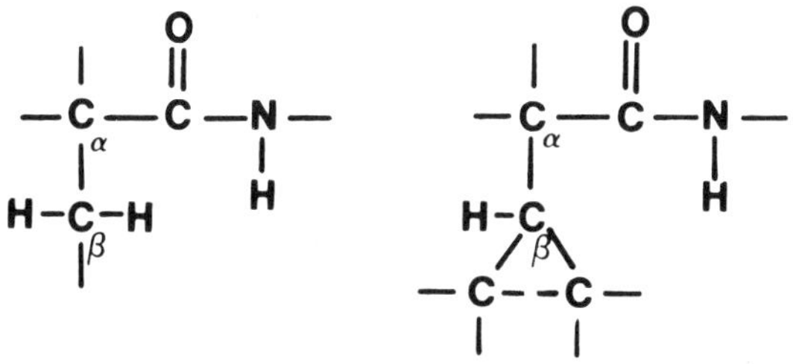

would be crowded in the α-helix conformation and is more sterically favored in the more extended β-sheet.
3. Amino acid residues with active groups on the C_β tend to form β-sheets. Thus the serine hydroxyl and cysteine sulfhydryl groups act as interchain hydrogen-bonding sites.

3.2 Application of Amino Acid Conformational Properties

The application of the previous observations to synthetic sequential polypeptides is generally self-consistent and extends fairly well to fibrous proteins.

Example 1. Silk Fibroin

The polydipeptide $(GlyAla)_n$ is thought to be the simplest model for the silk fibroin produced by the silkworm, *bombyx mori*. Writing down the conformational directing properties of the residues, we obtain the accompanying diagram.

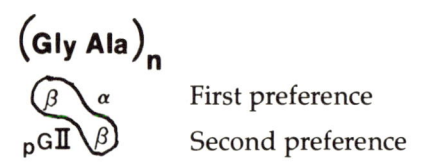

$(Gly\ Ala)_n$

$\beta \quad \alpha$ — First preference
pGII $\quad \beta$ — Second preference

The residues have in common the choice of β-sheet structure (though it is the second choice for alanine) and we would expect a β-sheet. (The third choice of Ala would be pGII, but since this is probably the second choice of Gly, the β-form is preferred.) The most common form of $(GlyAla)_n$ is indeed β-sheet, as predicted.

Glycine and alanine account for 74% of the amino acid residues in *bombyx mori* fibroin and we would expect the β-sheet conformation to be prevalent for the silk fibroin too, which it is. A more complete sequence is $[SerGly(AlaGly)_x]_n$. Since Ser is the next most common residue (12%) and it also favors the β-structure, the mutual influence of these residues consolidates the β-sheet conformation.

Example 2. Collagen

Without specifying the sequence of collagen, the main residues from amino acid analysis are glycine 33%, proline plus hydroxyproline 22%, and alanine 11%. Applying the conformational rules based on Table 3.1, we find

	Gly	Pro	Ala	
Conformation I	β	pPII	α	First preference
II	pGII	(pPI)	β	Second preference
III			pGII	Third preference

3. Prediction of Conformation

The only conformation capable of accommodating all three residues is polyproline II (pPII), which is, of course, the same as LH polyglycine II (pGII).

We would therefore expect the collagen structure to be based on the polyproline II helix—which it is.

Example 3. Tropomyosin

Tropomyosin does not possess a preponderance of any one amino acid but has Glu 25%, Ala 13%, Lys 13%, Asp 11%, Leu 9%. Most of these residues may be ionized and this poses an additional constraint, which we shall examine in more detail later. If we treat the residues in terms of their solid-state preference, then all are α-helix formers and we would expect the tropomyosin structure to be based on the α-helix, which it is.

From the preceding comments and examples, it appears that the conformational properties of amino acids in poly(amino acids), sequential polypeptides, and fibrous proteins are similar and thus we would be on firm ground extrapolating to nonsequential peptides, polypeptides, and proteins (globular proteins). Although at first sight there does seem to be a correlation between the concentration of α-helix forming residues and α-helix content of globular proteins (Figure 3.2), the correlation is deceptive. A detailed study of the relation between residue sequence and con-

Figure 3.2 Apparent correlation between α-helix directing residues (based on Table 3.1) in globular proteins and the percentage of α-helix derived from x-ray crystallographic or circular dichroism studies. Tm—tropomyosin; Mg—myoglobin; Hg—hemoglobin; Fib—fibrinogen; Lys—lysozyme; Ins—insulin; Rib—ribonuclease; Chy—chymotrypsinogen. △—x-ray; ○—circular dichroism.

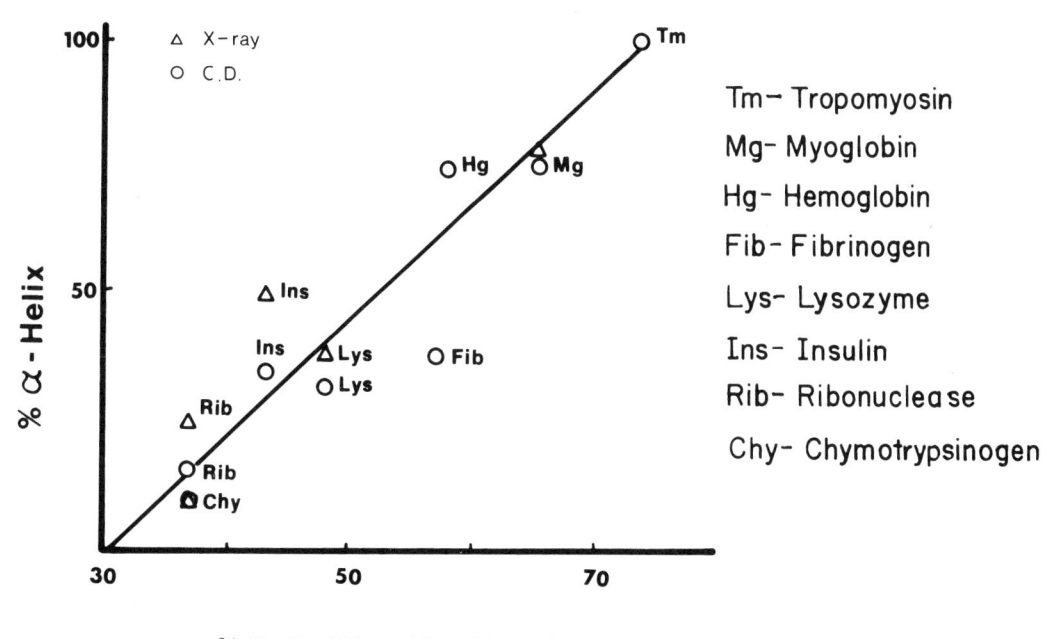

formational properties shows a disappointing lack of correlation in most cases—that is, the regions of α-helix in a globular protein do not necessarily contain only α-helix forming residues. Conversely, α-helix forming residues are often abundant in nonhelical regions of the protein.

An analysis of the inadequacies of the preceding approach shows that it is basically a first-order approach; that is, it depends on nearest neighbor interactions. The interaction between two similar residues in a poly(amino acid) may be quite different from the interaction of one of those residues with a different residue in a protein; that is, there is no real basis for assuming that the "conformation-directing property" (which is an assumption in itself) of a residue is the same in a poly(amino acid) and a protein.

In order to circumvent this problem, a great deal of effort has been expended on the peptide pair approach, encompassing the tenets of conformational energy calculations. The major assumption in this approach is that the energy of interaction of nearest neighbor peptide groups controls the chain conformation. Thus one could proceed, pairwise, along a polypeptide chain calculating the most favored arrangement of each pair, until the total chain conformation has been predicted. Unfortunately this approach is not entirely correct, mainly because interactions lying outside of the nearest neighbor residues clearly play a major role in determining chain conformation. As an example, it is clear from Figure 2.16 that residue no. 1 is hydrogen bonded to residue no. 4 in the α-helix, and so on.

3.3 The Chou–Fasman Approach

The approach proposed by Chou and Fasman for analysis of secondary structure in peptides and proteins uses elements of the residue conformer approach, but superimposes on it certain sequential constraints. Instead of the helix-forming or directing properties being derived from synthetic poly (amino acids) and polypeptides, they are derived from known structures of globular proteins; that is, the probability of amino acids falling in α-helical, β-sheet, and regions of irregular chain structure in most of the globular proteins of known structure (from x-ray diffraction) has been assessed. From this analysis it has been possible to assign breakers, formers, and indifferent for α-helical and β-sheet regions in proteins.

The conformational parameter for a residue lying in an α-helical region P_α is calculated from the probability:

$$f_\alpha = \frac{\text{No. of x residues in } \alpha \text{ regions}}{\text{Total no. of x residues}}$$

$$P_\alpha = \frac{f_\alpha}{\langle f_\alpha \rangle}$$

where $\langle f_\alpha \rangle$ is the average probability for each different residue lying in an α-helix region.
Similarly,

$$P_\beta = \frac{f_\beta}{\langle f_\beta \rangle}$$

3. Prediction of Conformation

Table 3.2 Assignment of Amino Acids as Formers, Breakers, and Indifferent for α-Helical and β-Sheet Regions in Proteins*

α-Helical parameters		β-Sheet parameters	
Residue	P_α	Residue	P_β
Glu$^{(-)}$	1.51 ⎤	Val	1.70 ⎤
Met	1.45	Ile	1.60 H_β
Ala	1.42 H_α	Tyr	1.47 ⎦
Leu	1.21 ⎦	Phe	1.38 ⎤
Lys$^{(+)}$	1.16 ⎤	Trp	1.37
Phe	1.13	Leu	1.30
Gln	1.11	Cys	1.19 h_β
Trp	1.08 h_α	Thr	1.19
Ile	1.08	Gln	1.10
Val	1.06 ⎦	Met	1.05 ⎦
Asp$^{(-)}$	1.01 ⎤	Arg$^{(+)}$	0.93 ⎤
His$^{(+)}$	1.00 ⎦ I_α	Asn	0.89
Arg$^{(+)}$	0.98 ⎤	His$^{(+)}$	0.87 i_β
Thr	0.83 i_α	Ala	0.83 ⎦
Ser	0.77 ⎦	Ser	0.75 ⎤
Cys	0.70 ⎤	Gly	0.75
Tyr	0.69 b_α	Lys$^{(+)}$	0.74 b_β
Asn	0.67 ⎦	Pro	0.55
Pro	0.57 ⎤ B_α	Asp$^{(-)}$	0.54 ⎦
Gly	0.57 ⎦	Glu$^{(-)}$	0.37 ⎦ B_β

*H, strong former; h, former, I, weak former; i, indifferent, b, breaker, B strong breaker.

Table 3.2 gives P_α and P_β values for 20 amino acids. In each case six classifications are made: H_α (strong α-helix former); h_α (α-former); I_α (weak α-former); i_α (α-indifferent); b_α (α-breaker); B_α (strong α-break). Similarly, assignments are made H_β . . . B_β for the β-sheet influence of peptide residues.

It is interesting to note that conformational directing properties in Table 3.2 are often distinctly different from those in Table 3.1 (e.g., Ser and Gly, which are β-formers in poly(amino acids), are found to be conformationally β-indifferent in globular proteins). Phe, Trp, and Arg, which are α-helix formers in poly(amino acids), would apparently prefer to be in a β-conformation in globular proteins and Asp, Trp, and Arg are not rated as likely α-helix formers in the Chou–Fasman scheme.

Apart from the frequency of occurrence of residues in α-helical or β-sheet segments of proteins, Fasman and Chou have examined the frequency of residues occurring in the β-bend (see Chapter 2). As explained previously, the β-bend may be treated as a tetrapeptide sequence, in which the first (N-terminal) and fourth residues are hydrogen-bonded together. Since each residue in a β-bend has a unique set of ψ,ϕ angles and is in a unique environment, it is appropriate to examine the frequency of occurrence in each of the four positions. These data are presented in Table 3.3. In this table, f is the frequency of occurrence of the particular residue in one of the four positions; for example, $f_i = 0.060$ for alanine means that, in the proteins surveyed, alanine occurred in the first β-bend position 6.0% of its total appearance in the proteins. The average occurrence of all residues in

Table 3.3 Frequency of Occurrence of Amino Acids in β-Turns

Amino acid	f_i	f_{i+1}	f_{i+2}	f_{i+3}
Ala	.060	.076	.035	.058
Arg	.070	.106	.099	.085
Asn	.161	.083	.191	.091
Asp	.147	.110	.179	.081
Cys	.149	.053	.117	.128
Gln	.074	.098	.037	.098
Glu	.056	.060	.077	.064
Gly	.102	.085	.190	.152
His	.140	.047	.093	.054
Ile	.043	.034	.013	.056
Leu	.061	.025	.036	.070
Lys	.055	.115	.072	.095
Met	.068	.082	.014	.055
Phe	.059	.041	.065	.065
Pro	.102	.301	.034	.068
Ser	.120	.139	.125	.106
Thr	.086	.108	.065	.079
Trp	.077	.013	.064	.167
Tyr	.082	.065	.114	.125
Val	.062	.048	.028	.053

Approximately 30% of all amino acid residues examined have been in β-turns.

the first, second, and so forth, positions is given by $\langle f_i \rangle$, $\langle f_{i+1} \rangle$, and so on, and is 0.086. Residues that are present in a given position in the turn substantially more than average, are underlined in Table 3.3.

The only residue that occurs in above-average probability in all four positions in the table is Ser. The high occurrence of glycine in bends comes as little surprise because of its large degree of rotational freedom. More surprising is the high occurrence of aspartic acid and serine amounting to 52% and 49%, respectively, of all such residues surveyed. The sampling is of course limited to a relatively small number of proteins and a wider sampling may show some variation in the probability function. However, the table already shows an interesting reverse correlation with the α-helix forming potential of Table 3.2.

The Procedure for Predicting Conformation (Chou–Fasman Method)

1. Write out the protein sequence with helix and sheet parameters underneath. As an example we shall use the first 50 residues of the (egg white) lysozyme sequence (since it possesses both α-helical and β-sheet regions). See Table 3.4.

2. Search for α-helix nucleation sites. Locate *four* helical residues (h_α or H_α) out of *six* residues along the polypeptide chain. Weak helical residues I_α count as $0.5h_\alpha$ (so that three h_α and two I_α residues out of six could also nucleate—initiate—a helix). Helix formation is not possible if the segment contains ⅓ or more helix breakers (b_α or B_α), or less than ½ formers.

3. Prediction of Conformation

For the lysozyme segment, helix nucleation can clearly be initiated at residues 7–12 (all H_α's) and 31–36 ($3H_\alpha$'s $2h_\alpha$'s and $i_\alpha = \frac{1}{2}h_\alpha$).

3. Extend the helical segment in both directions until terminated by a tetrapeptide with average P_α ($\overline{P}_\alpha$) < 1.00. Typical tetrapeptides causing *helix termination* will be b_4, b_3i, b_3h, b_2i_2, b_2ih, b_2h_2, bi_3, bi_2h, bih_2, i_4 and so on. The letters h and i in the preceding combinations may also be replaced by H and I.

Applying the P_α tetrapeptide criterion causes chain termination of α-helical segments between residues 4 and 5, between 15 and 16, between 26 and 27, and between 38 and 39. Thus there are two α-helical segments contained in chicken lysozyme 1–50 which are 5–15 and 27–38, using only the P_α tetrapeptide criterion. However, there is an hHib tetrapeptide that presumably could terminate the second helix at residue 35.

4. Proline cannot occur in the α-helix or at the C-terminal end. In addition Pro, Asp⁻, Glu⁻ prefer the N-terminal helical end. His⁺, Lys⁺, Arg⁺ prefer the C-terminal helical end. I_α assignments are given to Pro and Asp, near the N-terminus of its helix, and for Arg, near the C-terminus, if necessary, to satisfy the previous helix termination criteria.

Examination of our lysozyme helices shows that His is already at the C-terminal end of the 5–15 sequence, but Glu might prefer to terminate the sequence at residue 7, the N-terminus, as opposed to Arg, which prefers the C-terminus.

5. Any segment of six residues or more in a native (globular) protein with

Table 3.4 The Sequence of the First 50 Residues of (Egg White) Lysozyme*

	1	2	3	4	5	6	7	8	9	10
	Lys	Val	Phe	Gly	Arg	Cys	Glu	Leu	Ala	Ala
P_α	h	h	h	B	i	b	H	H	H	H
P_β	b	H	h	b	i	h	B	h	i	i

	11	12	13	14	15	16	17	18	19	20
	Ala	Met	Lys	Arg	His	Gly	Leu	Asp	Asn	Tyr
P_α	H	H	I	i	I	B	H	I	b	b
P_β	i	h	b	i	i	b	h	b	i	H

	21	22	23	24	25	26	27	28	29	30
	Arg	Gly	Tyr	Ser	Leu	Gly	Asn	Trp	Val	Cys
P_α	i	B	b	i	H	B	b	h	h	b
P_β	i	b	H	b	h	b	i	h	H	h

	31	32	33	34	35	36	37	38	39	40
	Ala	Ala	Lys	Phe	Glu	Ser	Asn	Phe	Asn	Thr
P_α	H	H	h	h	H	i	b	h	b	i
P_β	i	i	b	h	B	b	i	h	i	h

	41	42	43	44	45	46	47	48	49	50
	Glu	Ala	Thr	Asn	Arg	Asn	Thr	Asp	Gly	Ser
P_α	h	H	i	b	i	b	i	I	B	i
P_β	h	i	h	i	i	b	h	b	b	b

*The α-helix and β-sheet forming parameters are listed.

$\overline{P}_\alpha > 1.03$ and $\overline{P}_\alpha > P_\beta$ and satisfying the previous criteria is predicted as α-helical.

Our two helices meet these requirements and we predict then two α-helical regions in this segment of lysozyme: (1) residues 7–15; and (2) residues 27–35.

6. Search for β-sheet nucleation regions. Locate clusters of *three* β-formers (h$_\beta$ or H$_\beta$) out of *five*. Formation of a β-sheet is unfavorable if the segment contains one-third or more β-sheet breakers (b$_\beta$ or B$_\beta$) or less than one-half β-sheet formers.

Searching for β-sheet nucleation regions in our lysozyme sequence, we find regions 2–6 and 39–43 or 40–44. Residues 26–30 also fit the β-sheet nucleation criteria but already lie within a predicted α-helical region and are excluded.

7. Termination of β-sheet regions occurs when a series of four residues has $\overline{P}_\beta < 1.00$ and/or a sequence of b$_3$i, b$_3$h, b$_2$i$_2$, b$_2$ih, b$_2$h$_2$, bi$_3$, bi$_2$H, bih$_2$, or i$_4$ occurs. The preceding combinations apply if h = h$_\beta$ = H$_\beta$ or if i = i$_\beta$ = I$_\beta$. Extending the β-regions for lysozyme produces runs at 1–6 (terminated by an α-helix) and 38–43.

8. Any segment of five residues or longer in a native protein with $\overline{P}_\beta > 1.05$ and $\overline{P}_\beta > \overline{P}_\alpha$ and satisfying the preceding criteria is predicted as β-sheet.

Both regions of lysozyme fit the above criteria.

9. Prediction of β-turns. If the probability product $\prod_{i=1}^{i=4} f_j > 1.0 \times 10^{-4}$ the tetrapeptide sequence is predicted as forming a β-turn. A rapid-scan method of finding β-bends suggested by Chou and Fasman is to find four successive residues with $P_\alpha < 0.9$ (i$_\alpha$'s or poorer). Since a low P_α generally correlates with a high β-bend former, many of the bends may be located in this manner.

Regions in our lysozyme fragment that meet these criteria are: residues 19–23, 45–50, also with the exception of Phe[38], the region 36–40 seems

Table 3.5 Locating β-Turns in Lysozyme*

β-Bend position			Residue number			
	19	20	21	22	23	
	Asn	Tyr	Arg	Gly	Tyr	
1	.161	.065	.099	.152		
2		.082	.106	.190	.125	
3			.070	.085	.114	
	45	45	47	48	49	50
	Arg	Asn	Thr	Asp	Gly	Ser
1	.070	.083	.065	.081		
2		.161	.108	.179	.152	
3			.086	.110	.190	.106
	36	37	38	39	40†	
	Ser	Asn	Phe	Asn	Thr	
1	.120	.083	.065	.091		
2		.161	.041	.191	.079	
3			.059	.083	.065	

*Using f values from Table 3.3.

†In the last sequence (36–40), two of the residues strongly favor a β-bend when the bend is located at 37–40; the probability product is, though slightly less than the required product value for a probable β-bend of 1.0×10^{-4}.

3. Prediction of Conformation   59

Table 3.6 Comparison of Experimental (X-Ray) and Predicted (Chou–Fasman) Conformation of the N-terminal Region (50 residues) of Hen Egg White Lysozyme

	Predicted	Experimental
α-helix	7–15	5–15
	27–35	25–35
β-sheet	1–6	1–3
	38–43	38–46
β-bend	20–23	20–23
		36–39
	47–50	47–50

possible. Writing down these sequences and the probability factors, it is easily seen that residues 20–23 are favored to form a β-turn in the first sequence, and residues 47–50 are more likely to be involved than residues 46–49 (Table 3.5).

In summary, we have assessed three conformational characteristics: α-helix, β-sheet, and β-bend in the first 50 residues of hen egg white lysozyme. Residues not accounted for in this approach are assumed to be of irregular conformation. We can compare these data with experimental data derived from x-ray crystallography as shown in Table 3.6.

The agreement between prediction and experiment is clearly quite good, though it should be remembered that lysozyme is one of the proteins from which the parameters P_α, P_β, and f_i were derived. The accuracy claimed for the prediction of each of the three conformational states in the set of proteins used for parameter determination was 77%. Subsequently, peptides and proteins of initially unknown structure have been approached. Predictions based on this method are in fact presented in the following section and later chapters of this book.

3.4 Other Predictive Methods and Comparison of Data

Several other methods for predicting protein structure are in the literature, including approaches assigning helix probabilities (Lewis, Scheraga), tripeptide (ϕ,ψ) correlations (Wu, Kabat), information theory (Pain, Robson), and qualitative forms of conformational analysis (Ptitsyn). All of these other methods involve computer techniques and are thus more complicated to use. Although in some cases these other methods seem to achieve better results in isolated cases, the simplicity and effectiveness of the Chou–Fasman approach have brought it into common practical use. A demonstration of the relative effectiveness of the various methods is given in Table 3.7 for the protein adenyl kinase. The conformation was predicted by each method without knowledge of the x-ray crystallographic data and the predictions and experiment were then compared.

3.5 Basis for Qualitative Conformation Prediction

The Chou–Fasman approach has several shortcomings, but in spite of its simplicity of applications it appears to be surprisingly effective. The first problem inherent in the approach is the establishment of the probability

Table 3.7 Comparison of Predicted Conformation* by Various Methods for Adenyl Kinase†

	α-helix			β-sheet			β-turn		
	Correct	Incorrect	Missed	Correct	Incorrect	Missed	Correct	Incorrect	Missed
Chou and Fasman	70	14	35	20	29	4	28	10	18
Levitt and Robson	42	8	63						
Finkelstein and Ptitsyn	79	12	26	13	5	11			
Burgess and Scheraga	36	19	69				33	14	13

*Numbers of residues in conformation compared with x-ray numbers.
†It should be noted that the pattern of success or failure for the above methods is not necessarily repeated for other proteins and peptides.

factors. Since these are derived from a specific set of proteins, the values obtained are for the average environment in which the residues find themselves in those particular proteins. Environment affects the helix-forming ability, as may be demonstrated by considering the effect of polar or nonpolar solvents on conformation. Polyalanine can, for example, exist in solution in an α-helix in a weakly interacting solvent but may easily be rendered random in a strongly interacting solvent. On this basis the Chou–Fasman method is not particularly effective in predicting the solution conformation of synthetic polypeptides, extended peptides, or some fibrous proteins. (Since there is no provision for pPII conformation, the collagen structure is not predicted.) The ultimate example of this analysis is that if fragments of globular proteins are "snipped" from the main structure by enzymatic or chemical means, the conformation generally changes despite the fact that the sequence is constant and the predicted conformation remains the same.

Reversing the argument leads to an interesting conclusion: if we take a small peptide and predict its conformation, the predicted conformation may be more likely for the peptide bound to its substrate than free in solution. The reasoning here is that the probability of a residue lying in a specific conformational region is based on its average environment, which is biased towards a hydrophobic, highly interactive (residue/residue) state in large proteins. For small peptides, the environment is hydrophilic in solution but becomes more hydrophobic in the substrate-bound state.

Helix Prediction

The concept of α-helix nucleation at specific sites runs parallel to the concept that collagen helices are nucleated in regions rich in proline which induce the appropriate ψ,ϕ angles. Particularly interesting is the concept that four out of six helix formers are needed for nucleation. It is easy to see that since, in the α-helix, third nearest neighbors hydrogen bond, correlations of this type are required.

β-Sheet Prediction

It is more difficult to understand why there is a correlation along a chain in the β-conformation when the conformation is maintained by interchain hydrogen bonding to residues outside of the counting scheme.

β-Bend Prediction

The concept that certain residues occur more frequently in β-turns than would be expected on a random basis suggests that such residues lower the energy required to cause folding and are thus important in developing the tertiary structure of globular proteins.

It has, in fact, been possible to study in some detail the energetic factors causing folding by the conformational energy methods outlined previously.

3.6 Energetics of Chain Folding

One of the β-turns that is not predicted by the Chou–Fasman method occurs between two β-sheet regions in lysozyme at residues Gly^{54} Ile^{55} Leu^{56} Glu^{57}. Energy calculations have been carried out to assess why the chain folds at this point. The calculation starts with an extended β-helix in aqueous solution (one strand of β-sheet) and compares the energy of folding in three different places: Tyr^{53} Gly Ile Leu; the native fold; and Ile^{55} Leu Glu Asn^{58}. (See Table 3.8).

The calculation shows that for this particular fold it takes *energy* to go from an extended to a folded state. It is thus the driving force of residues away from the fold that causes the folding to be energetically stable. Nevertheless, the fold forms in the position dictated by the most favorable set of tetrapeptides in the region. This feature would certainly explain why not all folds are predicted correctly by the Chou–Fasman approach. If a chain needs to fold in a certain region for thermodynamic reasons, it will use the best set of tetrapeptides available.

3.7 Prediction of Tertiary Structure

Broadly speaking, there have been two schools of thought concerning the subject of polypeptide-chain folding in globular proteins. One holds that protein folding proceeds by equilibrium energy minimization, with the final structure being the "global" energy minimum; the other suggests that folding proceeds along paths chosen, not by the state of the final product, but by the kinetics of the various available pathways. These two approaches have been

Table 3.8 Change of Energy on Fold Formation (Res 53–58 Lysozyme)

Residues in fold	Energy of folding (kcals)
52–57	+11.5
53–58	+2.3
54–59	Very large*

*Due to steric interference.

termed "thermodynamic" and "kinetic," respectively. The basic tenet of the kinetic approach is that initiation of a chain-folded structure requires that a positive free-energy barrier be traversed. The energy for traversing this barrier is provided by thermal fluctuations. In this theory the driving force towards folding is the internal free energy arising from molecular forces, and the opposing forces are folding and desolvation energy.

Energy-Minimization Procedures

If the tertiary structure of globular proteins is determined by the absolute minimum free-energy conformation arising from interactions between all atoms within the molecule (and also surrounding solvent molecules, counterions, etc.), then the detailed methodology of conformational energy calculations should provide the detailed three-dimensional geometry. However, it is immediately obvious that the complexity of such a calculation is far beyond the reach of even the most sophisticated computer methods, except for peptides with very few residues. Thus systematic quantitative approaches of this type, elegant though they are, have often failed because of the need to make simplifying assumptions that are difficult to validate.

Kinetic Approaches

It is evident that as the protein evolves from its biologic synthetic source, the ribosome, one end is free to undergo thermodynamic fluctuations in conformation and folding while the other is attached to the ribosome. Although the protein may only "crystallize" into its three-dimensional geometry after syn-

Figure 3.3 Kinetic nucleation of three-dimensional structure in a nascent protein.

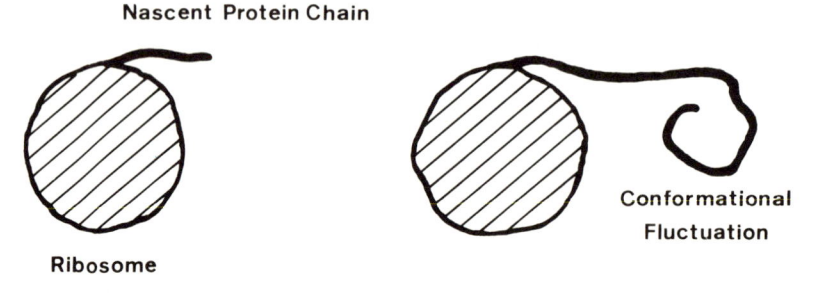

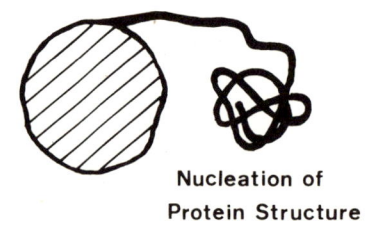

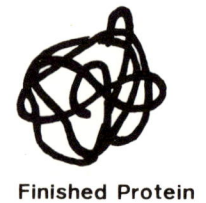

3. Prediction of Conformation

thesis is complete, such a process seems unlikely compared with the nucleation of a three-dimensional structure before chain completion. The remainder of the chain would then "wrap around" the nucleus in a stable structure. Such a process is shown schematically in Figure 3.3. Some caution should be exercised in regarding this procedure as a literal interpretation, since often a protein or its subunits are synthesized in precursor form and assemble only after transport through the cell membrane and after chain scission.

The kinetic approach implies that the final structure is determined by microthermodynamic chain fluctuations that nucleate a structure which, though stable thermodynamically, may not be the ultimate low-energy conformation for the total chain sequence.

If the preceding postulate is true, then:

1. Denaturation and refolding of a protein would not necessarily produce the same tertiary structure.
2. Conformational energy calculations seeking atomic resolution of maximum internal interactions may not be expected to predict the detailed structure.

The fact that commercial polymers produce chain-folded crystals, which are believed to form in accordance with microthermodynamic fluctuation theory (nucleation theory), suggests a parallel in the biologic situation.

Application of Tertiary Representation

If we admit that random chain fluctuations produce the folded tertiary structure of globular proteins, then it may be instructive to develop an approach that does not seek atomic resolution but that follows the gyrations of the polypeptide chain at a lower level of resolution. Such an approach is provided by a method devised by Dr. M. Levitt. It incorporates an energy-minimization procedure but differs in two important respects from previous approaches.

The residue side chains are treated in terms of an equivalent radius or volume which they occupy in the condensed state and instead of separate ψ,ϕ angles, a composite rotational function is used. The second innovation is that the folding process is simulated by molecular dynamics in which the molecule first heads towards a minimum energy conformation and then is randomly perturbed. We shall now examine, briefly, the method and type of data generated and study some of the implications.

Molecular Geometry

Side chains of peptide residues in proteins appear to take up atomic configurations very similar to those found in crystals of the individual amino acids. Thus it is suggested that side-chain rotations do not radically affect protein structure. The side chain is simplified by treating it as a single effective atom located at the centroid of the side-chain atom positions (C_α is treated as part of the side chain). Van der Waal's equivalent radii for the side chains are then as given in Table 3.9.

Rotation angles are treated in terms of two variables α and τ. These variables are depicted in the representation of four peptide residues shown in Figure 3.4.

Table 3.9 Effective Side-Chain Radii of Amino Acids

Amino acid side chain	r° (Å)	Amino acid side chain	r° (Å)
Ala	5.2	Trp	7.6
Val	6.4	Asp	5.0
Leu	7.0	Asn	5.0
Ile	7.0	Glu	6.0
Cys	6.1	Glu	6.0
Met	6.8	His	6.0
Pro	6.2	Ser	4.9
Phe	7.1	Thr	5.0
Tyr	7.1	Arg	6.0
Lys	6.0	Gly	4.2

Relating these parameters to the more familiar ψ and ϕ angles,

$$\alpha_i = 180° + \phi_{i+1} + \psi_i + 20°(\sin\phi_i + \sin\psi_{i+1}), \tag{3.1}$$

$$\text{and } \tau_i = 106° + 13°\cos(\alpha_i - 45°) \tag{3.2}$$

Molecular Force Field

The complete energy function is broken down into

$$V_{tot} = V_{vW} + V_s + V_{s-s} + V_H + V_H \tag{3.3}$$

V_{vW} is the van der Waal's term in which a (r^{-6}, r^{+8}) term is used instead of the more usual (r^{-6}, r^{+12}) (Lennard-Jones 6, 12 equation). V_s is the solvation energy term in which the approach of side chains displaces water, the solvation energy of side chains being derived from experimental data.

V_{s-s} is the specific interaction between half-cystine pairs (when their partners are known, otherwise the term is omitted). V_H is the hydrogen-bond interaction. V_N is an effective short-range nonbonded interaction.

Molecular Motion

The conformation of the polypeptide chain is perturbed in accord with the equations of molecular dynamics. In this approach the conformation of the molecule at time $t + \Delta t$ depends on the conformation, velocity, and operative

Figure 3.4 Angular variables α and τ for the Levitt model polypeptide. α is taken as zero for the *cis* configuration.

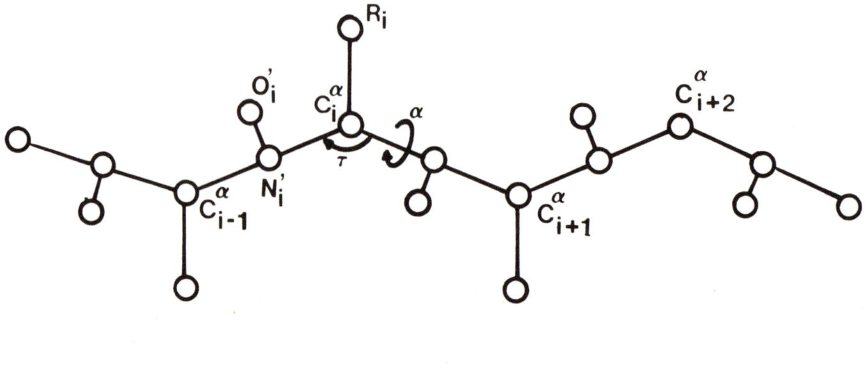

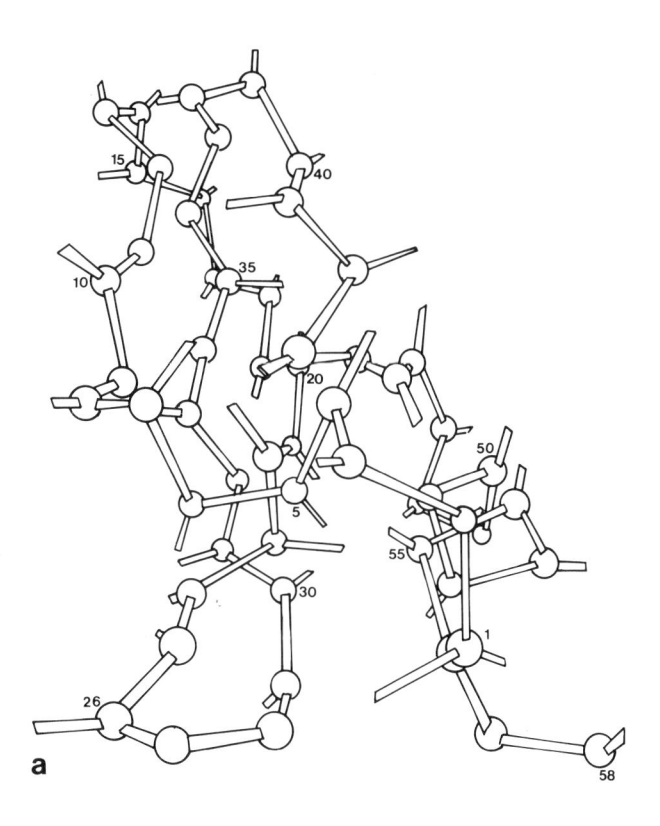

Figure 3.5 Representation of native pancreatic trypsin inhibitor using Levitt nomenclature: (a) native; (b) predicted.

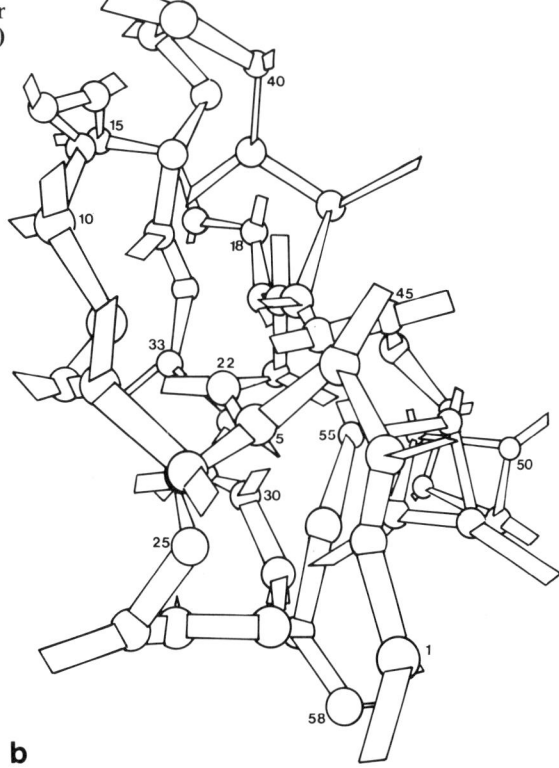

Figure 3.6 Computer assessment of the chain conformation for pancreatic trypsin inhibitor as it starts from an open structure and fluctuates towards the native structure (Figure 3.5).

---

forces at a slightly earlier time t. In a medium, such as water, inertial forces are damped out quickly and frictional forces, which depend on velocity, tend to balance the applied forces. As a result of applying the preceding concepts, a picture of the evolving (fluctuating) polypeptide structure may be obtained.

Application to Pancreatic Trypsin Inhibitor

Using the previous approximations, it is evident that some limitations on the accuracy of the predicted model of any protein will be produced. In Figure 3.5a we see a representation of the native form of pancreatic trypsin inhibitor (PTI),

3. Prediction of Conformation 67

iii

iv

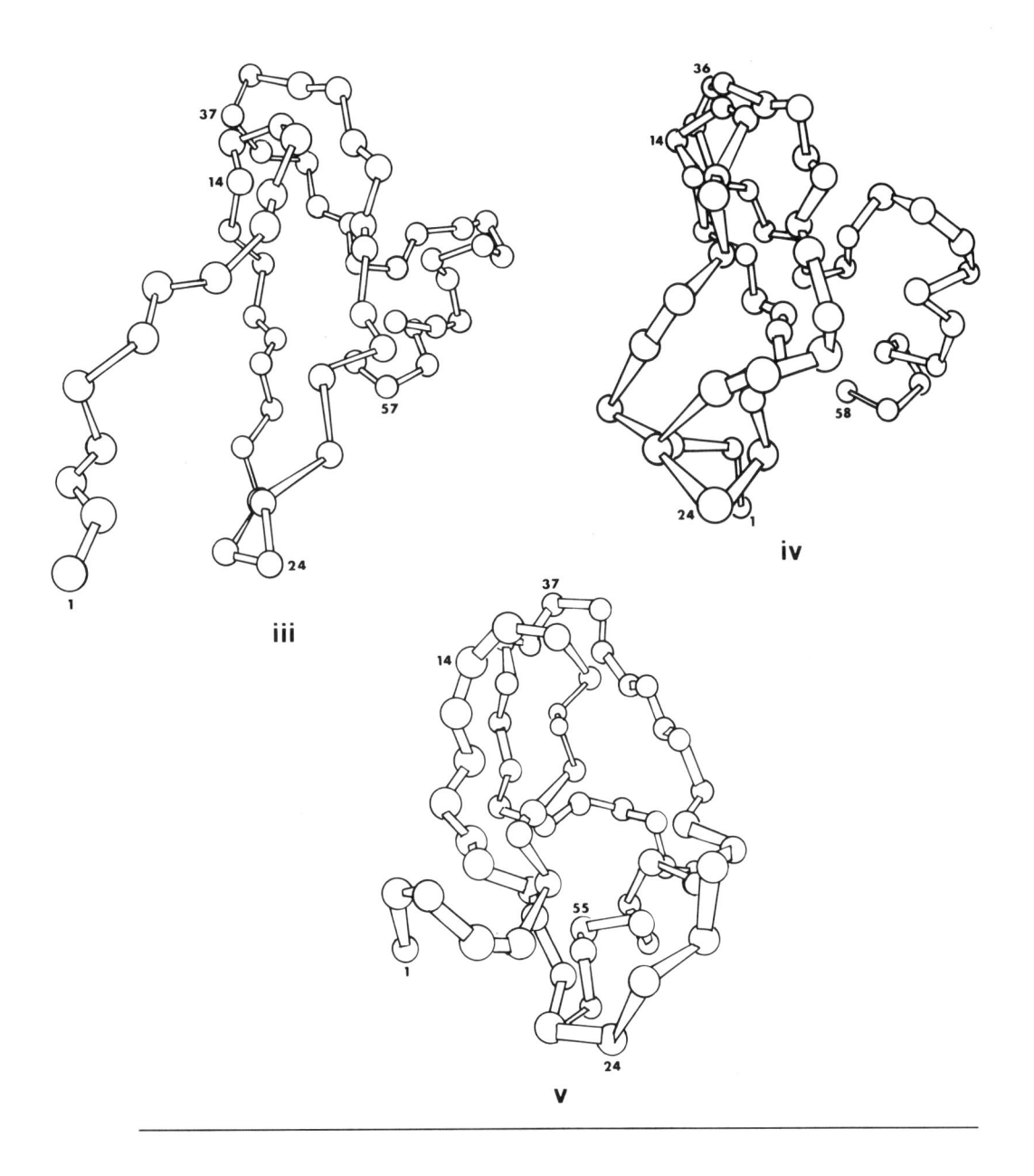

v

and in 3.5b is a similar structure, arrived at by starting with the native structure and using the aforementioned folding procedure to reach the nearest total energy minimum. The main ribbon follows the C_α backbone chain, while side ribbons join the side-chain centroids to the backbone. Although the detailed structures are different, several aspects are the same. The C-terminal region, in particular, contains an α-helical region, and most of the folding geometries are maintained.

Various starting conformations may be used to examine the folding process; presumably the more accurate the initial chain conformation, the better the folded structure. The series of diagrams in Figure 3.6 follows the computer-simulated folding of PTI from a relatively open structure to a near native structure.

The conclusion that has been reached in the preceding type of work is that the main features of folding are probably generated by a large number of long-range interactions and by thermal chain fluctuation. The fine detail of three-dimensional structure may be generated by a final crystallization involving detailed atomic interaction potentials.

Further Reading

Anfinsen, C.B. and H.A. Scheraga. Experimental and Theoretical Aspects of Protein Folding. *Advances in Protein Chemistry* 29:205–300, 1975.

Chou, P.Y. and G.D. Fasman. Empirical Predictions of Protein Conformation. *Annual Reviews of Biochemistry* 47:251–276, 1978.

Nemethy, G. and H.A. Scheraga. Protein Folding. *Quarterly Reviews of Biophysics* 23:239–352, 1977.

Schulz, G.E. and R.H. Schirmer. *Principles of Protein Structure.* Springer-Verlag, New York, Chapters 6 and 7, 1979.

II

EXPERIMENTAL METHODS

4

X-Ray Diffraction

The first experimental technique that we shall study in its application to polypeptides and proteins (though in a very simplified approach) is x-ray diffraction. This technique has undoubtedly provided more insight into the detailed structure of biologic macromolecules than any other, yet its concepts and theory are often difficult for the student to grasp. Rather than attempt a detailed description of the method, we will examine some very simple methods that would enable a student to make certain deductions about a synthetic polypeptide or fibrous protein, and also cover the concepts involved in more complex measurements.

4.1 Bragg's Law

We shall begin by making the tacit assumption that, as with inorganic crystals, polypeptide crystals (or crystalline powders) contain atoms that lie in regular planes. As one set of parallel planes is exposed to an x-ray beam, reinforcement or destructive interference will occur depending upon the inclination of the planes to the incident beam, as expressed by Bragg's Law.

Figure 4.1 shows that for constructive interference to occur, AO + OC must be an integral multiple of the wavelength; that is,

$$AO + OB = n\lambda = 2d\sin\theta, \tag{4.1}$$

where n is the order of the diffraction (generally 1), λ is the wavelength of x-ray radiation (with a copper source $\lambda = 1.542$ Å), and d is the separation of the planes.

As a result of the ordered array of atoms in crystals, constructive interference only occurs at discrete angles with respect to the atomic planes. Since x-rays are scattered by the electrons surrounding the atomic nuclei, the intensity of scattering is a function of the electron density.

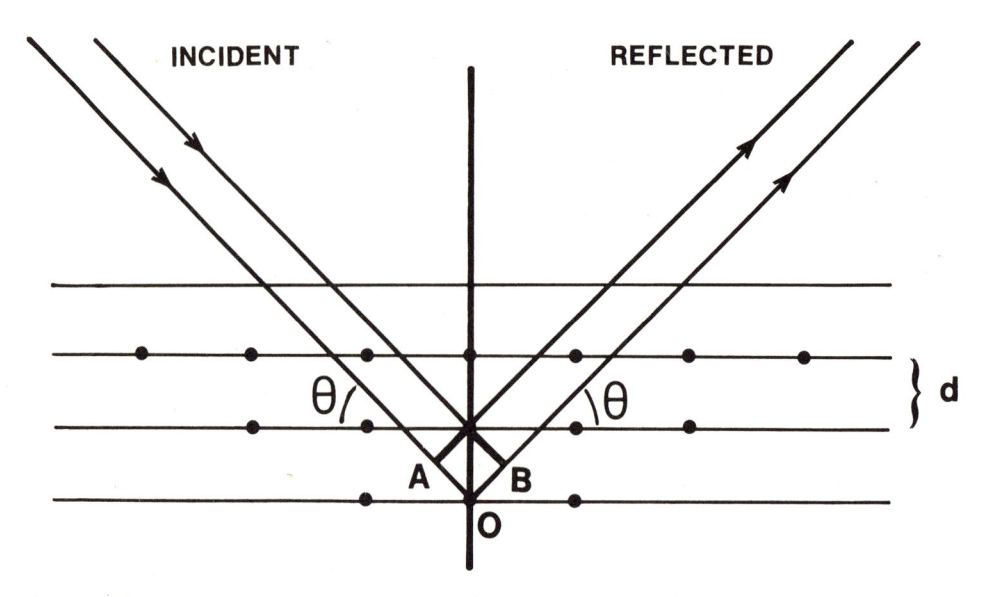

Figure 4.1 Diagram showing scattering of an incident x-ray beam by atoms arranged in crystal planes.

The relation between these planes of atoms and repeat structures in our crystalline sample is conveniently represented in terms of unit cells and Miller indices.

4.2 Two-Dimensional Unit Cells and Miller Indices

We have already seen that peptide residues repeat in a symmetrical manner in helices. Similarly, when polypeptides or proteins form a crystal, there is always a symmetrical relation between the molecules. The smallest element of this repeat structure is known as a "unit cell." If, for example, we represent a protein molecule by a dot, then the relation between these dots in two dimensions may be represented by only five different patterns such that each dot has identical surrounding. These are shown in Figure 4.2.

If we wish to construct two-dimensional unit cells, we could do so by drawing imaginary (dotted) lines bounding the smallest repeat unit. The choice of the location of the unit is arbitrary. In the point arrangement of

Figure 4.2 The five types of repeating arrangements of points that can be drawn in two dimensions such that each point is identically surrounded. **(a)** Monoclinic; **(b)** centered orthorhombic; **(c)** hexagonal; **(d)** orthorhombic; **(e)** cubic.

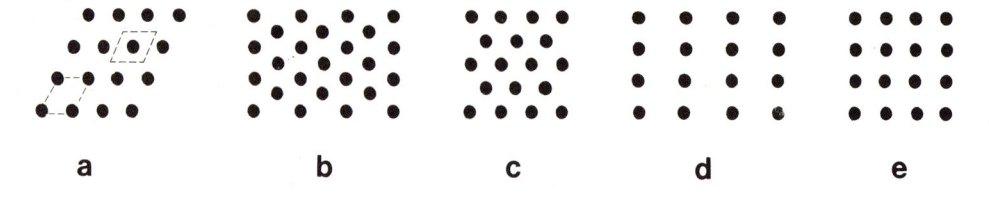

4. X-Ray Diffraction

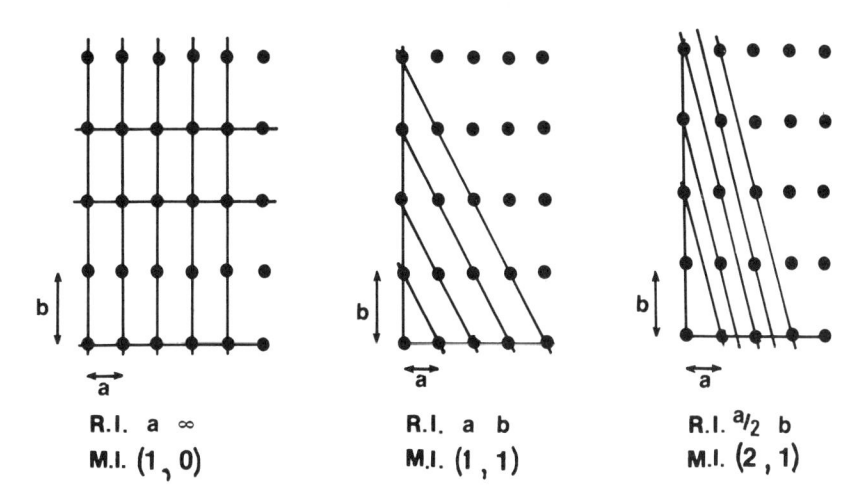

Figure 4.3 Two-dimensional representation of planes drawn through points, with intercept distances, and Miller indices.

Figure 4.2 (a), two equivalent "unit" cells are drawn, each of side a. There is one point per unit cell (either four points divided between four cells 4/4 = 1, or 1 isolated point).

If we now examine an array of lines drawn through a set of points using axes x and y, we can define these lines by their intercept with the axes in terms of the dimensions (unit cell) of a and b in the x and y directions. Figure 4.3 shows three sets of lines (two-dimensional planes) drawn at various angles with respect to the x and y axes. In Figure 4.3 (left panel) the planes are parallel to the y axis and intercept the x axis at intervals of the unit cell repeat a. These lines (planes) are parallel to each other and pass through all points. Their intercepts with the y and x axis are then a and ∞.

In x-ray crystallography it turns out that it is much more convenient to deal with diffraction data in terms of "reciprocal space" instead of real space. In terms of our intercepts, in reciprocal space they are $1/a$ and $1/\infty$ (= 0). To normalize to a standard system the values are multiplied individually by the appropriate unit-cell dimension (i.e., $a \times 1/a = 1$ and $b \times 1/\infty = 0$) and the resultant numbers are the Miller indices (1,0). For Figure 4.3(b) and (c) the resultant Miller indices are (1,1) and (2,1). Each set of indices refers then to a family of lines related by one unit-cell translation and the higher the indices, the closer together are the lines.

4.3 Three Dimensions and Crystal Systems

The preceding concepts may readily be extended to three-dimensional space. There are 14 different ways of arranging points in three dimensions such that each point lies in an equivalent environment. These are known as the Bravais lattices: there are three cubic, two tetragonal, four orthorhombic, two monoclinic, and one each of hexagonal, rhombohedral, and triclinic arrangements. Since rhombohedral and hexagonal lattices are closely related, there are sometimes said to be only six crystal systems. The unit-cell dimensions and axial orientations are given in Table 4.1.

Table 4.1 The Six Crystal Systems

System	Unit-cell constants	Angles between crystal axes (degrees)
1. Cubic	$a=b=c$	$\alpha=\beta=\gamma=90$
2. Hexagonal		
Hexagonal Divn	$a=b\neq c$	$\alpha=\beta=90, \gamma=120$
Rhombohedral Divn	$a=b=c$	$\alpha=\beta=\gamma\neq 90$
3. Tetragonal	$a=b\neq c$	$\alpha=\beta=\gamma=90$
4. Orthorhombic	$a\neq b\neq c$	$\alpha=\beta=\gamma=90$
5. Monoclinic	$a\neq b\neq c$	$\beta=90$
6. Triclinic	$a\neq b\neq c$	$\alpha=\beta=\gamma\neq 90$

$\alpha = \angle bc, \beta = \angle ac, \gamma = \angle ab$.

For the sake of simplicity, the internal symmetry of the lattice, which is the ultimate criterion of choice of crystal system, has been omitted. Extending the concepts of Figure 4.3 to three dimensions, we find that now *three* Miller indices define planes and may be represented by $(\bar{h}kl)$, which again are reciprocal unit intercepts. Several examples of planes with various Miller indices are shown in Figure 4.4. It can be seen that negative axial intercepts are indicated by (h, k, l).

Formulae for the distance of separation d of planes having Miller indices h, k, l may be derived for the various crystal systems. Five of the formulae are as follows:

Figure 4.4 Diagrams of planes having various Miller indices in a unit cell with edges *a, b, c.*

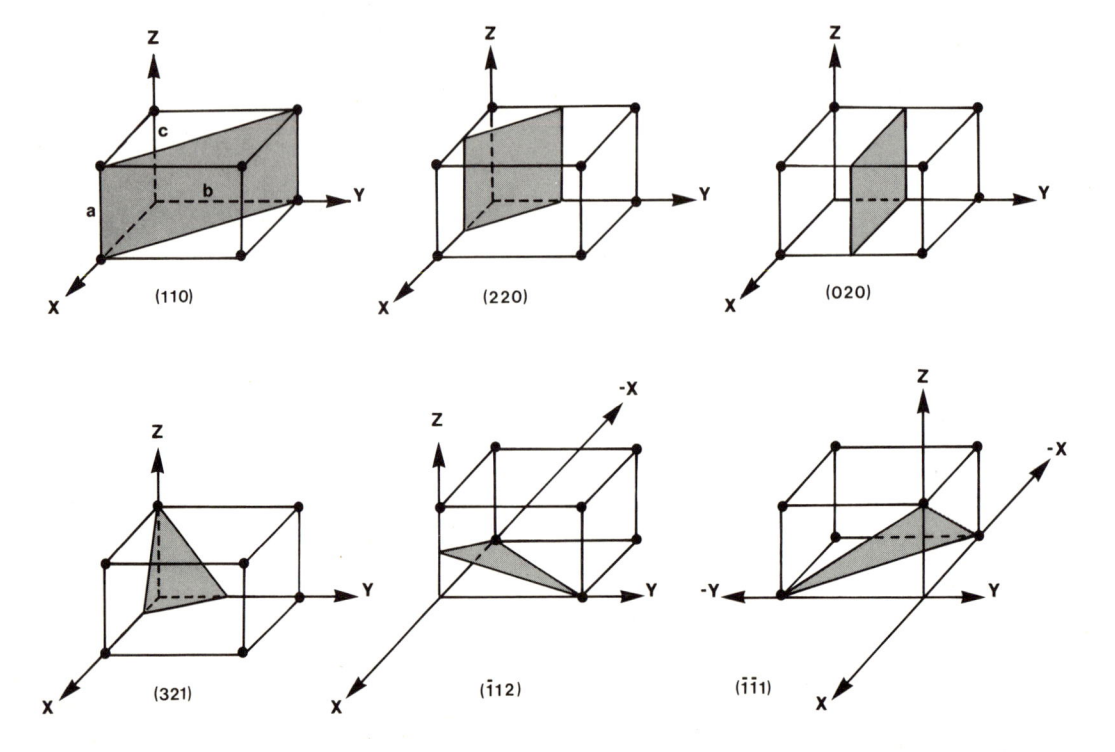

(110)   (220)   (020)

(321)   ($\bar{1}12$)   ($\bar{1}\bar{1}1$)

$$\text{Cubic } d_{hkl} = \frac{a}{\sqrt{h^2 + k^2 + l^2}} \tag{4.2}$$

$$\text{Tetragonal } d_{hkl} = \left[\left(\frac{h^2 + k^2}{a^2}\right) + \frac{l^2}{c^2}\right]^{-1/2} \tag{4.3}$$

$$\text{Orthorhombic } d_{hkl} = \left[\frac{h^2}{a^2} + \frac{k^2}{b^2} + \frac{l^2}{c^2}\right]^{-1/2} \tag{4.4}$$

$$\text{Hexagonal } d_{hkl} = \left[\frac{4}{3a^2}(h^2 + k^2 + hk) + \frac{l^2}{c^2}\right]^{-1/2} \tag{4.5}$$

$$\text{Monoclinic } d_{hkl} = \left(\frac{\frac{h^2}{a^2} + \frac{k^2}{b^2} - \frac{2hk\cos\gamma}{ab}}{\sin^2\gamma} + \frac{l^2}{c^2}\right)^{-1/2} \tag{4.6}$$

The other two formulae (rhombohedral and triclinic) are more complex and are usually used in conjunction with a computer. Simple methods of indexing polypeptide unit cells using some of the above formulae are known and are outlined in the following text.

4.4 Diffraction by Powders

Most specimens of synthetic polypeptides and native peptides and proteins become available as powders, which can be of varying degrees of crystallinity, unless special orientation or crystal-growing procedures are used. Let us suppose that the powder in question is microcrystalline and therefore has crystallographic planes that scatter x-rays such that reinforcement occurs according to the Bragg equation (4.1). The powder-diffraction method uses a film that is either in the form of a flat plate placed normal to the beam, or a cylindrical film wrapped around the sample such that a wider range of scattering angles may be observed. To understand the scattering from a very large number of small crystals it is instructive to examine first the scatter from one plane of one crystal. The diagram in Figure 4.5 shows a portion of the scattering geometry looking down the end of a cylindrical film. If the crystal plane is rotated about the axis 0 going into the plane of the paper, reinforcement occurs symmetrically in two positions ($\pm 2\theta$) in the forward direction. Reinforcement will also occur at any other angle that meets the Bragg criterion (equation 4.1). Since the beam is collimated to fine dimensions, if there were only crystal plane, as indicated, the diffraction pattern would consist of two points on the film, symmetrically arranged about A, the zero displacement point. However, the same plane from other crystallites will be rotated about the x-axis and will produce a ring of diffraction instead of two spots. Various planes in various crystals will produce "cones" of reinforced x-rays. Apart from the geometry of Figure 4.5, the "flat plate" camera is often used. Figure 4.6 shows powder diffraction with diffracted cones.

It is interesting to note that a very large number of crystallites produce a small number of detectable diffraction rings, in principle one ring for each (h, k, l) set.

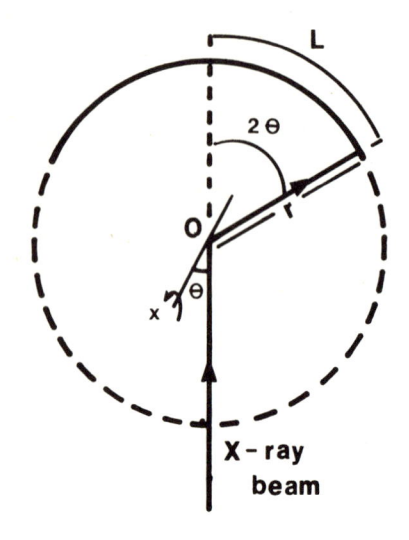

Figure 4.5 Arrangement of powder-diffraction system using narrow film placed at a uniform distance from diffracting material.

In Figure 4.5 the scattering geometry is such that:

$$\frac{2\theta}{360} = \frac{L}{2\pi r} \tag{4.7}$$

Since r is determined by the experimental geometry, the scattering angle θ may be calculated after L is measured. Hence from equation (4.1), d, the distance between scattering planes may be calculated.

In Figure 4.6 the scattering geometry is such that the ring radius L is given by:

$$\tan 2\theta = L/r, \tag{4.8}$$

and hence θ and d may be obtained as before.

Figure 4.6 Powder diffraction using flat film.

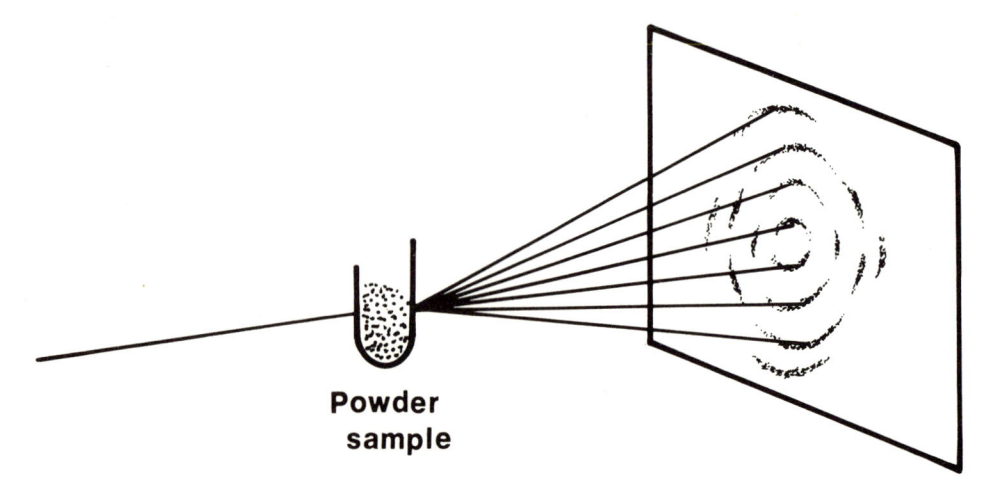

4.5 Obtaining Unit-Cell Parameters from Powder-Diffraction Patterns

We have assumed so far that x-ray diffraction is an all-or-none proposition; that is, that constructive interference produces a point or line of a certain intensity and destructive interference completely prevents the appearance of diffraction points or lines. In practice, differences in the scattering power of various atoms (electrons) causes definitive differences in line intensities, which can provide important information. Nevertheless for the time being we shall revert to our all-or-none situation in trying to solve powder-diffraction patterns.

Unfortunately, experimental powder-diffraction patterns from polypeptide materials are notoriously poor (i.e., there are relatively few rings, most of which are generally quite diffuse). Naturally, the fewer the lines and the more diffuse they are (generally caused by low crystallinity), the less information the diffraction pattern contains.

Most textbooks on elementary aspects of x-ray diffraction from powders deal with inorganic crystals in which a well-defined array of ions can be shown to produce constructive or destructive diffraction. In the current context, however, we are faced with polypeptide microcrystals for which we initially have no idea of the backbone geometry or atomic positions. A rapid and approximate method of interpretation is examined first. (Experience shows that synthetic polypeptide powders give, at best, 10 to 14 lines, more often 6 to 9.)

Indexing Diffraction Lines

Whereas in practice it is now usual to use a computer and computer-fit lines to a general diffraction equation of the form of equations (4.2) to (4.6), it is perhaps instructive to examine a simple noncomputational procedure for solving powder-diffraction patterns.

The equations (4.2) to (4.6) show that d_{hkl} is always related to the inverse square root of squares and products of the Miller indices h, k, and l. We need to identify these indices in order to extract the unit-cell dimensions. A convenient approach to indexing is to set up equations in terms of the inverse square of d_{hkl}. For example, from the Bragg equation (4.1) and equation (4.2),

$$\sin^2\theta = \frac{\lambda^2}{4a^2} [h^2 + k^2 + l^2], \tag{4.9}$$

for the cubic crystal system.

Now h, k, and l can only have values 0,1 $\bar{1}$, 2,$\bar{2}$, and so on, and thus $[h^2 + k^2 + l^2] = 0,1,2,3,4,5,6,8,9$, and so on.

Since θ is measured directly, a table of $\sin^2\theta$ values may be set up and a search made for ratios of 1,2,3, and so on.

No polypeptide is known to have a simple cubic cell, so an example of such an approach that is of more value in the current case is for a hexagonal cell. We have seen previously that helices tend to pack hexagonally, so that the hexagonal case would be expected to be quite common.

A schematic view of hexagonal packing for helices is shown in Figure 4.7.

Proceeding as before with the cubic cell, but using equations (4.1) and (4.5), all $\sin^2\theta$ values must obey the relation:

$$\sin^2\theta = A(h^2 + hk + k^2) + Cl^2 \qquad (4.10)$$

Permissible values of $(h^2 + hk + k^2)$ are 1,3,4,7,9 and so on.

For hk0 planes (i.e., $l = 0$),

$$\frac{\sin^2\theta}{(h^2 + hk + k^2)} = A. \qquad (4.11)$$

Example. Let us examine some data for a hexagonal system in which we have only minimal information (six lines).

Step 1. Set up a table of lines in order of increasing θ. Across the table set up values of $\sin^2\theta$ divided by 1,3,4,7,9, and so on, which are the only possible combinations of $(h^2 + hk + k^2)$.

Line	$\sin^2\theta$	$\dfrac{\sin^2\theta}{3}$	$\dfrac{\sin^2\theta}{4}$	$\dfrac{\sin^2\theta}{7}$	hkl
1	0.097	0.032	0.024	0.014	
2	0.112*	0.037	0.028	0.016	100,010
3	0.136	0.045	0.034	0.019	
4	0.209	0.070	0.052	0.030	
5	0.332	0.111*	0.083	0.047	110
6	0.390	0.130	0.098	0.056	

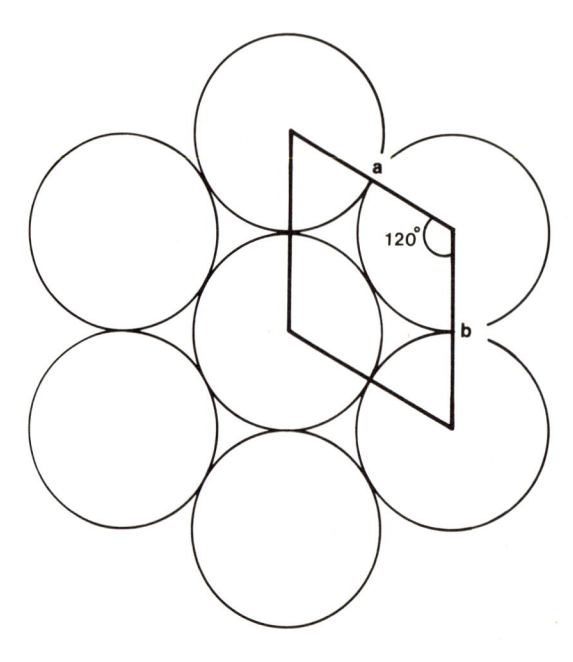

Figure 4.7 Polypeptide helices in hexagonal packing. Unit cell $a = b \neq c$ $\angle ab = 120°$. c is the fiber axis repeat (into the plane of the paper).

Now in order for equation (4.11) to be satisfied, there would have to be numbers in the first, second, third, etc. columns, that would be identical (if we had enough lines). Starting with 0.097 we find that this does not appear in the second column (or any other) and therefore, this value does not meet the criteria of equation (4.11). On the other hand, 0.112 of column one, and 0.111 of column two, (marked with asterisks) are equal within experimental error. This number (unless it is entirely coincidental) would also appear in columns three and four for higher line numbers if they were available. Since for the appropriate number in column one $(h^2 + k^2 + hk) = 1$, this can be satisfied by $h = 1$, $k = 0$ or vice versa. Therefore we assign line 2 (tentatively) to diffraction caused by the 100 or 010 plane of the crystal. In the second column, the appropriate number (0.111) corresponds to a solution where $(h^2 + k^2 + hk) = 3$—that is, $h = k = 1$.

If the preceding assignments are correct, then from (4.11), $A = 0.112$.

Step 2. We note that from (4.10),

$$\sin^2\theta - A(h^2 + hk + k^2) = Cl^2. \tag{4.10a}$$

Now $h^2 + hk + k^2$ must have values of 1,3,4,7,9 and so on, and $l^2 = 1,4,9,16$, and so on.

Set up a table of lines in order of increasing θ. Across the table set up values of $\sin^2\theta$, $\sin^2\theta - A$, $\sin^2\theta - 3A$, and so on.

Line	$\sin^2\theta$	$\sin^2\theta - A$	$\sin^2\theta - 3A$	hkl
1	0.097			002
2	0.112	0.000		(100),(010)
3	0.136	0.024		(101),(011)
4	0.209	0.097		(102),(012)
5	0.332	0.221		110,(103),(013)
6	0.390	0.278	0.054	004

One next searches this table to find numbers that are related by the possible values of l^2 (i.e., 1,4,9,16). In columns 1 and 2 the value of 0.097 appears which is approximately 4×0.024. 0.221 would be the $9 \times$ multiple; 0.390 the $16 \times$ multiple (the primary value would be perhaps 0.0245). We thus assign l values of 1,2,3, and 4 to the 0.024, 0.097, 0.221, and 0.390 lines.

For line 1, column 1, $l = 2$ ($l^2 = 4$) and the whole column corresponds to $h = k = 0$. For line 3, column 2, $l = 1$ and the column corresponds to h or $k = 1$, the other being zero. For line 4, column 2, $l = 2$ and again h or $k = 1$. For line 5, column 2, $l = 3$, h or $k = 1$. (Note here that two x-ray planes correspond to the same line.) Finally, line 6, column 1 gives $l = 4$ and $h = k = 0$. The number in column 3 evidently is not part of our sought-for series. In fact, setting up tables in this manner generates many numbers that have no particular relevance to the solution. The fact is, though, that all of the lines can be indexed in terms of a hexagonal unit cell and the cell constants can be evaluated.

$A = 0.112$ $\quad a = 2.66\text{Å} \quad C = 0.024 \quad c = 4.98\text{Å}$

This example actually applies to an inorganic system, but the calculation would have been identical with a polypeptide powder. In that case, if for example we were dealing with a polyproline II helix with a helix repeat of 3 × 3.12Å = 9.36Å, then the a dimension would equal the diameter of the helix and c = 9.36Å.

Of course if our tables showed no number correlations, it would be assumed that the material was not of the hexagonal class and another system would have to be applied.

The same approach may be used for other crystal classes.

Tetragonal Case

Here $\sin^2\theta = A(h^2 + k^2) + Cl^2$, (4.12)

where $A = \lambda^2/4a^2$ and $C = \lambda^2/4c^2$

$h^2 + k^2 = 1, 2, 4, 5, 8 \ldots$

and one proceeds as with the hexagonal case.

The orthorhombic unit cell is more difficult to deal with at this level because there are three constants in the basic equation:

$$\sin^2\theta = Ah^2 + Bk^2 + Cl^2 \tag{4.13}$$

Similarly, monoclinic cells are virtually impossible to deal with unless some concept of the axial angle β (or γ using c as the polypeptide axis) is available, and triclinic cells are not accessible by the preceding approach.

4.6 Crystalline Perfection

If we were dealing with a protein single crystal with a high degree of crystalline order, the Bragg diffraction angle θ would give rise to spots of infinitesimal thickness. We can approach such a situation experimentally for some inorganic materials, but particularly for polypeptide powders the powder-diffraction lines show considerable broadening. A schematic densitometer trace of a diffraction line is shown in Figure 4.8. The diffraction broadening is caused basically by the finite size of the crystallite producing the diffraction. (The crystallite that diffracts is almost always much smaller than the particle size of the powder under investigation.) A convenient algebraic relation exists between the half-width of the diffraction line and the crystallite width B, known as the Scherrer equation:

$$B = \frac{0.9\lambda}{L\cos\theta_0} \tag{4.14}$$

L is the angular spread of 2θ in radians (as shown in Figure 4.8). The calculated value of B is the thickness perpendicular to the plane corresponding to the diffraction.

In addition to crystallite size, lattice disorder also leads to line broadening. One mode of expressing this phenomenon is in terms of amorphous character

4. X-Ray Diffraction

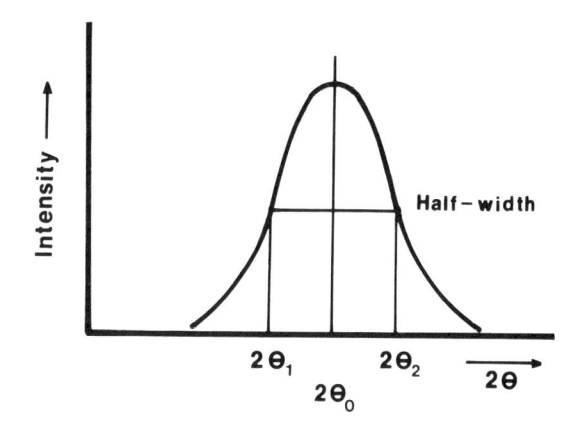

Figure 4.8 Diffraction broadening.

or degree of crystallinity. If a portion of the sample is amorphous, it gives rise to a broad background scattering, which occurs over all scattering angles. The ratio of crystalline to amorphous scattering is then given by:

$$\frac{\text{Crystalline}}{\text{Amorphous}} = \frac{\text{Integrated intensity of all lines}}{\text{Integrated background scattering}} \quad (4.15)$$

4.7 Preferentially Oriented Microcrystalline Specimens

In contrast to the randomly oriented crystallites examined by x-ray diffraction powder methods in the previous section, more information may be derived from the sample if it is a single crystal exposed to the beam and rotated about a major axis, or if orientation of microcrystals or fibers can be obtained.

The rotation of a single orthorhombic crystal about its c axis produces cones of discrete diffraction as the appropriate crystallographic planes cause "in-phase" reinforcement. (See Figure 4.9.) The rotation picture thus produced contains diffraction spots on discrete-layer lines (l = 0, 1, 2 . . .). Similar types of patterns are produced when oriented microcrystals or protein fibers are placed such that the x-ray beam interacts at right angles to the fiber or orientation axis. In this case the sample need not be rotated about its fiber axis since the various microcrystalline orientations or relative fiber planes are already present in random orientation in the other two dimensions.

The solution of the diffraction spots from a rotation or fiber pattern is similar to the methodology used to approach powder patterns. In this case, however, one of the Miller indices (l) is defined and each layer line can often be treated as a two-dimensional powder pattern. For example, if the unit cell were hexagonal with c axis perpendicular to the beam, then all spots on the l = 0 layer line should correlate with (hk0) planes derived from the $\sin^2\theta = (h^2 + hk + k^2)A$ formulation. The situation is similar for other layer lines and unit cells. Such methods are straightforward if rotation is about the c axis for tetragonal or hexagonal cases. For orthorhombic or monoclinic unit cells, graphical methods are sometimes used. For example with an orthorhombic cell it can be shown from equation (4.4) that:

$$\log d(hk0) = \log a - \tfrac{1}{2} \log \left[h^2 + \frac{k^2}{(b/a)^2} \right] \quad (4.16)$$

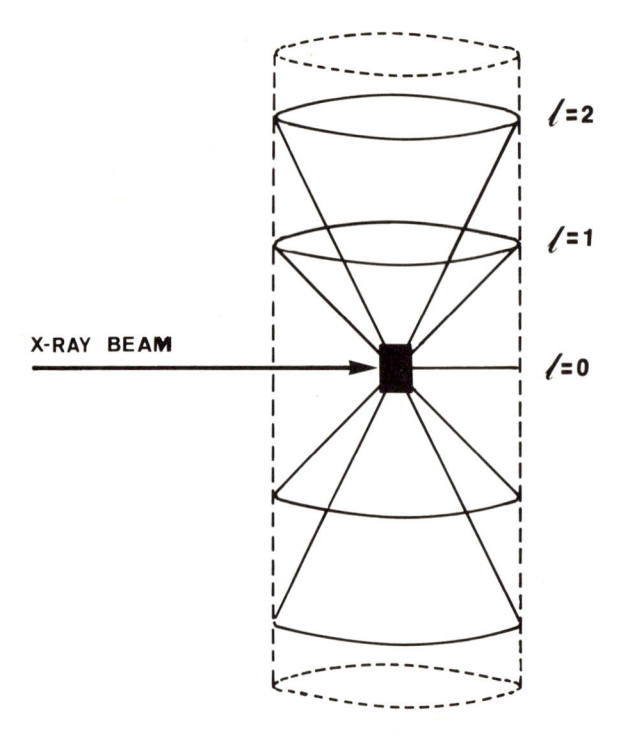

Figure 4.9 Cones of "constructive interference" for a rotated crystal showing layer lines.

A chart for indexing the reflections of a rectangular reciprocal lattice net can next be constructed by assigning pairs of values (0,1) (1,0) (1,1), and so on, to h and k, and constructing a curve of b/a versus d.

Measured d or $\log d$ values are then set up on the same scale and matched to the Miller indices as shown in Figure 4.10. The following two sections involve some advanced algebra and the student may wish to skip to the important result, namely equation (4.27) in section 4.10.

4.8 Intensity of Diffraction

The intensity of x-ray diffraction depends upon the electronic configuration of the scattering atoms and their relation to a given set of planes in the crystal. The intensity of scattering from hkl planes is given by:

$$I(hkl) = K(hkl) \mid F(hkl) \mid ^2, \tag{4.17}$$

where F (hkl) is the structure factor that embodies the atomic distribution and atomic scattering factors, and K is a correction factor based on experimental parameters such as beam polarization, an angular factor (Lorentz Factor), and absorption and temperature factors. In addition, K contains a multiplicity factor, which for single-crystal diffraction work is one, but in powder photographs may have a fairly high value based upon the simultaneous diffraction of various planes being displayed in the same place (e.g., 110, $\bar{1}10$, 101, $10\bar{1}$, etc.).

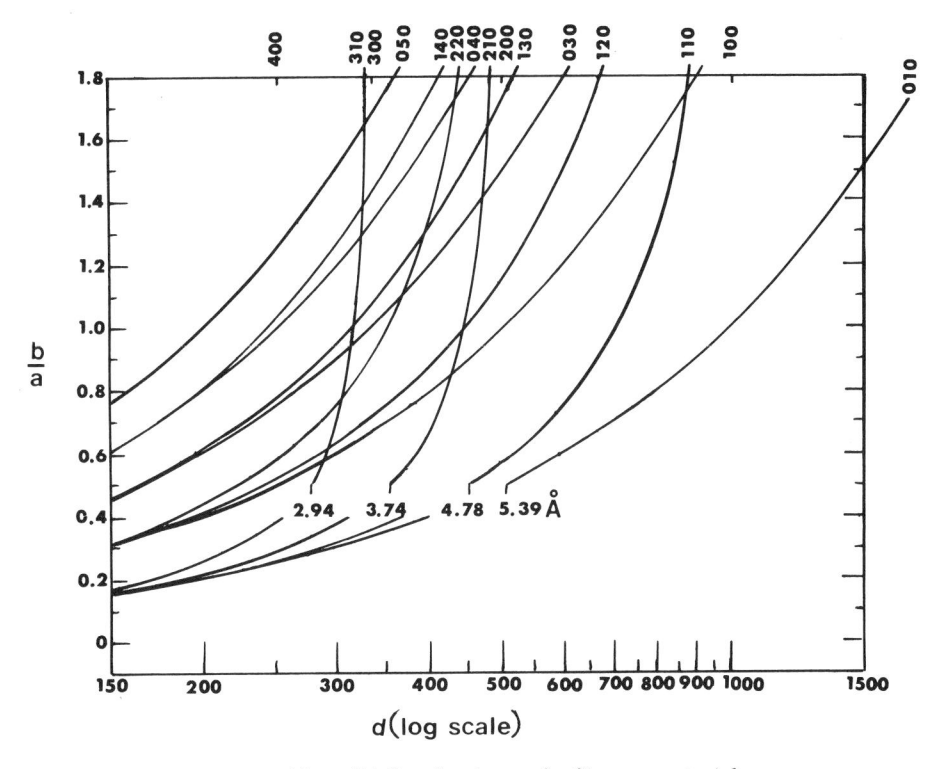

Figure 4.10 Indexing of four (hk0) reflections of a fibrous material.

The structure factor involves summation of the scattered waves from the various atoms within the unit cell with coordinates x_j, y_j, z_j, and is given by

$$F(hkl) = \sum_{j=1}^{n} f_j \exp[2\pi i(hx_j + ky_j + lz_j)]. \tag{4.18}$$

f_j is the atomic scattering factor.

The above expression may be separated into real and imaginary parts (remembering $\exp ix = \cos x + i \sin x$ and $\exp -ix = \cos x - i \sin x$),

$$F(hkl) = \sum_{j+1}^{n} f_j \cos 2\pi(hx_j + ky_j + lz_j) + \sum_{j=1}^{n} f_j i \sin 2\pi(hx_j + ky_j + lz_j).$$

The first and second terms on the right are designated A(hkl) and B(hkl) respectively, so that

$$|F(hkl)| = [A(hkl)^2 + B(hkl)^2]^{1/2}, \tag{4.19}$$

(eliminating i).

$F(hkl)$ is a vector of magnitude $|F(hkl)|$ and phase angle α given by $\alpha = \tan^{-1} B(hkl)/A(hkl)$.

Thus the presence or absence of diffraction lines or spots, as well as their intensities, may be derived from equations (4.18) and (4.19).

A simple exercise, for example, shows that for a simple (primitive) cubic cell, having one unique atom located at $x, y, z = (½, ½, ½)$, reflections from all of the hkl combinations are observed.

If the preceding approach is extended to other atomic arrangements, then certain reflections may be cancelled out. This observation may be generalized to state that if the unit cell contains a number of identical groups arranged in a symmetrical manner, then certain hkl reflections will be absent. For chain molecules, the most important symmetry data to be determined are N and t for the N_t helix. For the 00l reflections equation (4.16) reduces to:

$$F(00l) = \sum f_j \exp(2\pi i l z_j).$$

Reflections for the N_t helix are only observed when $l = qn$, where q is an integer; consequently the absence of all reflections except 00 ϵ, 002 ϵ, 003 ϵ and so on, suggests a helical structure with $n = \epsilon$. We will see how simple calculations make use of this type of information later.

Equation (4.19) may be written in a slightly different form, which treats the unit cell as a distribution of electron density instead of discrete atoms. Since the distribution is continuous and the summation is carried out in three dimensions, it can be rendered to the integral form

$$F(hkl) = (\text{const}) \int_0^a \int_0^b \int_0^c \rho(xyz) \exp 2\pi i(hx + ky + lz) dx dy dz \quad (4.20)$$

The density function can be transformed into the subject of the equation:

$$\rho(x,y,z) = \frac{1}{\text{const}, abc} \sum_{h=-\infty}^{\infty} \sum_{k=-\infty}^{\infty} \sum_{l=-\infty}^{\infty} F(hkl) \exp -2\pi i(hx + ky + lz). \quad (4.21)$$

We note that the previous type of transform converts a function in real space to reciprocal space or vice versa.

Since we wish to deal specifically with helical polypeptide chains, let us consider diffraction from a continuous (isolated) helix.

4.9 Diffraction by a Helical Chain

The most useful approach to a consideration of diffraction by polypeptide helices is to consider first the diffraction of a single, isolated, helical wire (Figure 4.11). As with the helix of Figure 2.13a, the wire will have a helical pitch of p and radius r. Equation (4.20) is still applicable but in place of the cartesian coordinates (x,y,z) it is necessary to substitute the coordinates for a continuous helix. In real space, the Cartesian coordinates may be replaced by cylindrical coordinates as shown in Figure 4.12. The replacement is as follows: $x = r\cos\psi$, $y = r\sin\psi$ and $z = z$, where ψ is the rotation angle about the z axis and z is the translation along the z axis. In reciprocal space we have the relations:

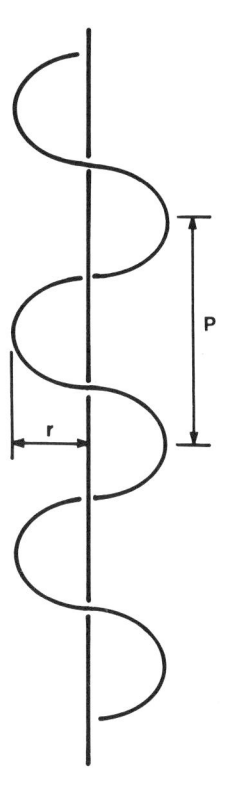

Figure 4.11 Projection of a helical wire of radius r and pitch p.

$$X = R\cos\psi$$
$$Y = R\sin\psi$$
$$Z = z$$

Finally we have a translation of $z = p$, when $\psi = 2\pi$ and $\psi = 2\pi z/p$. Using this last relation, the Cartesian coordinates x,y,z become $x = r\cos(2\pi z/p)$, $y = r\sin(2\pi z/p)$, and $z = z$.

Figure 4.12 Relation between Cartesian and cylindrical coordinate systems.

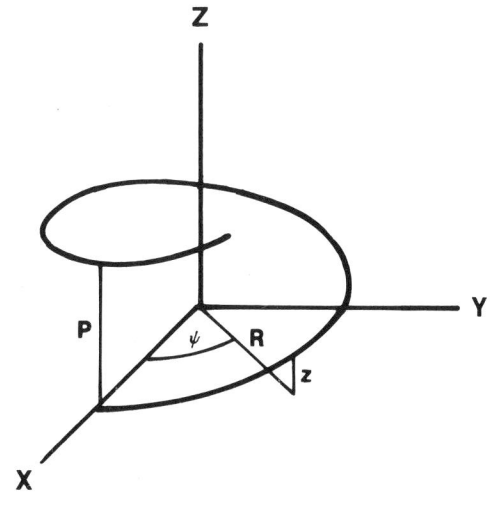

Equation (4.19) becomes on substitution and rearrangement:

$$F\left(R, \Psi, \frac{n}{p}\right) = \int_0^p \exp\left\{2\pi i \left[Rr\cos\left(\frac{2\pi z}{p} - \Psi\right) + \frac{nz}{p}\right]\right\} dz. \quad (4.22)$$

This integral is in the form of the so-called Bessel function J.

$$J_n(X) = \frac{1}{2\pi i^n} \int_0^{2\pi} \exp(iX\cos\psi)\exp(in\psi)d\psi. \quad (4.23)$$

$J_n(X)$ is the Bessel function of order $n(= 0,1,2 \ldots)$ and argument X. Tables of Bessel functions are available, so putting equation (4.22) in Bessel function form:

$$F\left(R, \Psi, \frac{n}{p}\right) = 2\pi J_n(2\pi Rr) \exp\left[in\left(\Psi + \frac{\pi}{2}\right)\right]. \quad (4.24)$$

Finally the intensity on the n-th layer line of the diffraction pattern is obtained from multiplying F by its complex conjugate, that is,

$$F^2 = I = [2\pi J_n(2\pi Rr)]^2. \quad (4.25)$$

It can be seen, therefore, that the intensity of diffraction of a defined continuous helix is dependent only upon the behavior of the Bessel function J_n.

We could, therefore, based on the solution of J_n from tables, construct an expected intensity pattern of x-rays diffracted from a continuous helix. Figure 4.13 shows a plot of F for Bessel functions of order 0 to 3. Since the scattered intensity is proportional to the square of the Bessel function, we would expect diffraction spots to follow the X cross arrangement of the Bessel function diagram.

When the previous approach is extended to a discontinuous helix with a repeat unit having atoms at coordinates (r_j, ψ_j, z_j) and scattering factor f_j, then for N (unit) repeats in t turns of the helix (repeat distance),

$$F\left(R, \Psi, \frac{1}{c}\right) = \sum^n \sum^j f_j J_n(2\pi r_j R) \exp\left\{i\left[n(\Psi - \psi_j + \frac{\pi}{2}) + 2\pi \frac{lz_j}{c}\right]\right\}, \quad (4.26)$$

where the summation is over the values of n defined by the selection rule

$$\boxed{l = Nm + tn.} \quad (4.27)$$

l is the layer line, m, an integer, and n the Bessel function order.

Equation (4.27) has the advantage of being both simple and very useful and should be noted as a VIE (Very Important Equation). The equation is of such central importance in fiber pattern interpretation that it is represented in heavy outline.

4. X-Ray Diffraction

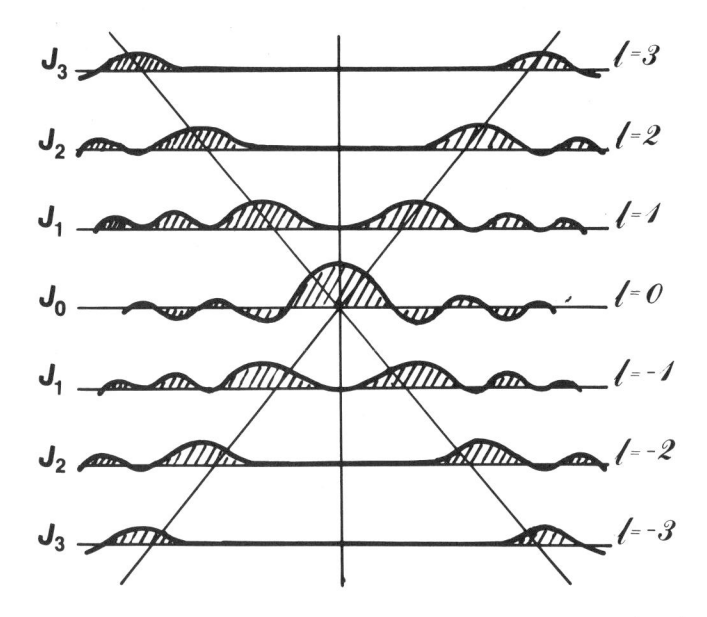

Figure 4.13 Theoretical transform of a continuous helix showing Bessel functions of order 0–3 and corresponding layer lines.

4.10 Calculation of Diffraction from a Discontinuous Helix

From the preceding discussion it would appear that diffraction is such a complex process that it would not be easy to interpret diffraction patterns from helical molecules. That such is not so will be shown in the following example.

Example. We wish to predict the fiber diffraction pattern of a single isolated (hypothetical) 7/2 helix.

The only equation required for such a procedure is the simple selection rule—equation (4.25). For the 7_2 helix N = 7 and t = 2, and we must evaluate m and n for the various layer lines l. Since the most intense diffraction results from minimum values of the Bessel function order n, we choose the layer line, arbitrarily put, n = 0, and search for an integral value of m. If none exists, we proceed to n = ±1, ±2, and so on. Table 4.2 presents values of n and m that satisfy equation (4.25) for a 7_2 helix (not all values are shown for clarity). Rarely are diffraction spots for n > 3 observable for a structure containing single helices.

Presenting Table 4.2 in diffraction form, we represent n = 0 by a heavy diffraction spot n = 1 as less dense, and so on, in accord with the values of J_n in Figure 4.14.

The most intense spots occur at the origin (l = 0, n = 0) and on the seventh layer line (l = 7, n = 0) on the "meridian."

The diffraction pattern is repeated in four symmetrical quadrants. Experimentally, intensities for the same n number also fall off from the origin; that is, the intensity of the n = 0 spot on the seventh layer line is less than that on the

Table 4.2 Solutions to the Helix-Diffraction Equation*

l (Layer line)	m (Integer)	n (Order of Bessel function)
0	0	0
1	1(−1)	−3(4)
2	0	1
3	1	−2
4	0	2
5	1	−1
6	0	3
7	1	0
8	0	4
9	1	1

*l = Nm + tn for a 7_2 helix.

zero*th*. (In practice the spot at the origin is obscured by undiffracted beam intensity.)

Consequently, we immediately see that the diffraction pattern of a 7_2 helix is characterized by a first meridional spot on the seventh layer line and off meridional (n = 1) on the second layer line. This phenomenon is easily shown to be true for a variety of helices, so for a 3_1 helix, the first meridional spot will

Figure 4.14 Representation of the diffraction from an isolated 7_2 helix (based on data of Table 4.2).

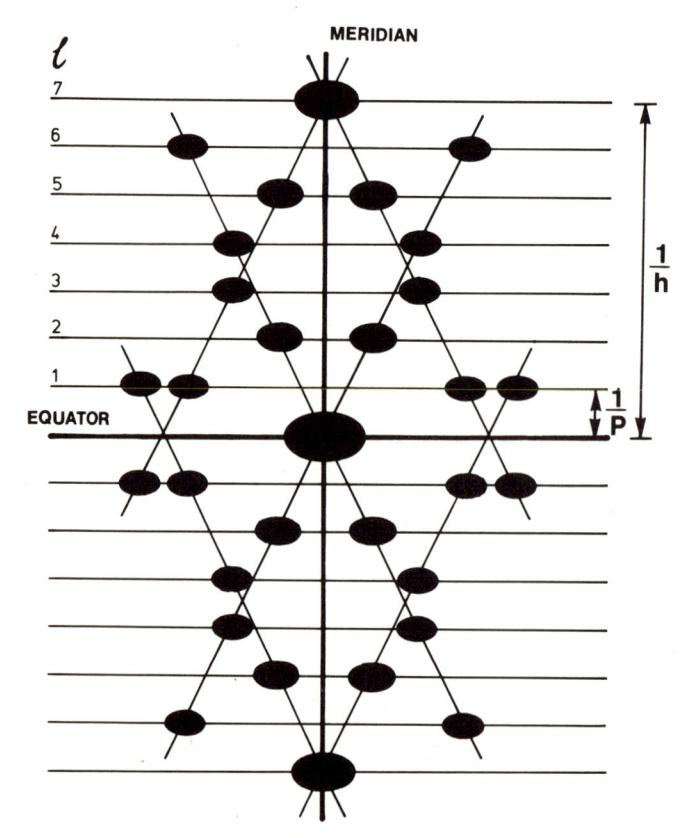

be on the third layer line, a major off meridional on the first layer line, and so on.

Furthermore, since the x-ray pattern is in reciprocal space, there is an inverse relation between the layer line positions and helix parameters. For the 7_2 helix the meridional reflection of the seventh layer line is $1/h$ Å^{-1} from the equator and the first layer line is the inverse distance of the helix repeat from the equator (h is the peptide repeat distance, see equation 2.12).

The α-helix in its generally accepted form has 18 residues in five turns of the helix. It should therefore produce a diffraction pattern in which the selection rule $l = 18m + 5n$ is obeyed. Close studies of diffraction patterns from poly-L-alanine and poly-γ-methyl-L-glutamate, both in the α-helix conformation, show slight distortions from the ideal; for the latter 29_8 or 69_{19} helices have been suggested. (In these cases instead of 3.6 residues per turn there would be 3.625 or 3.632 residues per turn respectively: the peptide repeat remains essentially unchanged.) Since all helices can presumably be distorted slightly, the nomenclature 18_5, 3_1, and so on, is said to refer to rational helices; that is, they represent the minimum number of residues and turns that approximate the helix.

4.11 Additional Complexities of Helical Arrays

Packing of Helices in Unit Cells

So far we have considered only the diffraction of an isolated helix. Evidently in the crystalline state helices are packed together in a three-dimensional array and the diffraction pattern becomes more detailed. For example, the equator contains information (h, k, 0) on the packing of helices and may be analyzed by the methods used for powder patterns with one index plane (l = 0) fixed. The basic pattern of the single helix is maintained (including the meridional rule).

For example, with poly-L-alanine we would expect for the α-helix and 18_5 (rational) helix a peptide repeat $h = 1.5$ Å and a helix repeat P of 27 Å. In fact, the layer line spacings are close to, but not quite exact orders of 27 Å. The best solution so far observed corresponds to a 47_{13} helix, which is a slight distortion ($18_{4.98}$) from the ideal helix. The measured peptide repeat is 1.496 Å and thus the helix repeat is 70.3 Å. Polyalanine helices pack hexagonally and since the helix is aligned with the c dimension of the unit cell c = 70.3 Å. The other dimensions are obtained from the fiber pattern in much the same manner as the rotation pattern was solved in section 4.8. Table 4.3 gives the diffraction spots observed on the equator for polyalanine. Since l = 0 these spots must conform to the relation (from equation 4.5).

$$\frac{1}{d^2(hkl)} = \frac{4}{3a^2} [h^2 + hk + k^2] \tag{4.28}$$

Since [h^2 + hk + k^2] can have values 1,3,4,6 and so on, we should search for values of $1/d^2$ (obtained from the photograph and Bragg equation) in the ratio 1:3:4:7. . . . This is a very simple procedure since the most intense diffraction spot arises from the lowest index plane (1,0). The experimental structure amplitudes F_m are also given in Table 4.3. The 13 lowest index planes have

Table 4.3 Equatorial Reflections for Poly-L-alanine

d	$1/d^2$	$F_m(exp)$	h, k	Ratio of $1/d^2$
7.41	.018(2)	1002	1 0	1
4.30	.054(6)	459	1 1	3
3.71	.072(8)	566	2 0	4
2.82	.126	475	2 1	7
2.47	.163(8)	372	3 0	9
2.14	.218	213	2 2	12
2.05	.237	135	3 1	13
1.85	.291	109	4 0	16
1.70	.346	259	3 2	19
1.62	.382	189	4 1	21
1.48	.455	134	5 0	25
1.42	.491	199	3 3	27
1.40	.510	125	4 2	28

been found, and from the data in the table and equation (4.28), $a = 8.55$ Å. Thus the diameter of the polyalanine α-helix is 8.55 Å.

Experimentally it is unusual to obtain such good data as that reported in the table. Deviations from perfect orientation low crystallinity and other experimental (and theoretical) factors combine to produce diffuse broad lines instead of the clear spots generally associated with nearly perfect inorganic crystals. In poorer fiber diagrams only a few of the most intense lines are seen.

We are now ready for the "shock" of looking at real experimental (smeared) diffraction patterns. The two patterns shown in Figure 4.15 are for polyalanine: the first corresponding to the axial arrangement of Figure 4.9; the second with the fiber axis tilted both with respect to the x-ray beam and the film. The patterns are overlayed with a grid indicating the hkl values of the major reflections.

The major equatorial reflections of Table 4.3 are noted in Figure 4.15a. The tilting of the specimen brings out the 1.5-Å meridional reflection in Figure 4.15b. Layer lines are set up as though the α-helix were a rational 18_5 helix with appropriate orders of Bessel function.

Coiled Coils

Unfortunately for the crystallography student, nature rarely uses the single, undistorted helix (though regular antiparallel β-sheets are common). Instead helices are generally intertwined in some manner. Two different types of intertwined helices may be defined; the first, such as DNA or some of the polysaccharides, involves two or more chains intertwined to form helices. These helices often possess a common linear axis as shown in Figure 4.16.

Diffraction complications arise from the interference between the two helices. For example, if two 3_1 helices form a double helix in which one chain with identical structure is displaced halfway between identical points on a second chain, then the odd layer lines predicted for a single strand are not observed and the first meridional reflection is on the sixth layer line with respect to the repeat for the single chain. Proteins, however, prefer a different type of helical intertwining that is described as a rope or supercoiled helix in Chapter 3. Examples of α-helices in rope structures are found in α-keratin, myosin and

4. X-Ray Diffraction

ℓ

18   1.5 Å
13
8
0   Equator

a

←1.5 Å

b

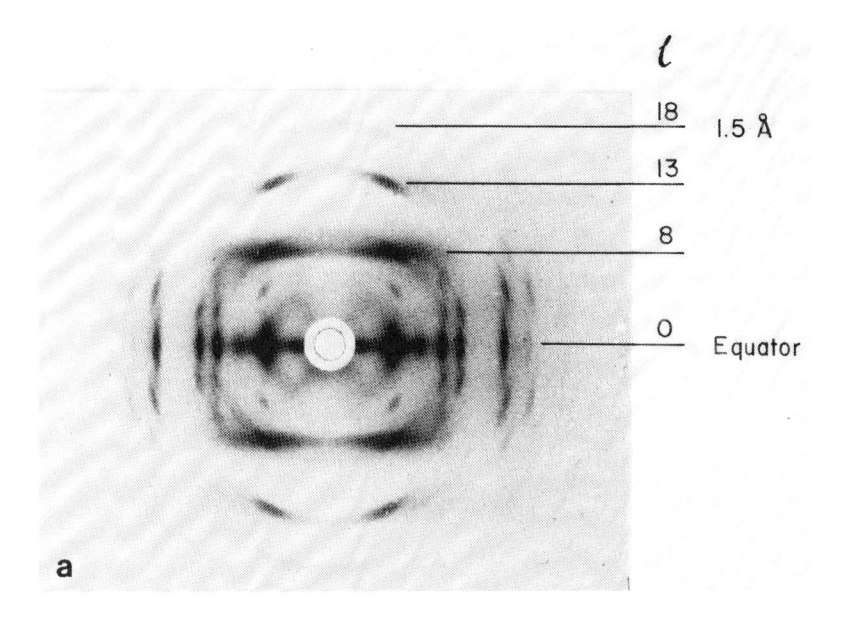

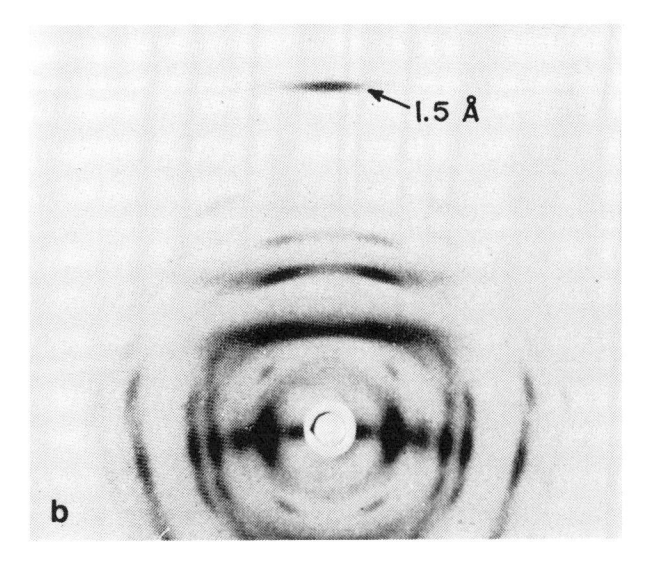

Figure 4.15 Fiber-diffraction patterns of α-helical poly-L-alanine: **(a)** fiber axis and cylindrical film axis vertical; **(b)** fiber tilted 31° to vertical.

tropomyosin, fibrinogen and fibrin. First let us define the parameters of a coiled coil shown in two-strand form in Figure 4.17. Previously we defined a helix having N residues in t turns. There is, in the rope structure, an additional level of helix formation; in other words each minor (α-) helix itself traces a (super) helical path. The superhelix repeat in Figure 4.17 is c. It is necessary therefore to define parameters referring to both minor and major helices. The coiled coil repeat contains N residues in repeat distance c. Over this distance c, there are t′ turns of the major helix (shown as one-half turn in Figure 4.17),

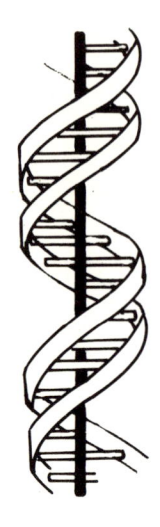

Figure 4.16 Schematic diagram of a nucleic acid (DNA)-type coiled helix. Both strands have a common axis.

and t turns of the minor (α-) helix. Crick has deduced that for this situation a structure factor equation analogous to equation (4.26), that is,

$$F(R, \Psi, l/c) = \sum_p \sum_q \sum_s J_p(2\pi r_0 R) J_q(2\pi r_1 R) J_s\left(2\pi \frac{1}{c} r_1 \sin\alpha\right)$$

$$\times \exp\left\{i\left[p\left(\frac{\pi}{2} + \Psi\right) + q\left(\frac{\pi - \Psi}{2}\right) + s\pi\right]\right\} \quad (4.29)$$

is appropriate where r_0 and r_1 are the radii of the major and minor helices and α is the angle of the slope of a tangent to the major helix. Instead of having one Bessel function J_n, we now have three, J_p, J_s, J_q. As before, it is not necessary for us to know the detailed derivation of such an equation but to see that its form resembles that for other scattering equations.

The selection rule for the orders of Bessel functions analogous to equation (4.25) is:

$$t'p + (t - t')q + ts = 1 + Nm, \quad (4.30)$$

where p, q, and s are orders of the appropriate Bessel functions and l and m are the layer line number and integer, as before.

Figure 4.17 A two-strand coiled coil or rope structure.

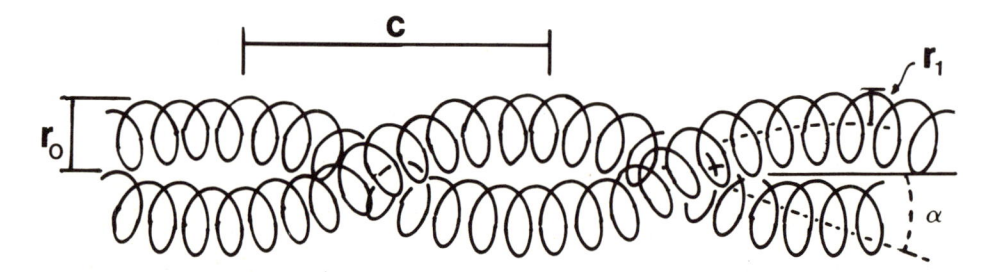

4. X-Ray Diffraction

Figure 4.18 "End-on" view of one helical component from a coiled coil.

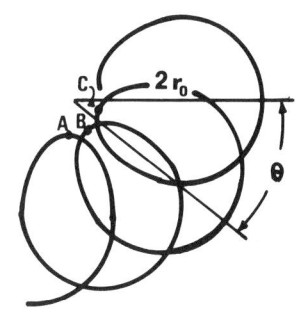

If we take a helix and twist one end with the other held fixed it will supercoil and might look something like Figure 4.18. In Figure 4.18 the radius of the superhelix is r_0 and successive loops of the helix are twisted through an angle θ. It is postulated that the superhelix forms in such a manner that points A, B, and C are equivalent. If we had an α-helix with a repeat sequence $(ZX_6)_n$ we would need to twist the helix such that the Zs come into equivalent positions. Looking down the end of an α-helix with such a sequence with 3.6 residues per turn and 100° twist per residue, we find that the first and eighth residues (Z) are 0.2 residues short ($= -20°$) of two turns (Figure 4.19). Thus if every seventh side chain is in an equivalent position on the inside of a coil, a supercoiling in the opposite sense of 20° per two turns must be executed.

The superhelix would repeat when the coil has completed a full turn ($\theta = 360°$); in other words, $360/20 = 18$ sets of two turns or 36 turns of the minor helix.

The repeat of the superhelix under the prescribed conditions thus consists of $18 \times 7 = 126$ amino acid residues. From Figure 4.17 the superhelix repeat c is given by $c = 126 \times 1.5 \times \cos 10° = 186$ Å.

Thus if two α-helices form a coiled coil with every seventh residue in an equivalent position, the superhelix repeat is 186 Å. We note that the peptide repeat h changes in this procedure to h cos 10°. Consequently *supercoiling has the effect of decreasing the peptide repeat distance.*

The values just quoted are in fact close to those observed for several fibrous proteins. This is not to say that other coiled coil arrangements are not possible.

Applying now the selection rule (4.30),

$$p + 35q + 36s = 1 + 126m \tag{4.31}$$

Figure 4.19 End-on radial projection of residues in an α-helix. Each residue progresses 100° around the circumference. First and eighth residues are 20° out of phase.

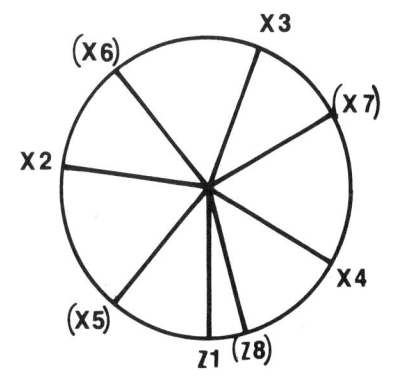

From equation (4.29) the Bessel functions J_p, J_q, J_s can be seen to relate to radial and translation functions. For reflections on the meridian $R = 0$; and thus the argument of J_p and J_q are zero, that is, they do not contribute to intensity. Thus for meridional reflections $p = q = 0$, and equation (4.31) becomes:

$$36s = 1 + 126m. \tag{4.32}$$

This equation can be reduced by putting $l' = 18l$ to

$$2s = l' + 7m. \tag{4.33}$$

This equation then applies to meridional reflections, the intensity being related to the order of Bessel function s (0 being the strongest, etc.). Solution of equation (4.33) as in Table 4.2 gives $s = 0$ for $l' = 0$ and 7, and $s = 1$ or -1 for $l' = 2$ or 5. Therefore, we expect the strongest meridional for $l' = 7$ ($d = 1.48$ Å) and next strongest at $l' = 2$ (5.17 Å) and $l' = 5$ (2.07 Å). The 5.17-Å reflection is observed but the 2.07-Å reflection is generally weak or absent. In addition to the meridional given by $p = q = 0$, reflections are expected: (1) with p small and $q = s = m = 0$ near equator; and (2) with $q = 1$ and $p = s = m = 0$ at 5Å near meridian.

Thus in supercoiling α-helices, many of the single helix features are maintained but slightly altered (decrease in peptide repeat) but instead of the peptide repeat being the only meridional reflection, the off meridionals near 5.1 Å move to the meridian and the helix diffraction pattern indexes as 7_2.

Similar procedures have been followed to solve the polyproline II coiled coil structure of collagen.

4.12 Single Crystals of Globular Proteins

X-ray diffraction from single crystals of proteins is, in many ways, the most important, yet the most abstract and complex of the techniques in the arsenal of the protein chemist. Since the student reading this text is unlikely to be presented with the problem of solution of single crystal patterns we shall confine ourselves to a short qualitative description of the information obtainable by this method. It is perhaps surprising that globular proteins form crystals at all (many have so far defied crystallization), but the crystals that have been formed range to millimeters in size. Solution of diffraction patterns of single crystals shows that all protein molecules (or sets of molecules) have the same orientation in the crystal and consequently the orientation forces and interactions exerted by each molecule causes an ordering effect on at least its near neighbors.

X-ray diffraction experiments seek to determine the atomic positions of at least the heavier atoms in a protein, such that the conformation and higher orders of structure may be determined. Since a fairly small protein molecule has one hundred peptides or more, and each peptide has 10 or more atoms, simultaneous diffraction from 1,000 or more uniquely positioned atoms will be producing the diffraction pattern. Often diffraction patterns are quite detailed, as shown in Figure 4.20 for the enzyme mitochondrial malate dehydrogenase (mMDHase).

4. X-Ray Diffraction

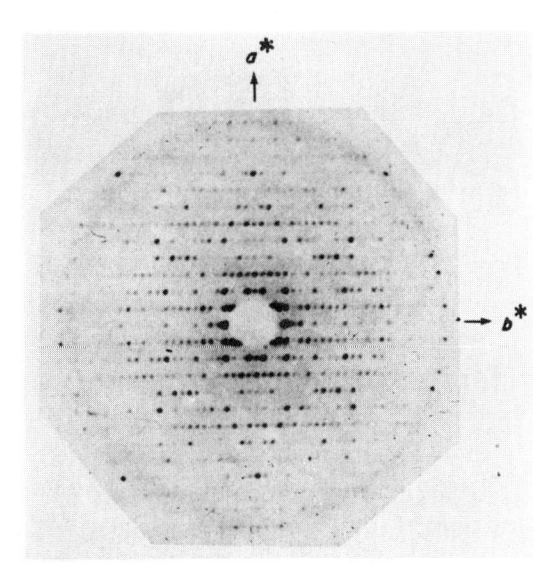

Figure 4.20 X-ray diffraction pattern for mMDHase crystal.

The atoms for different molecules are related by symmetry, are located in unit cells, and diffract in terms of lattice planes as before. However, in order to deduce the three-dimensional structure of the protein it is necessary to calculate the electron-distribution function $\rho(x,y,z)$ of equations (4.20) and (4.21). Thus if we can obtain the structure factor F (hkl) unambiguously, it is possible to locate atomic positions, provided enough experimental information is available. Unfortunately, as equation (4.17) shows, the intensity of scattering, which is the measured experimental variable, is related to the square of F (hkl). Thus although it is possible to establish the magnitude of F unambiguously it is not immediately obvious what the *sign* or phase of F is. This is known as the phase problem. Various approaches to this problem have been applied. The first that will be mentioned briefly is the Patterson synthesis. This method uses a Patterson function $P(x,y,z)$, which relates to the sum of the squares of F and thus avoids the phase problem. The Patterson function is written in a series (similar to a Fourier series):

$$P(xyz) = \frac{1}{V^2} \sum_{-\infty}^{\infty} \sum_{-\infty}^{\infty} \sum_{-\infty}^{\infty} F^2(\text{hkl}) \cos 2\pi (hx/a + ky/b + lz/c). \quad (4.34)$$

This series may be summed directly from the corrected intensities. A contour map is then established and the peaks on the map correspond to vector distances between pairs of atoms. If there are N atoms in the unit cell, there will be approximately N^2 Patterson peaks, and for protein molecules it is rarely possible to assign more than a few of the peaks to the correct pairs of atoms. Consequently the method of isomorphous replacement is used where the protein is crystallized with one or more heavy metal atoms in the unit cell. Since these atoms scatter strongly, it is relatively easy to locate these atoms and to assign the phase of these and related atoms. In some cases it is possible to make use of a pair of isomorphous crystals.

Since the diffraction pattern represents the atomic distribution in reciprocal space, it should perhaps be reiterated that reflections closest to the x-ray beam correspond to the largest interplanar spacings and those farther out reveal finer structural detail. Thus a first approximation to structure can be obtained by choosing a limited number of spots near the origin. It is usual to proceed by incorporating more spots in a series of refinements that produce higher and higher "resolution." It is usually possible to trace the polypeptide chain at 5 to 6 Å resolution, and at 3 Å or better it is possible to locate most of the non-hydrogen atoms.

In summary, the procedure for obtaining x-ray diffraction data for globular proteins (grossly simplified) is:

1. Obtain good single crystals. Determine unit-cell and other primary parameters.
2. Prepare crystals with heavy atom constituents.
3. Obtain diffraction patterns of isomorphous crystals.
4. Deduce heavy atom positions and phase of reflections.
5. Prepare electron-density map.
6. Build model and back-calculate diffraction density.
7. Refine model.

By such means several hundred protein structures have now been obtained and very valuable insight into structure and function relationships obtained. In most cases, crystals form with a heavy solvation shell, but nevertheless it is expected that some differences (though generally minor) in detailed structure will exist between proteins in the solid state and the same material in solution or bound in its native state. Several of the spectroscopic techniques outlined in this book are available for studying such problems, though most of these methods only provide qualitative comparative data. The power of the x-ray technique is its ability to provide unambiguous fundamental data.

4.13 Classifications of Protein Domain Structure Based on X-Ray Diffraction Analysis

We have already observed that various types of conformations (α-helix, β-sheet, irregular etc.) are built into the tertiary structure of proteins. Only very recently have various classes of proteins been established based upon conformation and the formation of domain structures.

Proteins with Antiparallel α-helix Structure

Up and down helix bundles

Several globular proteins (e.g., hemerythrin, tobacco mosaic virus protein, ferritin) contain antiparallel α-helices that lie essentially adjacent to each other. The helices are joined by folds as shown in Figure 4.21 for myohemerythrin. In this case there are four antiparallel α-helices forming the core of the protein.

Greek key helix bundles

By contrast with the adjacent antiparallel array of myohemerythrin, several globular proteins are known to have loosely packed α-helices in which pairs of helices tend to be perpendicular to each other. This crossover structure is demonstrated in Figure 4.22 for the β-chain of hemoglobin. Other proteins having such structure include myoglobin, thermolysin, and T4 lysozyme.

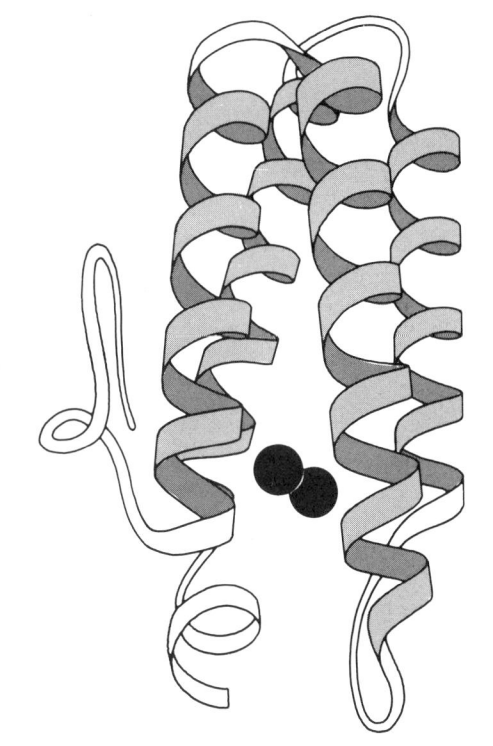

Figure 4.21 Antiparallel α-helix chain arrangement in myohemerythrin.

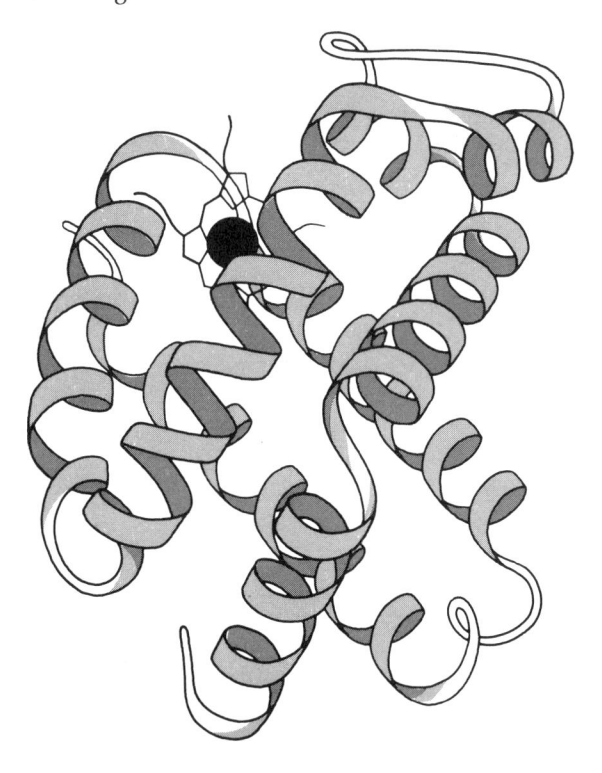

Figure 4.22 The crossover antiparallel α-helix (greek key) arrangement of the β-chain of hemoglobin.

Miscellaneous antiparallel α-helix

Not all antiparallel α-helical structures have the preceding types of tertiary structure. Some have only small regions of adjacent helices and other α-helices are widely separated. Such proteins include egg lysozyme, papain, and phospholipase.

Parallel α/β Structures

Considering the fact that parallel β-sheet structures have so far proven impossible to synthesize for simple poly(amino acids) and their derivatives, a remarkably large number of globular proteins contain regions (domains) of such structure. Invariably parallel β-sheet regions are twisted and are often found in conjunction with parallel α-helices. Two main categories may be described.

Singly-wound barrels

An interesting example of the parallel α-helix/β-sheet barrel structure is triose-P isomerase, shown in Figure 4.23. Each region of β-sheet is followed by α-helix such that the β-regions are adjacent in a twisted cyclic structure.

The barrel structure is relatively unusual but is also found in pyruvate

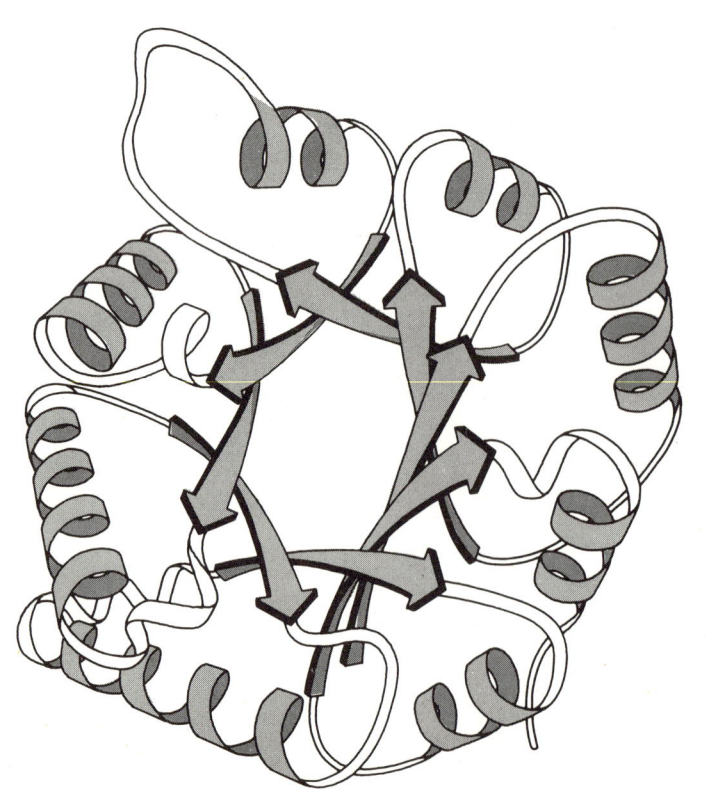

Figure 4.23 The parallel β-sheet barrel structure of triose-P isomerase.

4. X-Ray Diffraction

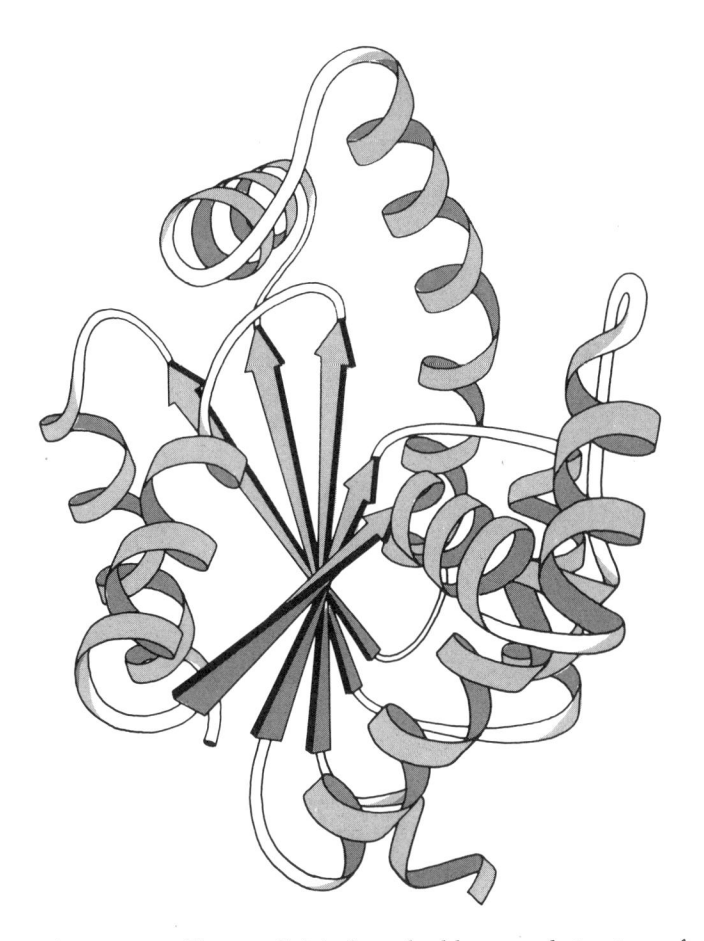

Figure 4.24 The parallel β-sheet doubly wound structure of adenyl kinase.

kinase. However, parallel β-sheet structures are commonly found in doubly-wound sheets.

Doubly-wound sheets

This classification of proteins includes around 20 known protein structures and consists of a twisted parallel β-sheet core along with peripheral α-helices, some of which are wound onto the sheet one way, and some the other. One example, adenyl kinase is shown in Figure 4.24. Among the more familiar proteins having similar structure are lactate and alcohol dehydrogenase, phosphorylases, flavodoxin, and subtilisin.

Miscellaneous α/β

The twisted β-sheet theme is a common one in globular proteins. The β-sheet can be parallel, antiparallel, or a mixture of both. Carboxypeptidase is an example of a protein with a mixed twisted β-sheet domain. Its structure is shown in Figure 4.25.

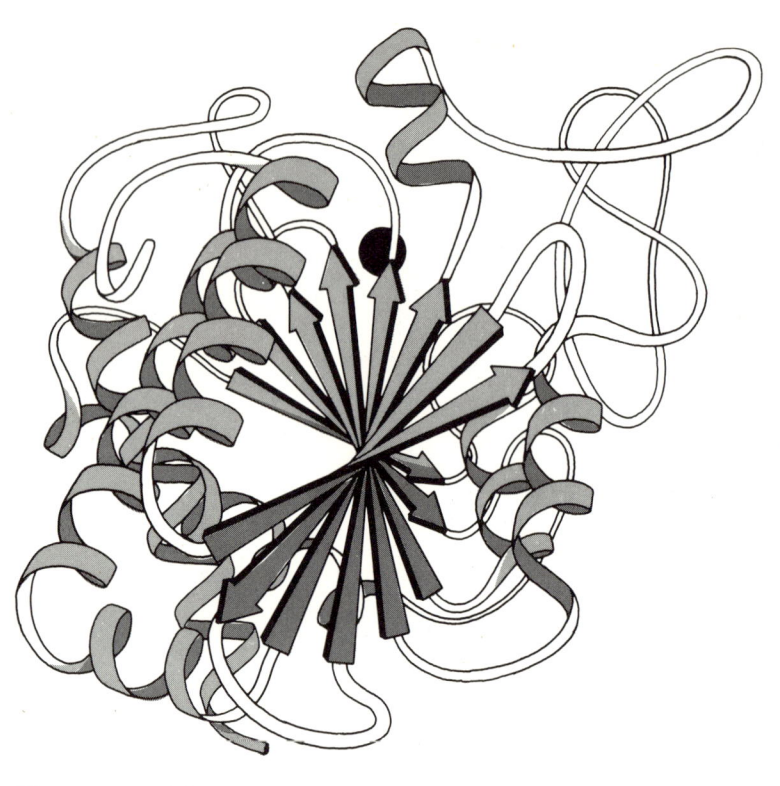

Figure 4.25 Schematic structure of carboxypeptidase showing the mixed parallel and antiparallel twisted β-sheet domain.

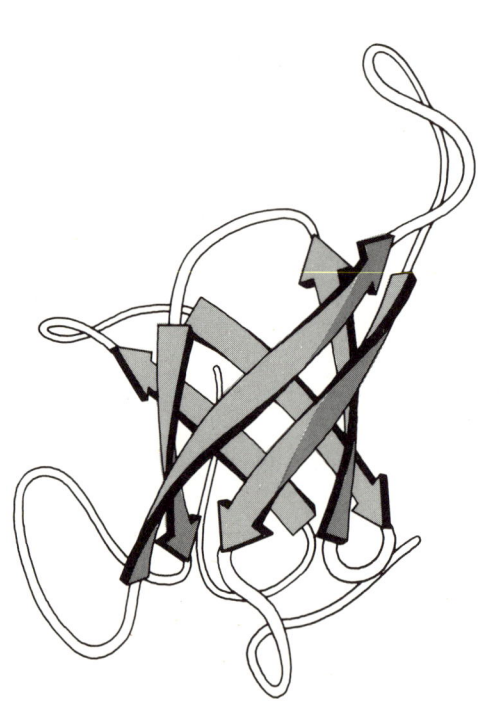

Figure 4.26 The antiparallel β-barrel domain of trypsin.

4. X-Ray Diffraction

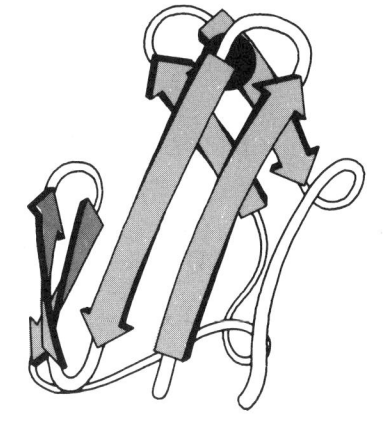

Figure 4.27 The β-sheet ligand cage of rubredoxin.

Antiparallel β-Sheet Structures

Globular proteins also contain the familiar antiparallel β-sheet structure, generally twisted, often into barrel domains. Such structures are commonly found in many proteins, including trypsin, chymotrypsin, elastase, immunoglobulins, and so on.

The structure of an antiparallel β-barrel is demonstrated in Figure 4.26 for one region of the trypsin structure.

Small Disulfide-Rich or Metal-Rich Structures

This class of proteins again contains familiar α-helix and β-sheet structures but the disulfide bridges or cage structures impose special constraints. The high sulfur content of snake neurotoxins and wheat germ agglutinin promote folded β-structures (toxin-agglutinin fold). Ion- or ligand-binding proteins can surround the nonprotein entity with α-helices, β-structure, or irregular structure. In rubredoxin there is an "up and down" antiparallel β-ligand cage, as shown in Figure 4.27.

Ferredoxin may by comparison be described as a "greek key" ligand that consists of both α- and β-chain conformations (Figure 4.28).

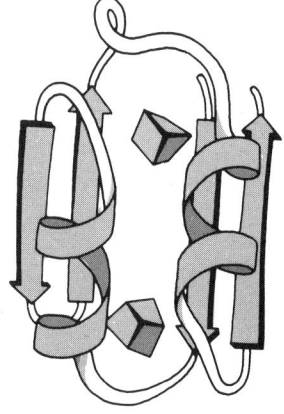

Figure 4.28 The greek key arrangement of chains in iron-binding ferredoxin.

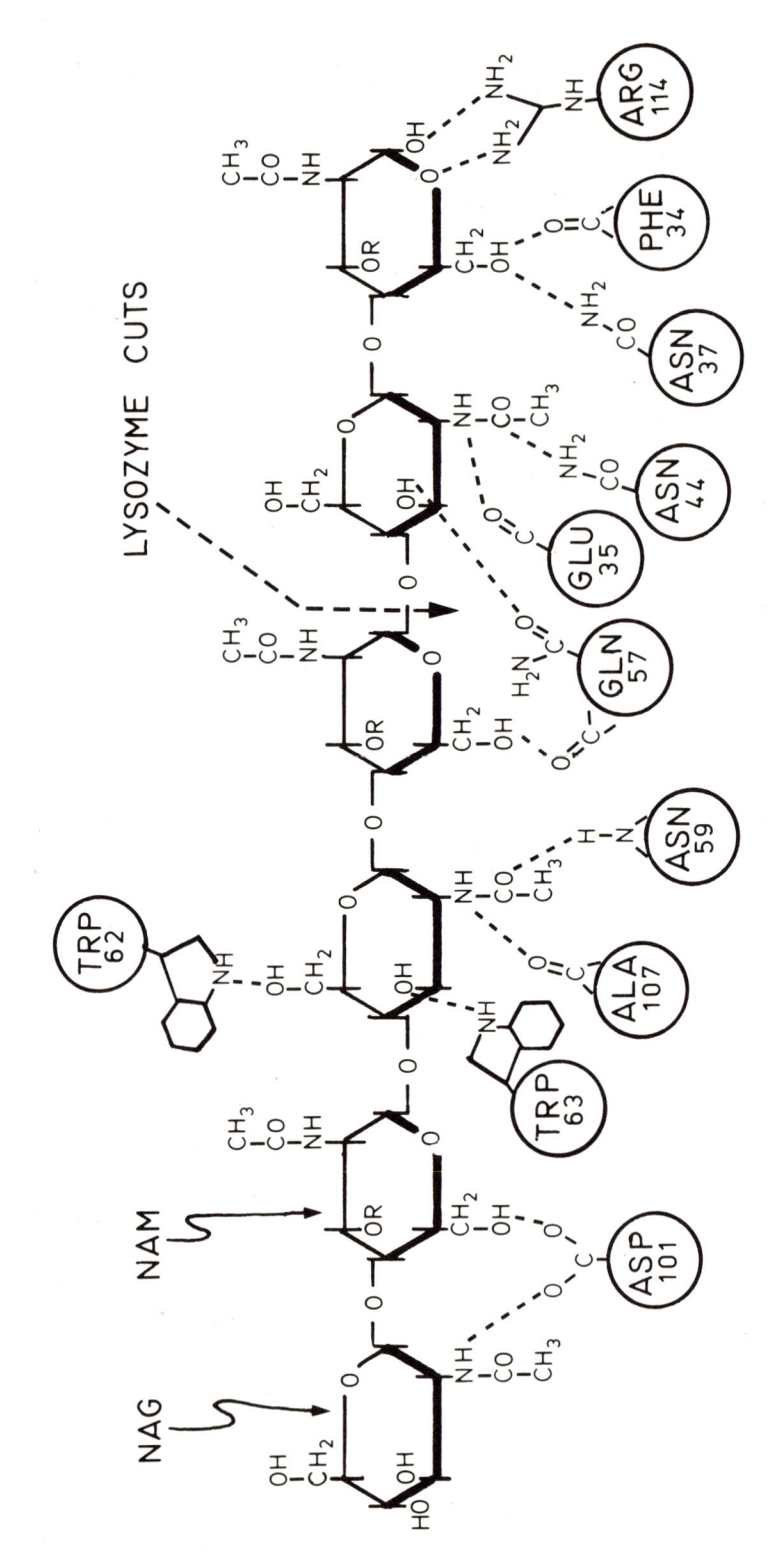

Figure 4.29 Amino acids involved at the active site of lysozyme and their proposed mode of binding to poly(NAG/NAM).

4.14 Active Sites

Many biochemists would argue that, beautiful though protein structures may be, it is their mode of functioning that is of prime importance. Consequently it behooves us to examine what type of information can be provided by x-ray crystallography concerning chemical activity. Normally enzymes are found to have clefts into which the substrate is inserted during enzymatic degradation. It is not, however, readily possible by studying the geometry of the cleft and the substrate separately to make an accurate analysis of how the enzyme functions and the substrate degrades. Since it is generally not possible to crystallize the enzyme with the substrate in position, more devious methods have been devised. In some cases it is possible to modify the substrate such that the enzyme complexes with the modified substrate but is unable to proceed with the chemical modification. One of the most thoroughly studied enzyme active sites is that of lysozyme. This enzyme was the first to have its tertiary structure determined by x-ray analysis and is a protein consisting of 129 amino acids (chicken egg white) or 125 residues (human). It is the former (egg white lysozyme) that has been studied most thoroughly. The enzyme

Figure 4.30 Schematic drawing of poly(NAG/NAM) located at the active site of lysozyme.

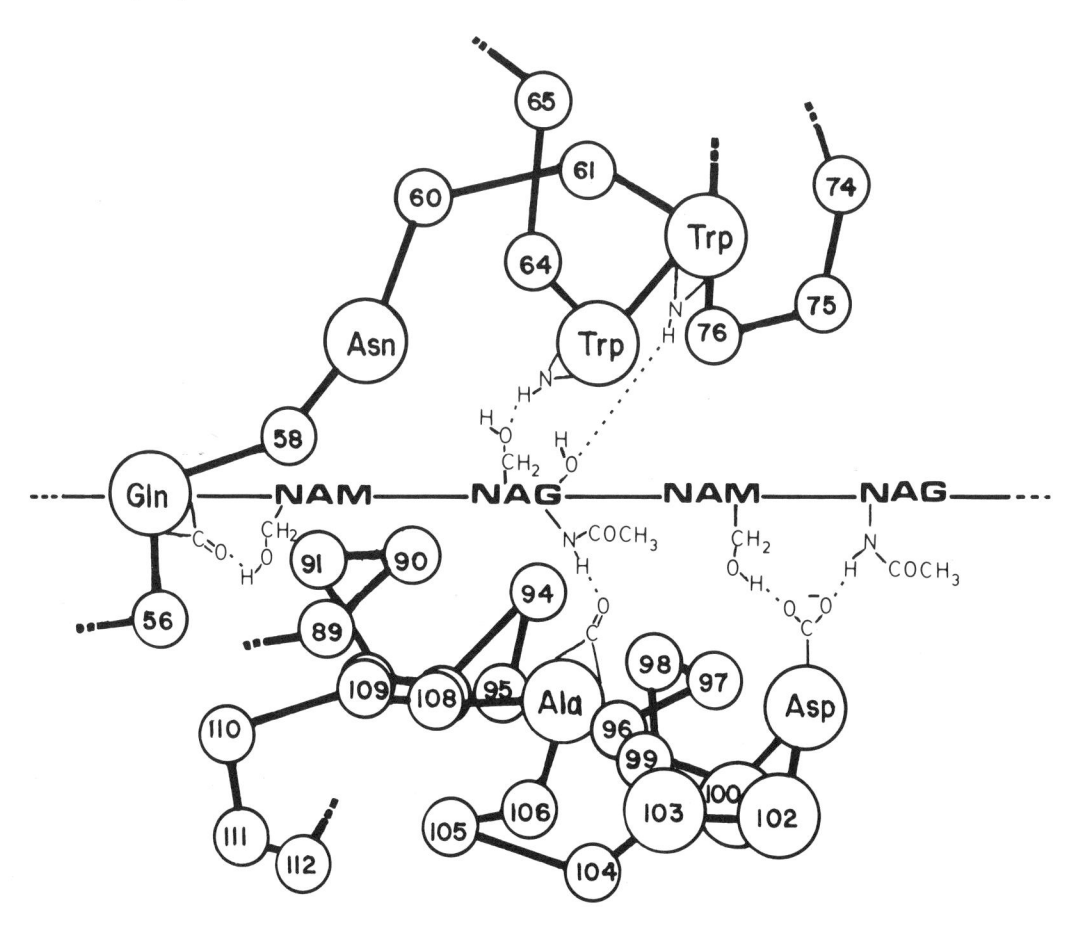

degrades certain polysaccharide structures, particularly the repeating polydisaccharide poly(N-acetyl-glucosamine, N-acetylmuramic acid) or poly-(NAG,NAM) in shorthand nomenclature. This polymer is found in bacterial walls that are degraded or "lysed" by lysozyme; hence the naming of the enzyme. In addition to poly(NAG,NAM), the (artificial) substrate poly-NAG is also cleaved by lysozyme and it has proven possible to crystallize the oligomer tri-NAG in the active site region of the enzyme. From the crystallographic and other evidence it has then proven possible to identify the binding mechanism and probable mode of action.

Figure 4.29 shows a schematic diagram of poly(NAG,NAM) and the lysozyme residues that are believed to play a role in binding and breaking the polymer bonds.

The suggested mechanism of action involves the binding of Glu^{35}-OH to the linking ether oxygen, O. The subsequent reaction proceeds as follows:

Of course the configurational strain on the polysaccharide chain induced by binding to other residues of the active site helps to overcome the energy barrier to the above process.

A schematic of tri-NAG located in the binding site of the enzyme is shown in Figure 4.30.

Further Reading

Alexander, L. *X-ray Diffraction Methods in Polymer Science.* Wiley, New York, 1969.

Blundell, T.L. and L.N. Johnson. *Protein Crystallography.* Academic Press, New York, 1976.

Cantor, C.R. and P.R. Schimmel. *Biophysical Chemistry,* Part II, Chapter 13. W.H. Freeman, San Francisco, 1980.

Dickerson, R.E. In *The Proteins,* H. Neurath, ed. Vol. 2, Academic Press, New York, 1964, pp. 603–778.

Kakudo, M. and N. Kasai. *X-Ray Diffraction by Polymers.* Elsevier, New York, 1972.

Matthews, B. X-ray structure of proteins. In *The Proteins,* H. Neurath and R.L. Hill, eds. Third edition, Volume 3, Chapter 4, Academic Press, New York, 1977.

Richardson, J.S. *The Protein Coloring Book.* Riverain Press, Durham, NC, 1979.

Richardson, J.S. *Advances in Protein Chemistry* 34, in press.

Schulz, G.E. and R.H. Schirmer. *Principles of Protein Structure,* Chapter 5. Springer-Verlag, New York, 1979.

Sherwood, D.H. *Crystals, X-Rays and Proteins.* John Wiley, New York, 1976.

5

Electron Diffraction and Microscopy

Electron microscopy is a tool that has advanced from its original objective of visualizing small objects to a powerful method for determining quantitative aspects of structure. The advantages of the methodology are associated with the short wavelength of the electron beam (roughly 0.05 Å) and the high electron-scattering power of matter. The disadvantage is that the method involves passing an electron beam (often of 100 KV) through a small sample (supported on a grid) in a high vacuum. These are naturally very harsh conditions, particularly for biologic samples, which usually undergo destruction in a few seconds. For the purposes of visualizing samples it is often standard practice to make replicas or to fix the material such that destruction of the organic material does not destroy the image. Some forms of quantitative electron microscopy mentioned in this chapter can, in fact, be carried out on fixed, stained, or replicated samples, but most require the careful and often tedious preparation of suitable native samples.

The questions addressed in this chapter are: (1) what information can be obtained concerning polypeptide chain conformation; and (2) what quantitative information can be obtained concerning chain arrangement and molecular arrangement? To approach these problems we shall study: (1) examples of electron diffraction; (2) quantitative staining procedures; and (3) an introduction to electron imaging.

5.1 Electron Diffraction

The wavelength of an electron beam in Å, accelerating potential V volts is:

$$\lambda = 12.27 \, (V + 0.978 \times 10^{-6} \, V^2)^{-1/2} \tag{5.1}$$

The electron-scattering power of biologic molecules is about 10^6 times

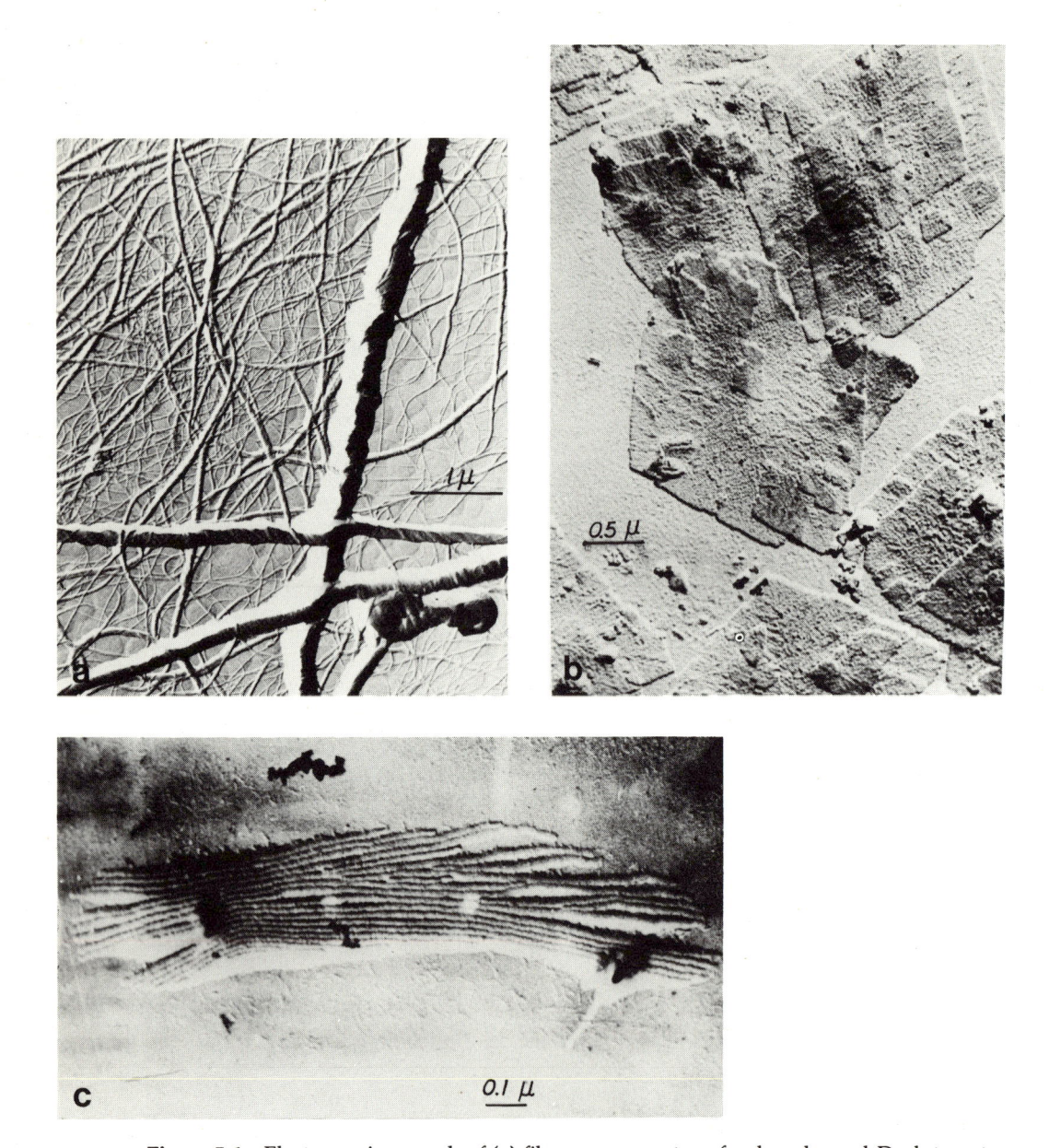

Figure 5.1 Electron micrograph of **(a)** fibrous aggregates of poly-γ-benzyl-D-glutamate; **(b)** single crystals of poly-O-acetyl-L-hydroxyproline; and **(c)** lamellae of poly-L-alanine.

greater than the x-ray scattering power, which means that the diffraction pattern from a sample some 500 Å thick may be collected on a photographic plate in a second or so. There are many different electron-scattering configurations of the microscope that may be put to use, but in the most refined, electron diffraction may be obtained from regions of the specimen that are 1,000 Å square or less. Consequently, the technique should be very useful if appropriate specimens can be prepared. One major disadvantage of electron diffraction from biologic specimens is that water of

5. Electron Diffraction and Microscopy 107

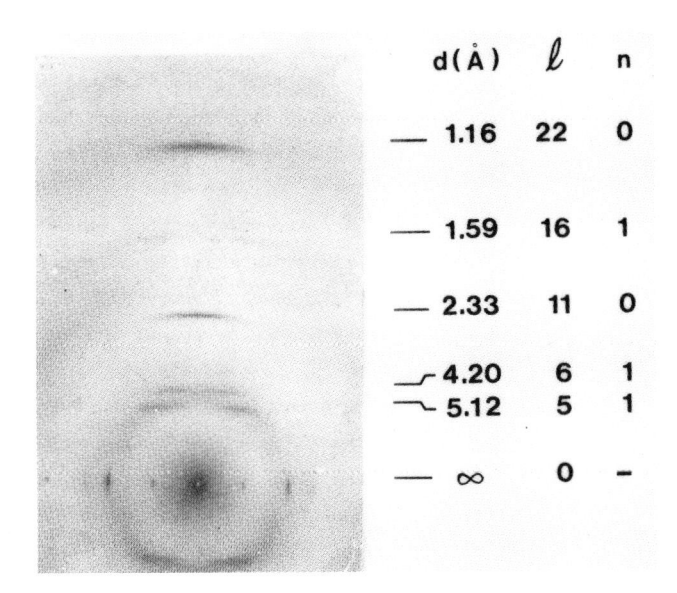

d(Å)	l	n
1.16	22	0
1.59	16	1
2.33	11	0
4.20	6	1
5.12	5	1
∞	0	–

Figure 5.2 Electron-diffraction pattern of oriented poly-γ-benzyl-D-L-glutamate.

hydration is almost immediately lost in the vacuum of the microscope; hence, if the structure is dependent on this water, it collapses very rapidly, thus preventing observation. Contamination caused by decomposition of material in the microscope also complicates quantitative microscopy.

Poly(amino Acids) and Polypeptides

Perhaps the simplest form of polypeptide chain analyzed by electron diffraction is that from synthetic poly(amino acids) and their modifications, and sequential polypeptides. Depending upon solvent crystallization or precipitation conditions, samples may be obtained that are ropelike or crystalline, the crystal dimensions being about 1μ in length or width and 100 Å in thickness.

Figure 5.1 shows various morphologies for (a) poly-γ-benzyl-D-glutamate, (b) poly-O-acetyl-L-hydroxyproline, and (c) poly-L-alanine. The electron diffraction is usually poor from type (a) precipitates, good for type (b), and fair to good for type (c). The same Bragg criteria need to be met in solving the diffraction patterns as for the x-ray work in Chapter 4. Some examples of diffraction from single crystals are as follows.

Fiber patterns

If the molecules lie flat on the surface and can be oriented by stroking or similar procedures, then fiber electron diffraction patterns are obtained resembling the x-ray fiber patterns of the previous chapter. Figure 5.2 shows the electron-diffraction pattern of poly-γ-benzyl-D-L-glutamate, a sequential polydipeptide which in this case, forms a π_{LD} helix. The layer lines l and order of Bessel function n are indicated. Evidently the material indexes (equation 4.27) as an 11_5 helix and is the $\pi_{LD}{}^4$-helix mentioned in Chapter 4.

Single crystal patterns

In Figure 5.1b and in all similar crystals examined to date, the chain axis of the polypeptide is perpendicular to the crystal face and will be parallel to the electron beam direction. This orientation gives single crystal reflections for hk0 (putting the *c* axis parallel to the electron beam).

Each spot then represents an hk0 plane in reciprocal space. Figure 5.3 shows such a pattern for polyglycine I.

Procedure for Solving Diffraction Pattern

1. Draw two lines through the origin 0 such that one passes through the spots closest to the origin, the second through the next nearest.
2. Draw parallel lines to the first two passing through the other diffraction spots and forming a grid.

We now have a representation of the reciprocal lattice. Since the angle between the two sets of lines is not 90°, we must have a unit cell in which

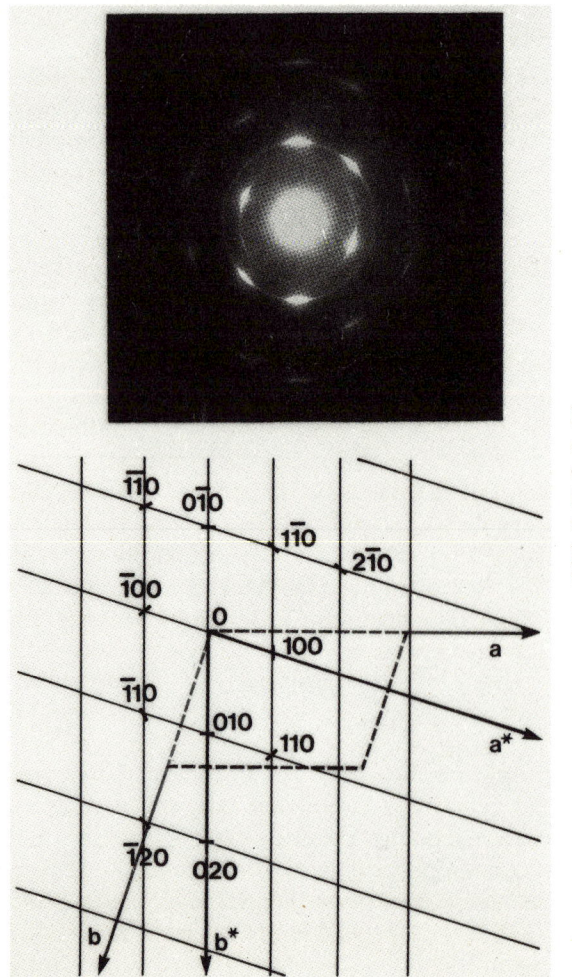

Figure 5.3 Selected-area electron-diffraction pattern of a single crystal of polyglycine in the polyglycine I (sheet) conformation. Below it is shown a diagram of the diffraction pattern along with the proposed unit cell.

5. Electron Diffraction and Microscopy

two of the axes in real space are not 90°, i.e., hexagonal, monoclinic, or triclinic.

3. Measure the angle between the two sets of intersecting lines. If the angle is 120° (or 60°), the system is hexagonal.

In this case the angle is 113° and hexagonal is ruled out. Triclinic cannot be ruled out at this stage but we shall proceed on the assumption that the system is indeed monoclinic.*

If the crystal had been inorganic, the monoclinic cell of Table 4.1 would have been chosen. However, it is common to identify the chain axis with the c unit-cell dimension. If this latter convention is used, axes a and b must be in the plane of the diffraction pattern and subtend (from the measured pattern) an angle of 113°. Similarly, the chain axis must be perpendicular to the crystal surface and parallel to the beam direction; hence the hk0 assignment. From equation (4.6),

$$d_{hkl} = \sin 113° [h^2/a^2 + k^2/b^2 - 2hk \cos 113/ab]^{-1/2}$$

We note that in this equation if $h = 1$ or $\bar{1}$ and $k = 0$ (or vice versa), there are two identical values of d_{hkl}. Since these are the lowest index planes, we would expect them to correspond to the largest possible values of d; they will correspond to the smallest distances from the origin. Thus the innermost two spots may be tentatively identified.†

4. Arbitrarily define axes a^* and b^*, which represent a and b in reciprocal space. If a^* is chosen, as in the diagram, the two closest spots to the origin become ($\bar{1}$00) and (100). [If b^* had been chosen instead of a^*, we would have had (0$\bar{1}$0) and (010) respectively.]

5. All vertical lines passing through the ($\bar{1}$00) point will have $h = \bar{1}$, and through (100) will have $h = 1$ (and through the origin $h = 0$). Similarly, all lines parallel to a^* will have the same values of k.

We can then set up the grid as shown with the spots lying at intersections identified by their Miller indices.

6. To check out these assignments and calculate unit-cell dimensions, it is necessary to measure the distance of the spots from the origin, convert this distance to a refraction angle θ (requires a knowledge of the diffraction geometry), and, by substituting in the Bragg equation, obtain $d_{hkl} = \lambda/2 \sin \theta$. The value of λ is calculated from equation (5.1). To a first approximation the measured distance on the film is proportional to θ, so that even if the diffraction geometry were not known, the ratio of a to b could be obtained. In the present case, the simplest method of procedure is to choose spots identified with h or k = 0, solve equation (4.6a) for a or b, then proceed to find whether higher index planes are accounted for by the equation. In the actual experimental data, $a = 4.77$ Å, $b = 3.67$ Å (and $c = 7.044$ Å from other information), $\gamma = 113°$. It is interesting to note that

---

*Since electron diffraction usually examines structure in only two dimensions, it is not possible to differentiate between monoclinic and triclinic unit cells unless information can be obtained by tilting the crystal.

†If the chains are arranged in a body-centered array, there will be forbidden reflections and the assignments will represent a simple multiple of the actual unit cell.

because of the values of a and b, the $0\bar{1}0$ and $\bar{1}10$ spots are essentially equidistant from the origin.

One final observation is that a representation of the size and orientation of the unit cell in real space may be drawn on the reciprocal lattice by drawing a line perpendicular to the axis marked b^* from the origin; this is the a axis of the unit cell. A second line drawn perpendicular to a^* from the origin represents the b axis of the unit cell. These new axes a and b are in the inverse ratio of a^* and b^* segments (between lattice points). The solution of the two-dimensional electron-diffraction pattern above does not directly indicate which unit-cell axis lies parallel to the chain axis (although we have assumed it to be the c axis). Confirmation of this assignment is readily achieved because the polyglycine I structure is known to have a peptide repeat h = 3.5 Å and a helix = unit-cell repeat of 2 × 3.5 = 7.0 Å (2_1 helix). Since this value does not correspond to the two calculated cell dimensions, it may be concluded that the chains lie perpendicular to the plane of the crystal and parallel with the c axis (and electron beam).

Fibrous Proteins

Relatively little electron-diffraction work has been reported for fibrous proteins, though the general area of electron diffraction from proteins is regarded as one of the current "frontiers" in protein structure measurements. The reason for this is that electron diffraction may be obtained from extremely small structures. In general, we may expect diffraction of the type

Figure 5.4 Fiber-diffraction patterns from silk *(Bombyx mori)* fibroin: **(a)** electron diffraction; **(b)** x-ray diffraction.

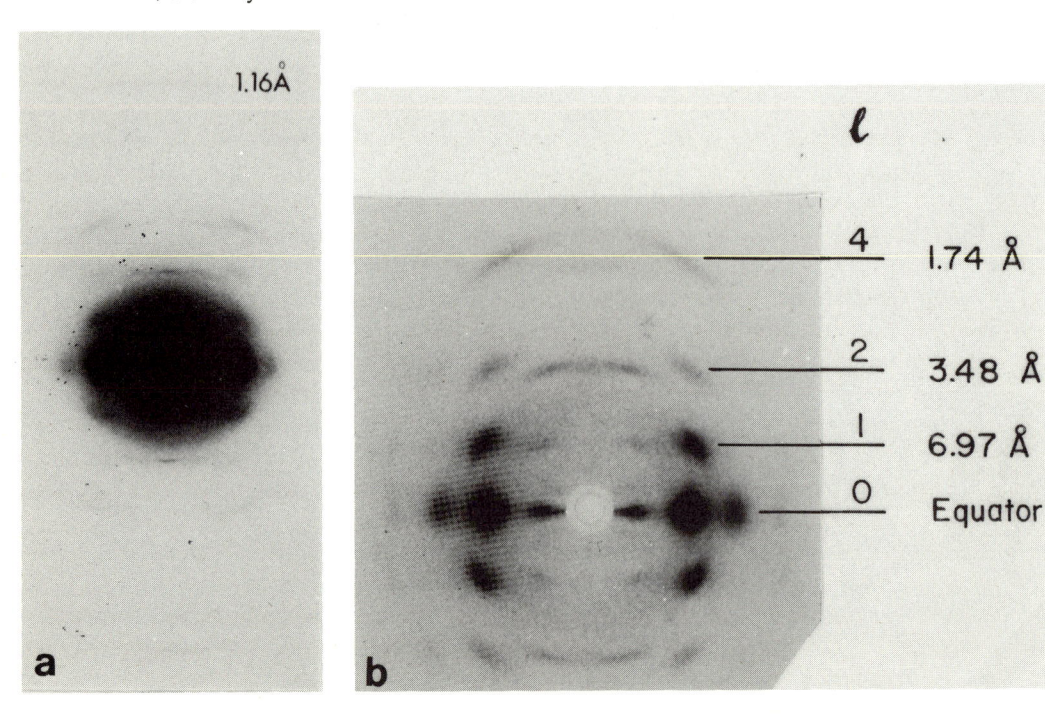

5. Electron Diffraction and Microscopy   111

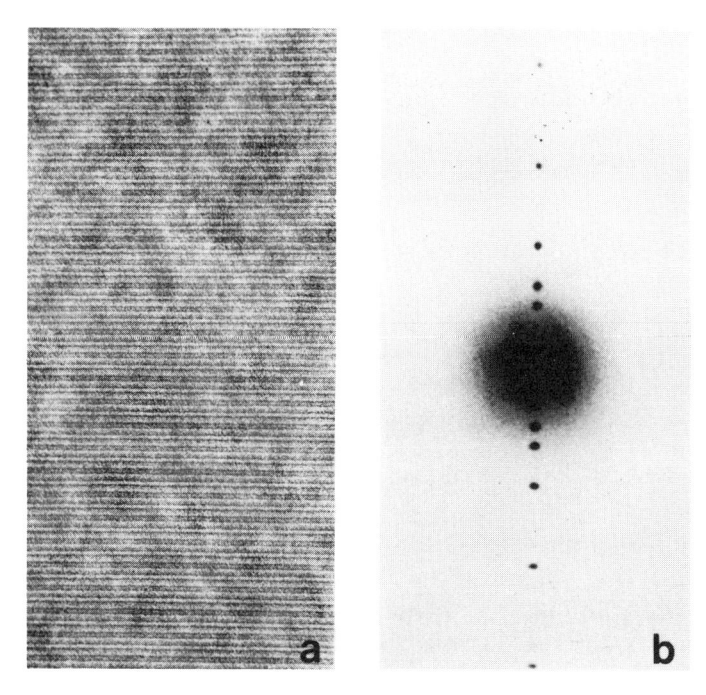

Figure 5.5 Electron micrograph of uranyl acetate–stained oyster muscle tropomyosin crystal: **(a)** major banding periodicity ~ 380 Å; and **(b)** low-angle electron diffraction from the same sample (after Tulloch and Fraser).

reported in the previous section for fiber patterns when dealing with fibrous proteins. Figure 5.4 shows the electron-diffraction pattern obtained from an 800 Å thick section of *Bombyx mori* silk fibroin. Figure 5.4b is an x-ray diffraction fiber pattern from the same material. Some of the features have been labeled so that the similarities (and differences) between the two diffraction patterns are evident. It can be seen that the low-wavelength electron-diffraction pattern leads to greater clarity in the high-index region.

Efforts to obtain diffraction from fibrous proteins in which water is an integral part of the structure have met with mixed success. In this case, research is directed towards fixation methods, freezing and building hydration cells that prevent the escape of solvent vapor and thus maintain the hydrated state. Collagen is an example where diffraction has been difficult to obtain; nevertheless, recent advances have enabled diffraction patterns to be collected.

Stained samples

The role of heavy-atom staining in obtaining quantitative structural data for fibrous proteins is discussed in some detail in section 5.2. The symmetrical deposition of heavy-metal atoms in a protein fiber does, in addition, allow low-angle diffraction (both x-ray and electron) to be obtained. Figure 5.5a and b show a tropomyosin tactoid (produced by precipitation of tropomyosin with a divalent cation) stained with uranyl acetate. Figure 5.5b shows the low-angle electron-diffraction pattern obtained from such a sample. The spots are simple orders of a primary diffraction $d = 380$ Å with Bragg reflections corresponding to $n\lambda = 2d\sin\theta$ having n = 1,2,3 . . . . The 380 Å repeat structure originates from the packing arrangement of tropomyosin molecules in the crystal. The fundamental repeat is different for different fibrous proteins and arises from molecular overlap such as that shown for paramyosin in Figure 5.7.

Globular Proteins

Notable advances in electron-diffraction studies of proteins, using hydration chambers, have been made in the past few years. Figure 5.6 shows a beautiful electron-diffraction pattern from the enzyme catalase. In this particular case the material was crystallized with glucose, which replaces waters of crystallization and "preserves" the crystal to some extent.

Once such a pattern is obtained, there remains the particularly difficult problem of reconstructing the crystal structure. As with x-ray diffraction from globular protein crystals, the phase of the diffraction is required as well as the spot intensity. For x-ray diffraction the assumption is generally made that for radiation of wavelength ~ 1.5 Å, there is only one scattering event as the beam passes through a crystal. For electron diffraction, however, this (kinematic) approximation will depend for its validity on sample thickness and the scattering power of atoms in the sample. For example, with metal samples, the scattering is multiple, particularly for thicker samples; and the diffraction intensities are related to the scattering factor by a power series. Fortunately, in the catalase case, it seems that the thin biologic sample does scatter kinematically. The phase problem is partially solved by a method devised by Klug and co-workers at Cambridge University in England. The details of the method will not concern us here but they involve the comparison of diffraction intensities (in reciprocal space) with direct microscopy of the crystals at very low-beam intensity (to minimize beam damage). The information obtained from high-resolution microscopy regarding the location of molecules in the unit cell is then fed back into intensity calculations. By this means the low-resolution "two-dimensional" structure may be evaluated. Electron-diffraction patterns from catalase and other proteins contain information that potentially allows 3 Å or so resolution (i.e., comparable with x-ray diffraction).

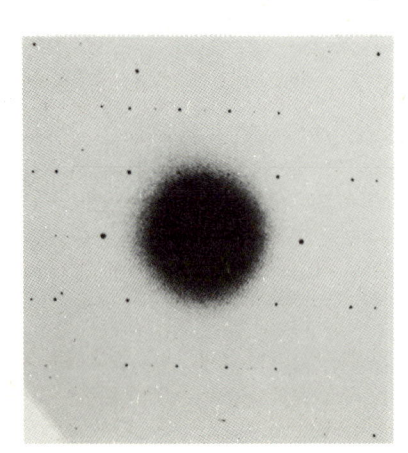

Figure 5.6 High-angle electron-diffraction pattern of catalase crystals immersed in glucose (after J. Unwin).

5. Electron Diffraction and Microscopy

5.2 Application of Electron Microscopy to Fibrous Protein

General Principles

A great deal of progress in understanding how molecules arrange themselves in fibers has come about from the use of electron microscopy. In the last few years in particular, it has become possible to stain the fibers in such a manner that a correlation of stain with primary sequence allows a very detailed understanding of molecular arrangement.

Two levels of sophistication are pertinent to this discussion. If nothing is known about the fundamental molecules except their length, then it is still possible in many circumstances to locate the molecules in the fiber. At the most sophisticated level, it is possible to locate clusters of residues in the molecules and to identify the most probable intermolecular forces.

There are generally three methods applied to the staining of fibers:

1. *Staining with phosphotungstic acid (PTA)*. Phosphotungstic acid interacts with basic groups, particularly arginine and lysine, leaving the heavy tungsten atom at these basic sites.

2. *Staining with uranyl acetate (UA)*. Uranyl acetate interacts with acidic groups, particularly glutamic and aspartic acid, such that the heavy uranium atoms "mark" the acidic sites.

3. *Negative staining*. If staining with UA is carried out at low pH, the carboxyl groups of glutamic and aspartic acids are undissociated and the proton only exchanges slowly with the uranyl ion, which preferentially accumulates in holes in the structure; this is called negative staining. Similarly, if a fibrous protein is exposed to PTA or UA at neutral pH for long periods of time, heavy atom complexes deposit in regions of low molecular density. In this latter case both positive and negative stains are produced.

Other stains commonly used are lead acetate and osmium tetraoxide, but for present purposes the PTA and UA staining methods will be studied in more detail.

Examples

Paramyosin

Paramyosin is a rod-shaped molecule, about 1,300 Å in length, that is found in molluscan muscles. It consists of a two-chain α-helix coiled coil in which the chains are parallel. When precipitated from solution with divalent cations it forms a number of different "paracrystalline" fibers. Four such micrographs are shown in Figure 5.7. The sample has been negatively stained with UA so that the regions in the fiber that have the lowest molecular density appear as dark bands. Beneath the micrographs are schematic drawings of molecular arrangements that would account for the negative staining pattern.

Let us consider first micrograph (a). According to the proposed molecular array, the molecules are all parallel to each other and there are half as many molecules in the dark band region as in the light region.

Actually if there were only dark and light bands, without any internal staining within the light bands, we would not be able to conclude whether the molecules were parallel or antiparallel. However, a close examination of the light region reveals that there are many electron-dense bands (in this case UA

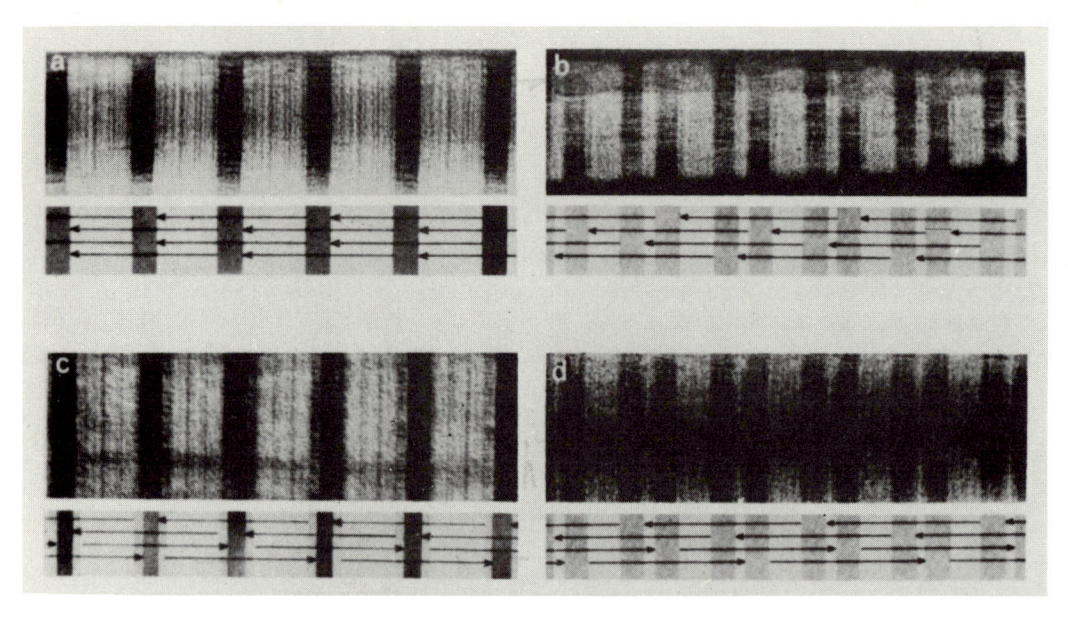

Figure 5.7 Electron micrographs of various forms of paramyosin precipitated with divalent cations and negatively stained with uranyl acetate. The darkest regions of the fibrous "tactoids" correspond to regions of lowest protein density. Correlation of the staining pattern and opacity enables the molecular arrangement to be established. Molecules are 1,275 Å long; the polar-banding pattern shows periodicity of about 725 Å (after Cohen et al.).

stained) that are not symmetrical about the midpoint of each light region. Since the stain is a molecular fingerprint that shows the distribution of acidic groups, the acidic groups must be asymmetric and therefore the molecules must be parallel.

It is not, however, possible to tell whether the C- or N-terminus of the molecules is at the position marked by the arrowhead. The light bands measure 725 Å and the molecular length, based on the preceding arguments, measures 1,275 Å.

In Figure 5.7b a different parallel arrangement of paramyosin molecules is shown that also gives rise to asymmetric fine structure.

Figures 5.7c and d, on the other hand, show antiparallel arrangements of molecules as revealed by the fine structure of the light bands. Figure 5.7c is particularly interesting in that an intermediate level of stain surrounds the dark region and is due, as shown, to an intermediate number of chains passing through that region.

One further question may be answered for these paracrystals and that is: why do parallel bands of stain fall within the light regions? To answer this question, we note that the bands stained with UA must correspond to clusters of glutamic and aspartic acid residues.

Since this type of paracrystal forms only in the presence of divalent cations, it seems likely that the divalent ions bridge the carboxylic groups, bringing them into register for staining purposes. The divalent ions themselves are somewhat electron-opaque and may contribute to the stain pattern. *Thus the molecules must assemble in arrays that optimize the ion-bridging interactions.*

Collagen

The details of the hierarchical arrangement of collagen molecules in tendon and connective tissue are discussed in detail in Section 3 (Chapter 13). The intent here is to investigate the type of information that has become available as a result of electron microscopic examination. Collagen molecules are known to consist of three coiled coil chains of polyproline II–type conformation, in parallel array. The molecules are rodlike with length approximately 2,800 Å. Sequential data are available that enable virtually all of the staining groups to be located within the molecule, thus providing us with the ability to carry out complete electron micrograph fingerprinting (quantitative electron microscopy) on the fibers.

Electron micrographs of native collagen fibers, stained with either UA or PTA show essentially the same banding pattern with a repeat of 640–670 Å. It is now generally agreed that 670 Å is a representative value for hydrated fiber but under the vacuum conditions of the electron microscope the fibers shrink. The native banding pattern shown below (Figure 5.8) is reminiscent of paramyosin, with the electron-opaque bands lying perpendicular to the fiber axis. X-ray diffraction shows the molecules to be aligned parallel with the fiber axis and therefore the problem presents itself of arranging 2,800 Å molecules such that they give 670 Å repeat patterns with detailed fine structure.

Figure 5.8 Native collagen fibrils PTA and UA stained (positive staining), showing 640-Å repeat bands with fine banding structure between the major bands.

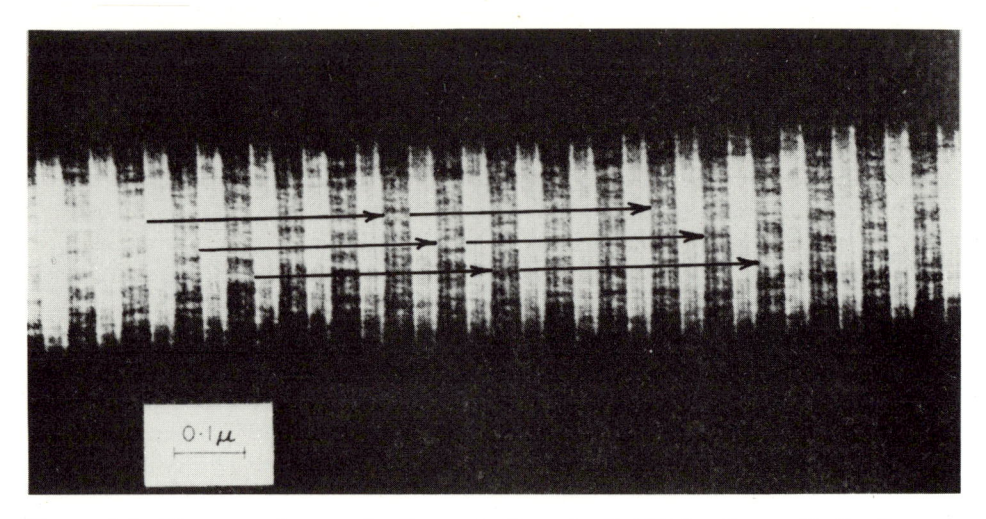

Figure 5.9 Electron micrograph of negatively stained (PTA) collagen fibrils showing the (schematic) location of molecules (~ 3,000 Å in length). The molecules must be parallel because of the asymmetric pattern of the positive stain pattern (Figure 5.8).

Negative staining of native collagen fibrils with PTA reveals an alternating dark- and light-banded structure with 640 Å repeat. Since the fine structure of the positive stain is asymmetric, the molecules must be in parallel array. The staggered array is shown in Figure 5.9.

By matching the sequence of the molecules with their position and staining pattern it has now been possible to account completely for the pattern based on the location of the groups Arg, Lys, Asp, Glu. The positive and negative residues are clustered in the same locations, thus accounting for the similarity between UA and PTA stain patterns. The driving force to fiber formation is, therefore, predominantly the need to form ion pairs.

5.3 Electron Imaging

The principles behind the reconstruction of three-dimensional protein structure have been introduced in section 5.1, where a two-dimensional reconstruction of catalase is examined. Electron imaging is a specific application of the general mathematical technique known as image reconstruction from projections. In its application to electron microscopy the process is sometimes known as Fourier microscopy. In its earliest applications electron micrographs were used as diffraction gratings, which upon illumination with visible light gave rise to optical diffraction patterns revealing some of the symmetry of the specimen. More recently, phase information has been obtained by scanning micrographs with a digital microdensitometer, which is then Fourier transformed (see Appendix A) for comparison with electron-diffraction data so that phase and amplitude data may be extracted.

In the most successful application of this technique to date, the protein component of *Halobacterium halobium* has been studied. (The organism is of particular interest because it can convert sunlight into energy without using chlorophyll.)

Samples of the membrane measuring about 1μ square and 45 Å thick (i.e., about 10^4 identical unit cells) were preserved from dehydration artifacts with

5. Electron Diffraction and Microscopy   117

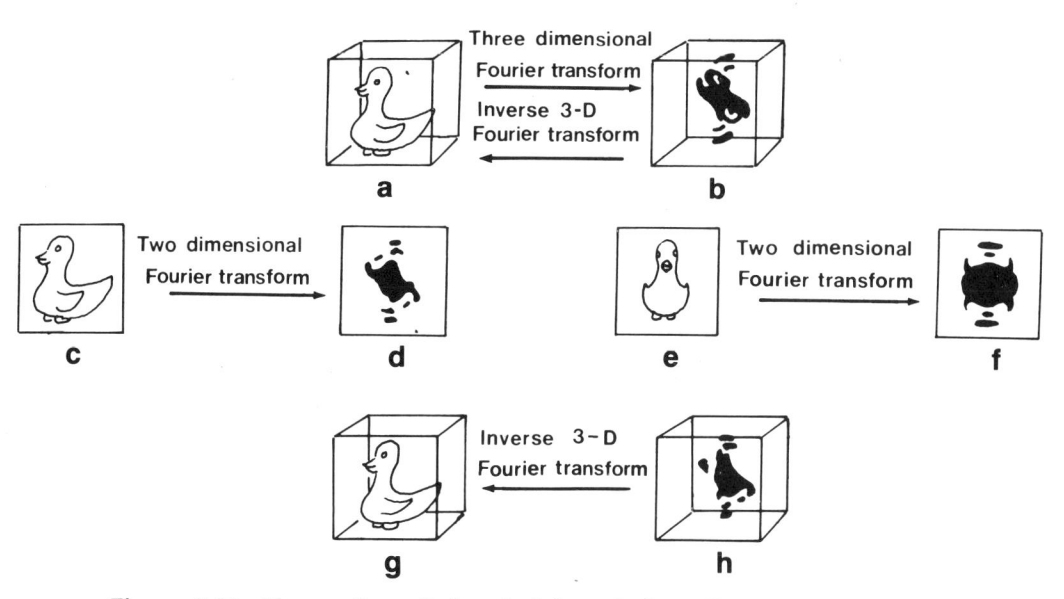

Figure 5.10 The mathematical principles of three-dimensional reconstruction. A three-dimensional duck (a) and its Fourier transform (b) are approximated as follows. Required are (c) a two-dimensional projection of (a); (d) the two-dimensional Fourier transform of (c); (e) another projection of the duck; and (f) its two-dimensional Fourier transform. The three-dimensional duck (g) is calculated from an approximate three-dimensional Fourier transform (h), which was reconstructed from the two-dimensional transforms (d) and (f). (*Source:* James Lake, New York University Medical School.)

glucose and illuminated by a defocused electron beam with a minimum intensity (0.5 electrons/ Å²) sufficient to image a single unit cell with good contrast. The probability of electron-beam radiation damage is thus very small for any unit cell. The resultant image transforms and diffraction patterns are reduced to number matrices with a densitometer and analyzed by Fourier analysis to produce a two-dimensional projection, in the beam direction, of the object transform. A three-dimensional object transform can be produced by tilting the specimen through known angles to produce a series of two-

Figure 5.11 A model of a single protein molecule in the purple membrane. The molecule is oriented perpendicular to the plane of the membrane. The α-helices are 35 to 40 Å apart. (*Source:* Richard Henderson and Nigel Unwin.)

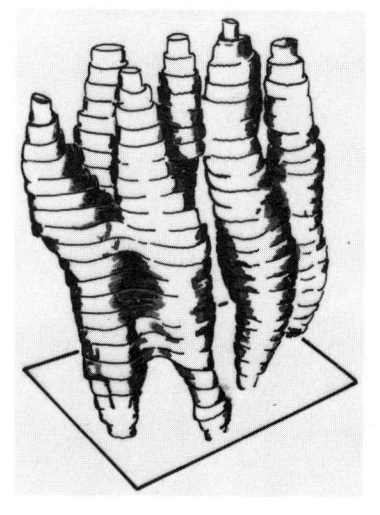

dimensional projections that contain the information to produce a three-dimensional transform for the average unit cell.

The process of transforming three-dimensional objects into two-dimensional Fourier-transformed diagrams, which are then reconstructed to the three-dimensional object, is shown in schematic form for a duck (a popular art representation!) in Figure 5.10.

The application of the Fourier conversion applied to the purple membrane protein is shown below (Figure 5.11). As can be seen, there are a series of (α-) helices spaced 10 to 12 Å apart which span the width of the membrane.

In summary, electron microscopy and diffraction techniques are of primary and growing importance in protein structure work. Furthermore, apart from the direct visualization of individual protein molecules (not discussed here), the detailed assembly of molecules in fibrous proteins, chain conformation, and crystal structure may be elucidated by electron microscopic methods. Such details are not generally accessible by other means.

Further Reading

Blundell, T.L. and L.N. Johnson. *Protein Crystallography*. Academic Press, New York, Chapter 18, 1976.

Crowthers, R.A. and A. Klug. Structural analysis of macromolecular assemblies by image reconstruction from electron micrographs. *Annual Review of Biochemistry* **44**:161–182, 1975.

Finch, J.T. Electron microscopy of proteins. In *The Proteins*, H. Neurath, and R.L. Hill, eds., 3rd ed., Volume 1, Chapter 6, Academic Press, New York, 1975.

Walton, A.G. and J. Blackwell. *Biopolymers*. Academic Press, New York, Chapter 4, 1973.

6

Infrared Spectroscopy

The application of infrared spectroscopy to biologic systems is a common method of obtaining structural information about polypeptides. Unfortunately, the information that may in principle be obtained has not been fully realized in practice because of limitations in experimental technique. New technical developments of the past few years have been changing this situation.

The basis of the technique is in molecular vibrations and how the vibration frequency of a particular chemical group is altered when that group is engaged in various types of intra- or intermolecular interactions.

6.1 Absorption of Radiation in a Vibrating System

Absorption of energy causes a transition in the energy state of a molecule, defined:

$$\Delta E = h\nu, \tag{6.1}$$

where h is Planck's constant and ν is the frequency of absorbed radiation. This relation may also be written in terms of the wavelength λ because

$$\nu = c/\lambda, \tag{6.2}$$

where c is the velocity of light.

In infrared spectroscopy, instead of the frequency ν, it is more common to use the wavenumber $\bar{\nu}$, which is defined as

$$\bar{\nu} = \frac{1}{\lambda} \tag{6.3}$$

and has the units of cm^{-1}.

Infrared spectroscopists use the reciprocal wavelength nomenclature because the numbers are convenient (10 to 10^5 cm^{-1} compared with 10^{11} to 10^{15} in the frequency nomenclature). The infrared absorption band shape is also conveniently expressed mathematically in terms of energy or reciprocal wavelength; that is the shape is symmetrical in reciprocal wavelength but not in wavelength.

Combining the previous equations,

$$\Delta E = hc\bar{\nu}. \tag{6.4}$$

In order for an atomic group to absorb infrared radiation, it must possess an electric dipole moment that changes during motion. Since dissimilar atoms or groups possess dipole moments, it is found that vibrational motions between dissimilar atoms lead to infrared absorption bands, whereas vibrations between identical atoms do not.

The necessary equations may be derived by considering the following situation.

Two atoms of mass m_1 and m_2 are joined by an elastic bond (e.g., a spring) such that they vibrate in harmonic motion with a maximum amplitude x_1 and x_2 (Figure 6.1). The displacement x from the mean at time t is given by:

$$x = x_2 \sin 2\pi\nu t. \tag{6.5}$$

The potential energy P of the system is:

$$P = \frac{1}{2} k x^2 = 2\pi^2 \mu \nu x^2, \tag{6.6}$$

where k is an elasticity constant and μ is the reduced mass $= \left(\frac{m_1 m_2}{m_1 - m_2}\right)$.

To develop the quantum mechanical equation required to produce the quantum energy levels one inserts P into the Schrödinger wave equation:

$$\nabla^2 \psi + \frac{8\pi^2 \mu}{h^2}(E - P)\psi = 0, \tag{6.7}$$

where ψ is the wave function and E the energy of the system. The solution of this equation is:

$$E = \frac{h}{2\pi}\sqrt{\frac{k}{\mu}}(v + 1/2), \tag{6.8}$$

where v is called the vibrational quantum number and has the values 0,1,2 . . . and k is the force constant acting between masses.

Incident energy is most easily absorbed or released by the vibrating atoms, by changing to the next higher or lower quantum energy state. This is expressed mathematically by the selection rule $\Delta v = \pm 1$. (Other changes are possible but unlikely.)

For a vibrational transition, then:

$$\bar{\nu} = \frac{1}{2\pi c}\sqrt{\frac{k}{\mu}}(v_2 - v_1). \tag{6.9}$$

6. Infrared Spectroscopy

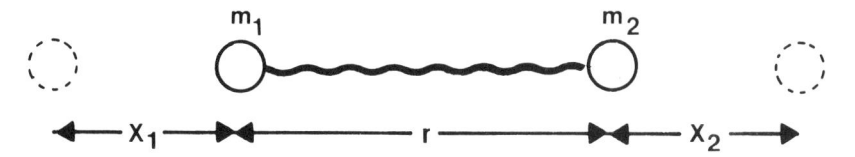

Figure 6.1 Schematic of two vibrating atoms.

Thus the vibrational spectrum of a simple harmonic oscillator consists of a single spectral line, since no matter what quantum levels are involved, $\Delta v = \pm 1$. This simple situation is not obeyed in real atomic systems because motion of atoms is not precisely harmonic; the anharmonic energy levels are instead given by:

$$E = hc[\bar{\nu}(v + \tfrac{1}{2}) + \bar{\nu}_x(v + \tfrac{1}{2})^2], \tag{6.10}$$

where ν_x is called the anharmonicity constant. Therefore, transitions between adjacent quantum levels are not exactly equivalent. Atoms are also subject to thermal effects and environmental influences from neighboring nonbonded atoms such that instead of getting a single infrared absorption line as predicted by equation (6.9), a band of definite width is produced. Furthermore, since the selection rule $\Delta v = \pm 1$ is also not valid for real systems, transitions corresponding to $\Delta v = \pm 1, 2, 3 \ldots$ can occur. This gives rise to bands of reduced intensity known as overtone bands which have frequencies that are approximately 2, 3 etc. times the frequency expected for $\Delta v = \pm 1$.

Whereas the calculation of absorption frequencies for diatomic molecules is relatively straightforward, polyatomic molecules provide a more complex problem. There are, in fact, $3N - 6$ ($3N - 5$ for linear molecules) vibrational modes for a polyatomic system of N atoms. In order to analyze such vibrations, the rules of symmetry and group theory may be applied (see Appendix B).

Polypeptides have such a large number of atoms and vibrational combinations, it might seem that there would be so many vibrational transitions that it would be impossible to interpret an infrared spectrum. However, as with x-ray diffraction, repeating symmetry in the structures considerably reduces the number of combinations that must be considered. Furthermore, not all transitions are of interest in structural studies; those that are most relevant are located within the peptide groups. There, chain vibrations are governed by simple rules derived from the relative motion of adjacent peptide groups.

6.2 Amide Bands

Several infrared absorbance bands have been shown to result from vibrations in the amide group of the polypeptide backbone. These are the major absorption bands of interest in polypeptides. Those in the infrared and so-called far infrared regions are labeled amide bands A, B, and I to VII; the near infrared bands are overtone and combination bands. These bands are described in Tables 6.1 and 6.2.

The origin of the amide I–III and V–VII bands is shown schematically in Figures 6.2 and 6.3. These assignments are based upon calculated vibrations that include interatomic forces.

Table 6.1 Amide Bands Falling in the Infrared and Far-Infrared Region of the Spectrum

Designation	Approximate Frequency (cm^{-1})	Description
Amide A B	3300 3100	Resulting from resonance between the first excited state of N—H stretch and second excited state of the amide II vibration.
Amide I II III	1650 1550 1300	Vibrations in the plane of the amide group amide I, mainly C=O stretch, amide II mainly N—H bend, amide III C—N stretch, N—H bend, etc.
Amide IV	625	Vibrations not localized to peptide group, rarely identifiable in spectra of polypeptides, mainly O=C—N in plane bend.
Amide V VI VII	650 550 200	Vibrations out of the plane of the amide bond: amide V mainly N—H bend, amide VI mainly C=O bend, and amide VII mainly C—N rotation.

Table 6.2 Amide Bands Falling in the Near-Infrared Region

Frequency (cm^{-1})	Description
6490	NH overtones for hydrogen-bonded solvent.
6620	1st NH overtone for protonated peptide.
6710	1st NH overtone for acid-solvated peptide.
4970	Combination (addition) of amide A and amide I.
4860	Combination of amide A and amide II.
4970	Combination of amide A and amide III.

As noted previously, electromagnetic radiation is absorbed only if, during the interaction, there is a change of dipole moment. Since the dipole moment of two atoms involves the product of effective charge on the atoms and the distance of separation, the dipole moment can be changed by changing the electronic distribution (e.g., by absorbance of ultraviolet light) or by changing the relative position of the atoms (e.g., by change of the vibrational quantum state by absorbance of infrared light) or by a combination of both. Since the new dipole moment is induced by absorbance of energy causing a transition between quantum states, it is called a transition moment. It is possible to measure the orientation of this "transition moment" in simple model crystals, the direction being, for example, approxi-

Figure 6.2 Atomic motions characteristic of amide vibrations in a plane. The arrows represent the direction of movement of the atoms.

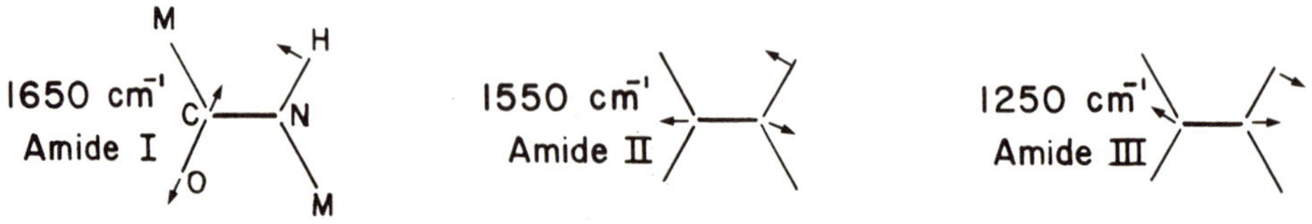

6. Infrared Spectroscopy

Amide V   725 cm⁻¹   **Amide VI**   600 cm⁻¹   **Amide VII**   206 cm⁻¹

Figure 6.3 Atomic displacements for amide "out-of-plane" vibrations.

Figure 6.4 Orientation of transition dipole on vibrational excitation.

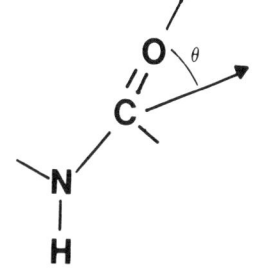

mately 20° from the direction of the C=0 bond for amide I. (See Figure 6.4.)

Of the 15 vibrational bands associated with the amide group in polypeptides, attention is mostly focused on the amide I, II, and V bands. This is because these bands are both prominent in polypeptide spectra and are *conformationally sensitive*.

6.3 Polarization of Radiation and Dichroism

By use of a polarizing filter which is often in the form of a fine wire grid, infrared radiation may be polarized such that the wave vector oscillates in only one plane, as indicated in Figure 6.5.

Figure 6.5 A plane-polarized infrared beam with wave vector E passes through two samples in which the transition moment is perpendicular ($\perp$) to the beam (A → B) and does not interact, or parallel ($\parallel$) to the beam (D → C) where absorption occurs.

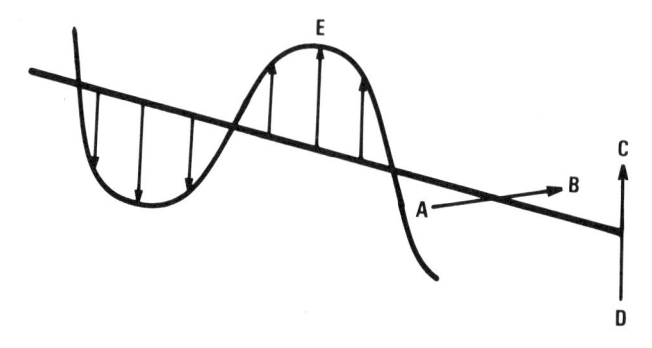

If the beam is parallel (∥) to the transition moment, maximum absorbance occurs and there is said to be a parallel dichroism. If the beam is perpendicular to the transition moment, no absorbance occurs. The direction of polarization relative to an absorbance band is written $\bar{\nu}_\parallel$ or $\bar{\nu}_\perp$. For helical polypeptides and fibrous proteins it is relatively easy to orient the material by stroking such that the helix axis is perpendicular or parallel to the polarized infrared beam. Consequently, infrared dichroic spectra may be obtained and studied in some detail. On the other hand, the polypeptide chains in globular proteins cannot be oriented in a regular (parallel or perpendicular) array and consequently no infrared dichroic information can be obtained.

6.4 Vibrational Analysis

We have so far made the simplifying assumption that the major absorption bands of interest in polypeptide structure analysis arise from simple atomic group vibrations located mainly within the peptide group. However, in order to obtain structural information it is necessary to evaluate the relative motions of chain segments.

Helices

For isolated helical chains it can be shown that the only chain vibrational modes that produce bands in the infrared spectrum are those for which the group motions in successive (peptide) units are related by a constant phase difference δ. The concept of phase is most easily presented in terms of groups vibrating in a chain as shown in Figure 6.6.

For a helical molecule, the phase difference is given by

$$\delta = 0, \pm t, \tag{6.11}$$

where t is the unit twist (see Chapter 2). For example, $t = 2\pi/3$ for a 3_1 helix, $\pi/2$ for a 4_1 helix, and so on.

Relating the vibrational frequency of a peptide group *in a helical chain* $\bar{\nu}(\delta)$ to that of the unperturbed group vibration $\bar{\nu}_0$ (that is, the frequency

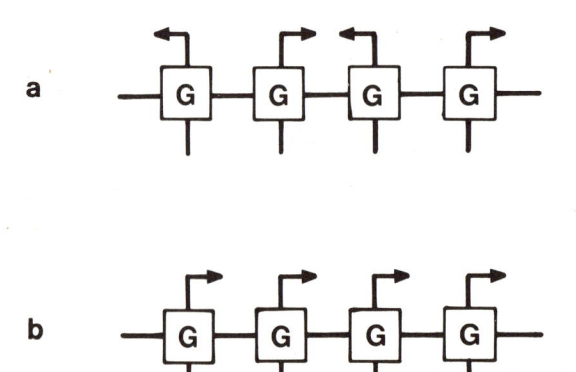

Figure 6.6 In **(a)** the adjacent groups G are vibrating in the opposite direction, as indicated by the arrows, and are 180° or π-radians out of phase ($\delta = \pi$). For case **(b)** the groups are vibrating in phase; that is, they are vibrating in the same direction at the same time ($\delta = 0$).

6. Infrared Spectroscopy

found in the random conformation), we have, following Miyazawa,

$$\bar{\nu}(\delta) = \bar{\nu}_0 + \Sigma_j D_j \cos(j\delta), \tag{6.12}$$

where D_j is an interaction coefficient that depends upon the coupling between the reference unit and the j-th unit along the chain. For example, the hydrogen bonding between a peptide unit and its third nearest neighbor will modify the vibrational frequency of the former in direct relation to the strength of interaction.

For an α-helix there will, in fact, be interactions between nearest neighbor and third nearest neighbor so that the vibrational expressions become:

$$\bar{\nu}_\parallel(0) = \bar{\nu}_0 + D_1 + D_3, \tag{6.13}$$

$$\bar{\nu}_\perp(\pm t) = \bar{\nu}_0 + D_1 \cos(\pm t) + D_3 \cos(\pm 3t). \tag{6.14}$$

The transition moment for which $\delta = 0$ is parallel to the helix axis and for $\delta = \pm t$ the moments are mutually perpendicular to the helix axis (Figure 6.7).

The parallel and perpendicular notations are included in equations (6.13) and (6.14). Vibrations corresponding to $\delta = \pm t$ have the same frequency and are said to be degenerate, so that for the α-helix, two absorption bands

$$\bar{\nu}_\parallel(0) = \bar{\nu}_0 + D_1 + D_3 \tag{6.13}$$

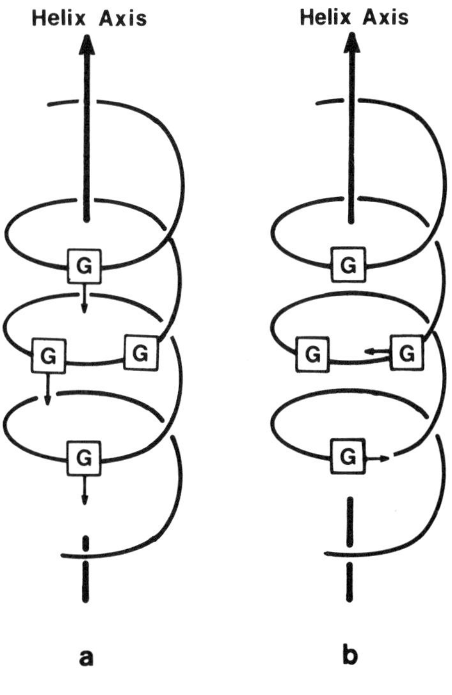

Figure 6.7 In phase ($\delta = 0$) motions of groups on a helix are shown in **(a)** with the vibration parallel to the helix axis. Vibrations in **(b)** are perpendicular to the axis ($\delta = \pm t$).

and $\bar{\nu}_\perp \left(\dfrac{2\pi}{3.6} \right) = \bar{\nu}_0 + D_1 \cos \left(\dfrac{2\pi}{3.6} \right) + D_3 \cos \left(\dfrac{6\pi}{3.6} \right)$ (6.14)

are expected. Note that since the transition moments of each differ in direction, they should be distinguishable by use of polarized light. We return to this point later.

The experimental band positions for helical polypeptides do indeed often show two bands with modes parallel and perpendicular to the chain axis.

It should be noted that the helical structure not only introduces a parallel and perpendicular mode of vibration but also shifts the frequency of the band. This is to be expected because peptide groups interact differently in different conformations (e.g., hydrogen bonding schemes are different, as is the atomic environment).

The frequencies observed for a polypeptide are clustered around the amide frequencies listed in Table 6.1 but show symmetry splitting as calculated by equations (6.13) and (6.14) and given numerically in Table 6.3. In most cases the $\bar{\nu}_\parallel$ amide I mode and the $\bar{\nu}_\perp$ amide II mode are much stronger and the associated $\bar{\nu}_\perp$ amide I and $\bar{\nu}_\parallel$ amide II infrared bands are often not readily detectable.

From the observed values of the $\bar{\nu}_\parallel$ and $\bar{\nu}_\perp$ modes, it should be possible to calculate the parameters D_1 and D_3 from equations (6.6) and (6.7a), provided that an unperturbed frequency ($\bar{\nu}_0$) can be estimated or calculated. One method of approaching this problem has been to look at amide bands in nonpolypeptide chains (e.g., nylon or simple amides). Values of $\bar{\nu}_0 = 1658$ cm^{-1} for amide I and $\bar{\nu}_0 = 1535$ cm^{-1} for amide II have been estimated from such sources. Calculated values of D are a measure of coupling strength or strength of interaction between appropriate peptide units, but details of such interactions for helices have not yet been fully evaluated.

Similar applications to other amide bands are feasible in principle, but the lower the frequency, the more complex is the vibration, and experimentally it is difficult to observe detailed band splitting.

Sheet Structures

Parallel β-sheet structures are found in certain globular proteins and provide a useful basis for application of the Miyazawa approach to vibrational analysis. In this case, it is necessary to consider both intra- and interchain vibrational coupling. Thus, including interchain terms, equation (6.12) becomes:

Table 6.3 Frequencies for Helical Polypeptides Showing Symmetry Splitting

Polypeptide	Conformation	Amide I $\bar{\nu}_\parallel(0)$ cm^{-1}	Amide I $\bar{\nu}_\perp(t)$ cm^{-1}	Amide II $\bar{\nu}_\parallel(0)$ cm^{-1}	Amide II $\bar{\nu}_\perp(t)$ cm^{-1}
[Glu(OMe)]$_n$	α_R	1654	1659	1519	1550
[Glu(OBzl)]$_n$	α_R	1652	1655	1518	1549
[Asp(OBzl)]$_n$	ω_L	1677	1670	1520	1538
[DGlu(OBzl)LGlu(OBzl)]$_n$	α_{DL}	1664	—	—	1550
[DGlu(OBzl)LGlu(OBzl)]$_n$	ω_{DL}	1676	—	—	1532
[DGlu(OBzl)LGlu(OBzl)]$_n$	π_{DL}	(1645)	—	—	(1540)
[Gly]$_n$	pGII	1644	1654	—	—
[Pro]$_n$	pPII	1660	—	—	—

6. Infrared Spectroscopy

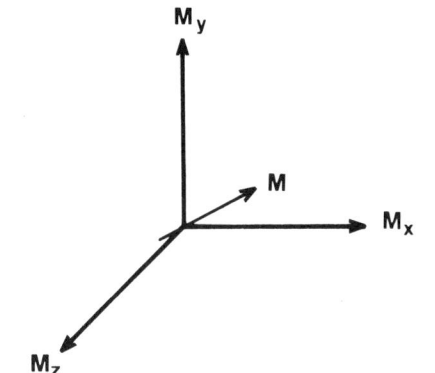

Figure 6.8 Vector decomposition of transition moment in three dimensions.

$$\bar{\nu}(\delta,\delta') = \bar{\nu}_0 + \sum_j D_j \cos(j\delta) + \sum D_i' \cos(i\delta') \qquad (6.15)$$

where δ' and D_i' refer to the interchain phase relation and coupling constant of the i-th residue. The interchain phase motion is simply described as the relative vibration motion of a peptide group in one chain compared with the peptide group under consideration in the adjacent chain. (See Figure 6.9.)

For infrared activity we have to consider only the motions of repeating groups of atoms (unit cells) that are in phase. This condition limits the possible values of δ and δ' to π or 0.

We actually need to consider the relative motion of transition moments, but let us break this vector quantity into three components acting along coordinate axes x, y, z, as shown in Figure 6.8. Looking at two chains of a parallel β-sheet we have the situation shown in Figure 6.9.

Figure 6.9 Transition moments in a parallel β-sheet.

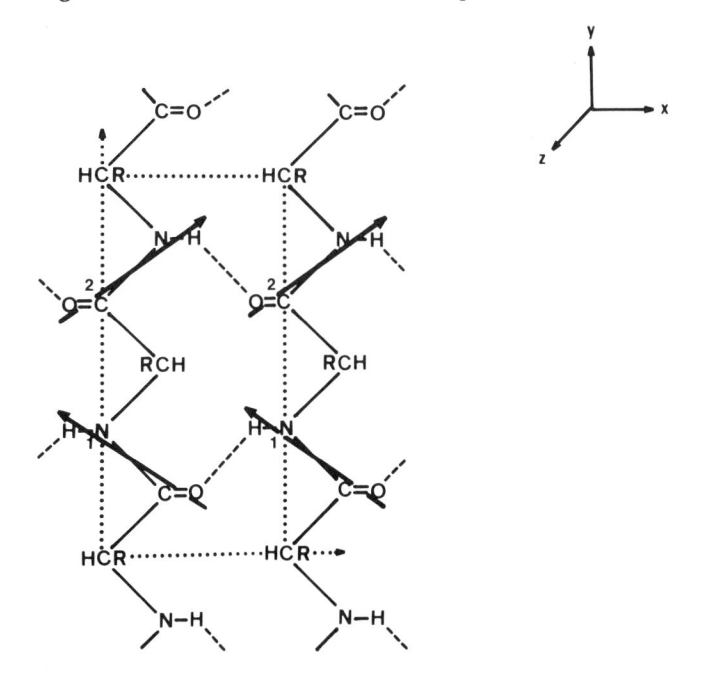

The unit contains two differently arranged peptides (a 2_1 helix) and there are two phase equations (equation 6.15). The vibrations in phase along the y-axis are shown in Figure 6.10a, in which the moments are rotated around the y-axis by 180°. The resultant is seen to be $2M_y$ along the y-axis and thus $\bar{\nu}(0,0)$ is the parallel $\bar{\nu}_\parallel$ vibration. The π-radian "out-of-phase" vibrations are shown in Figure 6.10b, in which the moments of group 2 are all $\pi = 180°$ different from the previous case. This leads to resultant components of $2M_z$ and $2M_x$ along the z and x axes. Since the resultants are at 90° to the chain axis, they are referred to as the $\bar{\nu}_\perp$ components. Since the second chain is a replica of the first, no interchain phase relationships are considered ($\delta' = 0$). Equation (6.15) then becomes:

$$\bar{\nu}_\parallel (0,0) = \bar{\nu}_0 + D_1 + D_1' \tag{6.16a}$$

$$\text{and } \bar{\nu}_\perp (\pi,0) = \bar{\nu}_0 - D_1 - D_1', \tag{6.16b}$$

assuming only nearest-neighbor coupling is significant. We would thus expect two infrared bands (for each of amide I, II etc.) for a parallel β-sheet chain arrangement (though this approach does not specify the relative intensities). As noted earlier, the parallel β-sheet occurs only in globular proteins so that the vibrational properties of the peptide groups cannot be established experimentally. Such is not the case, however, for antiparallel β-sheets, which are found in fibrous proteins and synthetic polypeptides.

In the antiparallel pleated sheet structure there are four amide groups in the polypeptide backbone (two in each chain), which make up a repeat unit (unit cell). A representation of the groups and their modes of vibrational motion is shown in Figures 6.11 and 6.12.

Figure 6.10 Vector decomposition of the transition moments of the two in-phase peptide vibrations for parallel β-sheet chain. **(a)** $\bar{\nu}(0,0)$; **(b)** $\bar{\nu}(\pi,0)$.

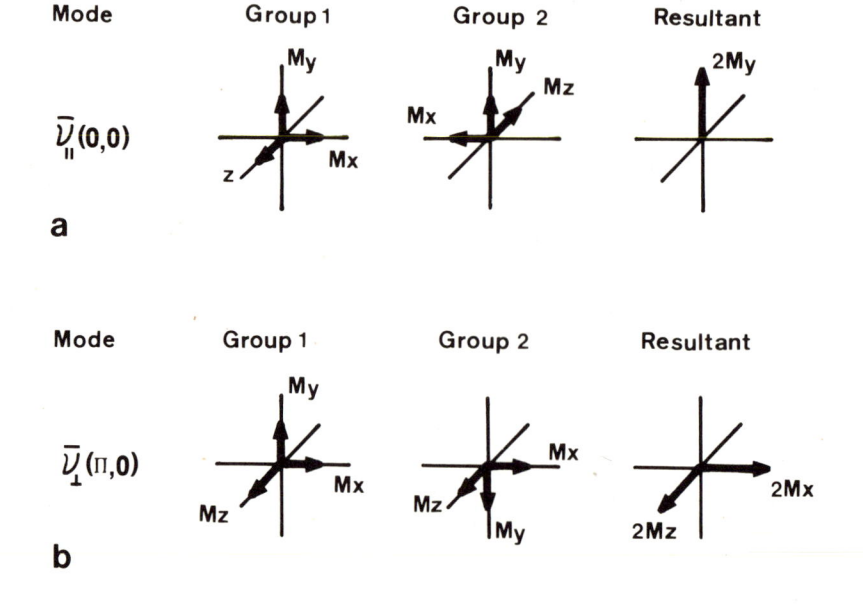

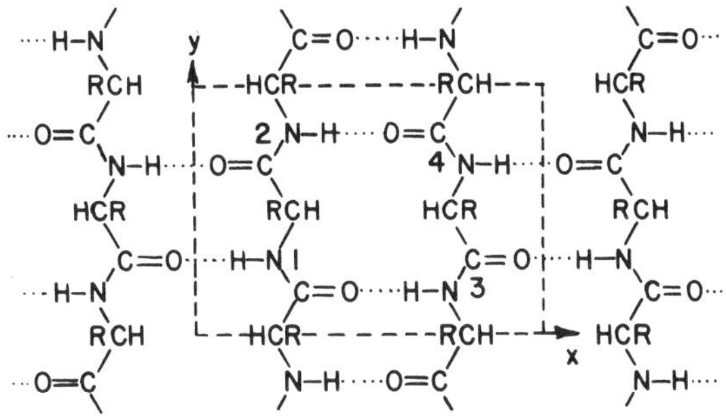

6. Infrared Spectroscopy

Figure 6.11 Representation of amide groups in an antiparallel β-sheet showing the phase-motion directions x and y.

The phase angle δ' between chains can now take on values of 0 or π. There are clearly four combinations of $\delta(0,\pi), \delta'(0,\pi)$. The phase relations are shown in Figure 6.11.

In this case there is no net transition moment for the $\bar{\nu}(0,0)$ mode so that there is no absorption. There are two transitions perpendicular to the chain axis and one parallel.

The coupling of these modes, shown schematically in Figure 6.11, is given from equation (6.15) (for nearest-neighbor groups) as

$$\bar{\nu}(\delta,\delta') = \bar{\nu}_0 \pm D_1 \pm D'_1. \tag{6.17}$$

Figure 6.12 Phase combinations of residues 1, 2, 3, and 4 of Figure 6.11.

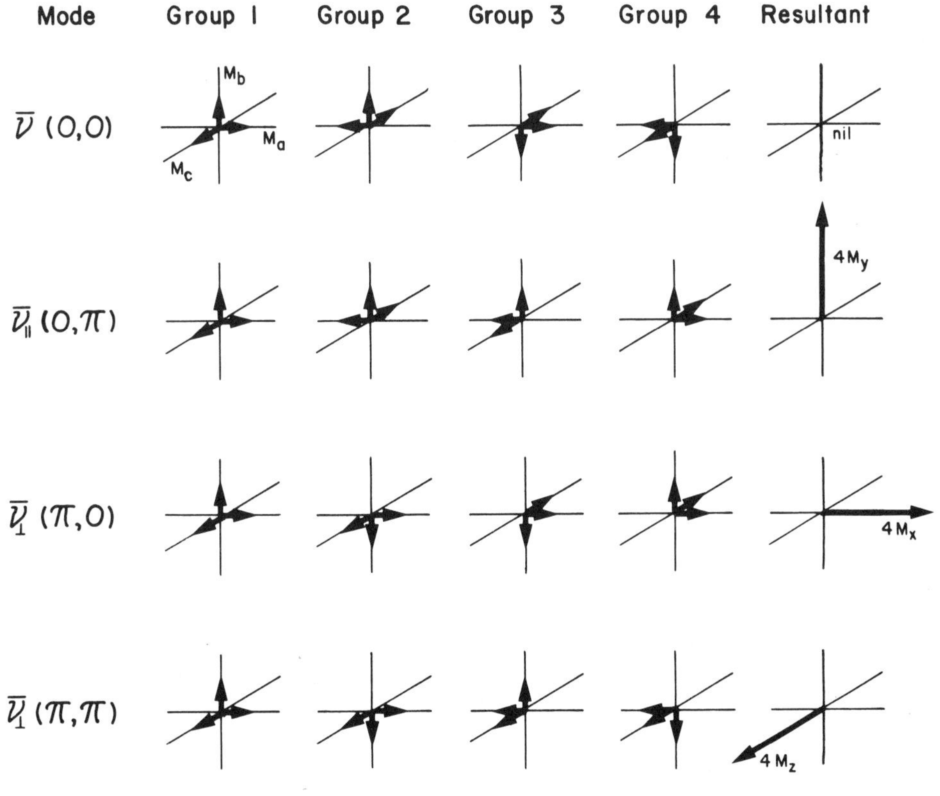

Table 6.4 Comparison of Theoretical and Experimental Values for the Vibrational Modes of Poly-L-alanine in its Antiparallel β-Sheet Arrangement

		Observed (cm^{-1})	Calculated (cm^{-1})
Amide I	$\bar{\nu}(0,0)$	(1669)*	1669
	$\bar{\nu}_{\parallel}(0,\pi)$	1695 (m)	1695
	$\bar{\nu}_{\perp}(\pi,0)$	1630 (s)	1630
	$\bar{\nu}_{\perp}(\pi,\pi)$	—	1701
Amide II	$\bar{\nu}(0,0)$	(1538)*	1535
	$\bar{\nu}_{\parallel}(0,\pi)$	1524 (m)	1523
	$\bar{\nu}_{\perp}(\pi,0)$	1558 (w)	1555
	$\bar{\nu}_{\perp}(\pi,\pi)$	1586†	1587

*Raman bands.
†Calculated from analysis of Fermi resonance A and B doublet.
s = strong, m = medium, w = weak.

Unfortunately it has not proven possible to extract consistent values of ν_0, D_1, and D'_1 from the experimental data (see Table 6.3) and it has been concluded that an additional next nearest term D'_2 must be added. In this model the four modes are represented as:

$$\bar{\nu}(0,0) = \bar{\nu}_0 + D_1 + D'_1 + D'_2, \tag{6.17a}$$

$$\bar{\nu}(0,\pi) = \bar{\nu}_0 + D_1 - D'_1 - D'_2, \tag{6.17b}$$

$$\bar{\nu}(\pi,0) = \bar{\nu}_0 - D_1 + D'_1 - D'_2, \tag{6.17c}$$

$$\bar{\nu}(\pi,\pi) = \bar{\nu}_0 - D_1 - D'_1 + D'_2 \tag{6.17d}$$

The detailed calculation of the amide I and II modes and their comparison with experimental values has been carried out for the β-sheet form of poly-L-alanine as shown in Table 6.4. The parameters used in the calculation (equations 6.17 a–d) were $\bar{\nu}_0 = 1673.6$ cm^{-1}, $D_1 = 24.2$ cm^{-1}, $D'_2 = 11.2$ cm^{-1}, and $D'_1 = 8.4$ cm^{-1} for the amide I, and $\bar{\nu}_0 = 1550.0$, $D_1 = -21.0$ cm^{-1}, $D'_1 = -5.0$ cm^{-1}, $D'_2 = 11.0$ cm^{-1} for the amide II bands. (Further subtle refinements involve additional constants which have, for present purposes, been incorporated in

Table 6.5 Major Amide Bands and Their Dichroism for Various Polypeptide Conformations

Conformation	Amide I (cm^{-1})	Amide II (cm^{-1})	Amide V (cm^{-1})
α_R	1652–1661 ($\parallel$)	1546–1555 ($\perp$)	610–620($\perp$)
α_L	1664–1668 ($\parallel$)	1555–1559 ($\perp$)	
β a$\parallel$ (stands for antiparallel)	1621–1637 ($\perp$)	1512–1545 ($\parallel$)	690–710($\parallel$)
ω	1667–1676 ($\parallel$)	1532–1538	
pGII	1644 ($\perp$)		
pPII	1660*		
π_{LD}	1645 ($\parallel$)	1540 (0)	
Random (irregular)	~1660 (0)	~1550 (0)	~650 (0)

*Value for the non-hydrogen–bonded solid state.

the three defined here.) Thus we see that unique absorption bands in the amide I and amide II region result if the polypeptide has the antiparallel β-sheet structure.

Before we turn to the examination of specific proteins by infrared spectroscopy we can see that, based on the principles explained in the previous pages, we may expect the *position* (frequency) of the bands, the *number* of bands in a given region, and the *dichroism* of the bands to be specific indicators of structure in polypeptides and proteins. The frequency and dichroism of the most intense bands for a number of different conformations are given in Table 6.5.

6.5 Dichroism and Conformation

As indicated in the preceding analysis, the vibrations have components parallel or perpendicular to the polypeptide chain, and it would be particularly useful if the orientation of any particular transition moment with respect to the molecular axis could be identified from an infrared spectrum. Although different polypeptide conformations produce infrared spectra in which the band positions are often shifted and typical of the conformation, it is the orientation of the vibration that serves to allow identification of the conformation with more confidence.

If a polarized beam is used, and the molecules in the sample are oriented, then it is possible to orient the polarizer (or the sample) so that the polypeptide chains are parallel or perpendicular to the beam polarization. Figure 6.13 shows such an experimental arrangement with the polarized light vector parallel to the chain axis of the sample. The spectra obtained by this approach are called dichroic spectra. Usually dichroic spectra are used qualitatively; for example, Figure 6.14 shows the polarized infrared spectrum of an oriented α-helical sample. The amide I band is clearly a $\bar{\nu}_\parallel$ mode since it absorbs more radiation when the chain axis is parallel to the plane of polarization. A similar argument may be made for the amide II band being $\bar{\nu}_\perp$.

When low-resolution infrared dichroic spectra* are obtained from polypeptides, relatively few vibrational components of the amide bands and only the most intense bands are seen.

It is generally necessary to examine high-resolution spectra along with Raman spectra (see next chapter) before the detailed vibrational assignments, such as those given in Table 6.3, may be obtained. For example, the 1695-cm^{-1}

Figure 6.13 Orientation of polarized beam and sample for obtaining dichroic spectra.

---

*Low-resolution spectra are obtained from conventional equipment by scanning rapidly or by including a relatively wide range of wavelengths in the incident radiation.

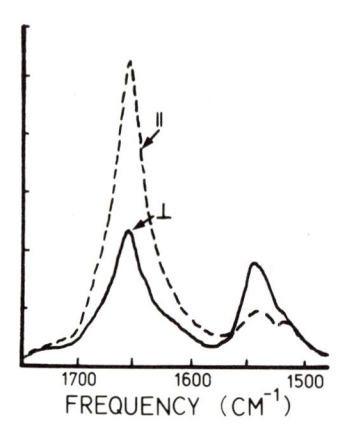

Figure 6.14 Infrared dichroism spectrum in the amide I and II region for the α-helix-rich, low-sulfur protein of wool keratin.

$[\bar{\nu}_\parallel (0,\pi)]$ band characteristic of antiparallel β-structure is much weaker than the 1630-cm^{-1} $[\bar{\nu}_\perp (\pi,0)]$ band and is often missed in low-resolution spectra.

The quantitative interpretation of dichroic ratios can be quite complex. In part this is caused by overlapping bands, background noise, and so on. When possible, dichroism is quantitated by the integrated band intensity, the dichroic ratio R_0 for a helix being defined as

$$R_0 = ([A_0]_\parallel + [A_0]_\perp)/([A_t]_\parallel + [A_t]_\perp) \qquad (6.18)$$

where $[A_0]_\parallel$ is the integrated area beneath the $\bar{\nu}_\parallel(0)$ band in the parallel ($\parallel$) spectrum, $[A_0]_\perp$ the integrated area in the perpendicular spectrum, and A_t refers to the $\bar{\nu}(t)$ mode.

For a helix with all amide groups equivalent, the angle α of the transition moment with respect to the helix axis can be shown to be:

$$\cot \alpha = \pm(\tfrac{1}{2} R_0)^{1/2}. \qquad (6.19)$$

This enables one to determine the orientation of the amide group with respect to the axis.

6.6 Amide A and B Bands

Whereas amide I and II bands have been used to identify conformation, the amide A and B bands contain information about hydrogen bonding of the peptide group. This is because the N — H stretching mode of the amide group is reduced in frequency when the amide is hydrogen bonded to a carbonyl group; the magnitude of the frequency change can be correlated with hydrogen-bond length.

The frequency of the $\bar{\nu}_{NH}$ stretch may be obtained from the amide A, B, and II frequencies $\bar{\nu}_A$, $\bar{\nu}_B$, and $\bar{\nu}_{II}$ by:

$$\bar{\nu}_{NH} = \bar{\nu}_A + \bar{\nu}_B - 2\bar{\nu}_{II}. \qquad (6.20)$$

The value of $\bar{\nu}_{II}$ required is the unperturbed $\bar{\nu}_0$ value (see equation 6.14). Thus for the α-helix, in which the experimental bands correspond to the $\bar{\nu}(0)$ and $\bar{\nu}(\pm t)$, a correction should be made for D_1 and D_3 (see equation 6.13 and 6.14).

6. Infrared Spectroscopy

Since these values are not known, the correction is not possible and use of the experimental value introduces a small error.

The value for an unperturbed N — H stretching frequency ($\bar{\nu}^0_{NH}$) is taken to be 3457 cm^{-1}. A useful equation linking the stretching frequencies and hydrogen-bond length b_H is:

$$\bar{\nu}^0_{NH} - \bar{\nu}_{NH} = 548(3.21 - b_H) \tag{6.21}$$

$$\text{or } b_H = 0.001825\bar{\nu}_{NH} - 3.1. \tag{6.22}$$

Table 6.6 shows some calculated bond lengths for helical polypeptides using this relation. We will see later that analysis of the amide A and B bands for collagen and its synthetic analogues provides interesting structural information relating to hydrogen bonding.

The amide A and B bands for the β-sheet structures have been examined in more detail. The measured frequencies $\bar{\nu}_A$ and $\bar{\nu}_B$ are related to the unperturbed frequencies by

$$\bar{\nu}_A = \tfrac{1}{2}([\bar{\nu}^0_A + \bar{\nu}^0_B] + s) \tag{6.23}$$

$$\bar{\nu}_B = \tfrac{1}{2}([\bar{\nu}^0_A + \bar{\nu}^0_B] - s), \tag{6.24}$$

where s is the observed splitting $\Delta\bar{\nu}$ between amide A and B. The intensity ratio of amide B to A is given by

$$I_R = \frac{I_B}{I_A} = \frac{s - s'}{s + s'}, \tag{6.25}$$

where $s' = \bar{\nu}^0_A - \bar{\nu}^0_B$ is the splitting between unperturbed bands. Use of these relations allows the calculation of $\bar{\nu}^0_A$ and $\bar{\nu}^0_B$. For poly-L-alanine in the antiparallel β-sheet conformation $\bar{\nu}_A = 3280$ cm^{-1}, $\bar{\nu}_B = 3072$ cm^{-1}, $\bar{\nu}^0_A = 3242$ cm^{-1}, and $\bar{\nu}^0_B = 3109$ cm^{-1}.

From symmetry considerations there are only two possible combinations of amide II bands that could lead to an unperturbed amide B band. These are

$$\bar{\nu}(0,0)_\parallel + \bar{\nu}(\pi,0)_\parallel \to \bar{\nu}^0(\pi,0)_B \tag{6.26}$$

$$\text{and } \bar{\nu}(0,\pi)_\parallel + \bar{\nu}(\pi,\pi)_\parallel \to \bar{\nu}^0(\pi,0)_B. \tag{6.27}$$

Substitution of the calculated vibrational frequencies from Table 6.4 gives $\bar{\nu}^0_B = 3090$ cm^{-1} from equation (6.26) or $\bar{\nu}^0_B = 3110$ cm^{-1} from equation (6.27). This latter value is very close to the experimentally derived 3109 cm^{-1} and suggests that it is the coupling of vibrations expressed by equation (6.27), which is the basis of the observed amide B.

Table 6.6 Hydrogen Bond Lengths Calculated from Amide A Bands

Polymer	Conformation	$\bar{\nu}_A$	$\bar{\nu}_B$	$\bar{\nu}_{II}$	$\bar{\nu}_{NH}$	$b_H(\text{Å})$
[Ala]$_n$	α_R	3293	3060	1545	3263	2.85
[Asp(OBzl)]$_n$	α_L	3308	3082	1561	3286	2.90
[Asp(OBzl)]$_n$	ω_L	3298	3075	1538	3301	2.92

Table 6.7 Positions of Side-Chain Vibrations in Infrared Spectra of Proteins

Group	Vibrational mode	Frequency (cm^{-1})
—NH$_2$	Antisymmetric stretching	3225
—NH$_2$	Symmetrical stretching	3370
—NH$_2$	Stretching and deformation	3050
—COO$^-$	Antisymmetric stretching	1410
—COO$^-$	Symmetrical stretching	1570
Acid C=O	Stretching	1700
Ester C=O	Stretching	1740
—SH	Stretching	2150
—CH$_3$	Bending	1440, 1370

6.7 Side-Chain Vibrations

In addition to the polypeptide amide bands, there are vibrations arising in the peptide side chains that produce bands in an infrared spectrum. Table 6.7 lists band positions for various amino acid side chains. This list is not exhaustive but is indicative of likely side-chain bands in a protein spectrum. These bands are generally insensitive to conformation but the band intensity tends to change with environment, and variation in ionization or hydrogen bonding will cause slight frequency shifts and/or band splitting.

6.8 Infrared Spectroscopy of Fibrous Proteins

Fibrous proteins behave in many ways like poly(amino acids), in that they often have long segments of relatively uniform conformation in which our concepts of vibrational symmetry are likely to prove valid. On the other hand, native fibers generally have more than one structural component, perhaps an amorphous matrix or local defects in the fiber.

Of the simplest fibrous proteins, the silks, most have an antiparallel β-structure (Chapter 11). The amide I and II region of the infrared spectrum of a section of *Bombyx mori* silk thread is shown in Figure 6.15.

In the regions of the spectrum where the silk fibers absorb most light that is polarized perpendicular to the fiber axis (full line in the diagram), the bands are said to have perpendicular ($\perp$r) dichroism. Two such major bands are seen in the spectrum, the amide A band at 3300 cm^{-1} and the amide I band at 1630 cm^{-1}. The bands at 1525 cm^{-1} (amide II) and at 1699 cm^{-1} (amide I) absorb light polarized parallel to the fiber axis preferentially and therefore show parallel dichroism. Comparison with Tables 6.4 and 6.5 shows that the structure must be antiparallel β-sheet with the following assignments: 1699 cm^{-1}, amide I, $\bar{\nu}_\parallel (0,\pi)$; 1630 cm^{-1}, amide I, $\bar{\nu}_\perp (\pi,0)$; 1525 cm^{-1}, amide II, $\bar{\nu}_\parallel (0,\pi)$. No other listed combination can account for this set of band frequencies and polarization. It may be noted that in addition to the amide I components already mentioned, there is a broad band centered at 1656 cm^{-1} that is taken to arise from amorphous, irregular-chain regions of the fiber. The ratio of the integrated amide I band areas gives an approximate ratio of β-sheet to amorphous form. (Found to be β-sheet = 70%, amorphous 30%.)

6. Infrared Spectroscopy 135

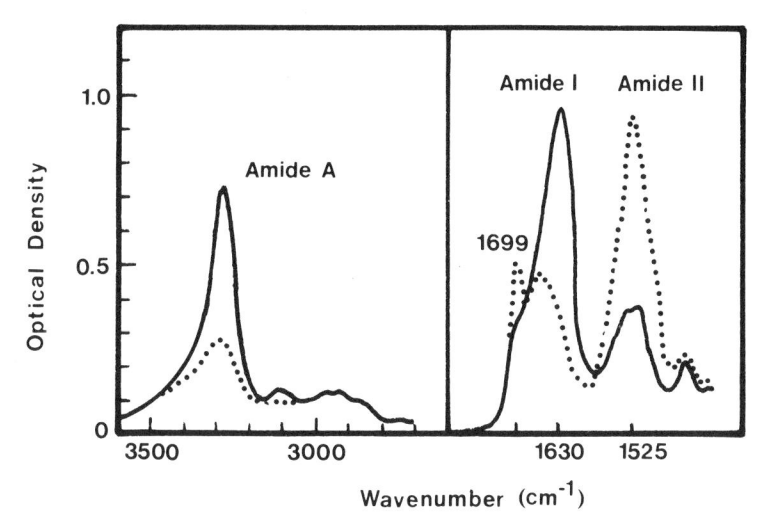

Figure 6.15 Infrared spectrum of a thin section cut from a thread prepared from a *Bombyx mori* silk gland. ⎯ Electric vector vibrating perpendicular to the fiber axis; .... polarization parallel to the fiber axis (Suzuki, 1967).

The amide I and amide II region of the infrared spectrum of porcupine quill is shown in Figure 6.16. In this case the amide I band shows parallel dichroism and is centered at ~ 1660 cm^{-1}. The main amide II band shows perpendicular dichroism and is at 1550 cm^{-1}. Comparison with Tables 6.3, 6.4, and 6.5 indicates that the predominant structure is α-helical with: 1660 cm^{-1}, amide I, $\bar{\nu}_{\parallel}$ (0); 1550 cm^{-1}, amide II, $\bar{\nu}_{\perp}$ (t).

Apart from the dichroism of the amide I and II bands, consistent with α-helix, there is a nondichroic component of the amide II and a broad band underlying the amide I, which probably originates from nonoriented polypeptide material. It is noteworthy that the α-helix in porcupine quill and hair (α-keratin) is probably in a coiled coil arrangement, but infrared spectroscopy is not currently capable of differentiating between such a chain arrangement and the isolated α-helix.

Figure 6.16 Amide I and II bands of a thin section of porcupine quill.

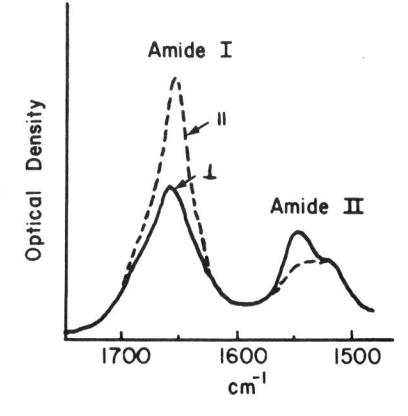

6.9 Spectroscopy of Globular Proteins

Globular proteins raise special problems and prior to the advent of Fourier transform infrared (FTIR) spectroscopy* a very limited number of globular protein spectra were accessible, particularly in aqueous solution, because of the interference of the very strongly absorbing bands originating from water. Spectra had to be obtained from D_2O solutions and studies concentrated on the amide I and amide II regions of the spectrum. However, the amide frequencies are affected by hydrogen–deuterium exchange in the peptide groups so that conclusions about structure were often ambiguous.

One paramount question answerable by infrared spectroscopy is whether globular proteins have the same structure in aqueous solution and solid state. Since the solid-state structure is sometimes known in considerable detail from x-ray diffraction, it is pertinent to ask how such a structure is related to the functioning of that protein in aqueous solution.

Hemoglobin

The detailed three-dimensional structure of hemoglobin is known (Chapter 15) from x-ray diffraction analysis of the protein in the solid state, and from circular dichroism (CD) spectroscopy in aqueous solution, the protein is found to be more than 80% α-helical. A similar conclusion is obtained from infrared spectroscopy.

Figure 6.17 shows FTIR spectra of hemoglobin in aqueous solution (with the water spectrum subtracted) and of a dried film. As can be seen, several bands are clearly defined, particularly the amide I and II bands and a band at 1106 cm^{-1}, which has been assigned to a $-NH_2$ or $-NH^+_3$ mode.

It is immediately obvious that because we do not have access to the dichroism of the amide I and II bands it is difficult to determine the conformation of the polypeptide chain. Comparison with Tables 6.3, 6.4, and 6.5 indicates that the structure cannot be β-sheet. Of the structures listed, it appears that hemoglobin should be α-helical (amide I at 1657 cm^{-1}, amide II at 1541 cm^{-1}). However, random or irregular structures also give bands at these frequencies, and without dichroic data these structures cannot be differentiated from α-helix. Usually a heat-denatured protein of random structure gives a set of broad bands, and irregular structures in intact proteins also give broad bands. Consequently the sharp narrow bands of hemoglobin suggest, but do not prove, that it is α-helical. Infrared spectroscopy is useful for globular proteins primarily for comparative purposes.

Hemoglobin shows a slight shift in amide I and amide II bands in water compared with the solid film. This change probably originates from hydrogen bonding of the solvent water molecules to the exterior peptide groups. Most of the other bands arise from vibrations in the peptide side chains, and the shifts in band position and band shape probably arise from differences of the environment of these side chains in the solid and solution state. The major shifts 1541 cm^{-1} → 1547 cm^{-1} (amide) and 1098 cm^{-1} → 1106 cm^{-1} are both caused by the solvent modifying the N — H vibration.

---

*The FTIR spectrometer produces an interference spectrum of the sample. A computer then (Fourier) transforms the spectrum into an infrared spectrum.

6. Infrared Spectroscopy 137

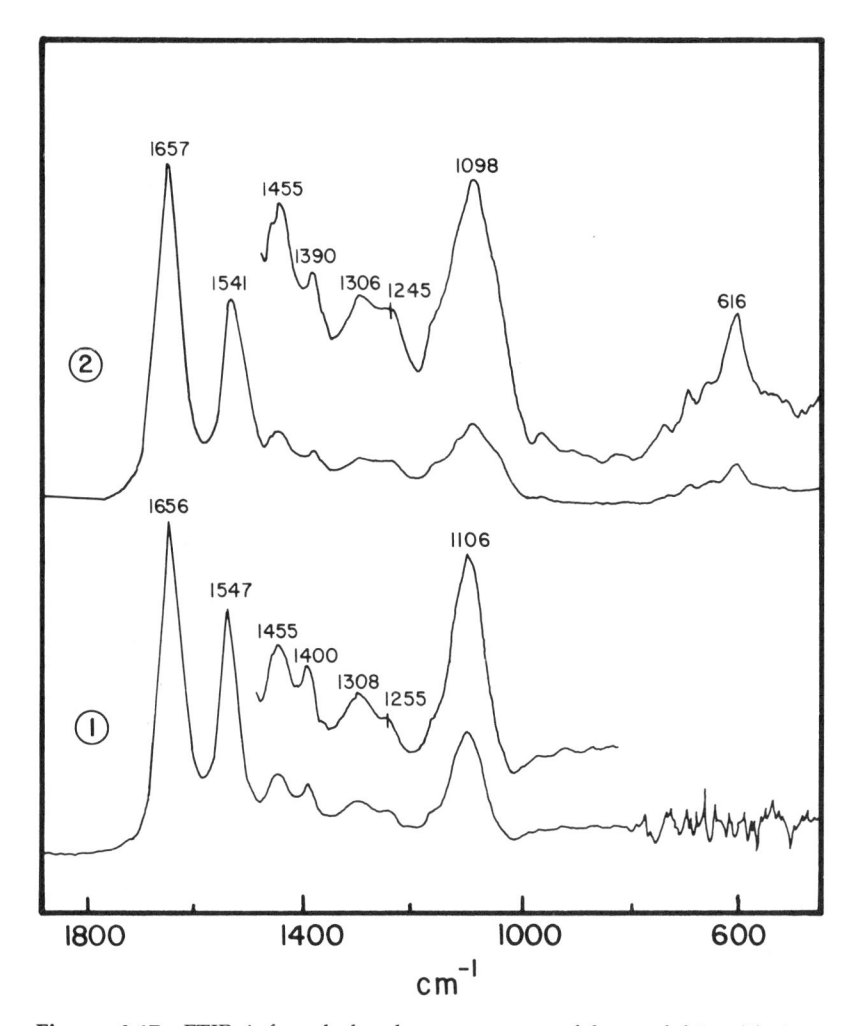

Figure 6.17 FTIR infrared-absorbance spectra of hemoglobin. (1) Aqueous solution (pH = 4.8); (2) dried films.

Other Proteins

Other proteins examined by FTIR spectroscopy include bovine serum albumin, ribonuclease, lactoglobulin, and alpha-casein. Since the conformations of these proteins are known to be different, it is of some interest to compare their spectral properties and the effect of solvent. Table 6.8 shows the positions of the amide I, II, and III bands for the five proteins plus hemoglobin, both in solid and solution. In terms of band position, the amide I bands appear in general where they are expected if based solely on the conformation. Thus 1653 cm^{-1} is in the region expected for α-helix and irregular conformation. However, despite supposedly similar conformation for ribonuclease and

Table 6.8 Comparison of Infrared Spectra for Globular Proteins in Solid-State and Aqueous Solution

Protein	Amide I (cm⁻¹) Solid	Amide I (cm⁻¹) Soln.	Amide II (cm⁻¹) Solid	Amide II (cm⁻¹) Soln.	$\Delta\nu$	Amide III (cm⁻¹) Solid	Amide III (cm⁻¹) Soln.	Conformation* α	β	I
Hemoglobin	1657	1656	1541	1547	(6)	1245	1255	75%		25%
Bovine serum albumin	1656	1655	1540	1548	(8)	1248	1247	55%		45%
Ribonuclease	1653	1656	1538	1548	(10)	1260	1260	10%	40%	50%
β-Lactoglobulin	1633	1632	1538	1551	(13)	1238	1247	10%	40%	50%
α-Casein	1656	1655	1538	1551	(13)	1242	1245			100%

*From x-ray or circular dichroism (CD) data.
Conformation: α = α-helix; β = β-sheet; I = Irregular.

β-lactoglobulin, the band corresponding to an α-helix or irregular structure seems to show up only for the former and the amide I band apparently corresponding to the β-form at 1633 cm⁻¹ only for the latter. Whether or not this is an experimental artifact is not known at this time. The amide II band surprisingly does not seem to be at all conformationally sensitive, occurring between 1538 and 1541 cm⁻¹ in all of the solid protein samples independent of their structure. This region is expected to be associated with the β-sheet conformation (though at the high end of the range—see Table 6.5).

The changes of amide II position on hydration are particularly interesting in that the frequency shift is greater for proteins possessing less α-helix. This is probably because the solvent probably does not have access to the inner regions (hydrophobic core) of a globular protein so that only a fraction of the residues become hydrated. Such an observation suggests that globular proteins having irregular chain structure on the surface are most prone to solvation effects that show up as shifts in the amide II band.

Other bands that shift with hydration include a stretching mode of the free carbonyl ($-COO^-$) found at 1390, 1394, 1393, and 1399 cm⁻¹ in the solid hemoglobin, BSA, ribonuclease, β-lactoglobulin, and α-casein respectively, and the $-NH_2$, $-NH_3$, $-CN$ combination mode at 1090 to 1100 cm⁻¹.

Table 6.9 FTIR Spectra of Ribonuclease

Film (cm⁻¹)	Solution (cm⁻¹)	Assignment
1653 vs	1656 vs ⎫	Amide I
1647 vs	1646 vs ⎭	
1538 s	1548 s	Amide II
1518 sh	1520 sh	Tyr
1450 m	1453 m	$\delta(CH_2)\ \delta(CH_3)$
1394 m	1405 m	$\nu_s(COO^-)$
1309 vw		$\nu_t(CH_3)\ \nu_w(CH_2)$
1260 sh	1260 sh ⎫	Amide III
1273 m	1240 m ⎭	
1089 ms	1101 ms	$\nu_t(NH_2)\ \nu_r(NH_3^+)$
		$\nu_s(C-N)$

s = strong, m = medium, w = weak, v = very, sh = shoulder, ν_s = stretching mode, ν_w = wagging mode, δ = deformation, ν_t = twisting vibration.

A table of resolved bands and their assignments is given for ribonuclease in Table 6.9.

From the spectra obtained so far, tyrosine is the only side-chain residue that appears to be identifiable in infrared spectra of globular proteins.

Further Reading

Elliott, A. *Infrared Spectra and Structure of Long-Chain Polymers*. Edward Arnold, London, 1969.

Fraser, R.D.B. and T.P. MacRae. *Conformation of Fibrous Proteins*. Academic Press, New York, Chapter 5, 1973.

Fraser, R.D.B. and E. Suzuki. Infrared methods. In *Physical Principles and Techniques of Protein Chemistry*, S.J. Leach, ed., Part B, Chapter 13, 1970.

Keighley, J.H. In *Introduction to the Spectroscopy of Biological Polymers*, D.W. Jones, ed. Academic Press, New York, Chapter 2, 1976.

7

Raman Spectroscopy

Raman and infrared spectroscopy are normally grouped together as one subject since they both depend upon absorption of radiation and are both forms of vibrational spectroscopy. The subjects are separated here for two reasons: (1) Since the selection rules for the Raman effect are different, the relative importance of amide bands, intensity relationships, origin of side-group effects, and so on, are distinctly different; (2) the literature on Raman spectroscopy of polypeptides has become quite extensive in recent years, mainly because the technique has the advantage over conventional infrared spectroscopy of providing structural information on proteins in aqueous solution. Water is generally a poor Raman scatterer but a strong absorber of infrared radiation; thus, until the very recent advent of the Fourier interferometric method, good infrared spectra of biologic molecules in aqueous solution were almost impossible to achieve. Raman spectra of biologic molecules in aqueous solution have, on the other hand, been relatively easily obtained.

Another advantage held by Raman spectroscopy until recently has been the fact that a single high-intensity (laser) monochromatic source could be used to excite the molecules into a higher vibrational state without the need for a source supplying widely different frequencies. Instead of requiring a permanent or induceable dipole for the interaction of incident radiation with a sample, as does infrared absorption, the Raman effect depends on the polarizability of a bond. Thus C—C bonds that give weak or nonexistent vibrational modes for infrared absorbance may give strong Raman lines.

Other advantages of Raman spectroscopy compared with infrared spectroscopy are:

1. The sample need not be transparent for Raman spectroscopy, since the Raman effect is a scattering technique rather than a transmission technique.

7. Raman Spectroscopy

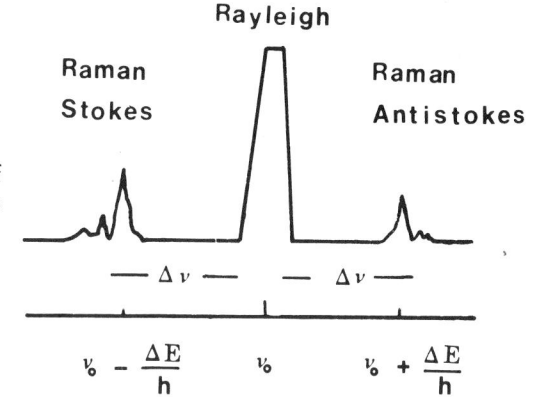

Figure 7.1 Schematic diagram of Rayleigh and Raman scattering. ν_0 is the frequency of incident light.

2. The accessibility of scattering due to homopolar bonds C—C, S—S, N—N, and so on, provides information on side chains, lipids, and so forth, which are inaccessible by the infrared method.
3. Glass containers may be used since glass is a weak Raman scatterer and small samples are often suitable for direct study.

One major disadvantage has been that many biologic samples show background "fluorescence"* that sometimes overwhelms the spectrum. Various methods of suppressing this effect are now known.

As explained in the previous chapter, the Raman effect originates in the emission of radiation from a vibrational excited state. A schematic diagram showing the relation between the frequency of the laser exciting line ν_0 and the Raman lines is shown in Figure 7.1. Incident light of frequency ν_0 is scattered elastically (or quasielastically) at, or close to, the incident frequency. (Small frequency shifts occur because of the molecular motion, which causes a "Doppler" shift in scattered frequency.) Inelastic scattering in which the molecule either accepts or gives up quantum energy from or to the incident photons gives rise to the Raman effect. The lines occurring as a result of the molecule giving up quantum energy appear at higher frequency ($\nu_0 + \Delta E/h$) and are called anti–Stokes lines, while those resulting from the sample accepting energy ($\nu_0 - \Delta E/h$) are called Stokes lines. The information contained in these two sets of satellite lines is the same and since the Stokes lines are stronger (but still possess only $\sim 10^{-6}$ to 10^{-9} of the intensity of the incident beam), it is these that are generally studied for Raman information.

The data obtained from infrared and Raman spectroscopy are complementary and are dependent upon the symmetry of the molecule in question. Thus in some cases there are vibrational modes that are Raman active but not infrared, or vice versa; there can also occur modes that are both infrared and Raman active. In terms of the appearance of the spectrum, if a mode were both infrared and Raman active, and perhaps had an infrared band at 1650 cm^{-1}, then it would appear in Figure 7.1 as

---

*The details of this type of fluorescence are not fully understood since normally fluorescent chromophores such as tyrosine, tryptophan, and phenylalanine do not interfere with Raman scattering.

$\Delta\nu_0 \pm 1650$ cm^{-1}. Thus the Stokes and anti–Stokes lines have similar appearance to an infrared spectrum.

7.1 Polarization of Raman Lines

We have seen previously that vibrations represented by moments parallel or perpendicular to polypeptide chains occur when the vibrational modes are stimulated by light absorption. Furthermore, infrared dichroism can be observed when the incident beam is polarized and couples with the orientation of the transition dipole.

The Raman situation is somewhat different in that the source (laser light) is inherently polarized and thus scattering from polypeptides can couple with the polarization or cause depolarization. In the usual Raman experiment, observations are made perpendicular to the direction of the incident beam, which is plane polarized as in Figure 7.2.

The "depolarization ratio" is defined as the intensity ratio of the two polarized components of the scattered light which are parallel and perpendicular to the direction of propagation of the (polarized) incident light. Nonsymmetric vibrations give depolarization ratios of 3/4, and symmetric vibrations from 0 to 3/4 depending upon the polarizability and symmetry of the vibrating group.

Raman polarization measurements are given the designation X(Z,Z)Y where the first capital letter (in this case X) indicates the direction of the light propagation, and the last capital letter, the direction of viewing the scattered light. Letters inside the parentheses denote the direction of incident and scattered polarization respectively. Referring to Figure 7.2, X(Z,Z)Y would represent polarized and X(Z,X)Y depolarized light, respectively.

7.2 Vibrational Analysis

As explained in the previous chapter on infrared spectroscopy, the behavior of groups of vibrating atoms may be treated in detail through group symmetry theory, or more simplistically and more conveniently through inter- and intrachain coupled motions as detailed by the Miyazawa approach.

For the case of the isolated helix, the Miyazawa approach provides for two vibrational modes corresponding to normalized frequencies $\bar{\nu}(0)$ and $\bar{\nu}(\pm t)$, where t is the unit twist of the helix. These are identified for the α-helix as in phase and $2\pi/3.6$ out-of-phase vibrations, having the representation A and E_1. The former is polarized parallel to the helix axis in the infrared spectrum and is the dominant band; E_1 is perpendicularly polarized.

For Raman spectra, three modes are derived from symmetry considerations: A_1, E_1, and E_2. The A mode is polarized; E_1 and E_2 modes are depolarized. The three modes correspond to phase differences between vibrations of equivalent atoms in adjacent repeat units of t = 0° (A species), 100° (E_1 species), and 260° (E_2 species).

Calculated frequencies of these three modes for the amide I band of α-helical poly-L-alanine are 1655 cm^{-1} (E_1, E_2), 1657 cm^{-1} (A) compared with

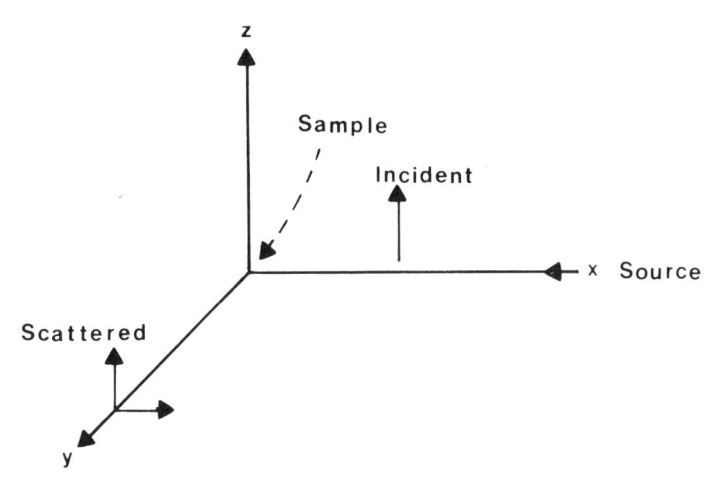

Figure 7.2 Polarization directions of incident and scattered beams in Raman spectroscopy.

the observed Raman line at 1659 cm^{-1} and infrared band at 1657 cm^{-1}. Similarly, amide II frequencies are predicted at 1541 cm^{-1} (E$_2$), 1542 cm^{-1} (E$_1$), 1545 cm^{-1} (A) compared with 1543 cm^{-1} (Raman), 1545 cm^{-1} (infrared). Despite the predicted line splittings, they are apparently unresolvable by either the Raman or infrared method.

For the antiparallel β-sheet structure the Miyazawa vibrational approach predicts, based on the general equation (6.15):

$$\bar{\nu}(\delta,\delta') = \nu_0 + \sum_j D_j \cos(j\delta) + \sum_i D'_i \cos(i\delta'),$$

that four vibrational modes will exist. As pointed out in Table 6.4, two of these modes are observable in the infrared for both amide I and II regions. For polyalanine these are: $\bar{\nu}_\parallel$ (0,π), 1695 cm^{-1}(m); $\bar{\nu}_\perp$ (π,0), 1630 cm^{-1}(s), and $\bar{\nu}_\parallel$ (0,π), 1524 cm^{-1}(m), $\bar{\nu}_\perp$ (π,0), 1558 cm^{-1}(w). Raman spectroscopy provides the symmetrical $\bar{\nu}(0)$ mode at 1669 cm^{-1} and 1538 cm^{-1}, respectively, and thus aids in the determination of coupling constants D$_j$ and D$_i$. Although very few coupling constants have been evaluated to date, it seems likely that it eventually will be possible to correlate such constants with fundamental atomic interactions and the conformational stability of the structure.

Apart from our ability to understand some fundamental aspects of vibration and vibrational symmetry, it is relevant to ask what information can be derived concerning the conformation and structure of proteins. In this regard it is often the relative intensity of the lines, as well as their frequency, which provides important insight. First, let us look at some general principles.

7.3 Amide Lines and Conformation

We have seen that three factors make infrared spectroscopy of polypeptide conformation particularly useful: the frequency of the bands, intensity of bands (particularly amide I and II), and vibrational splitting.

Table 7.1 Comparison of Infrared (IR) Bands and Raman Lines (cm^{-1}) for Some of the Main Polypeptide Modes

Conformation	Amide I IR	Amide I Raman	Amide II IR	Amide III Raman	Other Raman
α_R-helix	1652–1661	1655–1659 (s)	1546–1655	(1260–1245 w)	900–960 (s)
β-sheet	1621–1637 1690–1700	1667–1672 (s)	1512–1645	1229–1240 (s) (1270)	900–960 (w)
Random	~1655–1660	1665–1675 (s)	~1550	1243–1265 (m)	940–970 (s)

s = strong, m = medium, w = weak.

It turns out that whereas analysis of the amide I and III lines is most useful in Raman analysis, the amide II frequency line is generally weak or absent. A comparison of amide bands and conformational information is given in Table 7.1.

Of particular interest in the Raman spectra of polypeptides are the amide III and 900–960 cm^{-1} lines. The latter line seems to arise from a combination of C—C$_x$ stretching vibrations and is often quite intense. It can be seen that different combinations of these two lines occur for the three main conformations: α-helix, having a weak amide III and intense (C—C) line; β-sheet, having a strong amide III and no $\bar{\nu}$(C—C); and random, having both lines intense. These observations are then particularly useful for conformational studies of proteins.

7.4 Deuteration

Deuteration of polypeptides or proteins—that is, the treatment of the material with deuterium oxide—helps identify the lines by causing removal of labile hydrogen atoms (—OH, —NH, etc.) and replacement with deuterium, thus altering the vibrational frequency. For the amide I band, which involves mainly C=O stretching, relatively little effect of deuterium is expected, whereas the amide III line, involving >N—H bend, should change considerably. This effect may be demonstrated for polylysine and poly(glutamic) acid in water (neutral pH, coil form), as in Table 7.2.

The deuteration process also involves side chains (e.g., Tyr—OH, other —OH, His N—H, Trp N—H), and therefore helps identify the origin of the side-chain lines. In addition to spectral shifts, line intensities change. For example, in the lysozyme spectrum, changes in the 1180, 835, and 646 cm^{-1} region are ascribed to OH deuteration of tyrosine; 1580, 1553, 1337, 1014, 257, 574, and 544 cm^{-1} to N-deuteration of tryptophan, and so on.

Table 7.2 Effect of Deuteration on Amide I and II Raman Lines of Poly(amino Acids)

	Amide I H$_2$O	Amide I D$_2$O	Amide III H$_2$O	Amide III D$_2$O
Poly-L-lysine	1665 cm^{-1}	1660 cm^{-1}	1243 cm^{-1}	943 cm^{-1}
Poly-L-glutamic acid	1660 cm^{-1}	1656 cm^{-1}	1249 cm^{-1}	1006 cm^{-1}

7.5 Conformational Analysis of Proteins

From the preceding observations it is clear that Raman lines change in intensity and frequency with different conformations and it should be possible to combine this information in a manner such that the secondary structure of polypeptides and proteins may be evaluated. The following method, developed recently by Lippert and co-workers, uses initially the methodology that has been used with some success in circular dichroism spectroscopy (see Chapter 9).

Poly-L-lysine is chosen as a starting model because it can conveniently be converted between α-helix, β-sheet, and coil conformations in aqueous solution, simply by a choice of pH and temperature. The Lippert method involves a judicious combination of amide I and III intensities in water and deuterium oxide. The first evident problem in making a quantitative assessment of Raman spectra is that it is extremely difficult to obtain absolute intensities; published spectra use an ordinate axis of relative intensity. It is clear then that some internal standard is necessary. Since all proteins and almost all poly(amino acids) have C—C bonds in side chains of peptide residues, these homopolar bonds give rise to vibrations that show up as a rather intense Raman line in the 1440 to 1450 region. Although this line is not simply related to the number of C—C bonds, it does offer a convenient internal standard. Thus for polylysine the Raman spectrum in different conformations (Figure 7.3) may be calibrated in terms of the 1446 to 1448 line. Table 7.3 shows intensity ratios based on amide III in aqueous solution and amide I and III in D$_2$O compared with the 1446 to 1448 line. The reasons for this choice of line and wavelengths are complicated, but they include the need to avoid the Raman-scattering line of water at 1645 cm^{-1} and the need to provide three parameters to solve for three variables α, β, and coil content.

It can be seen from Table 7.3 that for each conformation, two of the three lines dominate. If we wished to examine a polypeptide with a mixture of α-helix, β-sheet, and coil, it is reasonable to suppose that the line intensity at any given frequency is composed of the sum of contributions from the various conformations. However, since the polypeptide will not, in general, have the same number of C—C bonds and the same standard (1447 cm^{-1}) line intensity as polylysine, it will be necessary to include a correction factor C^{poly}.

Then
$$C^{\text{poly}} I^{\text{poly}}_{1240} = f_\alpha I^{\alpha(\text{lys})}_{1240} + f_\beta I^{\beta(\text{lys})}_{1240} + f_c I^{c(\text{lys})}_{1240} \tag{7.1}$$

$$C^{\text{poly}} I^{\text{poly}}_{1632} = f_\alpha I^{\alpha(\text{lys})}_{1632} + f_\beta I^{\beta(\text{lys})}_{1632} + f_c I^{c(\text{lys})}_{1632} \tag{7.2}$$

$$C^{\text{poly}} I^{\text{poly}}_{1660} = f_\alpha I^{\alpha(\text{lys})}_{1660} + f_\beta I^{\beta(\text{lys})}_{1660} + f_c I^{c(\text{lys})}_{1660} \tag{7.3}$$

Table 7.3 Raman Spectral Intensities of Three Conformations of Poly-L-lysine Relative to the Spectral Intensity of the 1446–1448 cm^{-1} Line

Conformation	H$_2$O 1240 cm^{-1}	D$_2$O 1632 cm^{-1}	D$_2$O 1660 cm^{-1}
α-Helix	0.00	0.80	0.55
β-Sheet	1.20	0.33	1.22
Coil	0.60	0.21	0.70

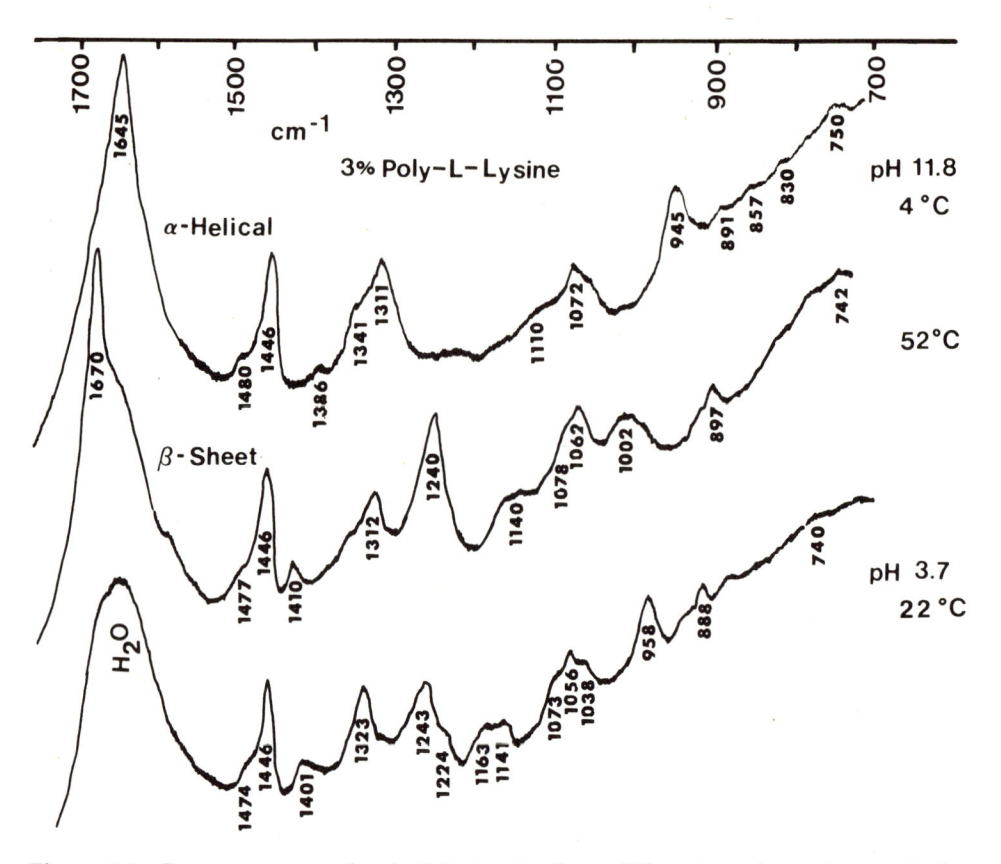

Figure 7.3 Raman spectra of poly-L-lysine in three different conformations: α-helix (pH 11.8, temp. = 4° C); β-sheet (pH 11.8, temp. = 52° → 35° C); coil (pH 3.7, temp. = 22° C).

where I_λ^{poly} is the relative intensity of the line at λ compared with the line at 1447 cm^{-1} for the polypeptide. Similarly for $I_\lambda^{\alpha(lys)}$, $I_\lambda^{\beta(lys)}$, $I_\lambda^{c(lys)}$ where the values of Table 7.3 would be used. f_α, f_β, f_c are the relative fractions of α-helix, β-sheet, and coil, respectively, so that:

$$f_\alpha + f_\beta + f_c = 1 \tag{7.4}$$

Equations (7.1)–(7.4) contain four unknowns: f_α, f_β, f_c and C^{poly}. The most convenient method of solving is to assume $C^{poly} = 1$, then obtain f_α, f_β, f_c. On addition these fractions will yield a number differing from 1, which will then be the true correction factor.

Application of the preceding methodology to proteins has the disadvantage that the irregular structure found in globular proteins should not be correlated with the random or coil forms found in polylysine. To circumvent this difficulty, a set of equations identical in form to equations (7.1)–(7.4) is used:

$$C^{prot} I_{1240}^{prot} = f_\alpha I_{1240}^\alpha + f_\beta I_{1240}^\beta + f_i I_{1240}^i \tag{7.1a}$$

$$C^{prot} I_{1632}^{prot} = f_\alpha I_{1632}^\alpha + f_\beta I_{1632}^\beta + f_i I_{1632}^i \tag{7.2a}$$

$$C^{prot} I^{prot}_{1660} = f_\alpha I^\alpha_{1660} + f_\beta I^\beta_{1660} + f_i I^i_{1660} \tag{7.3a}$$

In these equations I^x_λ is a standard relative spectral intensity for proteins referring to conformation $x(=\alpha, \beta,$ or $i)$ at wavelength λ. I^{prot}_λ is the measured intensity of lines at 1240 cm^{-1} in water and 1632 and 1660 cm^{-1} in D$_2$O for the protein in question. C^{prot} is the correction factor based on the relative intensities of amide bands compared with the 1447-cm^{-1} band for the protein in question and $\alpha, \beta,$ i refer to α-helix, β-sheet, and irregular conformations, respectively.

The problem is then to obtain appropriate values of I_λ. Such values have been obtained by incorporating data from the known (x-ray) structure of lysozyme and ribonuclease A and combining with the appropriate Raman spectra. The spectra for ribonuclease A in water and D$_2$O are shown in Figure 7.4. Standard spectral intensities obtained from the preceding approach are presented in Table 7.4.

Figure 7.4 Raman spectra of (5%) ribonuclease A solutions in H$_2$O and D$_2$O. Arrows point to the frequencies used for conformational assessment. The dashed line is drawn in so that intensities can be estimated. The sloping baseline indicates the presence of a low level of fluorescent impurity (after Lippert et al.).

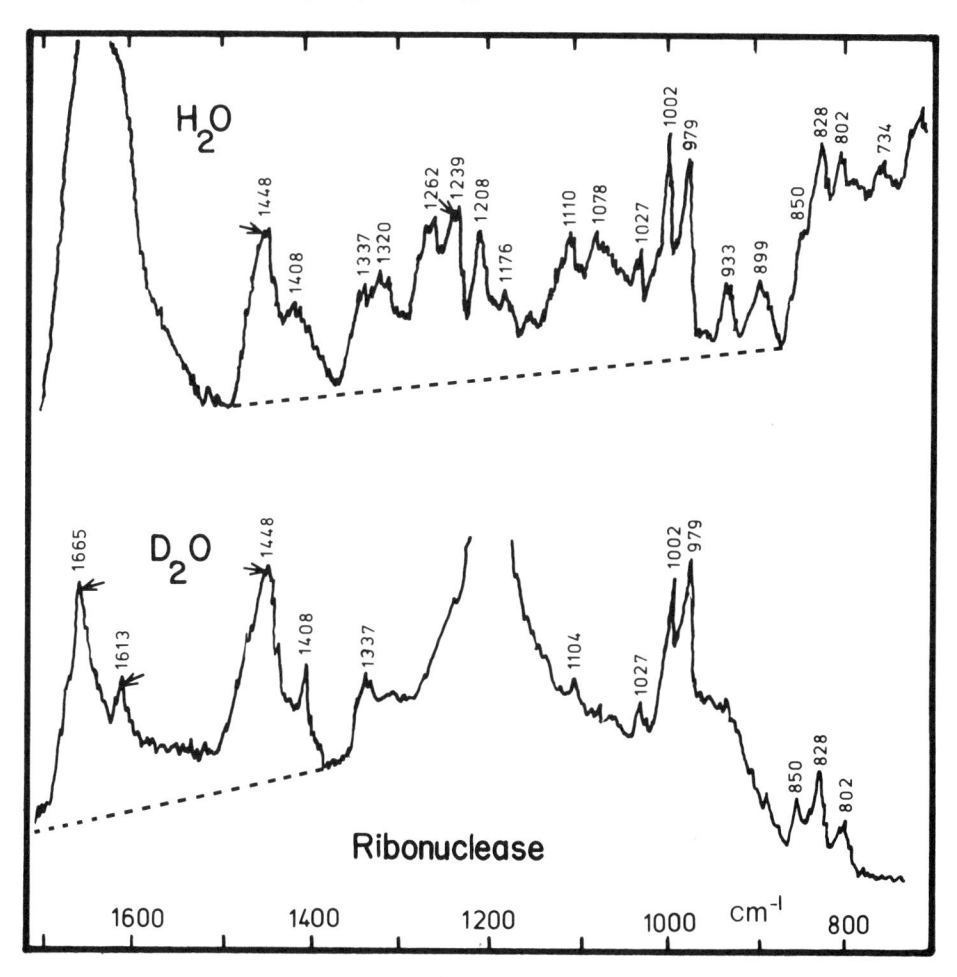

Table 7.4 Standard (Relative) Raman Spectral Intensities (I_x^X) for Proteins

	H$_2$O	D$_2$O	
Conformation (x)	1240 cm^{-1}	1632 cm^{-1}	1660 cm^{-1}
α-Helix	0.00	0.80	0.55
β-Sheet	1.20	0.72	0.88
Irregular	0.60	0.08	0.78

It can be seen that the α-helix parameters are identical with polylysine but that β-sheet and irregular/coil are changed. The reason for the β-sheet change is not entirely clear. The scale factors for lysozyme and ribonuclease A are found to be C^{prot} = 0.72 and 0.79, respectively.

The method has been used on both proteins in solution and in the solid state (crystallized from water or D$_2$O), and indeed Raman spectroscopy offers a sensitive method of determining subtle changes in environment.

Table 7.5 contains some relative spectral intensities of various proteins. It is noticeable that the relative intensities of the Raman lines from insulin in solution and in the solid state are substantially different. Although it is tempting to identify such changes with a conformational change, it is possible that a given conformation maintained in two different environments may give rise to different spectral intensities. Indeed the separation of environmental and conformational factors is not easy in Raman or other forms of spectroscopy. It is well worth pointing out that spectroscopic methods are *comparative* and *not absolute* methods of assessing structure and thus ambiguities often exist.

Given the preceding limitations, there does seem to be a good correlation between the Raman method and circular dichroism and x-ray estimates of protein conformation. Table 7.6 presents such a comparison based on the data of Tables 7.4 and 7.5 and the use of equations (7.1a)–(7.4a).

The range of values for x-ray data in Table 7.6 is caused by different methods of counting residues: for example, residues in β-turns may or may not qualify as β-sheet, distorted helices as α-helical, etc. CD ranges are generally caused either by different methods of estimation (see Chapter 9) or by variations based on measurements by different investigators.

Table 7.5 Raman Spectral Intensities of Proteins Relative to the Intensity of the 1446–1448 cm^{-1} Line

	H$_2$O	D$_2$O	
Protein*	1240 cm^{-1}	1632 cm^{-1}	1660 cm^{-1}
α-Chymotrypsin	0.86	0.36	0.88
Chymotrypsinogen	0.78	0.44	0.83
Bovine Serum Albumin	0.30	0.69	0.86
Concanavalin A	1.39	0.58	1.28
Pepsin	1.20	0.48	1.28
Pepsinogen	1.10	0.56	1.13
Insulin	0.75	0.55	1.10
Insulin (Solid)	0.60	0.65	0.91

*All protein spectra were measured in solution unless otherwise indicated.

7. Raman Spectroscopy

Table 7.6 Comparison of Protein Conformation Estimated by Various Methods

Protein and method	α-Helix	β-Sheet	Irregular
α-Chymotrypsin			
Raman	4	34	62
X-ray	3–9	22–50	47–70
CD	8–20	10–32	55–82
Chymotrypsinogen			
Raman	14	37	49
X-ray	6–14	22–50	44–65
CD	9–20	10–36	54–70
Bovine Serum Albumin			
Raman	60	0	40
CD	54–62	0–3	41–51
Concanavalin A			
Raman	0	48	53
X-ray	3	57	41
Pepsin			
Raman	4	36	60
CD	5–10	34–58	37–56
Pepsinogen			
Raman	9	39	52
CD	5–13	36–49	46–54
Insulin			
Raman	29	11	59
CD	20–30	10–25	47–55
Insulin (Solid)			
Raman	42	19	39
X-ray	52	6	42

7.6 Side-Chain Vibrations

Because of the many potential vibrational modes in the amino acid side-chain residues which involve Raman-active homopolar bonds (e.g., —C—C), individual residues often contribute more distinctly to the Raman spectrum than they do to the infrared spectrum of a protein. One popular method of identifying lines in the Raman spectrum has been to compare the spectrum of a protein with that of the equivalent mixture of its constituent amino acids. Two such spectra are shown of lysozyme and its constituent amino acids in Figure 7.5. Considerable similarity between the two spectra is apparent. Perhaps most prominent in the side-chain vibrations are lines originating from Trp, Phe, Tyr, His, and —S—S— bonds. Two rather interesting observations are: (1) The line positions do not vary as the residues are incorporated into the polypeptide chain, indicating localization of the vibrational motion within the side chains; and (2) the strong lines originating from the aromatic residues are out of proportion to their relative abundance in the protein. For example, although the relative abundance in the present case is Trp 6/129, Phe 3/129, Tyr 3/129,

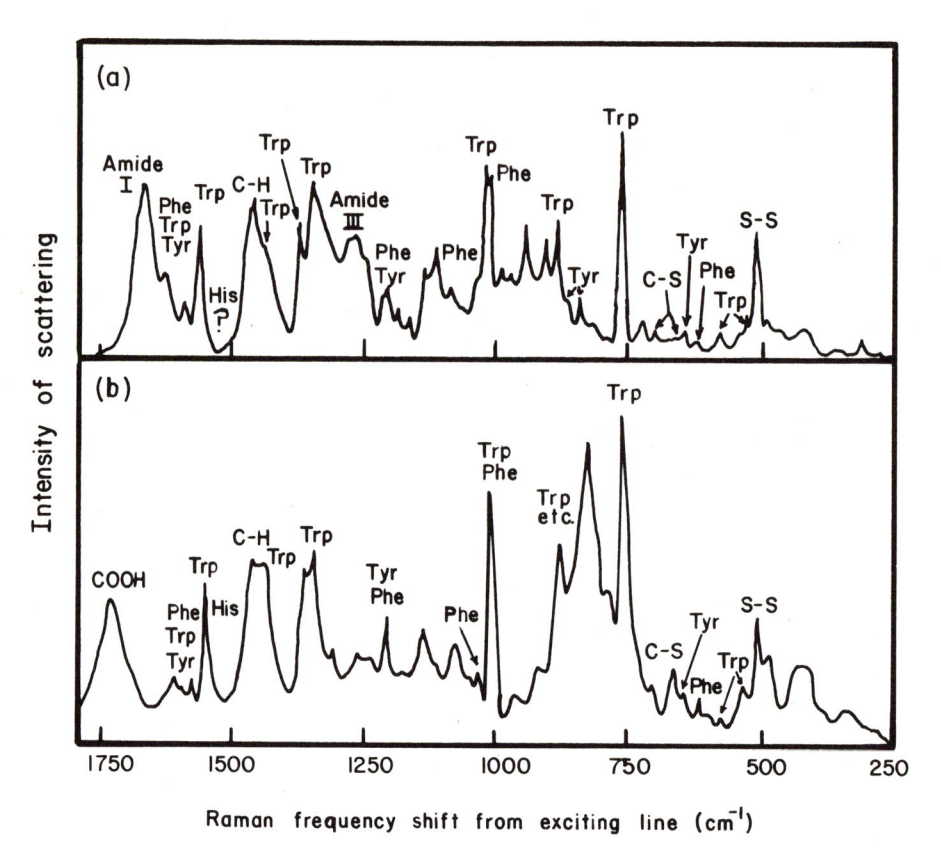

Figure 7.5 The Raman scattering spectrum of (a) crystalline lysozyme, and (b) amino acids mixed in the same ratio found in lysozyme.

His 1/129, —S—S— 4/129, the intensities for Trp and Phe in particular are the highest in the complete spectrum.

It is interesting to note that these residues are generally located within the hydrophobic core of the globular proteins and are often in the region of enzyme-active sites. The amide I and III bands are clearly evident and it is therefore possible to follow conformational aspects of both backbone and side-chain interactions. A survey of other proteins indicates that the same aromatic residues prominent in the lysozyme spectra are also prominent in other spectra, and of the other residues only proline (or hydroxyproline) are currently known to contribute to any extent. The pyrrolidine ring vibrations $\bar{\nu}(C—C)$ are generally found in the 800- to 900-cm^{-1} region.

7.7 Comparison of Proteins in Solid and Solution States

The only definitive method of assessing detailed protein structure known at present is that provided by detailed x-ray diffraction studies. However, those studies can only be carried out in the solid state, and although many of the protein crystals examined to date contain a large number of solvating molecules, there is some doubt that the solid-state structure is the same as the active form in solution. We have already seen that for one case—insulin—the

7. Raman Spectroscopy

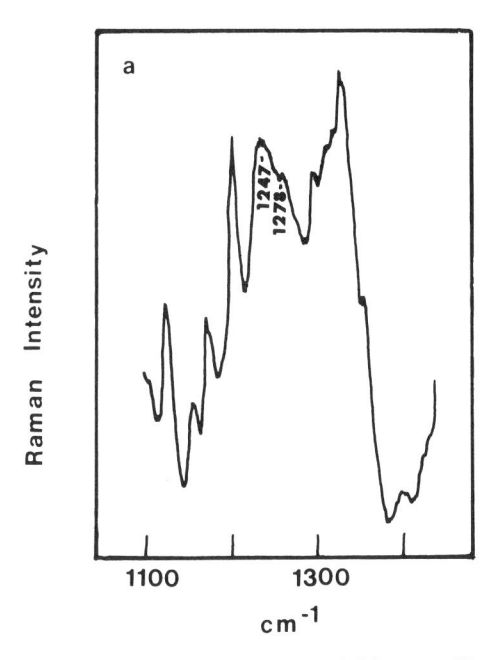

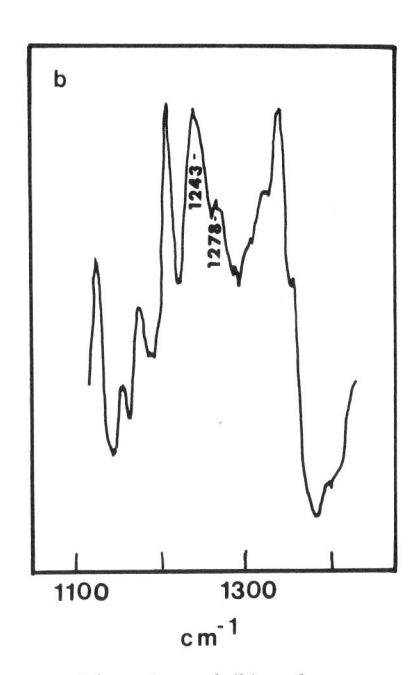

Figure 7.6 Raman spectra of **(a)** crystalline carboxypeptidase A, and **(b)** carboxypeptidase A in aqueous solution (pH 7.0, 60 mg/ml, 3M NaCl).

Raman amide lines have different intensity ratios in the solid and solution states.

Examples are known in which there are virtually no changes in spectral intensity observed on passing from solution to crystalline state. Such is true for ribonuclease, where the only significant change is the increase in intensity of the 664-cm^{-1} line, which is not associated with the backbone conformation. On the other hand, freeze-dried proteins, including ribonuclease, often show significant changes in Raman spectra compared with solution states, showing that the removal of normally bound water probably has a structural influence.

Another protein that appears to possess a different structure in aqueous solution and crystalline state is carboxypeptidase A. The amide III regions of the aqueous solution spectra and crystal spectra are compared in Figure 7.6. Although the observed changes could be interpreted through the application of the conformational approach spelled out in section 7.5, the details of environmental and structural changes accounting for the spectra are not known at this time.

7.8 Resonance Raman Spectroscopy and Structure/Mechanism Studies

One of the objectives of the structural/spectroscopy studies on proteins is to provide information concerning mechanism and function. Since the function of fibrous proteins is primarily mechanical, we may be more specific in seeking structural changes involved with the functioning of globular proteins. In this context we would like to know how enzyme and substrate interact, what changes in structure of the enzyme/substrate complex occur, and how the bond-breaking or joining processes occur. To this end, normal Raman and infrared spectroscopy are rarely, if ever, capable of the sensitivity necessary for such detailed studies and the newer method of resonance Raman spectroscopy is coming to the fore. In this adaptation of the Raman method, an exciting laser line that corresponds to an electronic absorption wavelength specific to the molecule in question is used. By this means, the normally inefficient light-absorption vibrational radiation process is greatly enhanced. For example, if the enzyme papain is reacted with a suitable chromophoric substrate as follows:

$$\text{PAPAIN} + (CH_3)_2N-\underset{\underset{\underset{\bigcirc}{C=O}}{NH}}{\bigcirc}-C=C-\overset{O}{\underset{}{C}}-OCH_3 \rightarrow (CH_3)_2N-\underset{\underset{\underset{\bigcirc}{C=O}}{NH}}{\bigcirc}C=C-\overset{O}{\underset{}{C}}-S-\text{PAPAIN} + CH_3OH$$

then it is possible to obtain spectra of the relatively long-lived intermediate state. Figure 7.7 shows the resonance Raman spectrum of the acyl-enzyme

Figure 7.7 Resonance Raman spectra of 4-dimethylamino-3-nitro-(α-benzimido)-cinnamoyl-papain (top) and substrate alone (bottom) in deuterated solution. (Kindly provided by Dr. Paul Carey.)

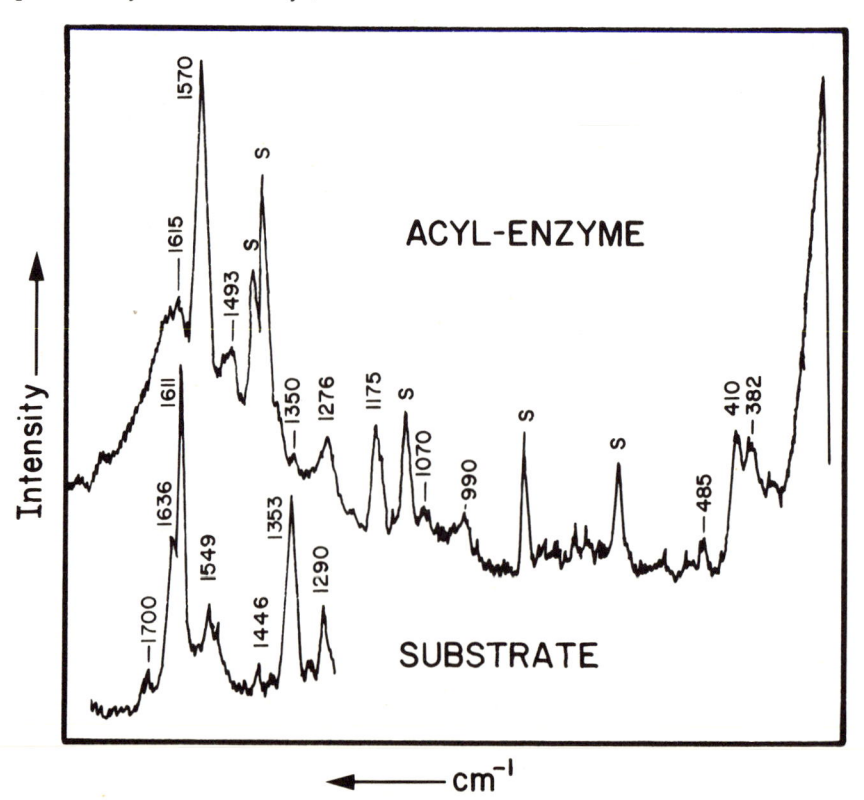

complex (4-dimethylamino-3-nitro(α-benzimido)-cinnamoyl-papain) and the substrate alone, using a 441.6-nm excitation laser line, which is in the region of the main electronic absorption band (above 320 nm) of the complex.

The 1636-cm^{-1} line in the substrate spectrum, which moves to 1570 cm^{-1} in the enzyme–substrate complex, involves the stretching vibration of the ethylenic C=C group. This large drop in the C=C line position in the active site means that this bond becomes more singular in character. This may be ascribed to the presence of a positively charged enzyme side chain located close to the substrate (an imidazolium side chain—from histidine—is a prime candidate). A probable intermediate form is then as follows:

$$(CH_3)_2 \overset{+}{N} = \underset{NO_2}{\underset{|}{\bigcirc}} = \overset{|}{C} - \underset{|}{C} = \overset{\overset{ImH}{\oplus}}{\overset{|}{\underset{|}{C}}} - S - PAPAIN$$

It appears then that certain types of important mechanistic information may be obtained from resonance Raman spectroscopy, including:

1. Effects of charge from protein environment;
2. Spectral differences of intermediates allowing kinetics to be measured directly;
3. Substrate distortion during catalysis can be detected (e.g., chymotrypsin reactions). This sort of information—that is, differences in the mode of action of trypsin and chymotrypsin enzymes—is not readily accessible by other more established methods (e.g., x-ray diffraction).

Further Reading

Koenig, J.L. Raman Spectroscopy. In *Introduction to the Spectroscopy of Biological Polymers*. D.W. Jones, ed. Academic Press, New York, 1976.

Van Wart, H.E. and H.A. Scheraga. Raman and Resonance Raman Spectroscopy. In *Methods in Enzymology*, Enzyme Structure, Part G. C.H.W. Hirs and S.N. Timasheff, eds., Vol. XLIX, Chapter 5, 1978.

Warshel, A. Interpretation of Resonance Raman Spectra of Biological Molecules. *Annual Reviews of Biophysics and Bioengineering* 6:273–300, 1977.

Woodward, L.A. *Introduction to the Theory of Molecular Vibrations and Vibrational Spectroscopy*. Oxford Univ. Press, New York, 1972.

8

Electronic Spectroscopy

8.1 Types of Spectroscopy

The ability of proteins to absorb quanta of light that promote electrons to higher (excited) levels has led to the development of several techniques that are useful for characterizing aspects of structure. Essentially one can measure the *absorbance* of monochromatic light by a sample or its *refractive index*. The incident radiation may be unpolarized, plane polarized, or circularly polarized. This combination of possibilities gives us six techniques in which various properties are measured. These are listed in Table 8.1. The symbols ∥, ⊥, L, and R refer to parallel, perpendicular, left, and right-hand polarized light.

In addition to these six methods, we have the possibility of irradiating the sample at one wavelength and examining the emission at a second; that is observing fluorescence or related phenomena. Before examining these techniques and the information they provide in more detail, we shall first examine the concept of polarized light, excitation of electrons, and optical activity.

8.2 Polarized Light

An electromagnetic wave is characterized by its amplitude, frequency or wavelength, and orientation. If we had a source of light that was "plane polarized" the electromagnetic wave would proceed down a linear axis with amplitude oscillations describing a sine-wave pattern (Figure 8.1). The radiation has two components, the electric field vector μ and the magnetic field vector m, which are mutually perpendicular and in phase. If one could follow the amplitude oscillations of, say, the electric field along the z axis, one would see at various times a sequence such that shown in Figure 8.2.

8. Electronic Spectroscopy

Table 8.1 Types of Electronic Spectroscopy

	Property measured	
Polarization	Absorption	Dispersion
None	Extinction Coef. ε	Refractive Index n
Linear	Dichroism $\varepsilon_\parallel - \varepsilon_\perp$	Birefringence $n_\parallel - n_\perp$
Circular	Circular Dichroism $\varepsilon_L - \varepsilon_R$	Optical Rotatory Dispersion $n_L - n_R$

The amplitude would, in fact, follow a sine-wave motion in the $\pm x$ direction. It is interesting to note that such an amplitude/time process can be generated schematically by imagining the amplitude vector as being composed of two vector elements of magnitude A, which are rotating in a circular manner such that they are in phase and have the same amplitude at all time t, but are rotating with constant angular velocity in opposite directions (left and right). The vector sum is given by the amplitude of sequence 8.2a. Thus plane-polarized light may be regarded as the vector sum of two equal circularly polarized field components. Of course, unpolarized light would have all vector components in the $x\,y$ plane simultaneously.

It is relatively easy to visualize the production of plane-polarized light where an absorbing material is coated on a transparent substrate in an oriented fashion such that if the light vector is in the direction required to promote an electronic transition, absorbance occurs but perpendicular to that direction light is transmitted. We are accustomed to sunglasses made from polarizing material and know that if two polarizers are used in series, light is either transmitted or blocked as the polarizers are rotated. The concept of circularly polarized light is, however, more difficult to assimilate. Circularly polarized light can in fact be generated by combining two plane-polarized waves that are of equal amplitude and frequency but have been *phase shifted* by one-quarter of a wavelength. (This is achieved physi-

Figure 8.1 Plane-polarized light of amplitude (related to intensity) A, wavelength λ, showing the electric field component μ and magnetic field component m.

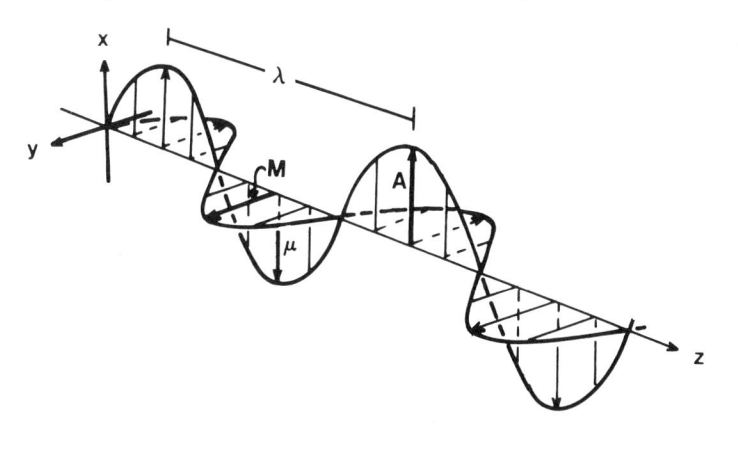

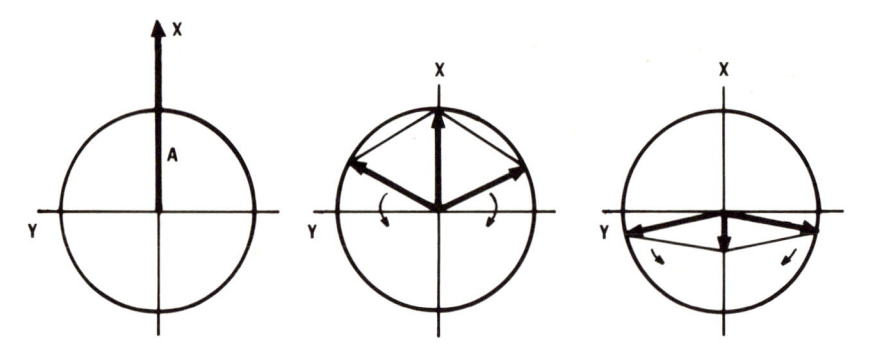

Figure 8.2 Decomposition of a plane-polarized wave vector into rotating components.

cally by a device known as a quarter-wave plate.) The result of this combination is shown in Figure 8.3, where the resultant vector is drawn in every quarter of a wavelength and is precessing (rotating) to the left. This then is left-hand circularly polarized light. If the light waves had been phase shifted by one-quarter of a wave in the opposite direction then right-hand polarized light would have resulted.

One other combination of waves will be of interest to us; that is the situation occurring when two field components are phase shifted and of unequal amplitude. In this case the resultant is elliptically polarized light, as shown in Figure 8.4.

Thus, in summary, two beams of plane-polarized light may be combined to produce plane-polarized, circularly polarized, or elliptically polarized light. If the planes of the two beams are perpendicular to each other, the possible combinations are shown in Table 8.2.

8.3 Absorption Phenomena

In the two preceding chapters on vibrational spectroscopy, we saw that the absorption of radiation in the infrared region of the spectrum (100 to 10,000 cm^{-1}) caused a change in the vibrational state of polypeptides. Many of the

Figure 8.3 Combination of two perpendicularly polarized light waves of equal amplitude but shifted in phase by one-quarter of a wavelength $\lambda/4$.

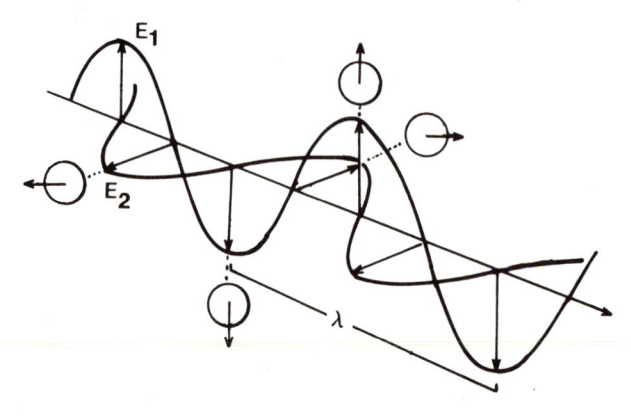

8. Electronic Spectroscopy

Figure 8.4 The result of combining plane-polarized light waves of the same wavelength and amplitude A_x (polarized in the x direction) and A_y (polarized in the y direction), which are quarter-wave shifted, is elliptically polarized light.

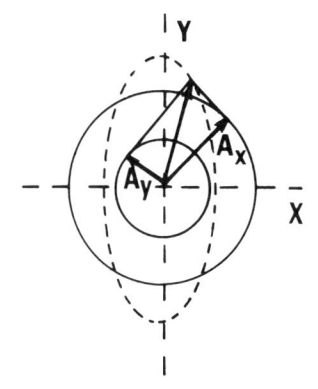

vibrational modes were localized in the peptide group. Visible/ultraviolet spectroscopy involves radiation of much lower wavelength (higher frequency and energy) and the absorption of light quanta involves a change in the electronic state of a polypeptide. Again the change of state often involves (electronic) transitions in the peptide bond.

Electronic Transitions

Electronic transitions within an atom, group, or molecule, obey the Planck relation:

$$\Delta E = h\nu \tag{8.1}$$

where ν is the frequency of the applied radiation and ΔE is the difference in energy level before and after the electronic transition. In order to know which electrons are involved in the transition, it is necessary to know something of the nature of the orbitals in which electrons can exist. Quantum theory tells us that for simple atoms, electrons are transferred from the ground state (low energy) to an excited state (higher energy) with the absorption of radiative energy. The electronic orbital for molecules or groups of atoms is often treated by molecular orbital theory. For two hydrogen atoms one has an energy diagram as shown in Figure 8.5. As the atoms are brought together the atomic orbitals combine to form bonding and antibonding molecular orbitals. In the hydrogen molecule case, absorption of radiation could, for example, promote an electron from the bonding (σ) orbital to the nonbonding (σ^*) orbital. This would be called a σ–σ^* transition. For proteins and polypeptides our major source of electron transitions is the peptide bond.

Table 8.2 Combination of Two Plane-Polarized Light Beams

Input polarization	Phase	Vector	Amplitude	Output	Schematic
Plane	In phase	⊥r	Equal	Plane polarized	
Plane	(¼) Shifted	⊥r	Equal	Circular	
Plane	Shifted	⊥r	Unequal	Elliptical	

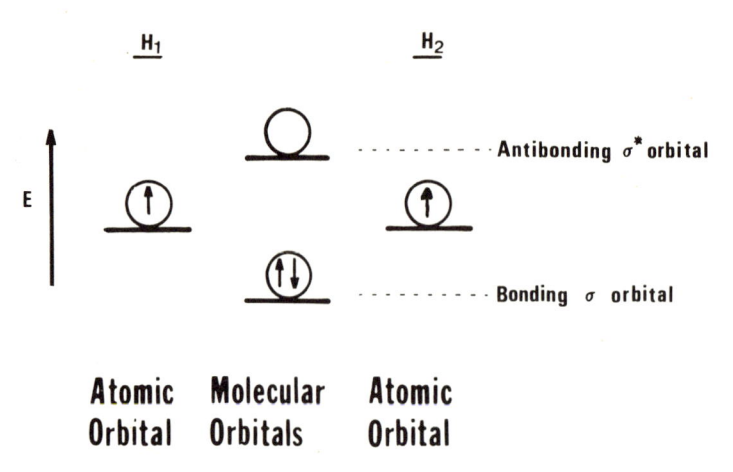

Figure 8.5 Combination of atomic orbitals to form bonding and antibonding molecular (σ) orbitals for the hydrogen molecule.

From Chapter 2 it will be remembered (perhaps!) that if the carbonyl bond electrons are treated as being delocalized, we can have four electrons (O(1), C(1), N(2)) traversing the bonding system of the three atoms O — C — N.

Molecular orbitals may be constructed for this system by a linear combination of atomic orbitals (LCAO) method. In this case, the wave function is based on a three-center molecular orbital (rather than two for hydrogen) and can be written:

$$\psi = c_1\phi_O + c_2\phi_C + c_3\phi_N \tag{8.2}$$

where ϕ_O, ϕ_C, and ϕ_N refer to the wave functions of oxygen, carbon, and nitrogen atoms, respectively. The c's are constants. Solution of such an equation in terms of energy levels reveals three orbitals often referred to as bonding, nonbonding, and antibonding (as shown in Figure 8.6).

The orbitals are labeled π_2 (bonding), π_1 (nonbonding), and π^* (antibonding). Since there are potentially four available π-electrons, these would occupy the lowest energy molecular orbital states (π_2 and π_1). Thus absorption of radiative energy could promote an electron from the π_1 to π^*

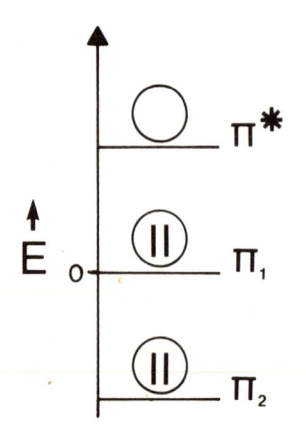

Figure 8.6 The three molecular (π) orbitals of the peptide bond.

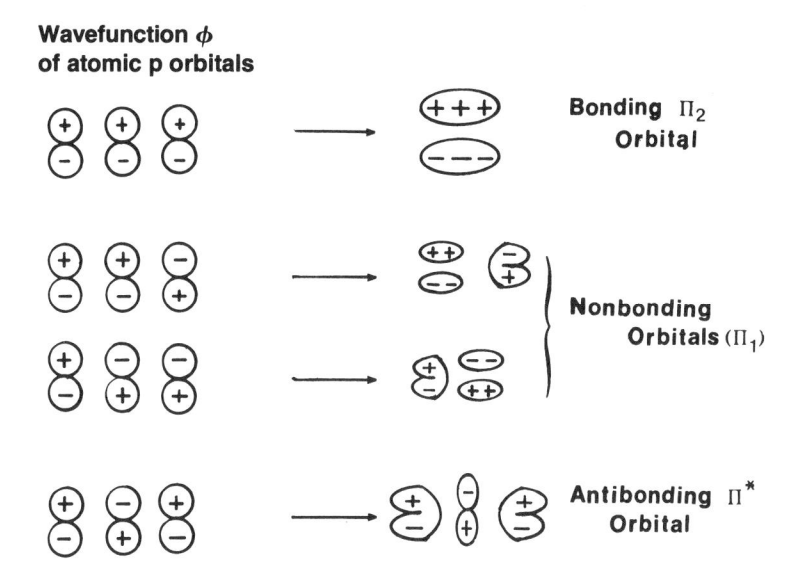

Figure 8.7 Representations of the combination of atomic p-orbitals to form molecular π-orbitals. (The dumbells on the left are the geometric representations of the wave function ϕ; ϕ^2 is the electron density.)

orbital, causing a π–π^* transition. Although the π^* and π_2 are equally separated from the energy level of π_1 (= 0) in Figure 8.6, the influence of the environment in real molecules lowers the energy level π_1.

A pictorial representation of the combination of atomic orbitals to give the three π-molecular orbitals is shown in Figure 8.7. It should be remembered that the + and − lobes of the wave function represent mathematical sign and not electronic charge.

The type of representation shown in Figure 8.7 will also be useful when we consider the electronic transitions in aromatic side chains. In addition to the delocalized π-electrons, the oxygen atom in the peptide group possesses nonbonding electrons in the p-orbital. It is possible to promote a nonbonding (n) electron in the π^* orbital, the so-called n–π^* transition. The relative energies (wavelengths) of these and two other related transitions to the antibonding σ^* orbital are shown in Figure 8.8. It can be seen that the lowest energy (highest wavelength) transition is the n → π^*, followed by π_1 → π^*, π_2 → π^*, and n → σ^*. The last two transitions occur in the vacuum ultraviolet range and are not accessible with conventional UV/visible spectrophotometers.

We shall consider the factors affecting the intensity of the ultraviolet absorption bands in a little more detail later.

Electronic Transitions in Side Chains

The electronic transitions enumerated above for the amide group of the polypeptide backbone are not the only transitions leading to UV/visible absorbance by proteins since a number of amino acid residues contain absorbing side-chain chromophores. Most of these absorbing side chains contain aromatic residues and it is therefore relevant to attempt to understand the

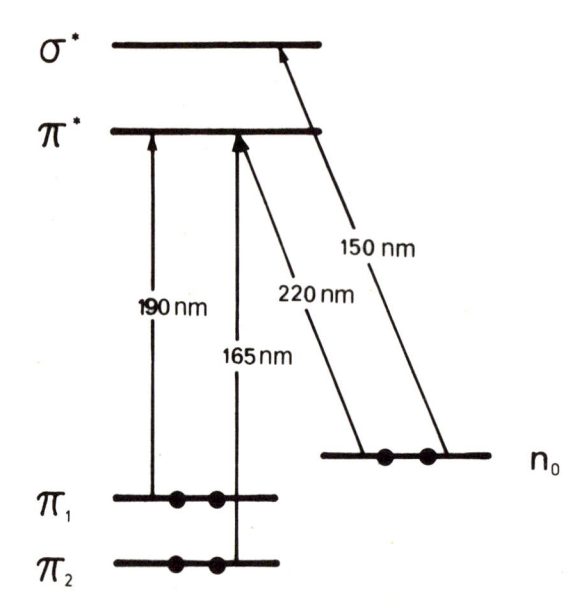

Figure 8.8 Electronic transitions and corresponding absorption wavelengths for an amide bond (based on simple amides).

electronic structure that gives rise to these transitions. Since the major residues we wish to consider (L-phenylalanine, L-tyrosine, and L-tryptophan) contain a single aromatic ring, we shall confine our consideration first to benzene. It is generally conceived that the six carbon atoms of the benzene ring are hybridized sp² and are σ-bonded to adjacent carbon atoms such that there is hexagonal arrangement with delocalized electrons traversing the π-system (Figures 8.9a and b). Compared with our peptide-conjugated

Figure 8.9 The p-electron orbitals of benzene. **(a)** σ-electron bonding; **(b)** π-electron system.

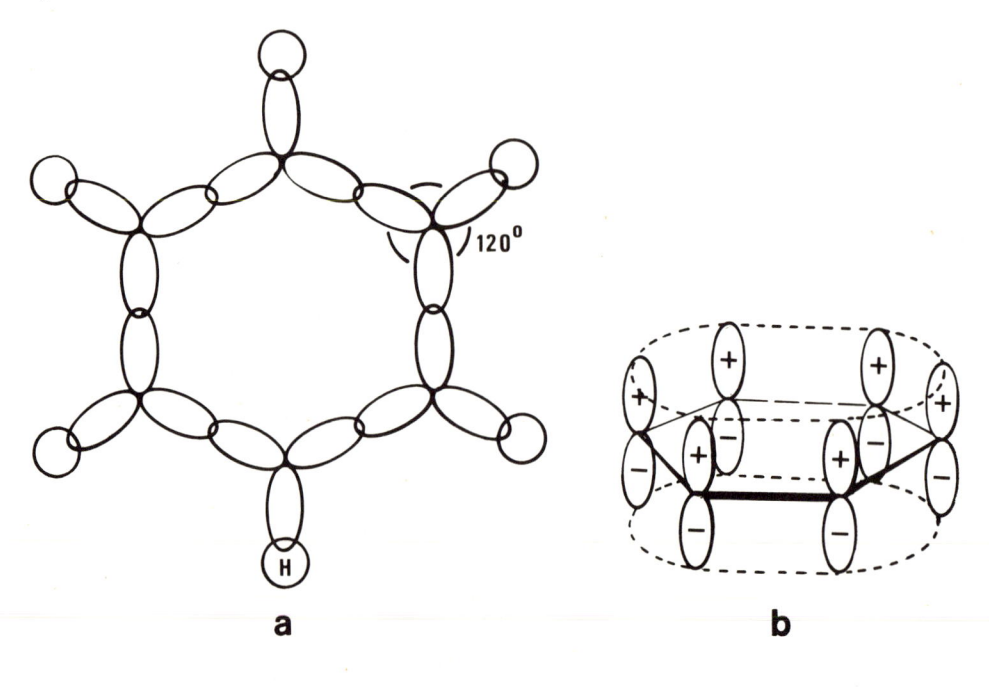

π-system, which contained (potentially) four electrons spread over three atoms, we have here six delocalized π-electrons spread over six atoms. Now we can proceed as before in setting up molecular orbitals from combinations of the atomic orbitals. In this case, since six atomic centers are involved, there will be six molecular wave functions corresponding to energy levels.

In this case, the molecular orbitals ψ_i are set up by combining atomic orbital wave function ϕ_i (in this case the six equivalent carbon p-orbitals) in the LCAO approach. There are six molecular wave functions identified with this approach, namely:

$$\psi(A) = \frac{1}{\sqrt{6}} (\phi_0 + \phi_1 + \phi_2 + \phi_3 + \phi_4 + \phi_5) \tag{8.3}$$

$$\psi(B) = \frac{1}{\sqrt{6}} (\phi_0 - \phi_1 + \phi_2 - \phi_3 + \phi_4 - \phi_5) \tag{8.4}$$

$$\psi(E_1 a) = \frac{1}{\sqrt{12}} (2\phi_0 + \phi_1 - \phi_2 - 2\phi_3 - \phi_4 + \phi_5) \tag{8.5}$$

$$\psi(E_1 b) = \frac{1}{2} (\phi_1 + \phi_2 - \phi_4 - \phi_5) \tag{8.6}$$

$$\psi(E_2 a) = \frac{1}{\sqrt{12}} (2\phi_0 - \phi_1 - \phi_2 + 2\phi_3 - \phi_4 - \phi_5) \tag{8.7}$$

$$\psi(E_2 b) = \frac{1}{2} (\phi_1 - \phi_2 + \phi_4 - \phi_5). \tag{8.8}$$

We can recognize the first two of these molecular wave functions as "bonding" and "antibonding" from a simplified overlap orbital picture as shown in Figure 8.10. The designations of A and B arise from consideration of the symmetry of the orbital distribution as follows. The benzene ring with six carbon atoms has a sixfold rotation axis C_6, meaning that a rotation of the ring by $2\pi/6$ brings the ring to an equivalent position. The

Figure 8.10 Representation of: **(a)** $\psi(A)$ bonding, and **(b)** $\psi(B)$ antibonding orbitals in benzene, showing the sixfold rotation axis C_6.

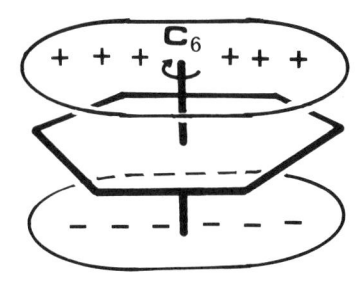

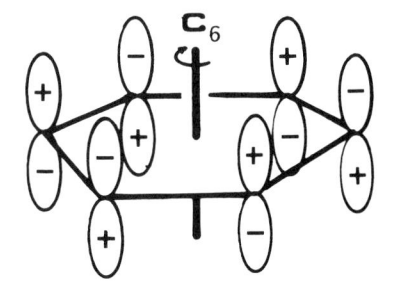

symbol A is given to indicate that the orbitals are symmetrical to such a rotational operation; that is, rotation by $2\pi/6$ does not change the orbital distribution. The same operation carried out for Figure 8.10b brings +'s to − positions and is thus antisymmetric with respect to the $2\pi/6$ rotation. B is used to designate this antisymmetric nature.

The orbital distribution may be represented schematically as in Figure 8.11, where the orbitals may be correlated with equations (8.3) through (8.8).

The symbols E, and so on, refer to the symmetry of the system, which is written in group theory nomenclature. For present purposes we note only that the $\psi(A)$ orbital has no nodal planes, $\psi(E_1a)$ and $\psi(E_1b)$ have one nodal plane, $\psi(E_2a)$ and $\psi(E_2b)$ have two nodal planes, and $\psi(B)$ has three. In a sense the number of nodal planes represents the energy level of the species, which may be understood by simple analogy with the wave motion in a rope held at two ends. The more energy that is put into shaking one end of the rope, the more waves or nodes are produced (and the shorter is the wavelength).

In the present case the energy state of $\psi(A)$ is the lowest and $\psi(B)$ the highest. This situation is shown in Figure 8.12. Also shown is the symmetry nomenclature for the various states (see Appendix B). Basically, A, B, and E are related to symmetry about the sixfold axis, the numerical subscript is related to symmetry about a twofold axis, and g and u refer to mirror symmetry through the plane of the benzene ring. An indistinguishable reflection through the plane of the ring is called g for even (German *gerade*) and a distinguishable reflection through the plane is u for uneven (German *ungerade*).

Figure 8.11 Orbital representations of equations (8.3) through (8.8). Lines are drawn where wave functions pass from + to −; that is, these are "nodal" lines.

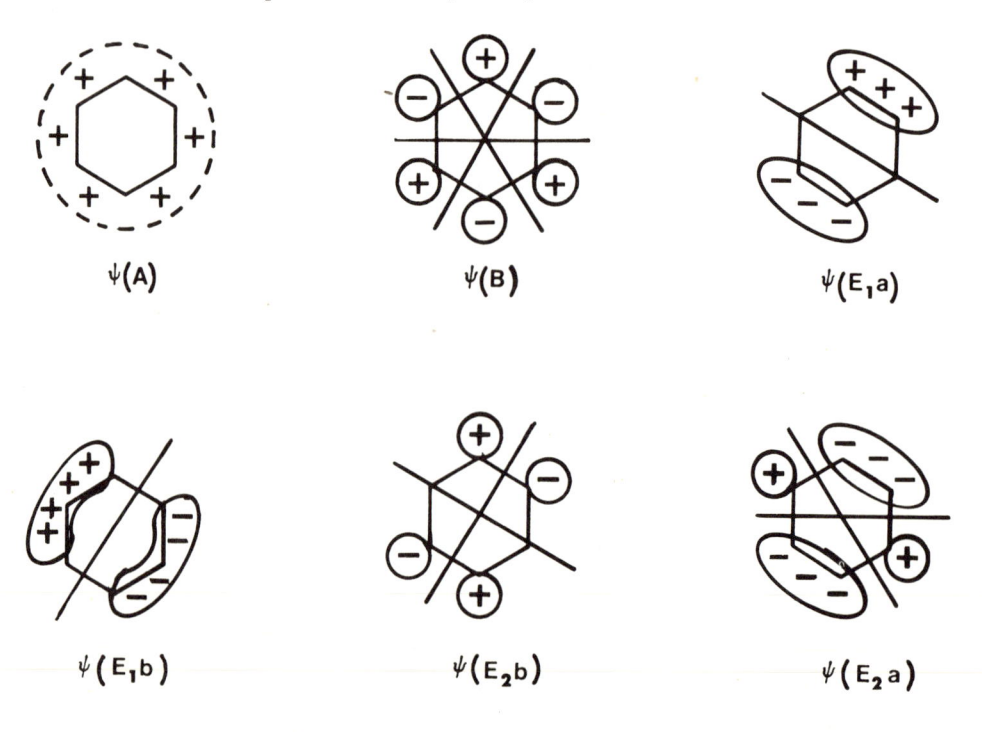

$\psi(A)$

$\psi(B)$

$\psi(E_1a)$

$\psi(E_1b)$

$\psi(E_2b)$

$\psi(E_2a)$

8. Electronic Spectroscopy

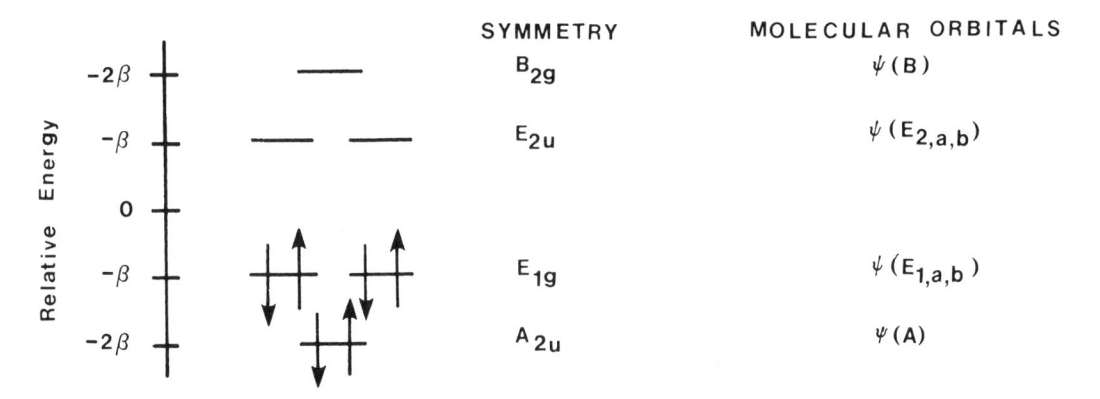

Figure 8.12 Molecular orbital energy level scheme for benzene.

Since six electrons are delocalized into the molecular orbitals (two into each orbital with spins paired), we may represent the energy levels as in Figure 8.12.

It is noticeable that two of the four energy levels are doubly degenerate; that is, they are each of equal energy. (This result is obtained from a quantum mechanical solution that is not spelled out here. The energy levels and their degeneracy are again a function of the symmetry of the system.)

If we now consider the possible electronic excitations and their symmetry, we have, for a one-electron transition, four possibilities as depicted in Figure 8.13.

We might expect then that the ultraviolet spectrum of benzene would correspond to such transitions, giving four bands. However, if our molecular

Figure 8.13 Molecular orbital energy level schemes showing four possible electronic transitions for benzene π-electrons.

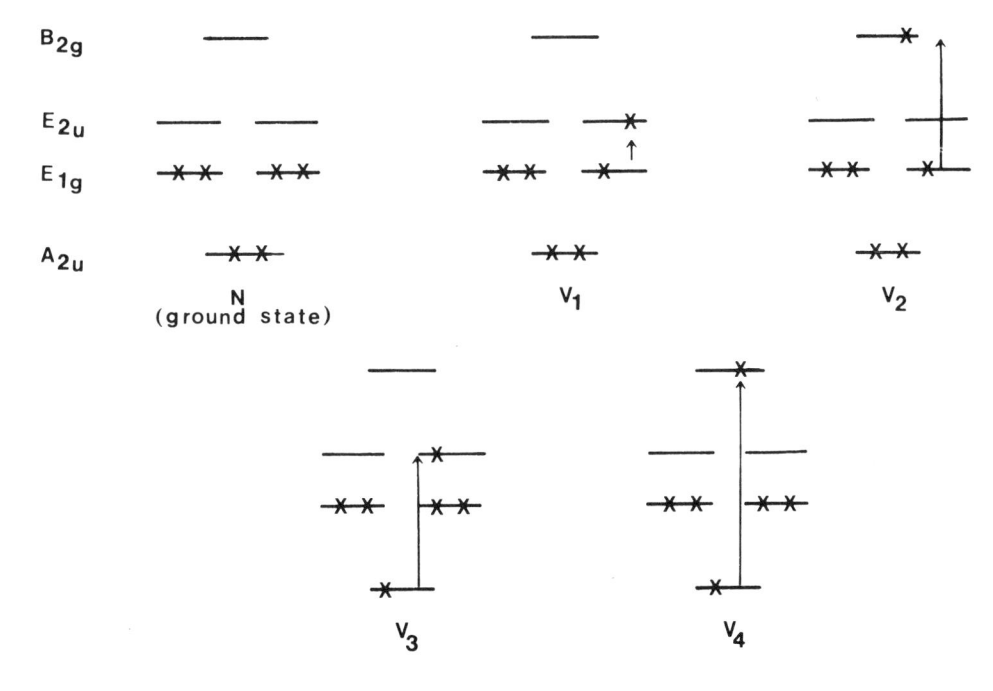

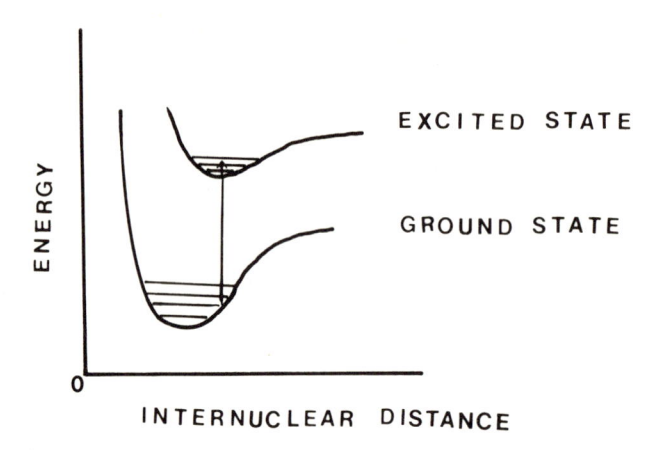

Figure 8.14 A vibronic transition given by simultaneous electronic and vibrational transitions.

orbital energy diagram (Figure 8.12) is precisely correct, two of these transitions (V_2 and V_3) are energetically equivalent and would be degenerate.

Because transitions can be represented, as in Figure 10.3, does not mean they are necessarily *allowed*. Symmetry criteria must be used and when these criteria are incorporated only one of the above transitions (N–V_1) is symmetry allowed. (There is a theorem—the Laporte rule—stating that for centrosymmetric systems only u→g or g→u transitions are symmetry allowed.)

Understanding the selection rules requires a detailed knowledge of group theory and its application to its symmetry of ground and excited states (see Appendix B). We will have to accept, however, that three of the four possible transitions are symmetry forbidden. However, in addition to the N–V_1 allowed transitions, two others can occur by what is known as *vibronic* transition. The vibronic transition involves a simultaneous change of electronic *and* vibrational quantum state, as shown in Figure 8.14. When vibronic transitions are also taken into account, the ultraviolet spectrum of benzene is expected to consist of three bands created by transitions of electrons in which the electron spin is the same in the ground and excited states (so-called singlet states). These bands are found experimentally at 255 nm (vibronic), 205 nm (vibronic), and 183 nm (allowed).

Unfortunately, there exist in the literature two different nomenclature systems: comparison of the Platt nomenclature and the molecular orbital (MO) nomenclature is given in Table 8.3 for the three major benzene bands. The superscript in the Platt nomenclature also refers to a singlet state (spin coupled).

Table 8.3 Molecular Orbital Group Theory and Perimeter Free-Electron Nomenclature (after Platt) for Benzene

Wavelength	MO nomenclature	Platt nomenclature
1830 Å	$^1E_{1u} \leftarrow {}^1A_{1g}$	$A \rightarrow {}^1B_{a,b}$
2050 Å	$^1B_{1u} \leftarrow {}^1A_{1g}$	$A \rightarrow {}^1L_a$
2550 Å	$^1B_{2u} \leftarrow {}^1A_{1g}$	$A \rightarrow {}^1L_b$

Figure 8.15 Phenylalanine (a) and tyrosine (b) showing symmetry axis.

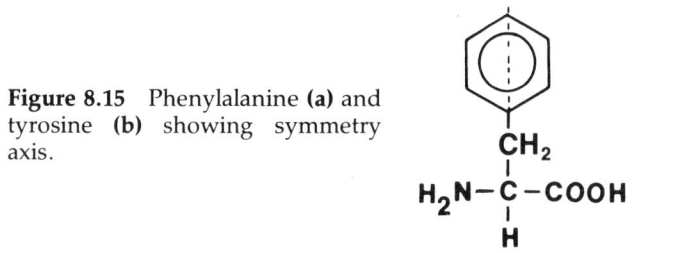

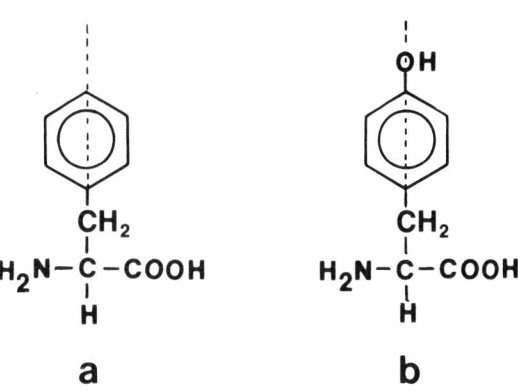

The introduction of substituents into the benzene ring, such as in phenylalanine or tyrosine, considerably reduces the symmetry (Figure 8.15). Instead of the sixfold C_6 rotation axis and twofold C_2 axes we now have only one representative twofold axis. The π-electron system of the aromatic ring is disturbed slightly but the predominant effects of substitution vest themselves through the symmetry. No longer are the $^1B_{1u}$ and $^1B_{2u}$ (^1L_a ^1L_b) transitions symmetry forbidden, so that more intense transitions are observed (although the total intensity is comparable with the vibronic transitions).

Table 8.4 presents the correlation between electronic transitions of benzene, phenol, and the aromatic amino acids phenylalanine and tyrosine. Tryptophan, which is more complex structurally, still shows several transitions that appear to be in common with phenylalanine and tyrosine.

Experimental Aspects of Absorption

Absorption of UV/visible radiation by proteins, as we have seen, involves the promotion of electrons to higher energy states. In dilute solutions the simple Beer–Lambert law holds:

$$2.303 \log_{10} \frac{I_0}{I} = \epsilon_\lambda l\, c \tag{8.10}$$

where I_0 is the intensity of the incident radiation; I, the intensity of transmitted radiation; l, the path length (cms); and c, the concentration in moles/liter. Then ϵ_λ

Table 8.4 Comparison of Absorption Bands (nm) and Assignments with Absorption Intensities (Extinction Coefficient $\varepsilon \times 10^{-3}$)

Benzene	ε	Phenylalanine	ε	Phenol	Tyrosine	ε(pH 7)	Symmetry assignment (MO)	(Platt)
183	60	188	58	193	193	47	$^1E_{1u}$	($^1B_{a,b}$)
205	8	206	9	223	223	8	$^1B_{1u}$	(^1L_a)
255	0.25	257	0.2	275	275	1.2	$^1B_{2u}$	(^1L_b)
360	very weak						$^3E_{1u}$	($^3B_{a,b}$)

The superscript [3] refers to a triplet state.

is the extinction coefficient at wavelength λ. Such an equation often does not apply at higher concentration or in systems where aggregates are formed.

Equation (8.10) may be written in the form:

$$A = \epsilon_\lambda l c \tag{8.11}$$

where A is the absorbance. Apart from identifying the extinction coefficient for various polypeptides and proteins, the procedure may generally be turned around to find concentration. If most of the absorbance arises from the polypeptides backbone, then the values

$$\varepsilon^{22°C}_{220nm} = 624 \text{ M}^{-1}\text{cm}^{-1}, \quad \varepsilon^{22°C}_{205nm} = 2.726 \times 10^3 \text{M}^{-1}\text{cm}^{-1}$$

are used. It is to be noted that at lower wavelengths the extinction coefficients become structure dependent, and without prior knowledge of structure the identification of very dilute concentrations of proteins or polypeptides becomes difficult or meaningless.

Absorption Intensity

When an electronic transition occurs there is a concomitant change in electronic distribution. A molecule or group in the ground state will have a different permanent dipole moment and polarizability than that in the excited state. Associated with the electronic transition, there is, therefore, a *transition dipole moment*. Furthermore, the transition dipole moment has direction and UV transitions show enhanced probability when the orientation of polarized light is parallel with the transition dipole.

The absorption intensity is related to the probability of an electronic transition. It will be no surprise to learn that the selection rules are again phrased in terms of quantum mechanics and group theory. Although the details for many of the situations that we would be likely to encounter with biologic molecules are complex, we shall proceed mainly on a qualitative or experimental level. A simple example of what is involved is provided by an examination of transitions in the aldehyde carbonyl group of (say) formaldehyde (Figure 8.16). As we have seen previously in the peptide amide group, the carbon atom forms a σ-bond between a hybridized sp^2 orbital and a p_x component of oxygen. One electron is delocalized into the p_y-orbitals, and nonbonding n-electrons lie in the p_z-orbital of oxygen. The symmetry of the electron-distribution changes as electrons are promoted into higher energy levels. The selection rules are based

Figure 8.16 Representation of carbonyl orbitals involved in electric dipole transitions.

8. Electronic Spectroscopy

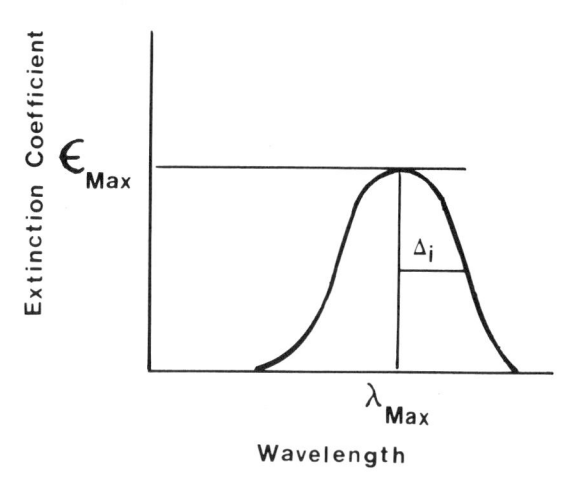

Figure 8.17 Typical UV absorption curve arising from an electronic transition λ_{max}.

upon a comparison of the symmetry of the ground and excited states (see Appendix B) and of the n–π*, n–π*, and σ–π* transitions, the n–π* transition is found to be forbidden.

Another, perhaps simpler, way of looking at this process is to realize that the probability of an electronic transition is given by the integral of a wave function product.

$$\int \psi^*_{ground} \, \mathbf{r} \, \psi_{excited} \, d\tau, \tag{8.12}$$

where **r** is a vector operator and $d\tau$ is a volume element. Without worrying how to apply the vector operator, the product must be zero for the n→π* transition since the wave functions (orbitals) lie in orthogonal (perpendicular) planes. Thus the n–π* transition is said to be an electric dipole–forbidden transition, as it is for the amide bond. Thus n–π* transitions should be absent (but are in fact weak) in experimental spectra.

We have thus been introduced to the concepts that electronic transitions are associated with changes in electronic distribution that may or may not be symmetry allowed. The probability of the transition is related to the experimental intensity and the change of electronic distribution can be represented by a transition dipole that has direction. Futhermore, the gain and loss in energy of electrons (the latter of which is considered in more detail later) produces an oscillating strength both experimentally and theoretically. Normally UV absorption curves are plotted with extinction coefficient ϵ (or some function thereof) against wavelength, as in Figure 8.17. However, the curve shape can be more neatly analyzed if the extinction coefficient is plotted as a function of energy or frequency. In this case, line shapes are generally treated as Gaussian in shape; that is, they conform to the equation

$$A_{\bar{\nu}} = \epsilon_{\bar{\nu}} lc = a \int \exp -\frac{\Delta \bar{\nu}^2}{\Delta_i^2} \, d\bar{\nu}, \tag{8.13}$$

where a is the amplitude, $A_{\bar{\nu}}$ and $\epsilon_{\bar{\nu}}$ are the absorbance and molar extinction coefficient at a wavelength corresponding to $\bar{\nu}$. Δ_ν is the wavenumber (cm^{-1})

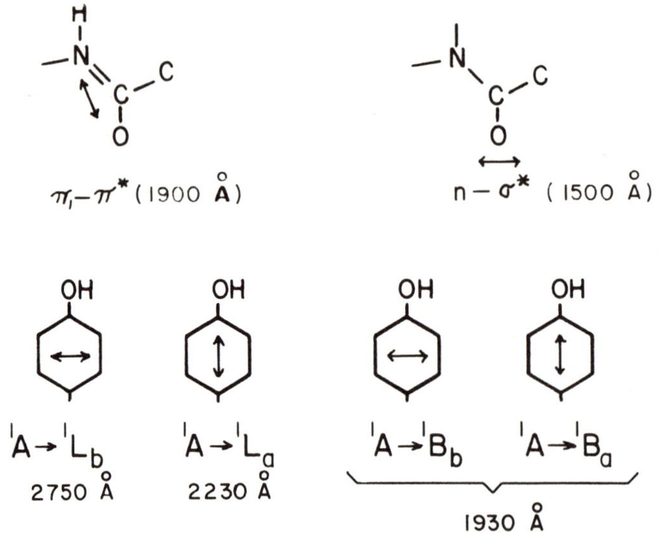

Figure 8.18 Some electronic transitions and their polarization for peptide and aromatic side-chain groups.

displacement from $\bar{\nu}_{max}$ and Δ_i is the band half-width when $A_{\bar{\nu}}$ is e^{-1} of its maximum. $\bar{\nu}$ is the frequency in wavenumbers (cm^{-1}).

The oscillator strength f is related to the extinction coefficient by:

$$f = 4.32 \times 10^{-9} \int \epsilon_{\bar{\nu}} \, d\bar{\nu}, \tag{8.14}$$

where the whole band area is integrated over all $\bar{\nu}$. The oscillator strength may be calculated from quantum mechanics (and symmetry considerations) and, as might be expected from the preceding discussion, the (forbidden) n–π* transition is ~ 1/60th of the strength of the π–π* (π_1–π*) transition of the peptide bond. The experimentally calculated oscillator strengths are a function of the electronic environment of the peptide group, which in turn is related to conformation: this phenomenon can be turned to good account in conformational analysis.

Transition Dipoles

One other feature of photon absorbance that should be considered before proceeding further is that of polarization. Plane-polarized light is absorbed preferentially in the direction of the transition dipole. Thus the direction of transitions of some isolated residues is shown in Figure 8.18. We shall see subsequently that in the milieu of the polypeptide or protein chain, interaction (coupling) with other residues can induce other polarization symmetry.

8.4 Ultraviolet Absorption and Conformation

So far we have considered (albeit at an elementary level) electronic transitions occurring in isolated residues or peptide bonds. If the absorption spectrum and intensity were dependent only upon the behavior of electrons in these groups,

8. Electronic Spectroscopy

we would be denied several powerful tools for examining the conformation of polypeptides and proteins.

The most often quoted example of the relation between the UV spectrum of a simple polypeptide and its conformation is the poly (amino acid), poly-L-lysine (Figure 8.19). Polylysine changes its conformation as a function of pH and temperature due to the side-chain ionization and degree of solvation. Thus with the same sample, the effect of these two parameters may be investigated. At neutral pH, polylysine is described as a coil, the conformation being dictated in part (Chapter 2) by the mutual repulsion of the positively charged ϵ-amino groups. The absorption maximum, associated with the peptide bond π–π^* transition is at 196 nm. At high pH and normal (25°C) temperature, polylysine becomes α-helical and the UV absorbance drops. This effect is known as *hypo*chromism. In addition to the decrease in absorbance the maximum absorbance band (n–π^*) splits and has two components at 209 and 191 nm.

At higher temperature, on standing at high pH, polylysine starts to aggregate and takes on an antiparallel β-sheet conformation. The maximum absorbance increases compared with the coil from (*hyper*chromism) and moves to 196 nm.

It is clear, therefore, that for these three different conformations (α-helix, β-sheet and coil), two effects predominate in the UV absorption spectrum, namely, a change in intensity of the band and its band position. These effects originate in the electronic interactions between the adjacent peptide groups and are said to originate from dipole or exciton coupling. It can be seen that the shorter the peptide repeat, the lower is the extinction coefficient, roughly in direct proportion between h(Å) and ϵ; that is, α:coil:β ~ 1.5:3.1: 3.5. In addition, the symmetry of the α-helix causes band splitting. This effect may be likened to the application of a magnetic field to proton or electron spins

Figure 8.19 Relative absorbance curves of poly-L-lysine in the three standard conformations: α = α-helix, β = β-sheet, c = coil.

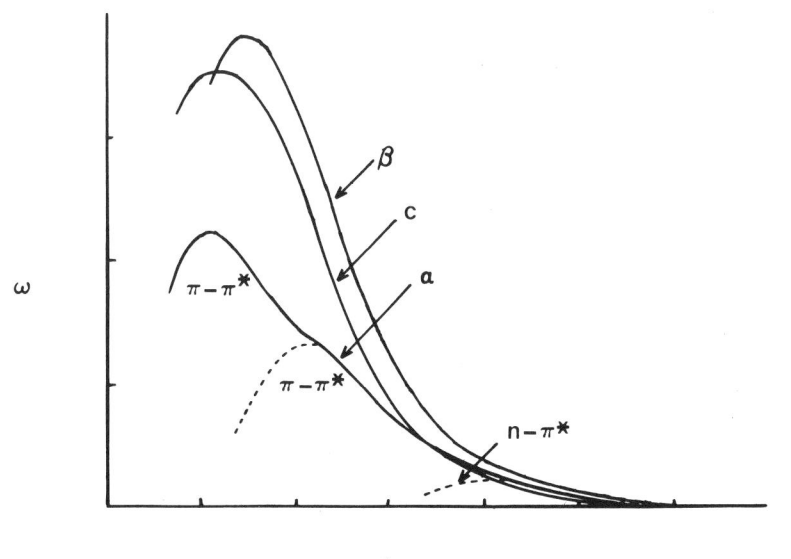

Table 8.5 Ultraviolet Absorption Bands for Polypeptide Conformations

Conformation	ε_{max}(nm)	Polarization	Assignment
α-helix	221	⊥r to helix axis	n–π*
	209	‖ to helix	π_1–π*
	191	⊥r to helix	π_1–π*
	155	‖ to helix	π_2–π*
β-sheet	219		n–π* (mainly)
(antiparallel)	196	Predicted 60% ‖	π_1–π*
	171	Predicted ⊥r	π_2–π*
Coil	216		n–π*
	201		π_1–π*
Polyproline II	226		n–π*
	215	‖ to helix	π_1–π*
	201	⊥r to helix	π_1–π*
	172		π_2–π*

which can cause spin splitting (see NMR, Chapter 11). Measurement of the dichroism of the α-helical π–π* bands shows an electronic polarization parallel to the helical axis (209 nm) and perpendicular to the axis (191 nm).

The very weak bands at 216 nm (α-helix), 216 nm (coil), and 219 nm (β-sheet) are associated with the forbidden n–π* transition. Since the nonbonding oxygen p_z-orbital electrons are involved and the oxygen atom is involved with solvent interaction, the exact position of the n–π* band is often influenced by the solvent or surrounding milieu. At lower wavelengths (not accessible by most conventional equipment) other peptide bands are observed. This information is collected in Table 8.5.

The table shows that there are distinct shifts for the different conformations and in principle the UV spectrum of a protein might be predicted to be the sum of all the conformational bands plus the side chains. Unfortunately the resolution in a UV spectrum of a protein is not adequate to predict conformation. All the information is "smeared" together. Recognizing this problem, biophysicists and biochemists have tended to concentrate in recent years on techniques capable of resolving the various transitions. Of these methods, those using polarized radiation have generally been considered the most applicable.

8.5 Linear Dichroism

Whereas the techniques of the next chapter—circular dichroism (CD) and optical rotatory dispersion (ORD)—are very useful spectroscopic methods applicable mainly to solutions of nonoriented molecules, UV linear dichroism is a method that depends for its success on the relative orientation of peptide dipoles in relation to a plane-polarized source. The method is likely to be useful, therefore, mainly for linear molecules; for example, fibrous proteins in the solid state.

When the linear dichroism of an oriented α-helical material is examined, as in Figure 8.20, the correlation between transition dipole orientation and absorbance is clearly indicated. In particular, the splitting and opposite polarization of the two π_1–π* bands at 191 nm and 209 nm are of note.

The use of this technique has been limited since essentially identical and more substantial information may be obtained from CD spectroscopy.

Figure 8.20 Comparison of the absorbance and linear dichroism spectra of α-helical poly-γ-ethyl-L-glutamate. ○---○---○ Conventional absorbance (optical density); ●—●—● the normalized transmittance $(T_\perp - T_\parallel)/(T_\perp + T_\parallel) = T_p$.

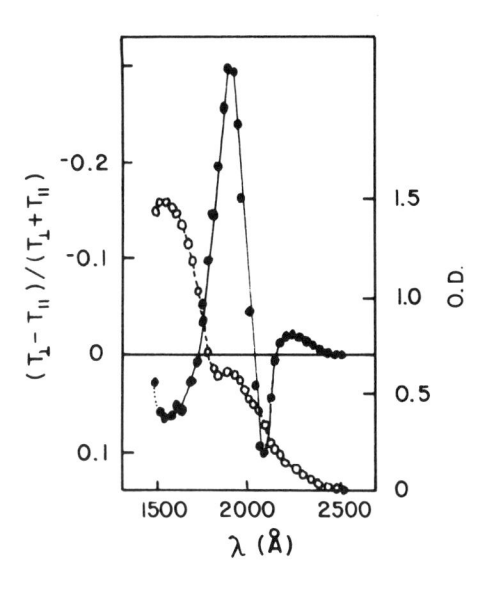

8.6 Topographic Mapping by Ultraviolet Spectroscopy

Apart from the ability of UV spectroscopy to provide information concerning concentration and to a limited extent, conformation of proteins, much recent work has been directed towards topographic mapping of aromatic residues and following the process of denaturation. Since the interaction of light with these residues is dependent both upon the electronic structure of the residue and the electronic environment, perturbation of the latter, either thermally or with probe solvents, produces a change in the spectrum. It has been concluded from an examination of proteins containing known ratios of exposed to buried tyrosine residues, that the contribution at 280 nm from buried tyrosine residues is approximately 12 times as great as that from exposed ones. This greater absorbance of buried residues is ascribed to restriction of the motional freedom of the residue, causing the phenol chromophore to experience a large asymmetric field. Such information may be turned to good account in working with proteins of unknown structure. Addition of solvents such as methanol and dimethyl sulfoxide to the native protein causes spectral changes in the tyrosyl and tryptophan residues on the surface, but not in the interior of the protein. Perturbants of various molecular sizes may be used to penetrate and probe the protein, thus providing information on surface topology.

Further Reading

Cantor, C.R. and P.R. Schimmel. Absorption Spectroscopy. In *Biophysical Chemistry*, Part II, Chapter 7. W.H. Freeman and Co., San Francisco, 1980.

Charney, E. *The Molecular Basis of Optical Activity: Optical Rotatory Dispersion and Circular Dichroism.* Wiley, New York, 1979.

Hofrichter, J. and W.A. Eaton. Linear Dichroism of Biological Chromophores *Annual Reviews of Biophysics and Bioengineering* 5:511–560, 1976.

King, G.W. *Spectroscopy and Molecular Structure.* Holt, Rinehart and Winston, New York, 1964.

9

Optical Rotatory Dispersion (ORD) and Circular Dichroism (CD) Spectroscopy

We have seen, in the previous chapter, that various types of light polarization are feasible and that absorbance or dispersion (refractive index) of a sample may be measured. We have also explored some of the electronic processes that underlie interaction with light, particularly plane-polarized light. Our attention now turns to the interaction of light with optically active compounds [amino acids → proteins] and how the optical activity may be used to good account in studying structure.

9.1 Optical Rotation

If monochromatic unpolarized light is passed through a slab of material, the velocity of light in the medium is reduced compared with that in vacuo, the ratio of the velocities being the refractive index. Since the emergent light has the same frequency but has been retarded, it is readily apparent that the emergent light has been phase shifted by an amount directly related to the refractive index and amount of material through which it has passed.

If the input beam is plane polarized and the medium is optically active, then the plane of the beam is rotated and the emergent wave is phase shifted (Figure 9.1).

Now the angle of rotation α will depend on the thickness of the sample ℓ, its specific rotation $[\alpha]$ and the concentration c (g/ml).

$$\text{Thus } \alpha = [\alpha]\,\ell c. \tag{9.1}$$

If rotation is towards the right (clockwise), looking towards the source, the material is said to be dextrorotatory, this rotation being given a positive

9. Optical Rotatory

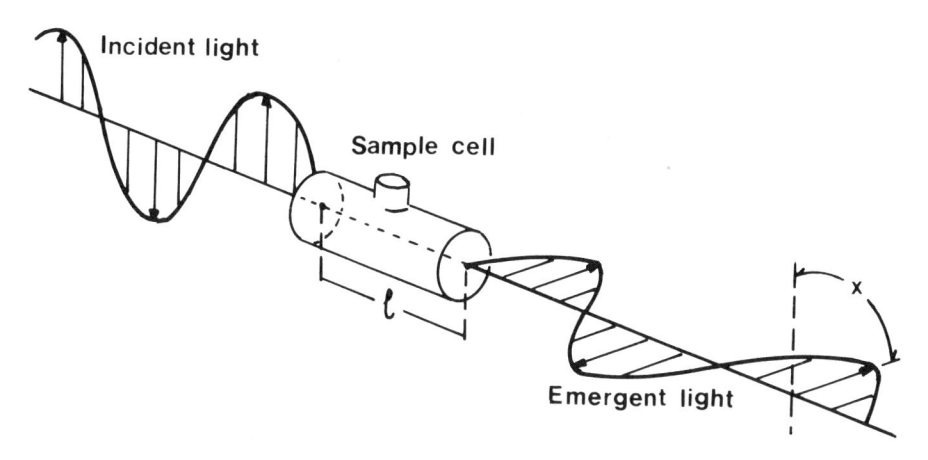

Figure 9.1 Optical rotation demonstrated by the rotation of a beam of polarized light on passing through an optically active sample.

sign by convention; and for anticlockwise rotation, the material is said to be levorotatory, the rotation being given a negative sign.

A more useful relation for macromolecular solutions expresses the molar rotation of solute (e.g., protein) [m'] in terms of [α], that is,

$$[m'] = [\alpha] \frac{3}{n^2 + 2} \cdot \frac{M}{100} \tag{9.2}$$

where M is the solute molecular weight and n is the refractive index of the medium. The refractive index term represents a small correction (Lorentz correction) for the changed polarizability of the system.

Since the molecular weight may not be known for a given protein it is common for M to be taken as the weight-average of amino acids (generally ~ 112) and then [m'] becomes the mean residue rotation.

The source of the optical rotation effect lies in the ability of light to interact with dipoles in the medium that are anisotropically distributed (e.g., helices, optically active centers, etc.). Such an effect is called dispersion and relies on cooperative behavior between active species in the medium. The effect is dependent upon the wavelength of incident light: the resonance interaction (dispersive effect) increases until very close to the electronic absorption frequency, where a complex interplay between absorbance and dispersion occurs. At the position of the absorbance band there is no net rotation. On passing through the wavelength of absorbance, the dispersive rotation appears again with opposite sign. The phase angle of the transmitted light has, in fact, passed through 180°.

The relation between dispersive rotation and absorbance is shown in Figure 9.2.

If the molar rotation becomes positive at high wavelength, then this is said to be a positive cotton effect; if the rotation goes from positive to negative at higher wavelengths, the effect is known as a negative cotton effect.

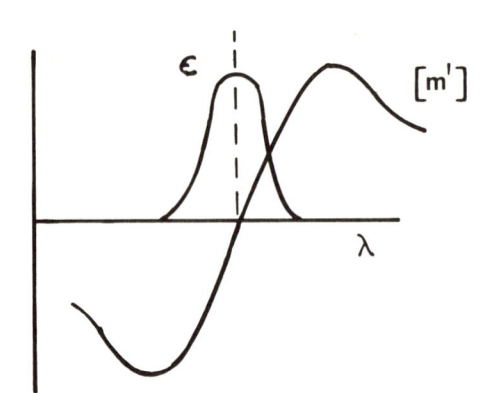

Figure 9.2 Behavior of molar rotation [m'] in the region of an electronic absorbance band.

9.2 Dispersive Effects and ORD

We can now correlate the three dispersive effects as follows:

1. If plane-polarized light is passed through an unoriented, non–optically active medium, there will be a change in light velocity in the medium as a function of the refractive index. At the wavelength of electronic absorbance, the refractive index will be one (1).
2. If the plane-polarized light is passed through an oriented non–optically active medium, there will be a difference of refractive index depending upon the relative orientation of the sample and the beam. The difference in refractive index at different orientations will disappear at the absorbance wavelength.
3. If plane-polarized light is passed through an optically active medium or solution, dispersive rotation will occur which is zero at the wavelength of absorbance. The phase angle of the emitted light passs through 180°.

Thus one major difference between the absorption of light and dispersion of light is that the absorption is a maximum at the absorption wavelength and the dispersion is zero.

Because absorptive effects are localized to regions of electronic transitions, whereas dispersive effects depend on molecular polarizability, techniques dependent upon the former effect generally have much narrower "band-widths." The examination of molar rotation as a function of wavelength is known as optical rotatory dispersion (ORD).

It is of some relevance to enquire what properties of a molecule or group lead to optically active properties. From a qualitative knowledge of organic chemistry we are aware that asymmetric carbon atoms possessing four different substituents have "optical" isomers that are not superimposeable. If, however, there are only three different substituents—say, C, N, and 2 Hs on one carbon atom, it loses its properties of optical isomerism or anisotropy, but gains a plane of symmetry. Such concepts clearly apply to the amino acids which are, with the exception of glycine, anisotropic and do not possess a plane or center of symmetry. In general, we can say that optically *active* molecules are asymmetric, and optically *inactive* molecules possess at least some symmetry elements (see Appendix B).

9. Optical Rotatory

Two molecules that are similar but possess quite different optical properties are coronene and hexahelicene, as shown in Figure 9.3. Coronene is clearly a highly symmetrical molecule, having many of the same symmetry properties as benzene. Hexahelicene, on the other hand, is highly asymmetric and is not superimposeable on its mirror image. However, we note that there is no specific carbon atom that we can point to that is asymmetric. In other words, optical activity can arise either at the atomic level or the molecular level.

Another more relevant example is that of glycine and polyglycine. As we have seen, glycine is optically inactive, yet when polymerized it can form a helix (3_1). Since the mirror image of a left-hand L-helix is a right-hand D-helix, it is obvious that whether the residues are optically active or not, the left-hand and right-hand helices are not superimposeable and are thus optically active. Thus glycine is not optically active, but the polyglycine helix is. However, since the left- and right-hand helices are energetically equivalent, solutions contain equal numbers of both and thus there is no net optical activity from a solution of polyglycine helices.

In molecules that are optically inactive, such as coronene, the electric transition moment μ, which is in the plane of the molecule (for the allowed π-electron transitions) is perpendicular to the magnetic transition moment m (see Figure 9.3). For optically inactive materials the vector product of $\mu \cdot m$ is zero. For optically active molecules (e.g., hexahelicene), the transition moments are not mutually perpendicular and thus the vector product is finite.

When considering the interaction of optically active molecules with plane-polarized light, it is convenient to think in terms of this plane-polarized light as resulting from two circularly polarized but opposite components, as explained in Chapter 8. If one component is absorbed preferentially, then the resultant plane of polarization shifts through the rotation angle α.

A schematic of the interaction of circularly polarized light with a helix is given in Figure 9.4. If the light is of the correct wavelength, it is able to

Figure 9.3 Two similar molecules that are (a) optically inactive (coronene), and (b) optically active (hexahelicene).

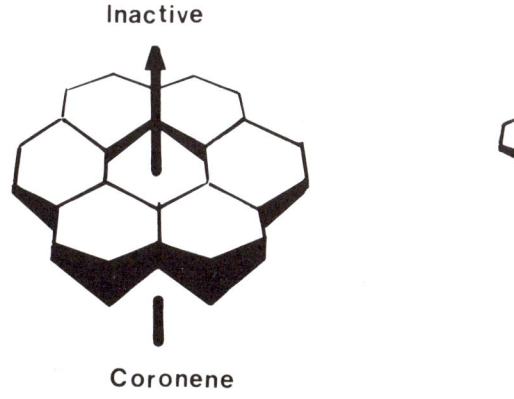

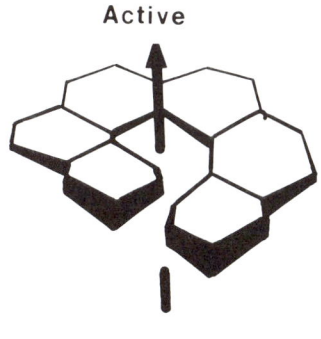

Inactive

Active

Coronene

Hexahelicene

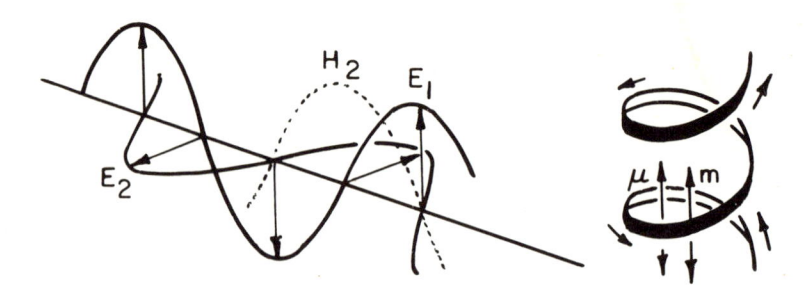

Figure 9.4 Interaction between a circulary polarized light wave and a helical molecule (after van Holde). The electric vector of the light is denoted by E and the magnetic field (90° out of phase) by H.

induce an electron displacement in the helix *in the direction of the helix;* that is, a transition dipole is created that has a vector component along the helix axis. Additionally, by the laws of electromagnetism, charge displacement in the helix creates a magnetic field in the core of the helix. We may say, therefore, that the light wave also interacts to induce a transient (transition) magnetic dipole moment. Since the helix has a sense to it (right-hand in Figure 9.4), the interaction with left- or right-hand circularly polarized is not equal. If we are considering plane-polarized light as composed simultaneously of left- and right-hand polarized components we can see that one will be absorbed preferentially on passing through an optically active medium and the direction of the polarization vector will change as shown in Figure 9.1.

The preceding discussion indicates (however crudely) that both plane-polarized and circularly polarized light will interact with optically active molecules in a discriminating manner; that is, there is a selective interaction of the sample with one component of the polarized light. This phenomenon is the basis of both ORD and CD spectroscopy.

Written in mathematical terms, the rotatory strength of a molecule is:

$$R = \text{Im}(\mu \cdot m), \tag{9.3}$$

where Im is the imaginary part of the vector product of the two transition moments μ and m.

The rotatory strength is analogous to the oscillator strength (equation 8.12) in non–optically active systems. There are thus two criteria for a system to produce an ORD or CD spectrum:

1. The incoming light must interact with the system to produce an electronic transition moment. This means that for absorption to occur there must be one or more groups capable of undergoing electronic transitions in the wavelength region of the exciting radiation. Dispersion (refraction) can occur at wavelengths further displaced from the absorption maximum, but still depends on the existence of electronic transitions.
2. There must be an optically active center in the neighborhood of groups that undergo electronic excitation.

Since it is the coupling between electron and magnetic transition moments (equation 9.3) that causes selective absorption and rotation of light,

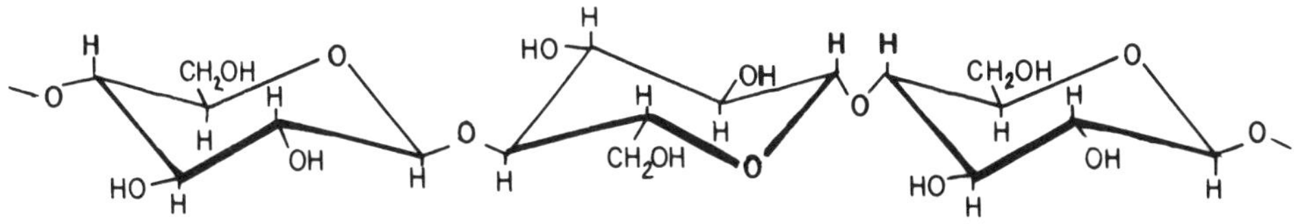

9. Optical Rotatory

Figure 9.5 The backbone of amylose [poly-α(1,4)-D-glucopyranose].

an optically active center must be present, otherwise μ and m are perpendicular and the product is zero.

In practice this means that such molecules as the plant polysaccharide amylose (Figure 9.5), though being very asymmetric and possessing both atomic and molecular asymmetry, does not have groups capable of undergoing electronic transitions in the near-UV region of the spectrum, and thus there is no absorptive (circular dichroism) spectrum. There are, however, electronic transitions (probably $\sigma-\sigma^*$) in the vacuum UV region and the coupling effect necessary to produce dispersive effects extends into the near-UV region. Hence optical rotation and ORD are observable.

On the other hand, the synthetic macromolecule, polystyrene (shown in Figure 9.6), although possessing groups that undergo electronic transitions in the near-UV region ($\pi-\pi^*$), is not optically active; therefore, $\mu \cdot m = 0$ (electric and magnetic transition moments are perpendicular) and there is no CD or ORD spectrum.

In an extension of the previous concepts to a system of optically active molecules, Rosenberg developed a quantum theory of optical rotation that presents the mean residue rotation of equation (9.2) in the form:

$$[m']_\lambda = \frac{96\pi N_0}{hc} \Sigma \text{Im}(\mu_i \cdot m_i) \cdot \frac{\lambda_i^2}{\lambda^2 - \lambda_i^2}. \tag{9.4}$$

The subscript i refers to the i-th absorption band (for systems possessing more than one); λ_i is then the absorption wavelength of the i-th band and λ the operating wavelength. The vector product is thus summed for all the electric and magnetic transition moments.

It turns out that equation (9.4) is identical in form to an equation that applies empirically to protein solutions, namely:

$$[m']_\lambda = a_0\lambda_0^2/(\lambda^2 - \lambda_0^2). \tag{9.5}$$

Equation (9.5) is known as the Drude equation. For random polypeptide chains in aqueous solution at $\sim 25°C$, $a_0 \sim -600$ and $\lambda_0 \sim 220$ nm.

It is evident, however, that neither equation (9.4) nor (9.5) has the form expected for the experimental curve shape of Figure 9.2, since when $\lambda = \lambda_i = \lambda_0$ the mean residue rotation should be infinite instead of zero as

Figure 9.6 The synthetic polymer polystyrene.

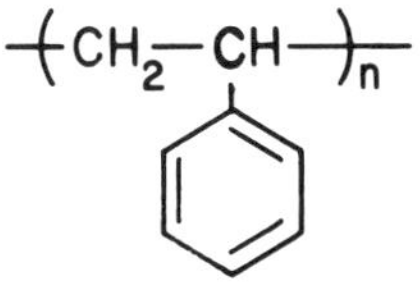

found experimentally. A comparison of the Drude equation with experimental data indicates that it applies only in a region well separated from the absorption wavelength (Figure 9.7).

No satisfactory equation has yet been derived that is applicable in the absorption region, but a more general equation covering a larger spectral region has been derived by Moffitt and Yang, their equation being:

$$[m'] = \frac{a_0 \lambda_0^2}{\lambda^2 - \lambda_0^2} + \frac{b_0 \lambda_0^4}{(\lambda^2 - \lambda_0^2)^2} \tag{9.6}$$

Since in practice, b_0 appears to be sensitive to conformation, the Moffitt–Yang equation has been used fairly extensively for estimation of protein structure in solution. Although certain aspects of the theory are thought to be unsatisfactory, it is found that b_0 has the values listed in Table 9.1 for different conformations.

If it is assumed that for structures other than α-helical, $b_0 = 0$, then the α-helical content of proteins and polypeptides may be estimated by:

$$f_{helix} = |b_0(\text{measured})/630|. \tag{9.7}$$

The rotator strength R may be extracted from the Drude equation or the first term of the Moffitt–Yang equation from the relation:

$$a_0 = \frac{96\pi N}{hc} \cdot R, \tag{9.8}$$

where h is Planck's constant, c the velocity of light, and N Avogadro's number.

In more recent usage, ORD has given way to the more general application of CD spectroscopy, except for certain special situations (e.g., where CD bands do not occur in the accessible region of the spectrum). ORD suffers from the disadvantage that various types of electronic transitions, including those from

Figure 9.7 Comparison of experimental and theoretical ORD curves.

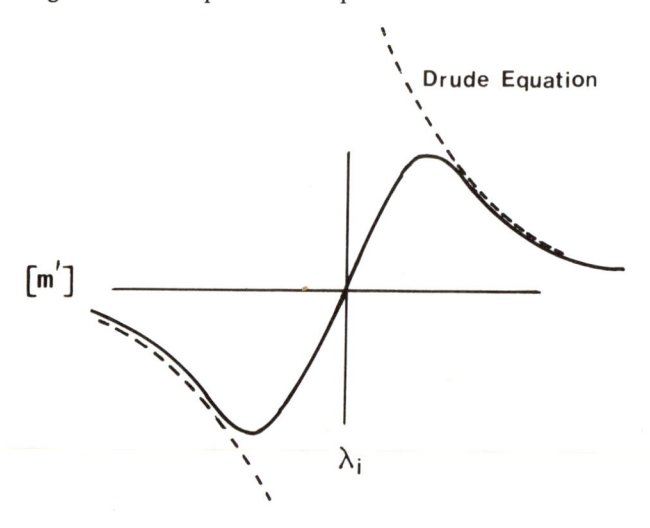

Table 9.1 Conformational Variation of the Moffitt–Yang Parameter b_0

b_0	Conformation	a_0
−630	Right-hand α-helix	+650
+630	Left-hand α-helix	
~0	β-sheet	~600
0	Random, irregular	−600
0	Polyglycine and polyproline helices	

peptide group and side chains, are "lumped together" in the broad bands. CD spectroscopy is, in principle, capable of resolving each set of electronic transitions and thus providing more detailed information.

9.3 Circular Dichroism Spectroscopy

As we have already seen, the technique of circular dichroism spectropolarimetry (CD spectroscopy) involves the application of circularly polarized light to the specimen in question (which is generally in solution). Circularly polarized light may be produced as shown in Chapter 8, by combining two mutually perpendicular plane-polarized waves that are phase shifted by one-quarter of a wave. Depending on which component is retarded, left- or right-hand circularly polarized light may be produced. In commercial instruments, the sample is exposed to alternating left- and right-hand circularly polarized light and the measured spectrum is a function of the difference. In Chapter 8 we saw that the result of passing circularly polarized light through a medium that selectively absorbs left- or right-hand polarized radiation is an emergent elliptically polarized beam. Circular dichroism is generally expressed in terms of ellipticity, which is given by:

$$[\theta] = 3300 (\epsilon_L - \epsilon_R) = \frac{\theta_{obs}(cms) \times f \times M}{10 \times c \times \ell} \tag{9.9}$$

where ϵ_L and ϵ_R are the extinction coefficients for left- and right-hand polarized light. The right-hand equation applies to experimental measurement. The curve "printout" produces bands θ_{obs} cm in height or depth; f is the instrument sensitivity factor; M, the mean residue molecular weight; c the concentration in g/ml, and ℓ is the path length in cm. The result is then expressed in degrees cm²/decimole.

Since both ORD and CD spectra depend on the existence of electronic transitions and coupling with optically active centers, it is not surprising that they are mathematically related. The ORD spectrum may be calculated from the CD spectrum by means of the Kronig–Kramers transform:

$$[m']_\lambda = \frac{2}{\pi} \int_0^\infty [\theta_\kappa(\lambda')] \frac{\lambda'}{\lambda^2 - (\lambda')^2} d\lambda'. \tag{9.10}$$

θ_κ is the ellipticity of the k-th transition, λ is the chosen wavelength, λ' is the wavelength pertaining to the integration.

The calculation is not entirely trivial; since θ and λ' vary, it is necessary to find some simplifying constraints. To convert to a common variable, use is made of the fact that the ellipticity is of Gaussian form:

$$[\theta_\kappa] = [\theta_\kappa^\circ] \exp - (\lambda - \lambda_\kappa^\circ)^2/(\Delta_\kappa^\circ)^2,$$

where $[\theta_\kappa^\circ]$ is the maximum ellipticity of the k-th band and λ_κ° is the corresponding wavelength. Δ° is the band half-width as previously described. If $\lambda_\kappa^\circ >> \Delta_\kappa^\circ$ (this is generally true with $\lambda_\kappa^\circ \sim 200$ and $\Delta_\kappa^\circ \sim 20$), then the integral of equation (9.10) may be rewritten in the form:

$$[m'] = \frac{2[\theta_\kappa^\circ]}{\sqrt{\pi}} \left[\exp \cdot (\lambda - \lambda_\kappa^\circ)^2/(\Delta_\kappa^\circ)^2 \cdot \int_0^{\lambda - \lambda_\kappa^\circ/\Delta_\kappa^\circ} \exp x^2 dx - \Delta_\kappa^\circ/2(\lambda + \lambda_\kappa^\circ) \right]$$

The procedure is then to set up a table of $[m']$ for a given λ, λ_κ°, θ_κ°, and Δ_κ. The integral is easily solved and hence an ORD curve calculated from a CD band (or set of bands) may be established. The reverse procedure may also be applied to obtain a CD curve from ORD data.

9.4 Utilization of CD Spectra for Determining Structure

An examination of CD data often begins with examination of the solution spectra of poly-L-lysine. As we have seen previously, this poly(amino acid) is particularly useful for reference purposes since it has three different conformations in aqueous solution depending upon pH and temperature. Figure 9.8

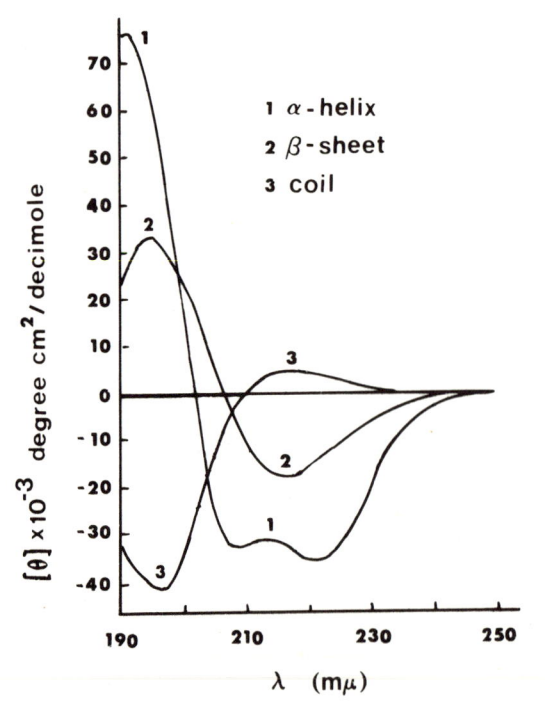

Figure 9.8 CD spectra of poly-L-lysine in the α-helical (curve 1), β-sheet (curve 2), and coil conformations (curve 3).

shows CD spectra of these conformations that have routinely been associated with the α-helix (high pH, low temperature), coil (neutral pH and normal temperature), and β-sheet (high pH, high temperature).

It can be seen that these spectra are distinctly different. The α-helix has three bands centered at 191 nm, 208 nm, and 222 nm. These wavelengths correspond to those previously noted for electronic transitions originating in the peptide bond. Particularly interesting is the magnitude of the 222-nm (n–π*) band, which is symmetry forbidden for linear dichroism but is allowed for the interaction with circularly polarized light. (This can be seen intuitively if it is remembered that the transfer of an electron from a nonbonding oxygen p-orbital to the π-system requires a circular rotating motion.) The two other CD bands correspond to the π–π* bands which are split due to the helix symmetry, and are oppositely polarized. The β-sheet shows only two bands: the first, being centered at 195 nm, originates from the π–π* peptide transition, and the 217-nm band originating again in the normally forbidden n–π* transition. The coil form also has two bands: the negative band at 197 nm corresponding to π–π*, and a small positive band, probably n–π* at 217 nm. The n–π* transitions are not normally seen in the absorbance spectra of polylysine β-sheet and coil.

Now it would be particularly convenient if the CD spectra of polylysine provided an absolute standard for estimation of structure in various polypeptides and proteins. Since each band is concentration dependent, it might be expected that a polypeptide having, say, 50% α-helix and 50% β-sheet structure would have a CD spectrum equivalent to adding one-half of the polylysine α-helix ellipticity to one-half of the β-sheet ellipticity as a function of wavelength.

In general, we would like to make the assumption that the ellipticity originating from various conformations in a molecule is additive, that is:

$$[\theta]_\lambda = {}^1[\theta°]_\lambda f_1 + {}^2[\theta°]_\lambda f_2 + {}^3[\theta°]_\lambda f_3 \ldots + {}^n[\theta°]_\lambda f_n, \qquad (9.11)$$

where ${}^i[\theta°]_\lambda$ refers to the ellipticity of 100% of conformation i at wavelength λ, and f_i refers to the fraction of conformation i in the sample. We note furthermore that:

$$f_1 + f_2 + f_3 \ldots f_n = 1. \qquad (9.12)$$

If a polypeptide exists as a combination of three conformations (e.g., α-helix, β-sheet, and coil), then it seems reasonable to use the polylysine data for the standard values of ${}^\alpha[\theta°]_\lambda$, ${}^\beta[\theta°]_\lambda$, and ${}^c[\theta°]_\lambda$. Indeed such an approach has been applied by Greenfield and Fasman. The standard reference ellipticities, based on the assumption that polylysine is 100% α-helical, 100% β-sheet, or 100% coil under the appropriate conditions, are given in Table 9.2.

Given Table 9.2 and equation (9.11), it should be possible to construct a predicted CD curve for any combination of the three conformations. Thus Figure 9.9 shows the computed curves for 0–100% α-helix and β-sheet. We note from these curves that there is a point (at approximately 200 nm) at which all the curves intersect. This is called an isodichroic point and is characteristic of a system with two variable conformational components.

Table 9.2 The Circular Dichoism Spectra of Poly-L-Lysine in the Three Reference Conformations

Wavelength (nm)	$^{\alpha}[\theta°]$ α-Helix	$^{\beta}[\theta°]$ β-Sheet	$^{c}[\theta°]$ Coil
190	74,800	22,400	−32,200
191	76,900 ± 8,400	25,300	−34,700
192.5	73,300	30,000	−37,500
195	64,300	31,900 ± 5,000	−41,000
197	44,300	30,000	−41,900 ± 4,000
200	14,300	24,300	−36,400
202	0	19,300	−25,600
205	−25,000	5,700	−14,500
208	−32,600 ± 4,000	−4,700	−3,400
210	−32,400	−10,800	−1,400
211	−32,100	−12,100	0
214	−31,000	−16,400	3,500
215	−31,400	−17,900	4,100
217	−33,100	−18,400 ± 1,800	4,600 ± 500
220	−35,300	−15,700	4,400
222	−35,700 ± 2,800	−13,800	3,900
225	−32,400	−11,400	2,700
230	−21,900	−6,400	800
234	−11,400	−3,600	0
238	−4,300	−1,400	−140
240	−3,300	700	−150
250	0	0	0

[θ] in deg cm²/dmole

It is to be noted that an isodichroic point occurs independent of the total number of conformations present, if only two are varied.

Now if we have a polypeptide CD spectrum, we are faced not with constructing a curve for assumed α-helix, β-sheet, and coil, but with the reverse procedure, which presents a few additional problems.

Greenfield and Fasman have noted, however, that the contributions of β-sheet and coil to the CD spectrum of polylysine at just below 208 nm are both small and equal (see Table 9.2). They therefore suggested that the α-helix content may be calculated from the relationship:

$$[\theta]_{208} = f_\alpha \, ^\alpha[\theta°]_{208} + f_\beta \, ^\beta[\theta°]_{208} + f_c \, ^c[\theta°]_{208}$$

where $^\beta[\theta°]_{208} = \, ^c[\theta°]_{208} = -4,000$, $f_\beta + f_c = 1 - f_\alpha$ and $^\alpha[\theta°]_{208} = -33,000$, so that $[\theta]_{208} = f_\alpha[-33,000] + (1-f_\alpha)[-4,000]$;

i.e., $f_\alpha = \dfrac{[\theta]_{208} + 4,000}{-29,000}$. (9.13)

Thus if the value of the ellipticity of a polypeptide is measured at 208 nm and inserted into equation (9.13), the fraction of α-helix is evaluated by a single measurement. This is known as the one-parameter equation.

In order to estimate the β-sheet and coil content, one now reverts to equations (9.11) and (9.12). Since the f_α and $[\theta°]$ values are known, there are two equations with two unknowns; hence f_α and f_c may be calculated.

[θ] × 10⁻³ Degree cm²/ Decimole — plot vs λ in mμ

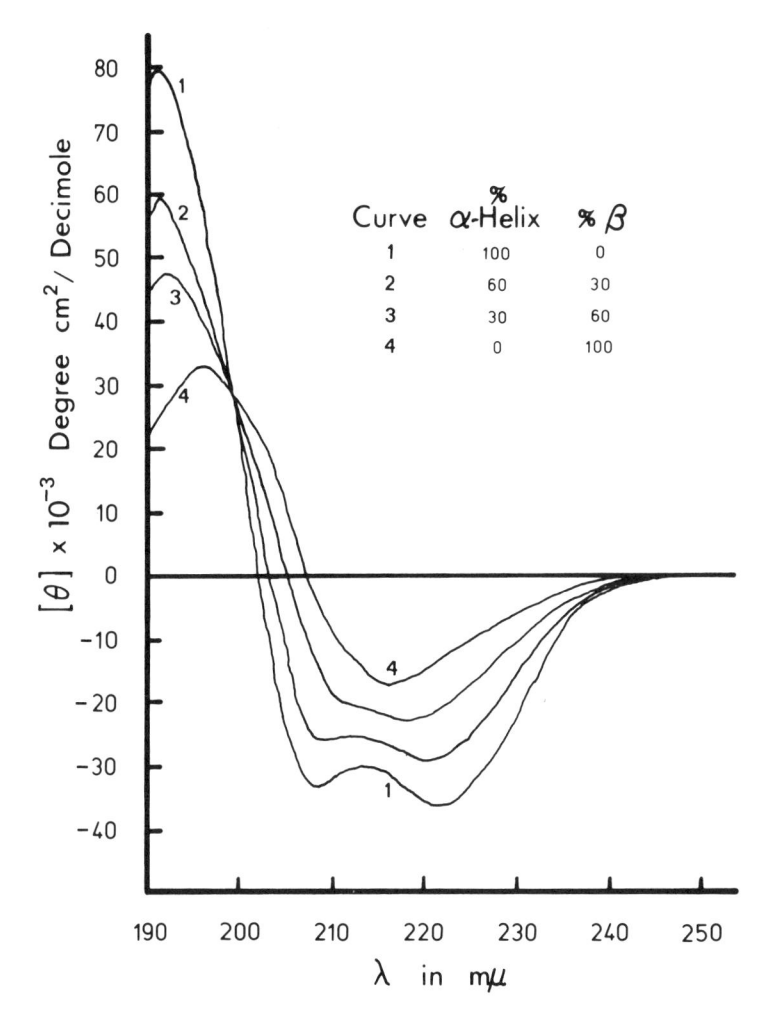

Curve	% α-Helix	% β
1	100	0
2	60	30
3	30	60
4	0	100

Figure 9.9 Calculated CD of poly-L-lysine containing 0% coil and varying percentages of α-helix and β-sheet structure as indicated.

9.5 Circular Dichroism and the Structure of Proteins

It is naturally tempting to extend the previous approach to evaluate protein structure. A rather naive approach would be to assume that the proteins contain only the conformations exhibited by polylysine and that they possess no other chromophores that contribute significantly to the CD spectrum, then proceed as before. This method is surprisingly successful for a number of proteins provided that the β and coil contributions are estimated from the CD data in the 208- to 222-nm region.

Table 9.3 shows a comparison of the conformation predicted for a few globular proteins, along with their known conformation as derived from x-ray diffraction data.

Although the correlation appears to be very good at first sight, there are certain disturbing discrepancies. For example, the CD curves of several proteins are not at all accurately represented by the above approach below 208 nm.

Table 9.3 Comparison of Protein Conformation by CD Spectroscopy and X-ray Crystallography

Protein	% α-helix X-ray*	% α-helix CD	% β-sheet X-ray	% β-sheet CD	% other X-ray	% other CD
Myoglobin	65–72	67	0	0–13	23–32	20–30
Lysozyme	28–42	29	10	11	48–62	60
Ribonuclease	6–18	12	36	38	46–58	50
Carboxypeptidase A	23–30	15	18	25	52–59	60

*The variability in x-ray percentages depends on the counting procedure and reference source.

Various suggestions have been made concerning the origin of this discrepancy, including the fact that various conformations in proteins are "buried" in the interior and do not interact with solvent as does polylysine. However, it is now fairly clear that certain assumptions inherent in the use of the polylysine model are not entirely appropriate. Some of these difficulties are:

1. The fluctuating coil conformation of polylysine is not representative of the fixed irregular structure of globular proteins.
2. The β-sheet conformation of polylysine in solution is thought to arise from chain folding and aggregation to form antiparallel arrays. The β-conformation in globular proteins is often of parallel chain nature, which apparently does not generate an identical spectrum. In any case, aggregation tends to increase apparent ellipticities and "standard" values for polylysine may be in error.
3. Even the α-helical spectrum of polylysine may not be entirely typical. Side-chain dipoles may influence the ratio of 208/222-nm bands. (Aggregation also affects this ratio.)
4. Globular proteins contain other, non–polylysinelike conformations, particularly the β-bend, that may contribute to CD spectra.
5. Aromatic residues may contribute significantly to the protein spectrum.

Again a word of caution is in order. Spectroscopic data do not *prove* the absolute existence of a conformation but only support it by comparison. It is conceivable that two different conformations will produce the same spectrum.

In order to circumvent the first of these problems, various approaches have been applied which extract $[\theta°]$ values from protein spectra, rather than from polylysine.

Figure 9.10 shows CD curves of three "standard" conformations constructed from the known conformation of three proteins (lysozyme, myoglobin, and ribonuclease) and their CD spectra. Curve I refers to irregular structure, that is, conformation other than α-helix or β-sheet. Comparison of this figure with Figure 9.8 shows qualitative similarities. The main difference, as expected, occurs with the I (irregular) and C (coil) curves. The I curve of Figure 9.10 includes β-bends, side-chain effects, and the "true" irregular spectrum. It appears that, at least for these three proteins, the use of $[\theta]_{208}$ for estimation of α-helix content is fairly sound. However, there has been a growing trend to use two (λ = 210,225 nm) or three (λ = 210,219,225 nm) wavelengths for simultaneous solution of equations (9.11) and (9.12) using either polylysine or protein $[\theta°]$s for conformational analysis.

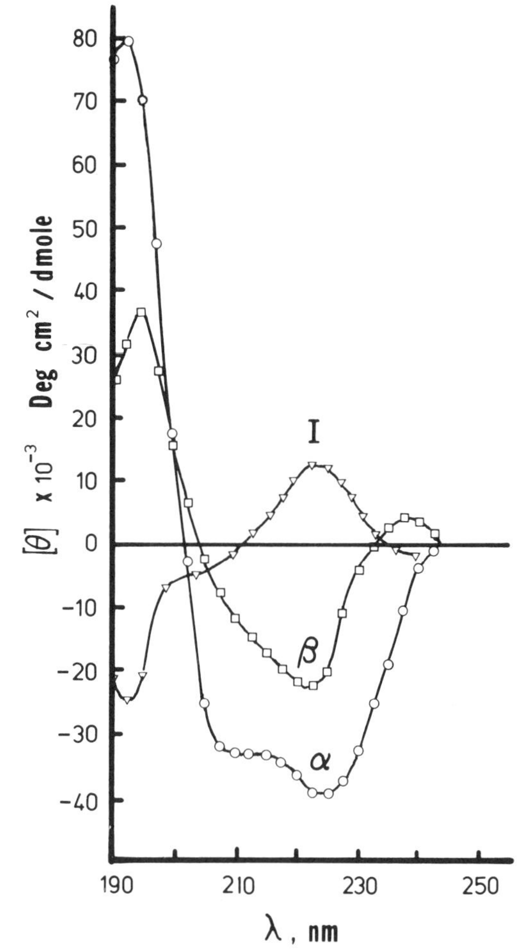

9. Optical Rotatory

Figure 9.10 Computed CD spectra of α-helix, β-sheet, and irregular conformations obtained from the known (x-ray) structure of lysozyme, myoglobin, and ribonuclease and experimental CD spectra curves of the proteins.

9.6 Circular Dichroism of the Polyproline Helix

One other major conformation exists that we have not yet examined; that is the polyproline II structure. Poly-L-proline forms a left-hand 3_1 helix for which the CD spectrum is shown in Figure 9.11. There are two components of the $\pi-\pi^*$ transition at 206 nm and 201 nm (not resolved) and a positive band at 226 nm which is probably $n-\pi^*$ in origin. Collagen, by comparison, consists of a three-chain coiled coil, each chain of which has a distorted polyproline II helix (see Chapter 13).

The thermal denaturation of proteins is readily followed by CD measurements; sometimes, by following various bands, details of the unfolding mechanism are revealed. In the collagen spectrum it is readily possible to follow the "melting" or denaturation process and a plot of ellipticity versus temperature for the collagen–gelatin transition yields a "melting temperature" T_m, which is defined as the temperature at which 50% of the initial structure is denatured. Figure 9.12 shows the thermal denaturation of collagen.

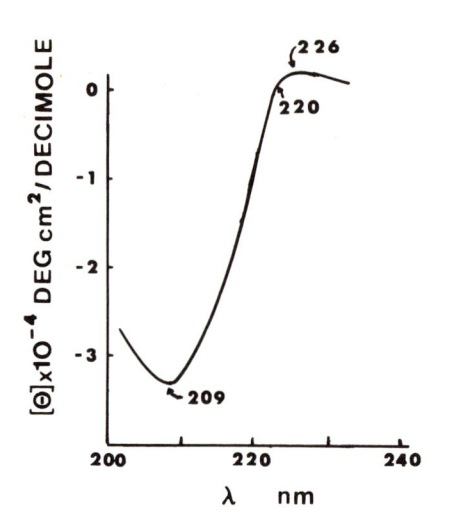

Figure 9.11 The CD spectrum of polyproline II in aqueous solution.

9.7 Effects of Side Chains

The effect of chromophoric side chains on CD spectra is particularly difficult to resolve. The reasons are several: the poly(amino acids) polytyrosine, polyphenylalanine, etc., are insoluble in water at neutral pH and even if they were not, some difficulty might be experienced in separating spectral contributions. As pointed out previously, there are two UV-absorption bands in the far-UV region at 192 and 225 nm for tyrosine that could conceivably interfere

Figure 9.12 Thermal denaturation profile of rat tail tendon collagen as measured by the change in ellipticity at 221 nm.

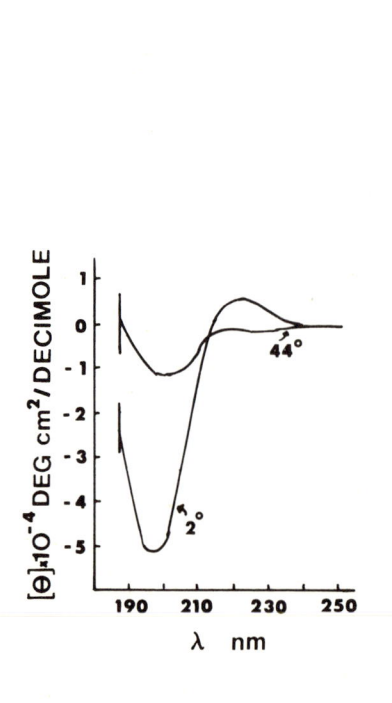

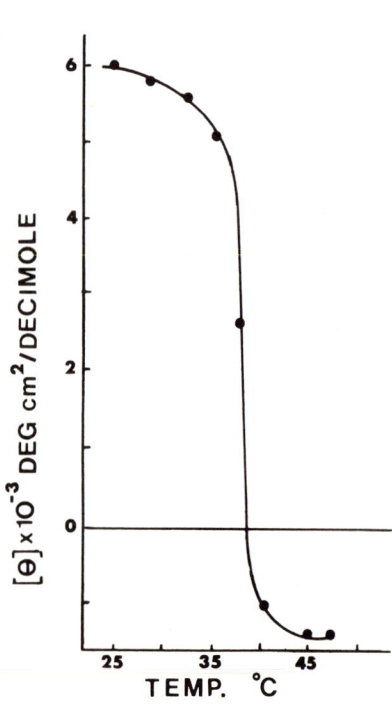

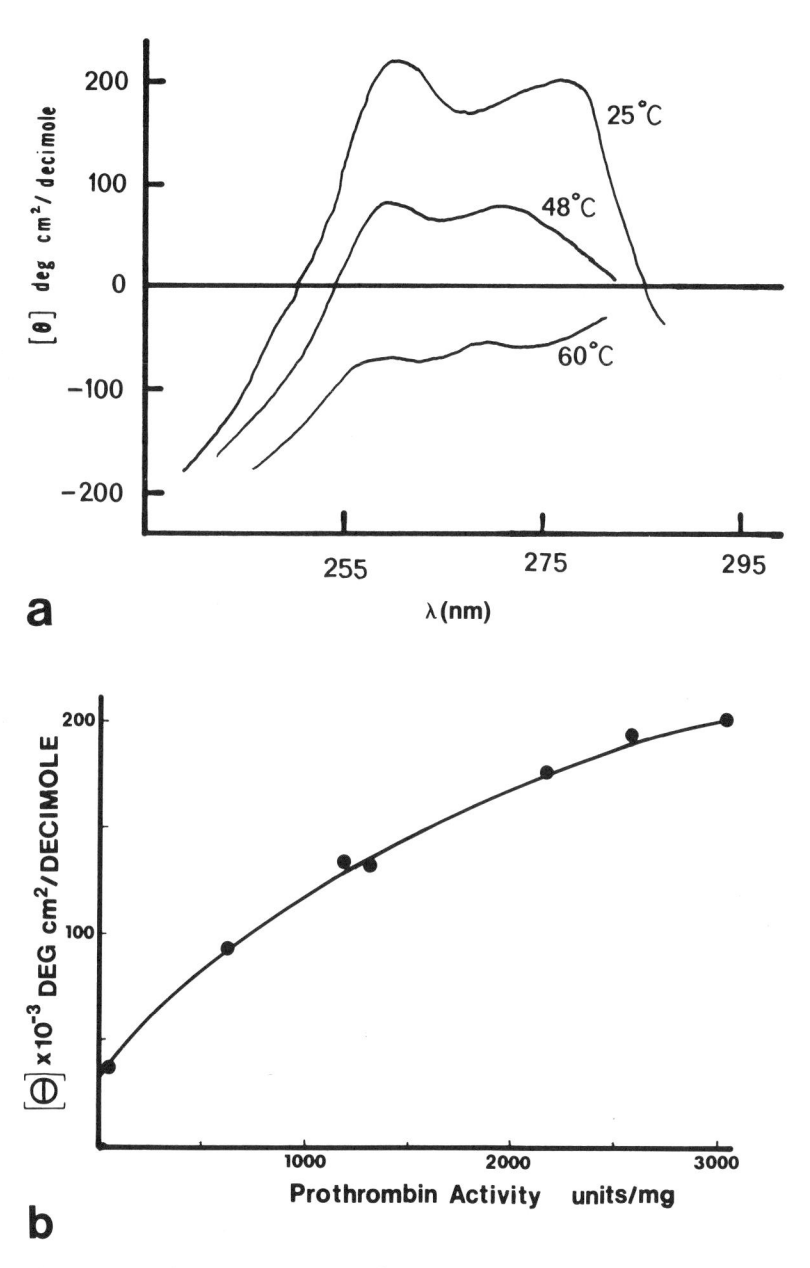

Figure 9.13 (a) CD spectrum of prothrombin in the high-wavelength region; bands originate primarily from tyrosyl side chains. (b) Correlation of 280-nm band ellipticity with prothrombin activity.

with the protein CD spectra. Two other bands at 277 and 282 nm appear in the UV region. Generally, the contribution of aromatic residues to the protein spectrum is negligible unless one or more of these residues is electronically perturbed by interaction with other aromatic residues or perhaps a peptide chromophore. Under such circumstances the aromatic contribution to the CD spectrum can become predominant and simple methods of estimating conformation can fail. This class of proteins is called CD abnormal; it may often be recognized by the unusual appearance of the spectrum, for example, positive ellipticity bands in the far-UV region. The CD-normal proteins generally have small aromatic bands in the near-UV region (240–350 nm) that can give important information on protein structure. An example of the change of aromatic CD band structure with temperature for prothrombin and its relation to prothrombin activity are shown in Figures 9.13a and b.

It can be seen that as the protein is heated, the band shape and position in the aromatic region of the CD spectrum changes as the protein denatures. This indicates a change of conformation (details unknown), particularly relating to the juxtaposition of aromatic residues. This change of structure appears to be the origin of the loss of prothrombin (blood-clotting) activity at higher temperatures.

In general, changes of conformation of proteins, globular and fibrous, may be readily followed by CD spectroscopy and often following the behavior of separate bands may give insight into the mechanism. However, details of structural change may be obscured by a number of factors including those mentioned above, and in particular, light scattering from aggregates. Nevertheless, CD spectropolarimetry remains one of the most convenient of methods for rapid assessment of the solution structure and behavior of proteins.

Further Reading

Adler, A.J., N.H. Greenfield, and G.D. Fasman. Circular Dichroism and Optical Rotatory Dispersion of Proteins and Polypeptides. In *Methods in Enzymology*, Enzyme Structure, Part D, C.H.W. Hirs, and S.N. Timasheff, eds., Volume XXVII, Chapter 27, 1973.

Charney, E. *The Molecular Basis of Optical Activity: Optical Rotatory Dispersion and Circular Dichroism*. Wiley, New York, 1979.

Imahori, K. and N.A. Nicola. Optical Rotatory Dispersion and the Main Chain Conformation of Proteins. In *Physical Principles and Techniques of Protein Chemistry*, S.J. Leach, ed., Part C, Chapter 22, 1973.

Sears, D.W. and S. Beychok. Circular Dichroism. In *Physical Principles and Techniques of Protein Chemistry*, S. Leach, ed., Part C, Chapter 23, 1973.

10

Electronic Emission: Fluorescence Spectroscopy

So far we have been concerned only with electronic transition processes that are involved in light absorption. We now turn to a consideration of what happens after electrons have been promoted to a higher energy state.

At room temperature and under the conditions of most light-absorption processes, the vast majority of electrons are in their lowest energy state, that is, the ground state. In this ground state the electrons are not only in their lowest electronic energy state, but are also in their lowest nuclear vibrational state and have their electron spins paired wherever possible. Transitions to a higher energy state can involve electronic states alone or simultaneous changes in electronic, vibrational (rotational), and spin pairing. We have already examined the so-called vibronic states (Figure 8.14) briefly. When the electron finds itself in the excited state it can either lose energy to the surroundings in the form of heat; or, if a pathway is available, can undergo various internal energy conversion processes, some of which lead to the emission of light.

Figure 10.1, which is known as a Jablonski diagram, summarizes many of these processes.

10.1 Internal Conversion Processes and Intersystem Crossing

On the right-hand side of Figure 10.1 are shown three electronic energy levels S_0, S_1, and S_2, which correspond to the electron states shown in the boxes. One of the two spin-paired electrons in the ground (S_0) state is promoted to the first excited state (S_1) or the second excited state (S_2) with its spin remaining the same. These states are described, as we have seen before, as singlet states. This term arises from atomic spectroscopy and is associated with the absence of effect of an applied magnetic field. The symbol "S" thus denotes singlet.

The light-absorption process can promote electrons from the ground elec-

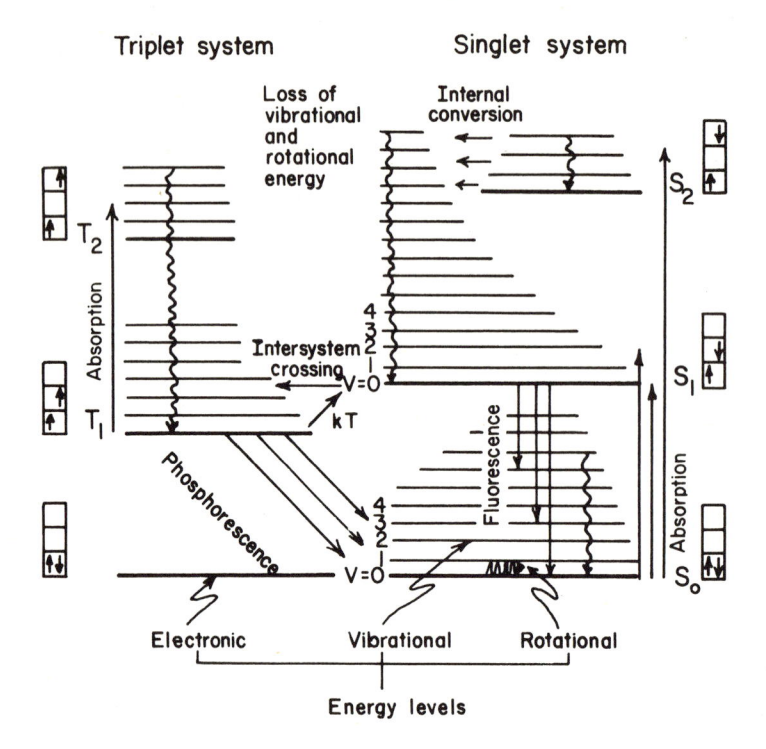

Figure 10.1 Jablonski diagram showing possible electron transitions in absorption and emission processes.

tronic and vibrational states to the first excited electronic state: (a) with no change of vibrational quantum state; (b) with a change in vibrational state; or (c) to a higher electronic and vibrational state, as denoted by the three long vertical arrows in the diagram.

Loss of energy can occur by a decay of vibrational energy through internal conversion in which (say) the second excited electronic level becomes a high vibrational energy level of the first excited state, and so on. Alternatively, if there is no vibrational state of equivalent energy, electrons may emit light (fluoresce) to reach a lower electronic state, of excited or ground vibrational states. The term internal conversion is applied, then, to changes between electronic and vibrational energy levels in the same (singlet) spin state.

We note that internal conversion can always occur in an excited vibrational state by decay to a lower vibrational state. Thus fluorescence only occurs by decay from the lowest v = 0 vibrational state of an excited electronic state.

It is also possible to attain *intersystem* crossing to triplet states. On the left-hand side of Figure 10.1 are shown boxes indicating the electron configuration corresponding to the first and second excited triplet (T) states. In this case the promoted electron changes its spin direction and is unpaired in the excited states. As we have seen in Chapter 8, promotion from a ground-paired state to a triplet state is a forbidden transition; however, absorption of energy can in some cases cause excitation of triplet states. The

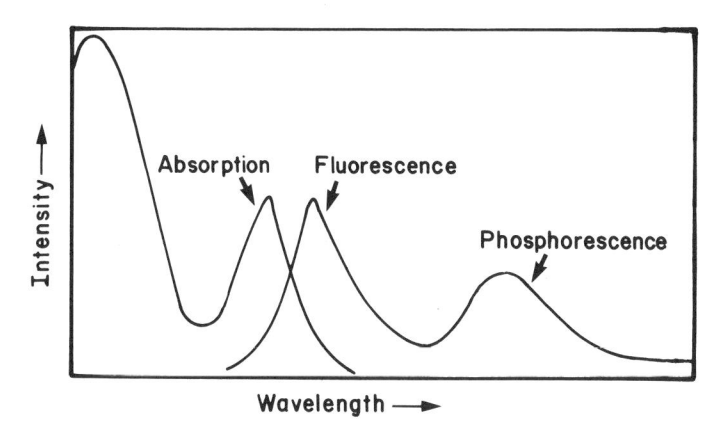

Figure 10.2 Relative spectral location of absorption, fluorescence, and phosphorescence bands.

triplet states are always of lower energy than their singlet counterparts and, as with the singlet states, there are various vibrational levels contained within the electronic levels. Intersystem crossing can occur when there are triplet vibronic states of equivalent energy to singlet vibronic states. In such intersystem crossing, the final loss of energy in returning to the ground state is phosphorescence, a light-emission process with a relatively long lifetime. Fluorescence, then, involves an $S_1 \to S_0$ transition, whereas phosphorescence involves a $T_1 \to S_0$ transition.

The frequency of the emitted light depends upon the vibrational state of S_0 to which the electrons return, and upon the energy of T_1 or S_1. In the situations of interest to us, absorption occurs at low wavelength (high frequency and energy), and radiation is emitted at higher wavelength, with phosphorescence lying at a higher wavelength than fluorescence. A typical spectrum is shown in Figure 10.2. Experimentally, a second difference is discernible between fluorescence and phosphorescence. The persistence time for emission after light absorption is determined by the stability of the S_1 and T_1 states. Fluorescence generally persists for approximately 10^{-8} seconds; phosphorescence may be emitted for several seconds.

Whereas fluorescence is a phenomenon that has found wide use in studying protein structure, the same is not true of phosphorescence. One reason is that loss of excitation energy to other surrounding molecules in their ground state (i.e., nonradiative energy loss), is particularly prevalent for the triplet state because of its relatively long life. Phosphorescence is generally observed only in cold rigid media.

10.2 Lifetime of States and Excitation Coefficients

In Chapter 8 we saw that electronic transitions can have high or low probability of occurring. Transitions with a high probability occur frequently and the lifetime of the excited state is small. Conversely, improbable transitions are associated with states having long lifetimes and low absorbance. Three features may be identified as underlying the transition probability.

1. *Symmetry-related transitions.* We have already noted (Chapter 8) that transitions may be symmetry forbidden. Transitions between states of the same symmetry are unlikely because they do not give rise to the linear translation of electron density necessary for the interaction with electromagnetic radiation.
2. *Quantum selection rules.* Singlet-triplet absorptive transitions have been noted as "forbidden." In general, it may be stated that transitions involving the simultaneous change of two quantum numbers (in this case electronic and spin) are improbable.
3. *Orbital overlap requirements.* Promotion of an electron from one orbital to another, with which it overlaps only slightly, is improbable. One example is the n–π^* transition, which we have already classified as "forbidden." Absorption is low, lifetimes are long, and fluorescence is negligible.

Evidently, the probability of fluorescent light emission is related to the probability of absorbance. The lifetime of the fluorescent process is defined as the time in which the population of the excited state falls to 1/e of its initial value. If fluorescence is the only means of depopulating the excited state, the lifetime τ_R can be represented:

$$\frac{1}{\tau_R} = 10^4 \, \epsilon_{max} \tag{10.1}$$

where ϵ_{max} is the extinction coefficient at the absorption maximum.

In practice other decay processes interfere, diminishing the "quantum yield" q_f and the lifetime, so that only the lowest singlet excited state provides measurable fluorescence. The quantum efficiency is defined as:

$$q_f = \frac{\text{Quanta emitted as fluorescence}}{\text{Quanta absorbed}},$$

$$q_f = \frac{k_f}{k_f + \Sigma_i k_i} = \frac{\tau}{\tau_R}. \tag{10.2}$$

k_f is the first-order rate constant for fluorescence decay and $\Sigma_i k_i$ the sum of the rate constants for all other processes depopulating S_1. The measured lifetime of fluorescence is denoted by τ.

The quantum yield may be determined directly, either by measuring or calculating the number of photons absorbed and emitted or by comparative methods. The latter approach is more commonly used, the quantum yield of the unknown being obtained from the relation

$$q_u = \frac{q_k F_u}{F_k}, \tag{10.3}$$

where k and u refer to known and unknown and F is the area under the corrected emission spectrum. Use of equation (10.3) requires that the absorbance of known and unknown solutions be the same at the excitation wavelength. In practice, it is more accurate to obtain quantum yields based on exciting the standard and unknown at their absorption peaks rather than at some wavelength where the solutions have equal absorbance.

The choice of fluorescence standard is usually made on the basis of similarity of the emission spectrum to that of the unknown. Useful standards are fluorescein in 0.1M NaOH ($q_f = 0.85$), emission maximum at 525 nm; quinine in 0.1N H_2SO_4 ($q_f = 0.54$), emission maximum at 458 nm; and tryptophan in 0.01M tris Cl$^-$ buffer, pH 7 ($q_f = 0.13$), emission max at 348 nm.

10.3 Origins of Intrinsic Fluorescence in Proteins and Polypeptides

Only three of the common amino acids fluoresce: namely, the aromatic acids phenylalanine, tyrosine, and tryptophan. Of these residues, tryptophan is usually the most important in terms of probing protein conformation because of its relatively high quantum yield.

Tryptophan

The excitation and fluorescence spectra of tryptophan are shown in Figure 10.3. The fluorescence-emission spectrum is obtained by using an appropriate excitation wavelength and then measuring the fluorescence intensity as a function of wavelength.

The fluorescence spectrum can be seen to consist of a broad structureless band centered at 348 nm with a half-width of 60 nm. *Excitation spectra* are determined by measuring the fluorescence intensity, usually at the wavelength of maximum emission, as a function of excitation wavelength.

The shape of the fluorescence spectrum and position of the maximum are determined by the indole ring. Various derivatives of indole have essentially the same fluorescence spectrum. In Chapter 8 we observed that the symmetry of the wave functions was such that electronic absorp-

Figure 10.3 Excitation (1) and fluorescence (2) spectra of tryptophan in aqueous solution.

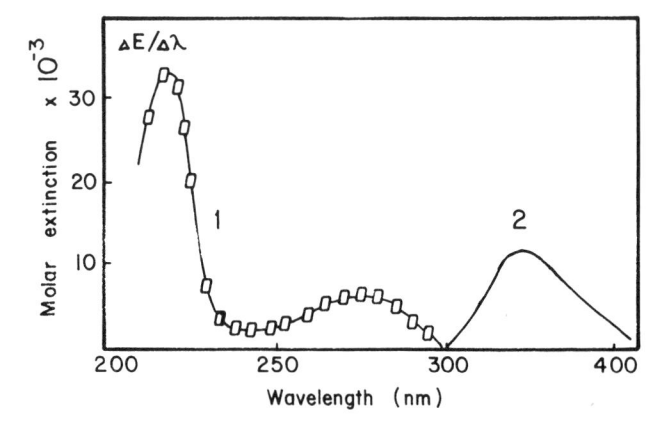

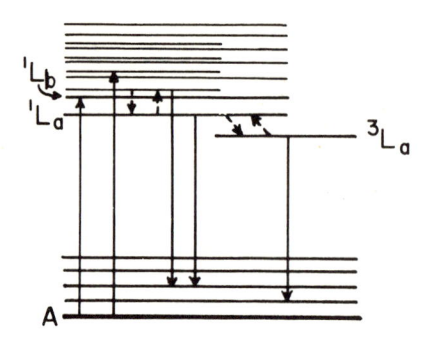

Figure 10.4 Electronic excitation and emission scheme for tryptophan.

tion occurred in preferred orientations that were associated with electronic oscillators because they produced charge displacement. In tryptophan it is possible to identify the direction of these oscillators by the use of polarized radiation and also to identify the fluorescence polarization, a feature that we will examine in more detail later. Based on such observations, it is possible to draw a Jablonski-type diagram (Figure 10.4) that shows absorptive transitions from the ground state (A) to first or second excited singlet states ^1L_a or ^1L_b and intersystem crossing to the triplet ^3L_a state with resulting fluorescence emission.

The "L" nomenclature is that of Platt and contrasts with the molecular orbital/symmetry theory of Chapter 8. In this case, the symmetry of the system is low so that all of the orbital transitions are symmetry allowed.

Figure 10.5 identifies the oscillator directions for absorption.

It can be seen that the ^1L_a and ^1L_b oscillators lie in the plane of the molecule, whereas the triplet oscillator lies almost perpendicular to the ring system.

Tyrosine and Phenylalanine

We have already examined the absorbance spectrum of tyrosine in some detail. Comparison of the near-UV absorbance and fluorescence curves for tyrosine and phenylalanine are shown in Figures 10.6a and b.

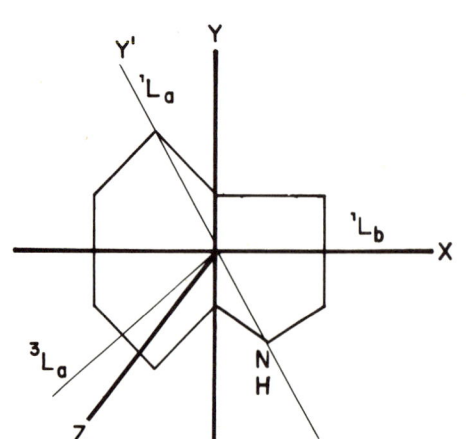

Figure 10.5 Orientation of oscillators in the indole ring corresponding to absorptive and emission transitions.

10. Electronic Emission

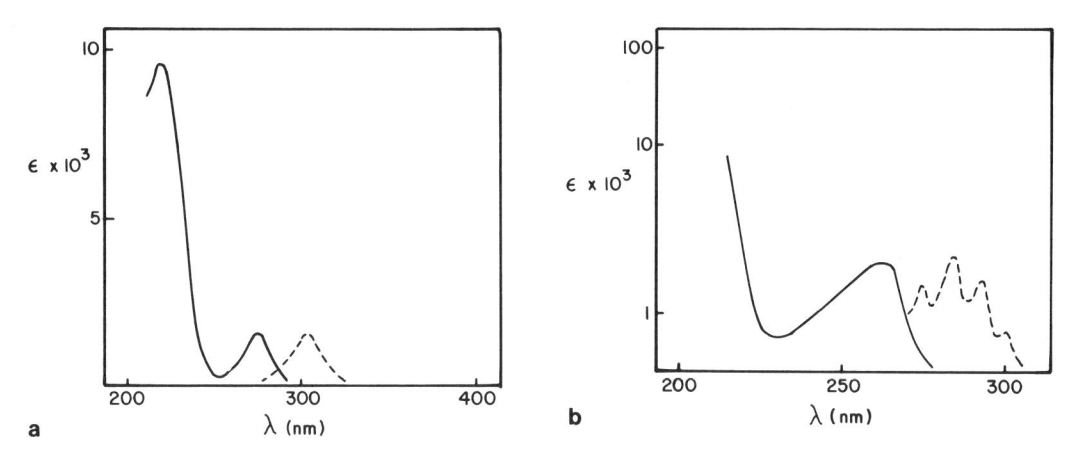

Figure 10.6 (a) Excitation and fluorescence of aqueous tyrosine solution. (b) Excitation and fluorescence of phenylalanine solutions (fluorescence in acetic acid/ethanol 1:4).

The fluorescence maxima lie at 304 nm and 285 nm, respectively, for tyrosine and phenylalanine compared with 348 nm for tryptophan. Thus, in principle, all three of the aromatic residues should be observable separately so that environmental changes of these residues in proteins in interactive systems or unfolding processes should provide interesting information. In practice, the phenylalanine fluorescence spectrum is unobservable in most proteins and tyrosyl fluorescence is small or overwhelmed by tryptophan fluorescence in proteins that contain both residues.

10.4 Uses of Fluorescence Spectroscopy

Fluorescence spectroscopy has found two main uses in protein chemistry, namely as a sensitive tool in assaying very small quantities of protein and following subtle conformational changes that perturb the environment of fluorophores, mainly tryptophan. To these may be added other uses based on the binding of other molecules, particularly extrinsic tags. The probing of separation distance between electron donor and acceptor, and features of surface mapping of proteins, may also be achieved by fluorescence spectroscopy. The efficacy of each method is dependent upon the high sensitivity of the fluorescence process. Let us first examine protein assay by fluorescence.

Fluorescence Assay of Proteins

Equation (8.10)—the Beer–Lambert Law—presented the relation between absorbance and protein concentration, that is:

$$\epsilon_\lambda cl = 2.303 \log_{10}\left(\frac{I_0}{I}\right)_\lambda. \tag{8.10}$$

The fluorescence observed at a constant exciting-light intensity is equal to the light absorbed multiplied by a factor ϕ', which is dependent upon the quantum yield of fluorescence and an instrumental factor. We can write, therefore, for the fluorescence intensity at wavelength λ

$$F_\lambda = \phi' \left(\frac{I_0 - I}{I_0} \right)_\lambda = \phi'(1-\exp-\epsilon_\lambda cl); \qquad (10.4)$$

expanding the exponential, for small absorbance:

$$F_\lambda \sim \phi' \epsilon_\lambda cl. \qquad (10.5)$$

Thus, at low concentration, the Beer–Lambert Law holds with F_λ/ϕ substituted for the absorbance. A calibration using a combination of absorbance and fluorescence enables the evaluation of ϕ' and ϵ_λ (the molar extinction coefficient) and hence for unknown solutions the concentration c may be established. Depending upon the quantum efficiency, it is often possible to assay protein solutions two or more orders of magnitude more dilute than by the absorption method.

At higher concentrations of protein, light is absorbed so strongly that excitation by the input light barely reaches the center of the cuvette. Since the detector is focused on the center of the cell, filtration of the input signal reduces its amplitude and it is possible that high concentrations of protein actually appear to have lower fluorescence intensity than at higher dilution.

Conformation/Environment

Although it is not possible to determine conformation by fluorescence spectroscopy, it is possible to detect small changes in conformation that manifest themselves in environmental changes of tryptophan or tyrosine residues. The secondary and tertiary structures of proteins bring different chromophores into close proximity, which produces physical interactions between them that can lead to energy migration. Thus the quantum efficiency of residues in a protein is, in general, different from that for the same residues in a denatured protein or isolated in solution.

Class A proteins

The fluorescence of proteins that contain no tryptophan residues, but contain tyrosine and phenylalanine, is attributable only to the presence of tyrosine. Proteins of this class include tropomyosin, ribonuclease, ovomucoid, nucleohistones, and insulin and are often termed "class A" proteins. The fluorescence spectrum has a maximum at 304 nm and, unlike the tryptophan band, the tyrosine band does not change in position with conformational change, but only in intensity. In general, the tyrosyl groups increase their quantum yield as a protein is denatured. The groups are said, therefore, to be "quenched" in the intact protein. Quenching seems to arise from several sources.

Quenching by an adjacent peptide bond. It is readily shown that the quantum yield of tyrosine is affected by adjacent peptide bonds by studying the fluorescence of small synthetic peptides. Table 10.1 compares the effect of various side-chain modifications on the quantum yield.

10. Electronic Emission

Table 10.1 Quantum Yields of Fluorescence for Some Tyrosine-Containing Peptides

$$\text{HO}-\langle\text{C}_6\text{H}_4\rangle-\text{CH}_2-\text{CH}(\text{R}_2)-\text{R}_1 \quad \text{Tyrosine-type compound}$$

Compound	R$_1$	R$_2$	Quantum yield
Tyrosyl	—COO$^-$	—NH$_3^+$	0.21 (neutral pH)
Tyrosine	—COOH	—NH$_3^+$	0.056 (acid pH)
Leucyltyrosine	—COO$^-$	NH$_3^+$—CH(CH(CH$_3$)$_2$)—CONH—	0.103
Tyrosylglycine	—CONH—CH$_2$—COO$^-$	—NH$_3^+$	0.074
Glycyltyrosine	—COO$^-$	NH$_3^+$CH$_2$CONH	0.070
(Lys tyr lys)			0.020

It can be seen that N- or C-terminal substitution greatly affects the quantum yield. The value of the quantum yield (q_f) in class A proteins is in the region of 0.03.

Involvement of tyrosyl—OH group in hydrogen bonding. It can be shown that phenolic—OH groups will interact with amides to form —OH $\cdots$ O = C$\langle$N$\langle$ type hydrogen bonds. Such an effect can account for the low fluorescence yield of tyrosines buried in the core of a globular protein such as bovine pancreatic ribonuclease. A second quenching mechanism arises through the interaction of ionized carboxyl groups to form hydrogen bonds, viz.

$$-\text{OH} \cdots \text{O} = \text{C}\langle^{\text{O}-}$$

In the electronically excited state of tyrosine, the phenolic group is believed to become partially ionized, causing the fluorescence quenching.

Table 10.2 shows that for four proteins, the tyrosyl fluorescence yield decreases with increasing carboxylate content, suggesting that interaction between tyrosine and the ionized carboxyl groups (of aspartic and glutamic acids) underlies, in part, the fluorescent quenching.

Another example of the role of environment on tyrosine quenching is for the random copolypeptide Tyr^4:Glu^{96}. At neutral pH, this polypeptide is in a coil conformation with carboxylates ionized. The relative quantum yield (on the same scale as Table 10.1) is 0.038. At low pH ($\sim$ 3.0), the polypeptide becomes α-helical and nonionized, the quantum yield rises to 0.08.

The denaturation of class A proteins leads then to an increase in fluorescence.

Table 10.2 Tyrosyl Fluorescence and Carboxylate Concentration for Proteins

Protein	Relative quantum yield	COO$^-$/tyrosine ratio
Zein	6.0	1.0
Insulin	3.7	1.14
Ribonuclease	1.7	1.4
Human serum albumin	1.0	1.65

Table 10.3 Fluorescence Behavior of Some Class B Proteins

	Aqueous solution		Urea (8 Molar)	
Protein	Fluorescence maximum	Relative quantum yield	Fluorescence maximum	Relative quantum yield
Lysozyme	341 nm	6.00	350 nm	4.1
Trypsin	332 nm	8.10	350 nm	13.8
Trypsinogen	332 nm	8.70	350 nm	14.8
Chymotrypsin	334 nm	9.50	350 nm	20.4
Chymotrypsinogen	331 nm	7.20	350 nm	20.0
Human serum albumin	339 nm	7.40	350 nm	5.2
Bovine serum albumin	342 nm	15.20	350 nm	7.4
Ovalbumin	332 nm	12.10	340 nm	13.1
Fibrinogen	337 nm	14.00	350 nm	14.0

Class B proteins

Tryptophan-containing proteins are called "class B" proteins. In general, proteins containing phenylalanine, tyrosine, and tryptophan show only a tryptophan maximum in their fluorescence spectrum. Hence the fluorescence spectrum of proteins is not an additive spectrum of the aromatic amino acids. The class B proteins have a fluorescence maximum in the 328- to 342-nm region of the spectrum. An examination of tryptophan fluorescence under a wide variety of solvent conditions and microenvironments has led to the concept that changes in the quantum yield and fluorescence maxima on denaturation of proteins are due predominantly to a change in local dielectric properties. Interactions often occur inside globular proteins that enhance the tryptophan fluorescence so that the protein will show fluorescence quenching upon denaturation. When the tryptophan residues are fully converted from the hydrophobic environment of a protein core to the hydrophilic environment expected under the denaturing influence of 8 Molar urea, the fluorescence maximum moves up to 345–350 nm—that is, the same region as tryptophan itself. Table 10.3 compares the fluorescence behavior of several class B proteins.

10.5 Fluorescence Depolarization

Let us now consider in more detail what happens if the incident light in a fluorescence experiment is plane polarized. If the polarized fluorescence is measured at right angles to the incident beam as shown in Figure 10.7, certain interesting deductions may be made concerning the properties of the fluorescing molecules.

Suppose that the incident light is polarized in the yz plane, as shown in the figure, and that it interacts with an oscillator in the sample. Preferential absorption occurs in oscillators parallel to the polarization direction and also in oscillators distributed about that direction. This process is known as "optical selection." If the fluorescence oscillator is parallel to the absorption oscillator and polarization is measured in the y direction of the xy plane, the polarized intensity is:

$$p = \frac{I_y - I_z}{I_y + I_z} \tag{10.6a}$$

10. Electronic Emission

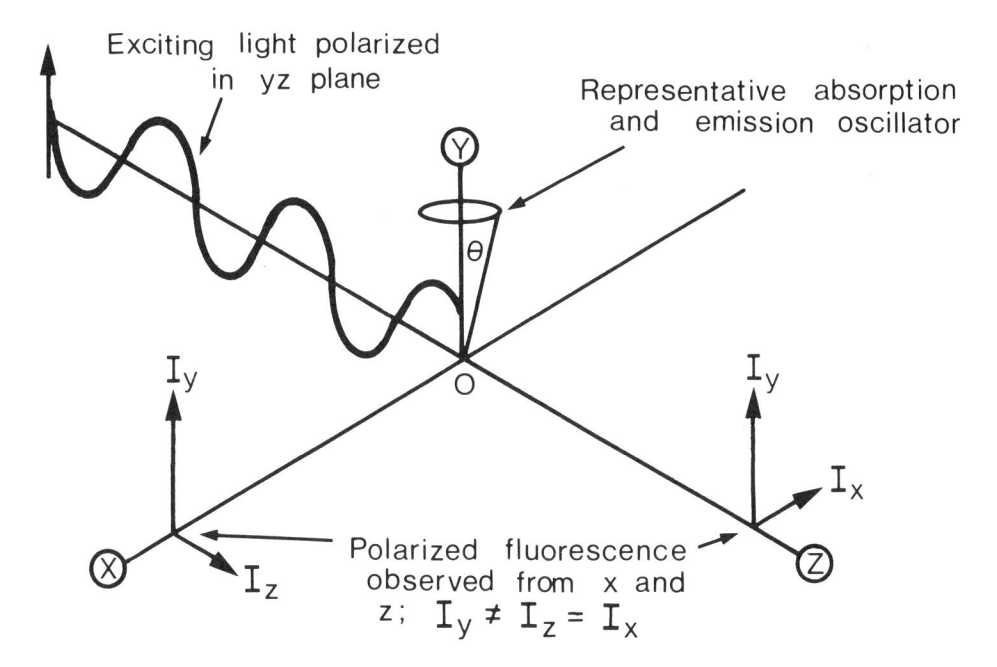

Figure 10.7 Experimental configuration used for measuring polarized fluorescence.

It would seem that the maximum value of p would be unity. However, absorption is selective and can never be totally oriented parallel to the y axis. Consequently the maximum observable value of p is 0.5 and there is a component of fluorescent intensity in the xz plane. The minimum observable value of p is −0.33, occurring when the absorption and emission oscillators are orthogonal (90° to each other).

If the molecules undergoing irradiation do not rotate during the lifetime of the excited state and the oscillator responsible for absorption is also responsible for emission—that is, there is no energy exchange—then it is possible to follow p values as a function of exciting wavelength. By this means the relative orientation of absorption and emission oscillators can be followed for various absorption transitions.

If, on the other hand, significant rotation of the molecules occurs in the time framework of fluorescence, depolarization occurs. A depolarization ratio D_p may be defined as −p of equation (6.6). Further, if we put $I_y = I_\parallel$ and $I_z = I_\perp$, the depolarization ratio is given by

$$D_p = \frac{I_\perp - I_\parallel}{I_\perp + I_\parallel}. \tag{10.6b}$$

In terms of a rotational relaxation constant ρ and excitation lifetime τ a relation can be written

$$\left(\frac{1}{D_p} \pm 1/3\right) = \left(\frac{1}{D_p^\circ} \pm 1/3\right)\left(1 + \frac{3\tau}{\rho}\right). \tag{10.7}$$

D_p° is the intrinsic depolarization—that is, that which the solution would have if no rotation had occurred.

Generally, τ, ρ, and D_p° are not known for protein solutions, but since $\rho = \frac{1}{2}\theta$ where θ is the rotational diffusional coefficient for spherical molecules, ρ may be replaced (assuming spherical molecules) by the approximate relation $\rho \sim 3\eta V/RT$, where η is the viscosity of the medium and V the molar volume. Then

$$\left(\frac{1}{D_p} - 1/3\right) = \left(\frac{1}{D_p^\circ} - 1/3\right)\left(1 + \frac{RT}{V\eta} \cdot \tau\right). \tag{10.8}$$

Hence $1/D_p$ is plotted as a function of T/η (T or η varied), then D_p° and τ may be evaluated. Alternatively, if the fluorescent lifetime τ is known, the molecular volume may be calculated. Equation (10.8) is known as the Perrin equation.

In practice it is not easy to find circumstances under which changes of T do not affect molar volume or protein shape, so that viscosity is usually the chosen variable. A comon method of varying the viscosity of a protein solution is the addition of sucrose or glycerol. Figure 10.8 shows fluorescence depolarization of bovine serum albumin (BSA) solutions, at two different pH values, plotted according to the Perrin equation. The slope of the plot is $\frac{R\tau}{V}$, so that changes in the fluorescent lifetime or molecular volume would result in a changed slope. Since it is known that albumin changes conformation at about pH 4.5 and carboxyl groups become protonated at this pH, thus altering their potential interaction with active chromophores, it seems probable that both effects are present.

Another simple use to which depolarization measurements may be put is concerned with protein binding. Evidently if a fluorescing protein molecule undergoes aggregation, either with itself (e.g. in subunit assembly) or in complexation with other molecules, then the rotational relaxation changes and

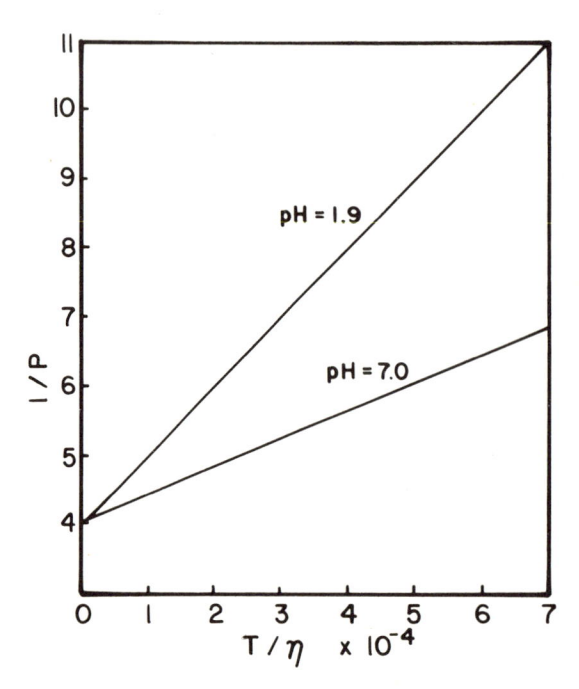

Figure 10.8 Fluorescence depolarization of BSA plotted in accord with the Perrin equation (equation 10.8).

a change in depolarization ratio occurs. Analysis of depolarization ratio versus component concentration can yield complexation ratios.

10.6 Fluorescent Lifetime

We have seen that an indirect method of measuring the fluorescent lifetime τ, which is the time required for the emission intensity to drop to 1/e of its original value, is that involving the use of the Perrin equation. Another indirect method involves the measurement of quantum yields under different conditions, that is,

$$\frac{q_1}{q_2} = \frac{\tau_1}{\tau_2}. \tag{10.9}$$

Such a relation might apply, for example, to a solution at different temperatures or with a material dissolved in two different solvents. This indirect method suffers from serious limitations and cannot be used for comparing different compounds of unknown fluorescence-decay mechanisms.

The direct measurement of fluorescence lifetime has come into its own in the last decade. The time dependence of fluorescence-emission anisotropy after pulsed excitation yields more information about the emitter than does the static measurement. Aspects of rotational diffusion, kinetic measurements involving the first singlet state, and quenching mechanisms due to "static" or "dynamic" mechanisms emerge from pulsed excitation experiments. A static quencher acts on the fluorescent molecule in the ground state, effectively reducing the concentration of potential fluorescent states. We have already seen that factors influencing the phenolic —OH group of tyrosine modify the apparent quantum yield and thus act as static quenchers. Dynamic quenching involves only the excited state where collision processes, particularly with O_2 or I^-, decrease the quantum yield. For static quenching the lifetime of the fluorescent state τ is not changed, whereas in dynamic quenching the decrease in quantum yield is proportional to a decrease in τ.

For measurement of fluorescent-decay processes, the sample is irradiated with a flash of light lasting approximately a nanosecond (10^{-9} seconds). The time dependence of polarized fluorescence is believed to be comprised of five exponential terms of which no more than three are independent (see the rate constants of equation 10.2).

Because of the complex energy-exchange processes that occur in protein fluorescence and quenching, the most straightforward method of obtaining shape or structural data from protein molecules by fluorescent-decay methods involves complexation with fluorescent dye molecules. The dye may then be excited by a light pulse and the resultant decay may be followed by modern instrumentation. The use of fluorescent markers is known as *extrinsic* fluorescent spectroscopy and may be compared with the use of the native fluorescent peptide residues, which involves intrinsic fluorescence. Details of extrinsic fluorescence will be examined later, but for our present purposes we will suppose that the extrinsic probe acts as an independent fluorescing entity that rotates as part of the host protein. The sample is excited with a light pulse polarized in the y direction (Figure 10.7) and the intensity of emission is

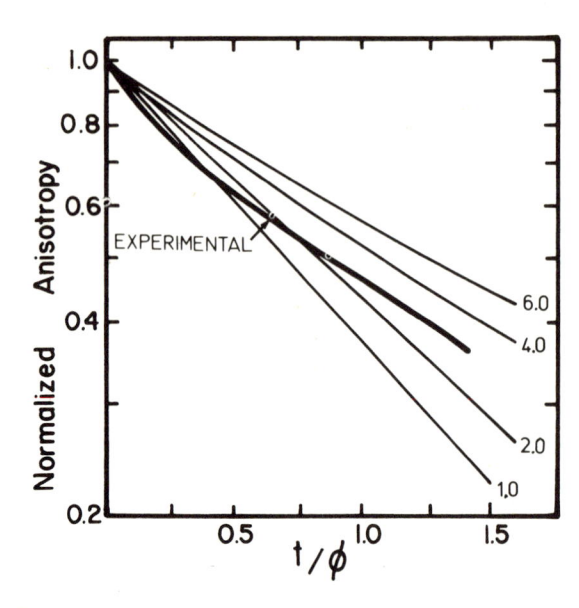

Figure 10.9 Comparison of observed and calculated plots for emission anisotropy of IgG. The experimental data correspond to $\phi = 53$ nsec and $A_0 = 0.32$ in equation (10.9).

measured as a function of time in the y polarization plane $F_y(t)$. Similarly the process is repeated in the x plane, yielding a decay function $F_x(t)$. The anisotropy function $A(t)$ that relates to the shape of the molecular complex is then given by

$$A(t) = \frac{F_y(t) - F_x(t)}{F_y(t) + 2F_x(t)} \tag{10.10}$$

If the complex can be represented by a rigid sphere, equation (10.10) reduces to

$$A(t) = A_0 \exp - t/\phi \tag{10.11}$$

where ϕ is the rotational diffusion time of the molecular complex being equal to $\rho/3$ (see equation 10.7). The anisotropy function is independent of the lifetime of the fluorescent dye and depends only on the rotational diffusion coefficient.

If a protein molecule is approximated as a rigid ellipsoid (see Chapter 12), it is possible to calculate the anisotropy function for ellipsoids of various major and minor axis ratios and to compare experimental data with these calculations.

Figure 10.9 shows a comparison of data for an immunoglobulin (IgG) molecule with the calculated values for various rigid ellipsoids. Since the immunoglobulin does not correlate with any of the predicted curves and it may be shown by the methods of Chapter 12 that it is highly anisotropic, it may be concluded that the molecule is probably quite flexible.

10.7 Energy-Transfer Processes

So far we have seen that changes in fluorescence spectra of proteins represent rather unspecific changes in the environment of the fluorophore. Fluorescence spectroscopy may, however, be used in a more quantitative manner if the mode of transfer of energy from the chromophore is known. Since all of the aromatic residues (and the peptide bonds and other groups) absorb, but only tryptophan fluoresces in group B proteins, it is evident that there are mechanisms for transfer of fluorescent energy, some of which have already been explored. The importance of energy-transfer studies lies in the ability of the researcher to extract information concerning the separation of donor and acceptor groups in a molecule (or in separate molecules). The process of energy transfer may be compared with the resonance of two tuning forks: when the first is caused to vibrate, the second begins to vibrate at the same frequency by resonance. Similarly, "induced resonance" or energy transfer occurs between donor and acceptor groups that have appropriate fluorescence and absorption spectra. If group A is the fluorescence donor and B is a nonfluorescent absorber in the same spectral region, the fluorescence of A will be quenched by B as a function of the inverse sixth power of their separation distance. If the fraction of photons that A transfers to B is x, then

$$1 - x = [(R_0/R)^6 + 1]^{-1}, \tag{10.12}$$

where R is the distance of separation and R_0 is a "characteristic" distance that corresponds to an energy transfer of 50%. Typically R_0 is ~ 20 Å. It is given by the Förster relation

$$R_0 = \sqrt[6]{\frac{1.66 \times 10^{-33} \times \tau J_{AB}}{n^2 \times \bar{\nu}_0^2}}. \tag{10.13}$$

$\bar{\nu}_0$ is the mean of the peak positions of the donor emission and receptor absorption bands. J_{AB} is the overlap area (overlap integral) of the donor and receptor bands, and n is the refractive index.

It can be seen from equations (10.12) and (10.13) that fluorescent quenching at a given separation distance is enhanced by the degree of overlap of donor emission and acceptor absorption bands and by a long fluorescence lifetime τ. Fluorescent quenching is often significant over distances up to 100 Å.

An example of the preceding approach is the application to small peptides.

The peptide ACTH (adrenocorticotropic hormone) has 24 peptide residues with tryptophan at residue 9, tyrosine at positions 2 and 23, and phenylalanine at position 7. Under normal circumstances the fluorescence spectrum of ACTH shows the typical single 348-nm band of tryptophan emission. It is possible to incorporate into ACTH an N^ϵ-dansyl group on Lys[21]. The primary sequence of the dansylated peptide is shown in Figure 10.10.

The dansyl group acts as an excellent energy acceptor and at the same time does not affect the biologic activity of the ACTH. The fluorescence spectra of

Figure 10.10 The sequence of ACTH showing the Lys(Dns)[21] structure.

```
         1                                              10
         Ser — Tyr — Ser — Met — Glu — His — Phe — Arg — Trp — Gly —
                                                        20
         Lys — Pro — Val — Gly — Lys — Lys — Arg — Arg — Pro — Val —
                     24
         Lys — Val — Tyr — Pro
```

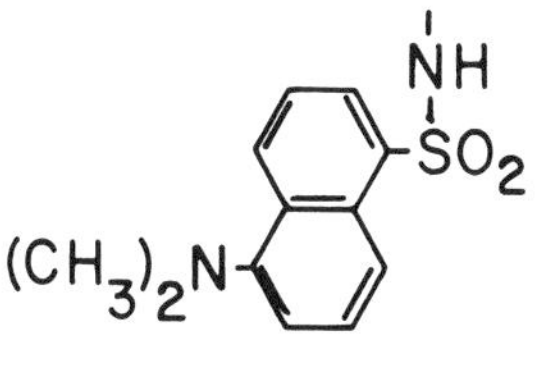

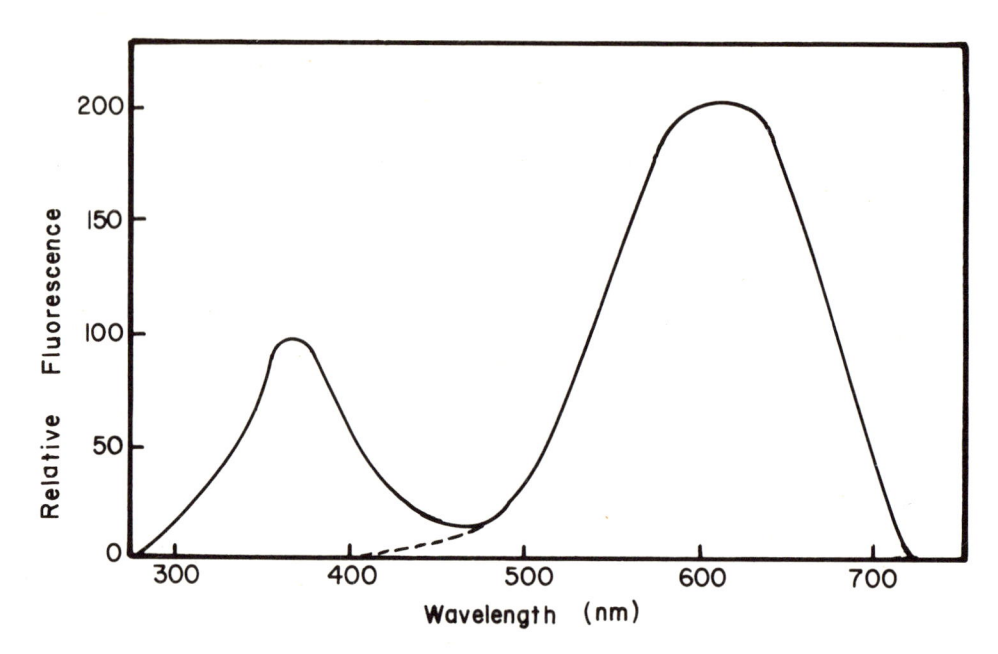

Figure 10.11 Fluorescence-emission spectra of [Lys(Dns)²¹]-ACTH$_{1-24}$ (———) and [Lys(Dns²¹]ACTH$_{11-24}$ (-----). Excitation wavelength = 293 nm (after P. Schiller).

Lys(Dns)²¹–ACTH and the fragment Lys(Dns)²¹ ACTH$_{11-24}$ which contains no tryptophan are shown in Figure 10.11. There is only a slight overlap of fluorescence bands, the separation being useful for this type of experiment.

Table 10.4 contains a collection of data obtained for the modified ACTH molecule. It can be seen that over a range of conditions that would be expected to cause changes in ordered conformation and tertiary structure, if any were to exist, the tryptophan/dansyl separation remains in the 20- to 26-Å region. Theoretical calculations indicate that a random chain of modified ACTH would have the tryptophan and dansyl groups separated by 29 Å. Consequently, it seems probable that ACTH has little, if any, ordered structure in aqueous solution.

A similar approach may be applied to proteins where the fluorescent donor and absorption receptor (usually a dansyl group or similar probe) can be clearly identified. In many proteins, however, the mechanism of energy transfer and internal fluorescence quenching is extremely complex.

Table 10.4 Experimental Parameters and Calculated Intergroup (Tryptophan/Dansyl) Distances for Modified ACTH

Solvents, conditions (25°C)	J_{AB}[cm⁶/mol] × 10¹²	R[Å]
H$_2$O	3.02	20.1
Sodium phosphate, 0.1M, pH 7.0	3.04	24.5
Sodium phosphate, 0.1M, pH 7.0, 70°C	3.17	24.8
6M Guanidine · HCl	3.22	23.6
8M Urea	3.20	26.0

Sometimes it is possible to use combinations of all the preceding methods, along with solvent and temperature variation, to probe surface topology of proteins and their unfolding mechanism. Again the analysis of the fluorescent processes and their meaning can be extremely complicated.

10.8 Complexation Processes

A wide range of complexation processes has been followed by fluorescence spectroscopy; considerable emphasis has, for example, been placed on studying the interaction between drugs and biologic macromolecules. When a drug binds to a protein the process may be followed by the perturbation of the fluorescent groups within the protein—that is, intrinsic fluorescence—or by the behavior of a fluorescent group in the drug or by the displacement of a fluorescent dye. The wavelengths of maximal activation and emission, quantum yield, fluorescence lifetime, and degree of polarization are all parameters that are sensitive to subtle changes in the environment of the fluorophore. Consequently the complexation of protein with a second entity can usually be detected by changes in these parameters.

In the simplest case we may imagine the interaction between macromolecule A and complexing agent (drug, dye, other entity) B as being given by a simple equilibrium reaction

$$A + B \xrightleftharpoons{K_a} AB$$

from which

$$K_a = \frac{[AB]}{[A][B]} \tag{10.14}$$

Let us suppose that the complexing agent becomes fluorescent on binding and the fluorescence may be measured at a wavelength at which neither macromolecule nor complexing agent alone fluoresces. As increments of the agent are added an increase in extrinsic fluorescence is observed, as in Figure 10.12.

Figure 10.12 Increase in extrinsic fluorescence of a protein/complex with increase in complexing agent (the concentration of protein is held constant).

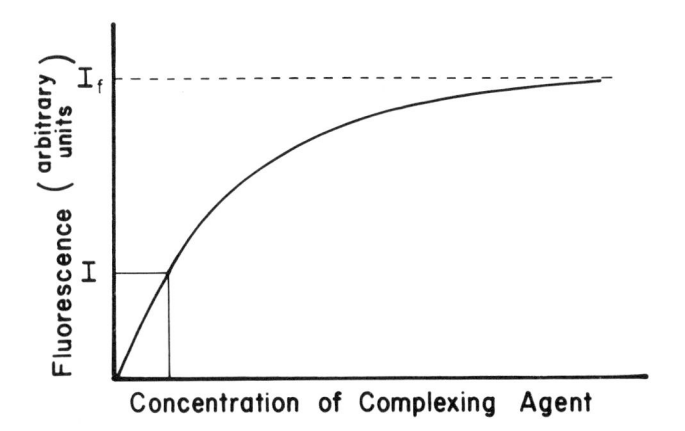

It is assumed that the fluorescence intensity is proportional to the concentration of complex [AB] formed so that

$$\frac{I}{I_f} = \frac{[AB]}{[A_t]} \tag{10.15}$$

where A_t is the total concentration of protein.

The concentration of free complexing agent is

$$[B] = [B_t] - [AB] = [B_t] - Q[A_t], \tag{10.16}$$

where $Q = I/I_f$ and the concentration of free protein is

$$[A] = [A_t] - [AB] = A_t(1 - Q). \tag{10.17}$$

Hence the association constant from equation (10.14) is

$$K_a = \frac{Q[A_t]}{([B_t] - Q[A_t])(1 - Q)A_t} . \tag{10.18}$$

Since $[A_t]$ and $[B_t]$ are known and Q is obtained from the fluorescence plot, K_a is readily obtained.

The preceding approach is not limited to the complexation of one molecule of B with one of A if additional complexation of B molecules with A causes an equal increase in fluorescence intensity and has the same association constant.

The number of binding sites per molecule of protein A can, under such circumstances, be obtained from a Scatchard plot. If N_B is the number of moles of complex B bound per mole of protein

$$N_B = \frac{[B_t] - [B]}{[A_t]},$$

then the Scatchard equation is

$$\frac{N_B}{[B]} = K_a(n - N_B), \tag{10.19}$$

where n is the number of binding sites per protein molecule. Hence a plot of $N_B/[B]$ versus N_B (both quantities obtainable from the fluorescence data, Figure 10.12) gives a slope of K_a and an intercept on the abscissa of $N_B = n$. Figure 10.13 shows a schematic plot of equation (10.19) for two different association constants.

An interesting example of fluorescence enhancement with drug/protein binding is for the enzyme erythrocyte carbonic anhydrase, which is inhibited by sulfonamides. One particular sulfonamide derivative, 5-dimethylamino-naphthalene-1-sulfonamide (conveniently called DNSA), shown in Figure 10.14, forms a highly fluorescent complex with bovine erythrocyte carbonic anhydrase, which may be analyzed by the preceding approach. The analysis

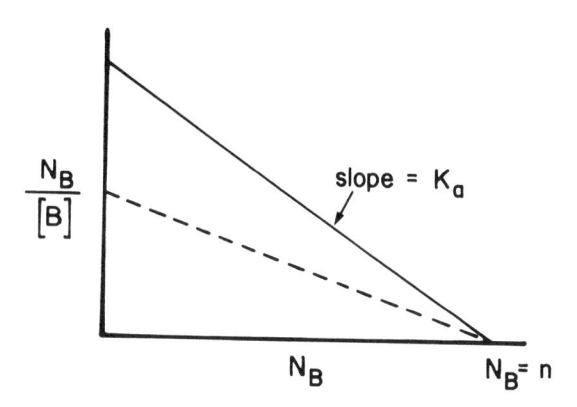

Figure 10.13 A Scatchard plot of binding data: N_B is the number of moles of complexing agent bound per mole of protein; K_a is the association constant; n, the total number of binding sites per protein molecule; and [B] is the concentration of free complexing agent.

reveals that one mole of DNSA binds to the protein at pH 7.4 with a $K_a = 2.5 \times 10^{-7}$ M. The fluorescence-enhancement curve is shown in Figure 10.15; however, it is evident that the enhancement of DNSA fluorescence at 468 nm is accompanied by a quenching of the tryptophan fluorescence at 336 nm. Evidently some of the energy absorbed by the tryptophans of the protein is transferred in the DNSA-activation process.

When a protein complexes with a drug or other molecule, such that various sites are used with variable quenching or enhancement of fluorescence, the analysis is much more complicated but may be approached in a similar manner if one set of sites is filled preferentially.

Next we turn to our analysis of the interaction between protein A and complexing agent B, where the intrinsic fluorescence of the protein (tryptophan and/or tyrosine) is quenched.

A number of drugs that bind to proteins cause fluorescence quenching by the energy-transfer process. Table 10.5 contains a partial list of such complexes.

If the protein can be titrated with complexing agents such that when all sites are occupied there is a fraction of the original fluorescence remaining, the titration curve looks like that shown schematically in Figure 10.16.

Analysis of the reaction may be accomplished by using equations (10.15)–(10.19) with $Q = (I_0 - I)/(I_0 - I_f)$. Hence the equilibrium constant and number of binding sites n may be obtained as before.

An example of intrinsic fluorescence quenching is shown in Figure 10.17 for thyroxine complexation with albumin.

Figure 10.14
DNSA (5-dimethylaminonaphthalene-1-sulfonamide).

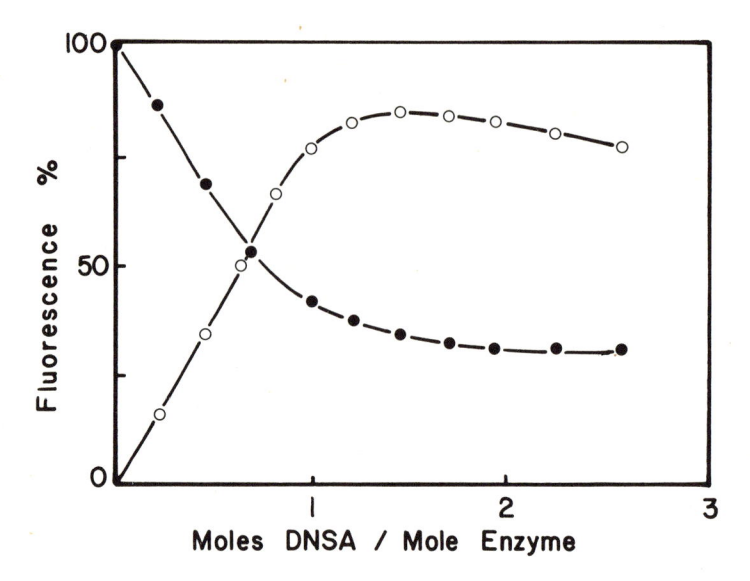

Figure 10.15 Fluorescence titration of carbonic anhydrase with DNSA. Tryptophan fluorescence (excitation at 280 nm, emission at 336 nm) -•-• is quenched. DNSA (excitation at 320 nm, emission at 470 nm) -o-o is enhanced.

In this case it seems probable that several thyroxine molecules (T) combine with one albumin molecule so that a series of equilibrium reactions may be written:

$$A + T \underset{}{\overset{K_1}{\rightleftharpoons}} AT$$
$$AT + T \underset{}{\overset{K_2}{\rightleftharpoons}} AT_2$$
$$AT_{n-1} + T \underset{}{\overset{K_n}{\rightleftharpoons}} AT_n.$$

The intensity of fluorescence (assuming a linear relation between fluorescence intensity and concentration) is given by

$$I = x_0 I_0 + x_1 I_1 + \ldots \sum_{n>1} x_n I_n, \qquad (10.20)$$

where x_n is the mole fraction of protein complex containing n thyroxine

Table 10.5 Proteins That Show Fluorescence Quenching on Interaction with Drugs

Drug	Protein
Dansyl sulfonamide	Carbonic anhydrase
Sulfonamides	Carbonic anhydrase
Warfarin, dicoumarol	Serum albumin
Thyroxine	Serum albumin
Steroids	Serum albumin
Bilirubin	Serum albumin
Chlorpromazine	Myosin

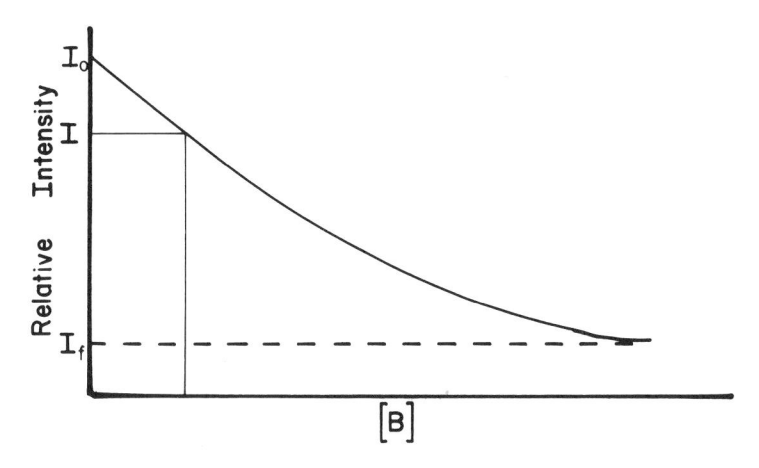

Figure 10.16 Fluorescence quenching of protein by complexing agent B.

molecules. So that

$$x_0 = \frac{[A]}{[A_t]} = \frac{[A]}{[A] + [AT] + [AT_2] + \ldots}. \tag{10.21}$$

Substitution of the equilibrium constant expressions yields

$$\frac{1}{x_0} = 1 + K_1[T] + K_1K_2[T]^2 + \ldots \tag{10.22}$$

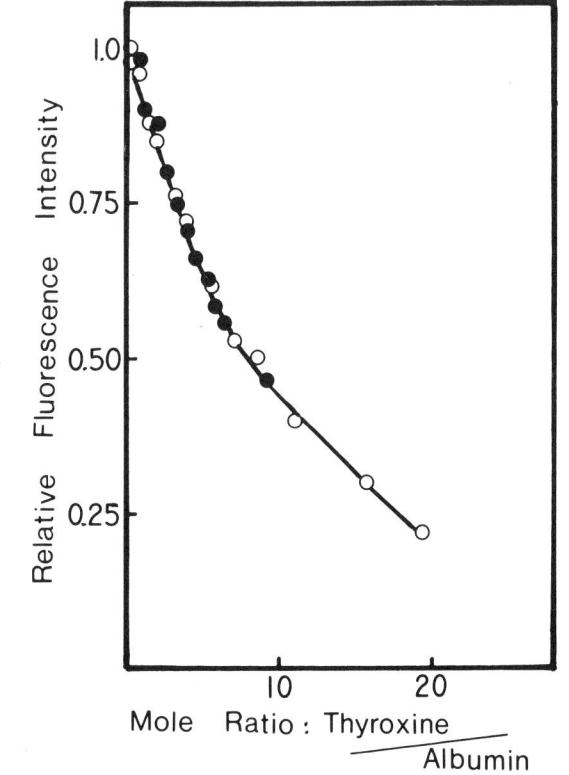

Figure 10.17 Fluorescence quenching of BSA by thyroxine.

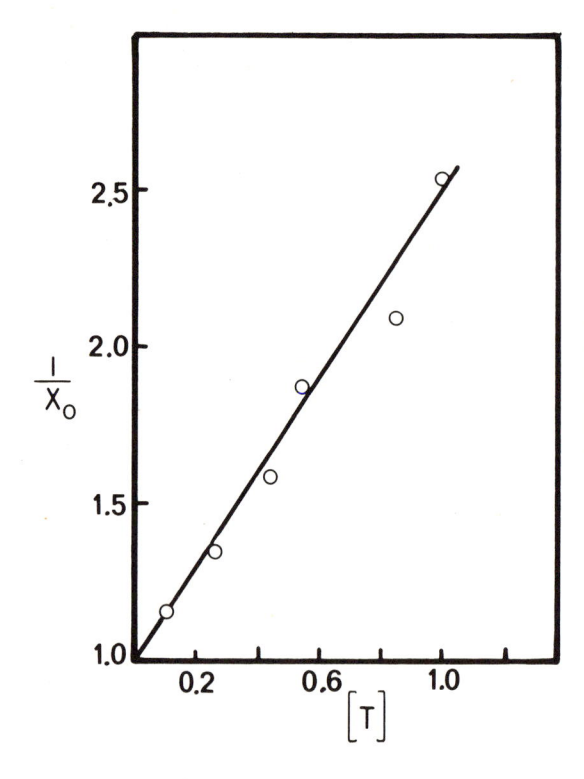

Figure 10.18 Fluorescence quenching of BSA as a function of the mole ratio of thyroxine, plotting the inverse mole fraction of free BSA (x_0) versus the concentration of thyroxine [T] (see equation 10.20).

The mole fraction of uncomplexed material is given by

$$x_0 = \frac{I - I_f}{I_0 - I_f}$$

in the initial region of quenching. In this region the complexed thyroxine may be assumed to be in the form of AT; consequently the free-thyroxine concentration [T] may also be estimated. Thus a plot of $1/x_0$ vs. [T] yields the constant K_1 as shown in Figure 10.18. Higher order equilibrium constants K_2, K_3, and so on may sometimes be estimated by an iterative process that first calculates K_1 and then proceeds to a correction for AT_2, K_2, and so on.

10.9 Extrinsic Fluorescence

We have already seen that the fluorescence of protein complexes may either originate from amino acid residues in the protein (intrinsic fluorescence) or from the complexing molecule (extrinsic fluorescence). There are, in fact, a large number of complexing agents known that fluoresce when combined with protein; the agents are known as fluorescent probes.

The probes are usually, as might be expected, highly aromatic in character. Figure 10.19 shows the molecular structure of a few probes in common use. Most of the probes bind to proteins predominantly by hydrophobic bonding. One exception is DNS (1-dimethylaminonaphthalene-5-sulfonate), which binds to the —NH_2 groups to produce the DNSA sulfonamides (Figure 10.14).

10. Electronic Emission

ANS
Anilinonaphthalene sulfonate

TNS
Toluidinonaphthalene sulfonate

PERYLENE

DNS
5-Dimethylaminonaphthalene-1-sulphonyl

DPH
1,6-Diphenyl-1,3,5,-hexatriene

Figure 10.19 Some common extrinsic fluorescence probes.

Figure 10.20 Extrinsic fluorescence of DPH bound to the blood-clotting enzyme, Hageman factor. ——— after activation; —•— prior to activation; ---- blank (activating agent (ellagic acid) + DPH) (after Rippon, Ratnoff, et al.).

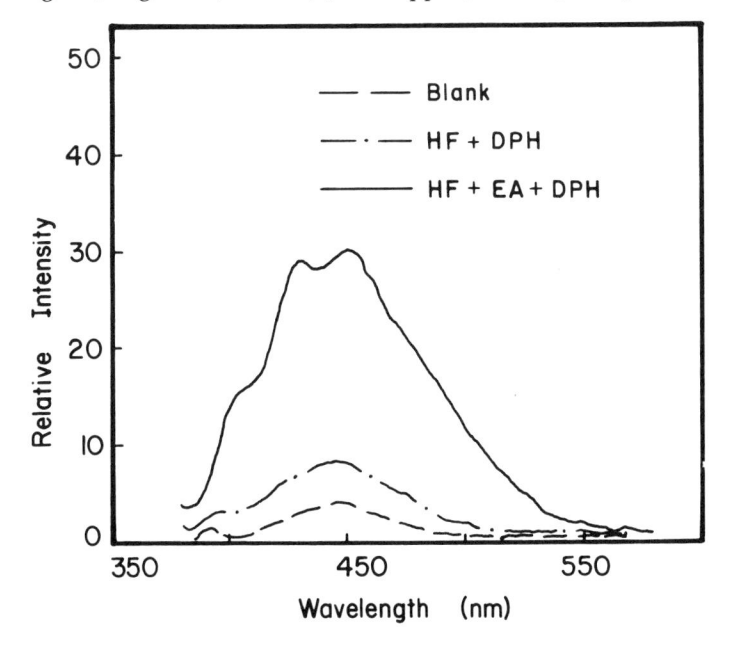

As an example of the use of an extrinsic fluorescent probe we will consider activation of the blood-clotting enzyme, Hageman factor (blood factor XII—see Chapter 15). The zymogen is activated by treatment with ellagic acid and Figure 10.20 shows that Hageman factor adsorbs much more DPH probe after activation than before, indicating that the activation process involves the development of hydrophobic binding sites. As the protein becomes more hydrophobic it is able to "turn on" the fluorescence of the extrinsic fluorophore.

More sophisticated applications of extrinsic probes include the use of depolarization measurements and cell microviscosity measurements where probes are buried in the cell membrane.

In summary, intrinsic and extrinsic fluorescence measurements provide a basis for detection of subtle structural changes, protein shapes, separation distances of certain groups, and hence some conformational information. In addition, considerable information may be derived concerning the electronic structure and electronic processes in peptides and proteins.

Further Reading

Brand, L. and B. Witholt. Fluorescence measurements. In *Methods in Enzymology,* C.H.W. Hirs, ed. Volume XI, Chapter 87, Academic Press, New York, 1967.

Cantor, C.R. and P.R. Schimmel. Fluorescence spectroscopy. In *Biophysical Chemistry,* Part II, W.H. Freeman and Co., San Francisco, 1980, pp. 433–465.

Chen, R.F. and H. Edelhoch, eds. *Biochemical Fluorescence,* Vols. 1 and 2. Marcel Dekker, New York, 1975 (Vol. 1); 1976 (Vol. 2).

Chen, R.F., H. Edelhoch, and R.F. Steiner. Fluorescence of proteins. In *Physical Principles and Techniques of Protein Chemistry,* S.J. Leach, ed. Part A, Chapter 4, 1970.

Guibault, G.G. ed. *Fluorescence: Theory, Instrumentation and Practice.* Marcel Dekker, New York, 1967.

Parker, C.A. *Photoluminescence of Solutions.* Elsevier, New York, 1967.

Schulman, S.S. *Fluorescence and Phosphorescence Spectroscopy: Physicochemical Principles and Practice.* Pergamon Press, Oxford, 1977.

Stryer, L. Fluorescence energy transfer as a spectroscopic ruler. *Annual Review of Biochemistry* 47: 817–846, 1978.

Udenfriend, S. *Fluorescence Assay in Biology and Medicine,* Academic Press, New York, 1969.

11

Nuclear Magnetic Resonance

11.1 The Resonance Phenomenon

In the three preceding chapters, we have considered how proteins and their constituents interact with light. We now turn to the interaction of molecules with a magnetic field. Two techniques that have proved of value in this area are nuclear magnetic resonance (NMR) and electron paramagnetic resonance. In the former technique, it is the atomic nucleus, rather than the surrounding electrons, that provides the basis of the phenomenon.

Both protons and neutrons have a spin quantum number of ½ and, depending on how these entities are paired in a nucleus, it may or may not have a net nonzero spin quantum number (I). Nuclei with even numbers of protons and neutrons (e.g., ^{12}C, ^{16}O, ^{32}S) all have I = 0.

Nuclear magnetic resonance spectroscopy is most often concerned with nuclei having I = ½ (i.e., ^{1}H, ^{13}C, ^{19}F, ^{31}P). When I = ½ there is one net unpaired spin in the nucleus, and the spinning charged nucleus behaves as a magnet. This nuclear magnet has a magnetic dipole moment directed along its axis of spin. The dipole moment is a vector quantity and is termed the magnetic moment, μ*. For I = ½ the charge distribution on the nucleus is spherical; however if I ≥ 1, then the charge is asymmetric and the nucleus is said to possess a quadrupole moment. The quantities μ and I are related by $\mu = \gamma \hbar I$ where γ is known as the magnetogyric ratio of the nucleus and $\hbar$ is Planck's constant divided by 2π.

The allowed energy levels for a spin of quantum number I in a magnetic field of strength H_0 are given by:

$$E = -\gamma \hbar H_0 m \qquad (11.1)$$

---

*Note that in previous chapters the group magnetic moment was termed m and the electric moment μ.

where m is the nuclear spin angular momentum quantum number that can have values of I, I − 1, ..., −(I − 1), −I. When I = ½, m = ±½, corresponding to alignment of the magnetic moment with, or opposed to, the magnetic field. If I = 1, m has values of 1, 0, and −1 corresponding to alignment of the nuclear magnetic moment with, perpendicular to, or opposed to the field. Figure 11.1 shows the quantized energies of nuclei in a magnetic field for different spin angular quantum numbers. It can be seen from this figure that for I = ½ in the presence of a magnetic field, there is a difference in energy between the spin orientations of $\Delta E = \gamma \hbar H_0$. If, therefore, an external energy source were provided, energy would be absorbed if it corresponded to a value that could promote the nuclear spin from the lower spin state to the upper spin state.

The energy required to flip the spin states is derived in the NMR experiment from a superimposed radiofrequency source, the required frequency ν being related to the magnetic parameters by the Planck equation (for I = ½).

$$\Delta E = h\nu = \gamma \hbar H_0 \tag{11.2}$$

The relative population of the upper and lower spin states (±½) for protons may be assessed when it is realized that the energy of separation is approximately 4×10^{-19} erg for a 14-K gauss field. From the Boltzmann relation the population of lower and upper states is given by:

$$\frac{N_L}{N_u} = \exp\left(\frac{\Delta E}{kT}\right) = \exp\left(\frac{\gamma \hbar H_0}{kT}\right). \tag{11.3}$$

At room temperature the ratio is approximately 1.00001 and thus the population of the two states is almost equal. The population difference increases with increasing field strength.

To consider the nature of the magnetic resonance phenomenon in more detail, we must examine the relationship between the magnetic dipole moment of the spinning nucleus, and the direction of the applied field. Figure 11.2 depicts a spinning proton in an external magnetic field H_0. It can be

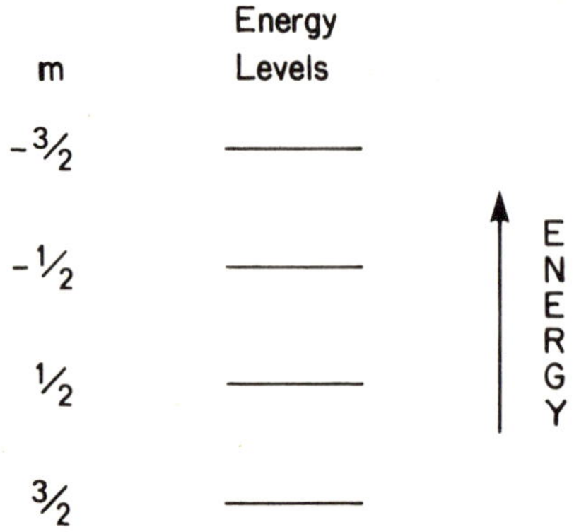

Figure 11.1 Quantized energies of nuclei in a magnetic field.

11. Nuclear Magnetic Resonance

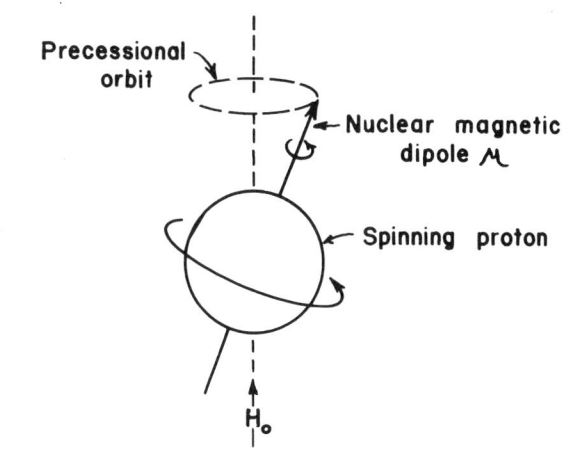

Figure 11.2 Proton precessing in a magnetic field H_0.

seen that the dipole axis precesses in a circular orbit about the axis of the magnetic field. In order to achieve the magnetic resonance phenomenon, it is necessary to achieve an alignment of the precessional orbit with the field of a radiofrequency oscillator coil. This arrangement is shown in Figure 11.3. Reviewing the proton magnetic resonance (PMR) experiment then, the protons are subjected to a strong uniform magnetic field (approximately 14-K gauss for 60-MHz spectra). Protons are aligned both with and against the field (thermal disorder causes most of the protons to be improperly aligned). The electromagnetic frequency is applied such that its field H_1 is at right angles to the main magnetic field H_0. The oscillator produces a linear field along its axis which can be resolved into two oppositely rotating (cir-

Figure 11.3 Oscillator generates rotating component of magnetic field H_1 (from Silverstein, et al.).

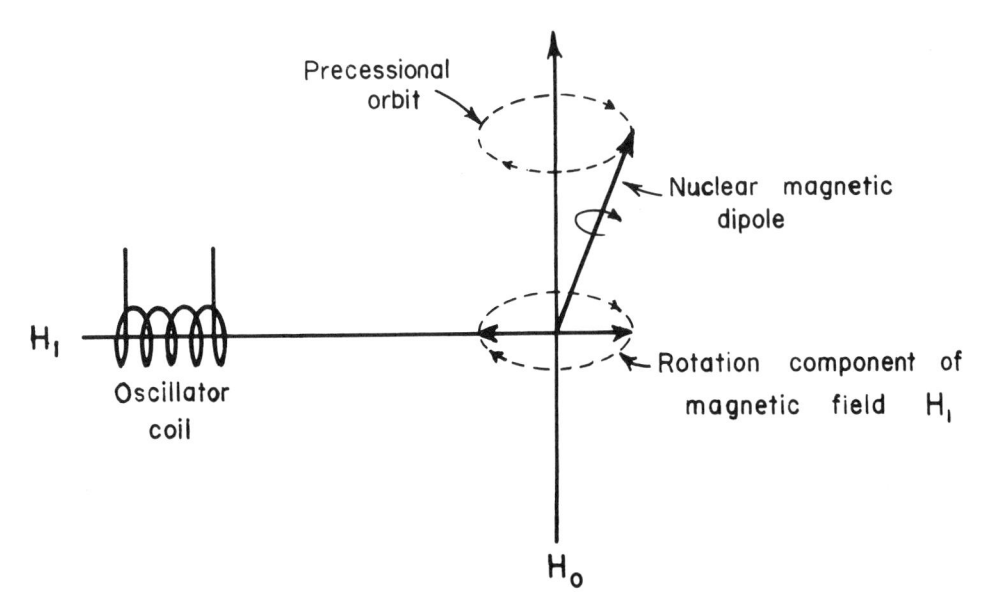

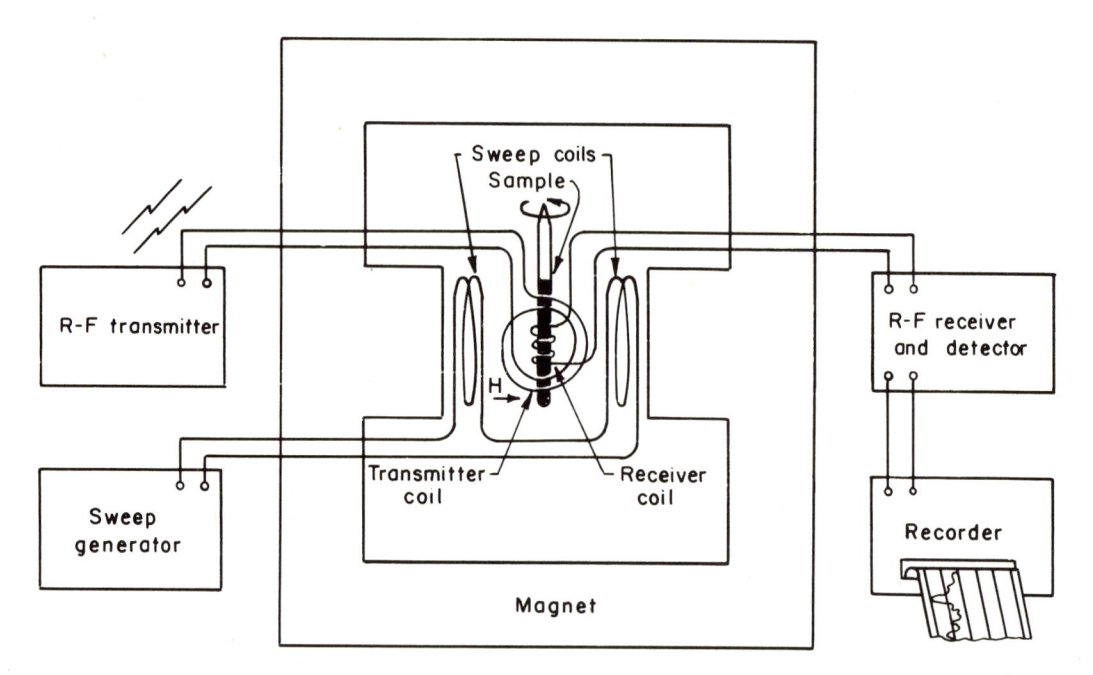

Figure 11.4 Schematic diagram of an NMR spectrometer.

cular) components, in the same manner as that which we analyzed for the circular rotating vectors of plane-polarized light in Chapter 8. One of the components is rotating in the same direction as the precessional orbit of the proton magnetic dipole.

When the rotating magnetic field is brought into resonance with the angular velocity ω_0 of the precessing proton, energy is absorbed and the nucleus flips to its higher energy level. A schematic diagram of equipment used in NMR spectroscopy is shown in Figure 11.4.

11.2 The NMR Experiment

The detection of the rather feeble NMR absorption of the radiofrequency energy is achieved by connecting a very sensitive radiofrequency receiver to a coil wrapped around the sample. The sample is commonly contained in a 5-mm diameter cylindrical tube with a volume of approximately .5 ml. This tube is spun at a rate of 20 to 100 revolutions/second (in most conventional instruments) by an air turbine. Spinning the sample helps to average out static magnetic field inhomogeneities in the plane perpendicular to the spinning axis. In the conventional mode of operation, the radiofrequency ν is maintained constant (60 MHz, 100 MHz, or 220 MHz in most instruments) and the magnetic field is altered so that the resonance condition is achieved.

As with the absorption of light that promotes electronic transitions, we need to identify what happens to the nuclear spin after it achieves its higher energy quantum state. The loss of energy is called spin relaxation and two methods are usually defined.

11.3 Relaxation Processes

Whereas the absorption of radiofrequency energy leads to the promotion of nuclear spin to a higher energy, the reverse process, namely decay or relaxation, can remove the nuclei from the upper spin state. The equation describing the probability of a transition from the lower to upper spin state is similar to that for UV transitions (equation 8.12) and involves a matrix having elements of the wave function for the two states. The relaxation time is inversely proportional to the probability of an upward or downward shift provided that the mechanism of energy loss involves transfer of the energy to the surrounding matrix. This mechanism is known as spin-lattice or longitudinal relaxation and is described in terms of a time τ_1. The influence of relaxation effects on NMR line shapes leads to some important applications in NMR spectroscopy. The lifetime of a given spin state Δt influences the spectral line width via the Heisenberg uncertainty principle, that is,

$$\Delta E \Delta t \sim \hbar, \tag{11.4}$$

Since $\Delta E = h\Delta\nu$ and $\Delta t = \tau$, the lifetime of the excited state, the range of frequencies giving rise to resonance is $\Delta\nu \sim 1/\tau$. In this case a resonance (energy absorbance) band or line half-width at half-height ($\delta = 1/\tau$) occurs in which τ combines all of the relaxation processes affecting the line width. There are actually various means by which the upper spin state may relax, that is, revert to the unperturbed state. In liquids or solutions the spin-lattice or longitudinal relaxation process predominates and τ_1 values lie in the range 10^{-2} to 10^2 seconds. For solids the energy is transferred from one high-energy nucleus to another with eventual loss of phase coherence. Interaction between nuclear moments in a solid may be looked upon as one nuclear spin of fixed orientation providing a nuclear field for a close neighbor of fixed (but not necessarily the same) orientation. The net influence of these local organized field effects in a solid is a broadening of the resonance line. The relaxation time τ_2 is called spin-spin or transverse relaxation. In a liquid or solution these effects are averaged out. Other factors may also affect line width through modification of the relaxation processes (e.g., inhomogeneous external fields), but these factors are usually controlled in the PMR experiment to be negligible. For larger nuclei (e.g., carbon) there are additional contributions to the relaxation process from chemical-shift anisotropy, which originates from the shielding effects of molecular electrons. For nuclei with spin ≥ 1 there are electric quadrupole moments that can interact with the surrounding electronic environment to produce inhomogeneous fields.

NMR measurements on biologic samples have rarely been carried out in the solid state until recently because of the low resolution normally encountered with solids. However, new methods outside of the scope of this text are becoming available to produce high-resolution NMR spectra of crystalline solids, and undoubtedly structural comparisons between proteins in the solid and solution states will be forthcoming shortly. For present purposes we will consider only NMR solution spectra in which the magnetic resonance lines have a Lorentzian shape, the shape function g(ω) being given by

$$g(\omega) = \frac{\tau_2}{\pi} \cdot \frac{1}{1 + \tau_2^2 (\omega - \omega_0)^2} \tag{11.5}$$

The width of the band between the points where absorption is half its maximum height is $2/\tau_2$ in radians/second. In units of Hertz (Hz), the full band-width at half-height is given by $1/\pi\tau_2$ (assuming only spin/lattice relaxation).

Although relaxation times may be obtained from NMR line widths, this is rarely done in modern practice. The detailed methods will not be spelled out here but are explained in detail in the further reading material. τ_2 may be obtained from line widths if $\tau_2 \ll \tau_1$ and it is less than about 1 second. Since line widths of proteins or polypeptides in solution are generally in the 5- to 10-Hz range, reasonably accurate measurements of τ_2 can often be obtained by this method. However, for $\tau_1 < 2$–3 second and $\tau_2 > 1$ second, pulse techniques are used in which a strong radiofrequency field is applied for a short time τ', during which no relaxation occurs. The resulting signal decays ("free-induction decay") with a time constant τ_2 that contains inhomogeneous field effects. In order to eliminate these effects a method known as spin-echo pulsing is used (the reader is referred to the book by Pople, et al., pp. 46–49). τ_1 is obtained by applying a second pulse at right angles to the first. If τ_1 is >2–3 seconds, it may be obtained directly by following the growth of the signal immediately after insertion of the sample into the field or after saturation with high field. A succession of scans provides a record of the change in signal intensity.

As an example of the application of relaxation times, let us examine the results of ^{13}C NMR experiments conducted on a series of pentapeptides of the form GlyGlyXGlyGly. In Figure 11.5 the pentapeptide carbon atoms are listed with $N\tau_1$, where N is the number of directly bonded H atoms. (Relaxation occurs predominantly through ^{13}C—^{1}H dipole-dipole relaxation.)

Five points should perhaps be made regarding these $N\tau_1$ values for the pentapeptides:

1. The shortest $N\tau_1$ values are found for the α-carbon in the central residue with increasing $N\tau_1$ toward the ends;
2. Values of $N\tau_1$ are identical for $C\alpha_2$ and $C\alpha_4$ in each case;
3. The C_β of alanine shows a long $N\tau_1$ compared with other protonated C atoms;
4. $N\tau_1$s for lysine atoms increase with distance from the C_α atom;
5. C_β, C_δ, and C_ϵ atoms in tyrosine are equal within experimental error.

The interpretation of these data are that rotational reorientation occurs more rapidly at the ends of the peptide than in the middle, and the methyl group of alanine undergoes particularly rapid reorientation. There is increasing rotational freedom with distance from the C_α atom in lysine and rotation is most rapid in tyrosine along the aromatic ring twofold axis.

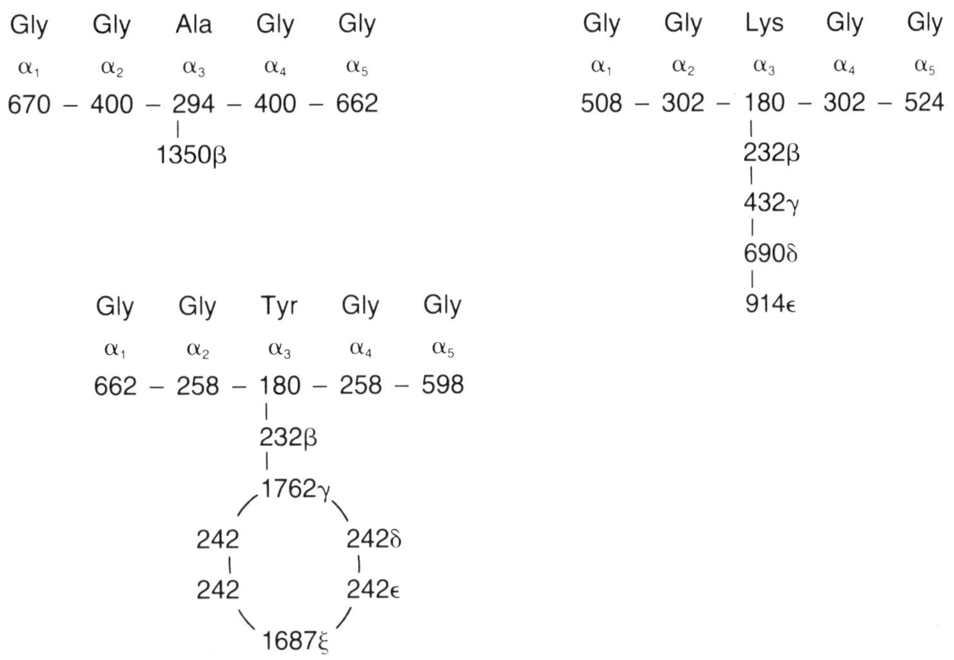

```
Gly    Gly    Ala    Gly    Gly              Gly    Gly    Lys    Gly    Gly
α₁     α₂     α₃     α₄     α₅               α₁     α₂     α₃     α₄     α₅
670 –  400 –  294 –  400 –  662              508 –  302 –  180 –  302 –  524
              |                                             |
             1350β                                         232β
                                                            |
                                                           432γ
                                                            |
                                                           690δ
                                                            |
       Gly    Gly    Tyr    Gly    Gly                     914ε
       α₁     α₂     α₃     α₄     α₅
       662 –  258 –  180 –  258 –  598
                     |
                    232β
                     |
                   ,1762γ.
                242       242δ
                 |         |
                242       242ε
                   \     /
                   1687ξ
```

Figure 11.5 $N\tau_1$ values for the C-atoms of three pentapeptides.

11.4 Chemical Shift

It has been mentioned that atoms resonate at a characteristic frequency. The remarkable power of NMR spectroscopy lies in the fact that the magnetic field defining the resonance frequency is the local microscopic field at the nucleus and is thus dependent on the local environment. For example, a particular nucleus may be shielded from the full effect of the external magnetic field by a neighboring nucleus which may add to, or counteract, the external field. This means that a series of isotopically identical nuclei in the same molecule may experience different effective magnetic fields because of the internal "molecular" field, and will thus appear at differing resonance positions characteristic of their nearest neighboring atoms. The difference between the applied and effective fields is defined as:

$$\delta = \frac{H\text{ effective} - H\text{ applied}}{H\text{ applied}} \times 10^6 \tag{11.6}$$

and is known as the chemical shift, measured in parts per million (ppm). Use of this ratio makes the measurement independent of field strength. This allows identification of the types and number of nuclei surrounding the nucleus of interest, providing evidence for primary structure.

An example of the chemical shift may be demonstrated for the low-resolution PMR (proton NMR) spectrum of ethanol shown in Figure 11.6.

The proton on the oxygen atom responds to the lowest applied field and is said to be least shielded—that is, there is at least an opposing local field from

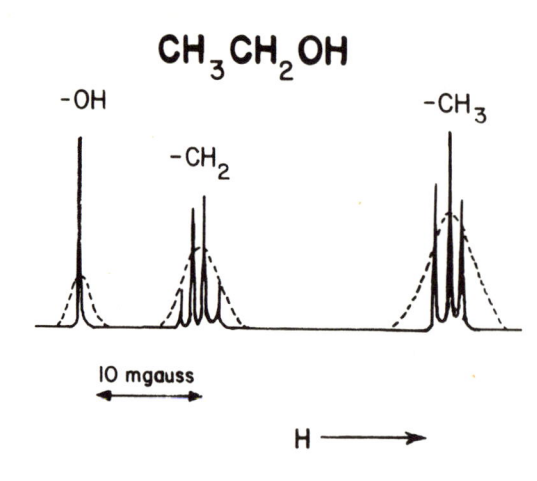

Figure 11.6 Low-resolution proton NMR spectrum of ethanol.

neighboring nuclei. The shielding coefficient (σ) increases for the methylene and methyl protons, which consequently resonate at higher external field strength. Additionally, the area under the absorption peaks is in the ratio of the number of hydrogen atoms associated with the —OH, CH_2, and CH_3 radicals, that is, 1:2:3.

Since accurate measurement of the effective and applied field is difficult, in practice a reference material is used which is used as a calibration agent. Typically tetramethylsilane (TMS) has been used as a common PMR reference material in nonaqueous solvents and the salt $(CH_3)_3SiCD_2CD_2CO_2^-$ is often used in aqueous solvents.

Although NMR instruments sweep the magnetic field, it is convenient to calibrate the scale in frequency (cycles per second or Hertz). The TMS is located at a position of $\delta = 0$ and the separation of resonance peaks is given by Δ cps. However, the shift depends on the shielding coefficient and the field strength, that is, σH_0. Thus the apparent shift will change based on the frequency of the probe (usually 60 MHz, 100 MHz, or 220 MHz). To overcome this problem and obtain peak positions that are independent of field strength, the chemical shift is defined as:

$$\delta = \frac{\Delta \times 10^6}{\nu_0}, \tag{11.7}$$

where ν_0 is the probe frequency. If the sample resonance peaks occur at lower field strength than the reference peak, Δ is defined as being positive in sign.

11.5 Spin-Spin Splitting

A further effect on the appearance of the resonance lines arises from the interaction of the magnetic fields of adjacent nuclei. For example, if the carbon-13 resonance of a methyl group is observed, a quartet of lines is evident. The small energy differences revealed by these nonequivalent signals have an intensity ratio of 1:3:3:1. This is explained by a tabulation of the possible spin orientations of the protons as shown in Figure 11.7. These spin components interact with a parallel spin component of the carbon-13 nucleus

11. Nuclear Magnetic Resonance   221

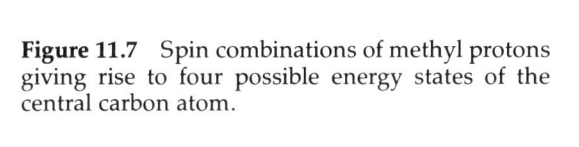

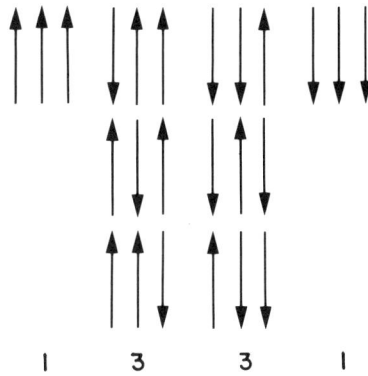

Figure 11.7 Spin combinations of methyl protons giving rise to four possible energy states of the central carbon atom.

to raise or lower the energy. Since no net energy is lost, the pattern is centered about the resonance position of the carbon-13 atom without proton coupling. Proton-proton coupling also takes place over extended distances (four to five bonds) and can be the source of additional structural and conformational information. However, fine structure is absent for proteins and larger peptides because of overlapping contributions. We will examine in section 11.9 how spin-spin splitting may be used to determine aspects of structure.

11.6 Free-Induction Decay Spectra

An advance of some importance in the biologic field is the application of Fourier-transform NMR. This process makes use of a broad radiofrequency pulse of short duration that perturbs the Boltzmann distribution of nuclei. The resulting free-inductive decay in which each type of nucleus transfers energy to the environment by τ_1 and τ_2 processes in order to assume the original thermal distribution of spins, can be transformed into a frequency spectrum (see Appendix A). Since information from the entire range of frequencies is present as an additive set of exponential decay curves, scanning times have fallen from several minutes to a second or less. By recording a series of pulses and adding the results, a tremendous signal-to-noise advantage may be exploited (a net gain in signal-to-noise ratio of $\sqrt{N}$ for N scans) in an amount of real time equivalent to a single continuous wave frequency (or field) sweep observation.

Further advantages accrue. Since the free-induction decay curve is dependent on relaxation times, manipulation of various pulse programs (changes in time and phase) results in signals of differing relaxation times having different amplitudes, thus allowing simplification of complex spectra and easier peak assignment. The Fourier-transform method requires an applied magnetic field that is stable with time; a pulsed, high-power radiofrequency transmitter; and a computer for ease of transformation. Nearly all high-resolution work on biopolymers in the last few years has utilized this method.

There is sufficient natural abundance of ^{13}C in amino acids, peptides, and proteins to obtain ^{13}C NMR spectra. However, in some cases enhancement of the signal has been achieved by use of residues containing enriched ^{13}C.

Figure 11.8 shows that all of the α, β, γ, and δ atoms of gramicidin-S [cyclo(-Phe-Pro-Val-Orn-Leu-)$_2$] are identified in the ^{13}C NMR spectrum by their distinctive chemical shifts.

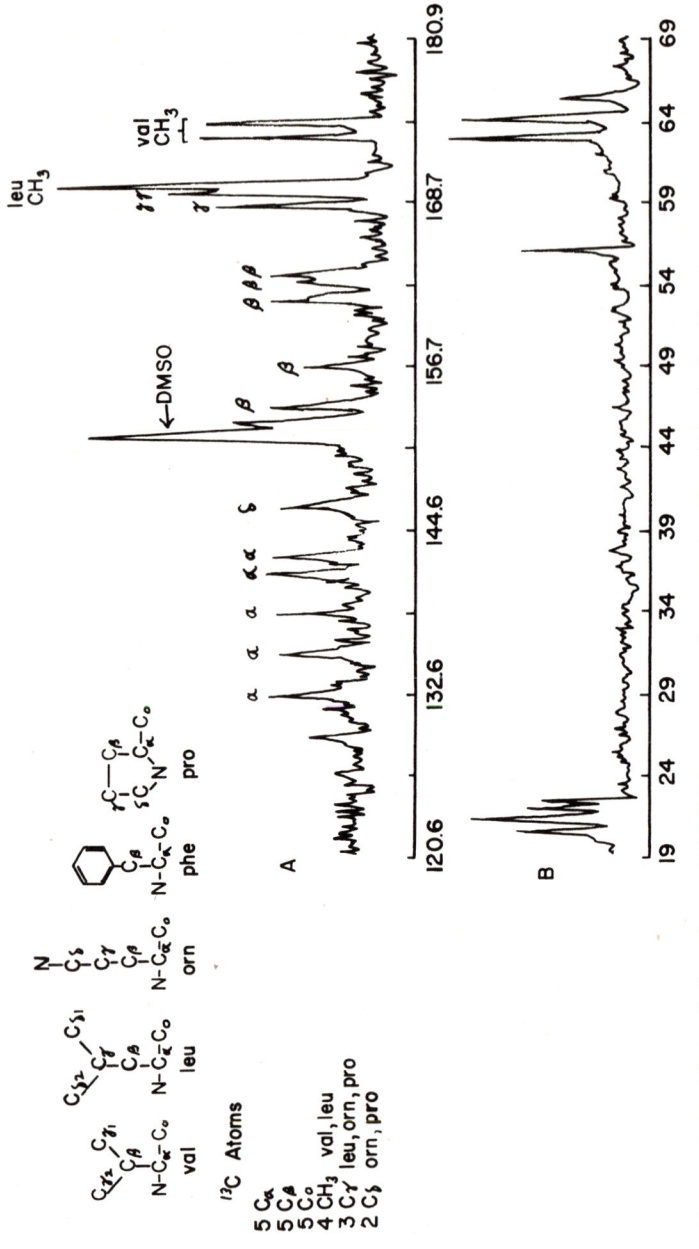

Figure 11.8 Natural abundance ^{13}C NMR spectrum of gramicidin-S in dimethyl sulfoxide.

As larger peptides, polypeptides, and proteins are studied, it becomes progressively more difficult to resolve the chemical shifts of all of the nuclei. Since increased frequency of the probe causes a greater separation of the resonances, many of the published spectra use 220-MHz or higher radiofrequency oscillators.

11.7 Multiple Resonance Effects

Interaction between local magnetic fields can be used to obtain further information. The two general areas of importance are multiple resonance experiments and paramagnetic effects.

If one recalls that interaction of nuclear spins of neighboring atoms can cause splitting of NMR signals, it is obvious that modification of the interaction between nuclei (a) and (b) would cause changes in the multiplicity of the spectral lines associated with them. One way in which this can be done is by spin decoupling, in which a saturating radiofrequency signal is applied to the

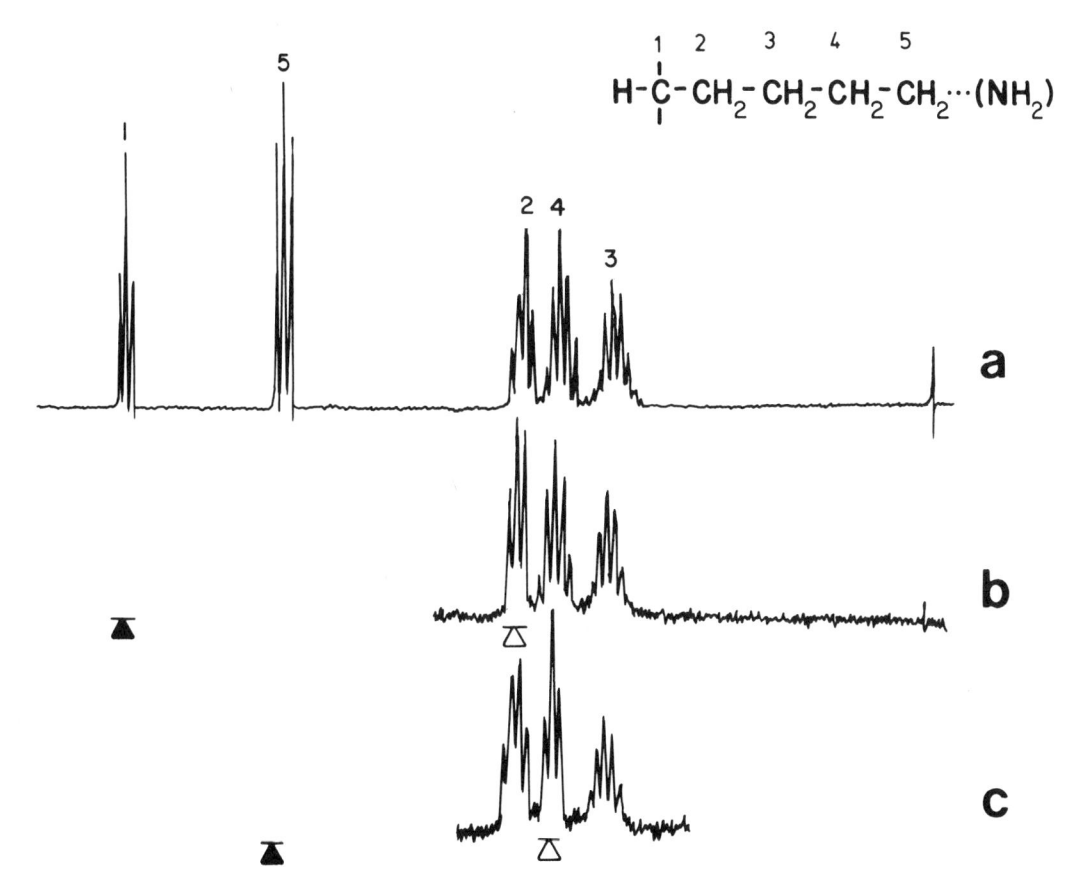

Figure 11.9 Spin decoupling of lysine. Spectrum **(a)** shows the undecoupled spectrum, while **(b)** and **(c)** show the effect produced (at △) by irradiation at ▲, and thus identifying the groups that are coupled to those of peak (1) (a) (the α-CH) and peak (5) (the ϵ-CH$_2$) (after Knowles, et al.).

spectral line of nucleus (a). This alters the Boltzmann distribution, resulting in zero net difference in spin population. The coupling effect of the local field of nucleus (a) virtually disappears, and the spectral line of nucleus (b) is simplified, thus identifying the spatial relationship of the two nuclei. Complex proton spectra can be easily simplified by a stepwise application of this process.

Figure 11.9 shows the NMR spectrum of lysine and the equivalent decoupled spectrum, using the preceding approach.

Another effect is seen during irradiation of one of a pair of coupled nuclear spins. Irradiation of nucleus (a) can cause changes in the population of the spin states of nucleus (b). The observed enhancement in amplitude of the spectral line of nucleus (b) is called the nuclear Overhauser effect. It is of particular interest in carbon-13 spectroscopy, since irradiation of the entire proton spectrum removes proton–carbon coupling and increases the relative amplitude of those carbons carrying protons. Selective irradiation of portions of the proton spectrum during observation of the carbon spectrum allows identification of carbon–hydrogen bonds and unequivocal assignment of resonances. The removal of proton–carbon splitting is standard procedure in observation of carbon-13 spectra; collapse of the spectral lines to singlets increases the apparent signal-to-noise ratio.

11.8 Paramagnetic Effects

Paramagnetic effects arise from the interaction of a nuclear spin with the magnetic moment of an unpaired electron, resulting in changes in either chemical shift or relaxation parameters, or both. Study of proteins by this method is carried out by incorporation of a stable free radical into the system, or by use of a paramagnetic metal ion in the case of a protein with a metal ion-binding site. The electron magnetic moment is about 10^3 times as strong as a nuclear moment and sets up a strong magnetic field in its immediate environment. The interaction of this moment with nuclear moments can be mathematically represented and contains a dependence on r^{-6}, where r is the distance from the paramagnetic center to the observed nucleus. Observation of the chemical shift and relaxation times of a nucleus in the presence and absence of the paramagnetic entity allows the determination of a difference spectrum, consisting only of those nuclei affected, and a consideration of the effect on relaxation times allows the calculation of distance r.

As an example of the application of paramagnetic effects, we shall consider the protein concanavalin A. This protein is one of a class of "lectins" whose function is to bind to terminal sugar residues such as those at membrane surfaces and on a host of glycoproteins such as immunoglobulins. Concanavalin A has two identical subunits of 237 amino acid residues each. Two metal ions must be bound per monomer or saccharide binding does not occur.

The incorporation of manganese II into concanavalin A and its complexation with α-methyl-D-pyranoside enables the τ_1 values of appropriate carbons to be obtained and a picture of the binding site established (Figure 11.10).

The paramagnetic spin-spin relaxation time τ_{1p} is related to the distance from the manganese ion by a sixth-power relation (i.e., $\tau_{1p} = Kr^6$), where K is a complex parameter derived from the resonance frequency.

11. Nuclear Magnetic Resonance

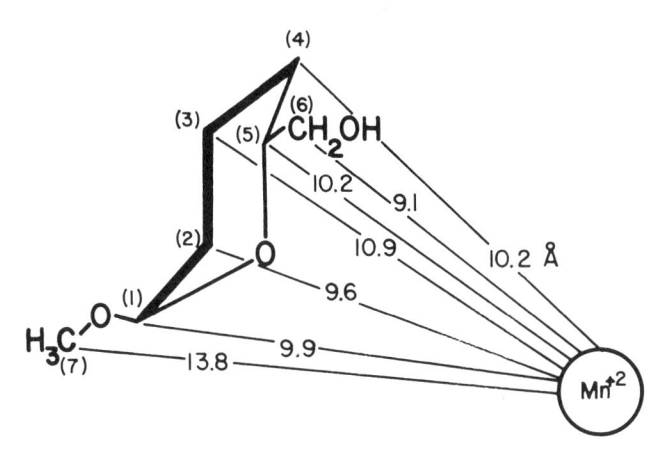

Figure 11.10 Distances from the Mn(II) site of concanavalin A to the seven carbons of α-methyl-D-glucopyranoside.

11.9 Structural Information for Proteins and Peptides

Most of the NMR data pertaining to peptides, polypeptides, and proteins have provided information on local configurational and kinetic properties, rather than an overall structural picture. In principle, there could be sufficient information in the ultimate high-resolution spectrum to obtain all coupling constants and hence all bond-rotation angles. In practice much of this information is "smeared" together, as it is in other forms of spectroscopy such as infrared and Raman. As noted above, much of the more recent methodology has been directed to promotion of some parts of the spectrum by enhancement techniques or by suppression of the remainder. Local perturbation effects, such as those of paramagnetic ions, are reminiscent of the fluorescent energy-transfer studies, both identifying distances of separation with an inverse distance to the sixth power. Nevertheless, for certain situations NMR is capable of providing broader information. We shall examine several situations: namely, peptide conformation, β-turn structure, isomerism, unfolding, and local structural effects that are difficult to approach by other methods.

Coupling Constants and Peptide Conformation

The spin-spin coupling of nuclear moments causes splitting of the resonance bands. In most protein spectra, even at high resolution, this splitting is generally obscured by the overlapping of many adjacent resonance bands. Consequently, the conformational information extractable from the spin-spin splitting is unavailable for most larger polypeptides. The coupling constant J is related to the electron density surrounding the nuclei in question and the bond character, but for present purposes, the most interesting aspect is that bond rotation causes a change in atomic configuration, which in turn changes the separation distance and nuclear coupling.

For example, in an ethylenic derivative

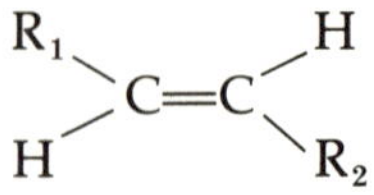

$$\begin{array}{c} R_1 \\ \diagdown \\ C=C \\ H \diagup \diagdown R_2 \\ \end{array}$$

the coupling—that is, the mutual magnetic interaction—between the two protons will be different in the *cis* and *trans* configurations. With a rotatable C—C bond it may be shown that the coupling constant J_{HH} is related to the rotation angle by:

$$J_{HH} = A + B \cos \theta + C \cos 2\theta, \tag{11.9}$$

where experimentally A, B, and C have been found to be 7, −1, and 5 Hz, respectively. Similarly, for peptides, the coupling between —NH and —CH protons may be represented approximately by:

$$^3J_{NH-CH} = A \cos^2 \theta + B \cos \theta + C \sin^2 \theta \tag{11.10}$$

where A = 7.9, B = 1.5, and C = 1.4 Hz. The nomenclature $^3J_{NH-CH}$ refers to the fact that in a peptide the two coupled protons are separated by three bonds (H—N—C$_\alpha$—H), and hence the superscript 3. (The same is of course true in the ethanol examples previously quoted. However, since coupling effects can occur over longer distances the superscript nomenclature is preferred.) Hence a favored ϕ (N—C bond rotation, see Chapter 2) may be obtained from measured peptide coupling constants ($\theta = \phi - 120°$ for a D residue and $\phi + 120°$ for an L residue).

An example of the application of such an approach is for the cyclic pentapeptide malformin A (Figure 11.11a).

By examining a number of linear and cyclic derivatives of malformin, and by use of solvent, temperature variation, and decoupling experiments, it has proven possible to identify all of the proton resonances as shown in Figure 11.11b. The results have been interpreted as in Table 11.1.

Although the C$_\alpha$—C rotation angles (ψ) are not obtained from the NMR data, a combination of the ϕ-angle data and the formation of the disulfide S—S bridge essentially defines the total conformation of the molecule.

Analysis of Denaturation Mechanism

A popular mode of approach to the structural aspects of globular proteins is to follow the behavior of aromatic groups, particularly aromatic protons, which are generally shifted away from the nonaromatic protons in the NMR spec-

Figure 11.11a Formula for the peptide malformin A.

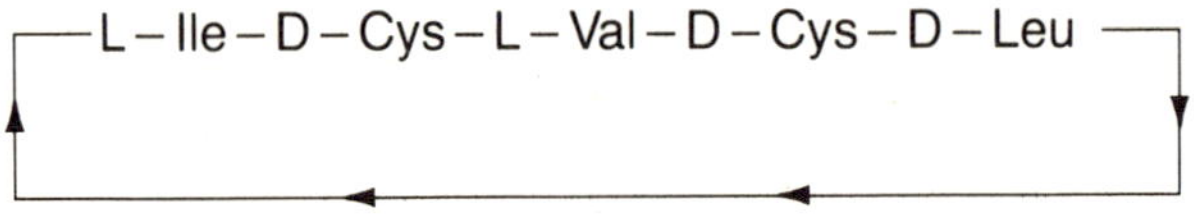

L – Ile – D – Cys – L – Val – D – Cys – D – Leu

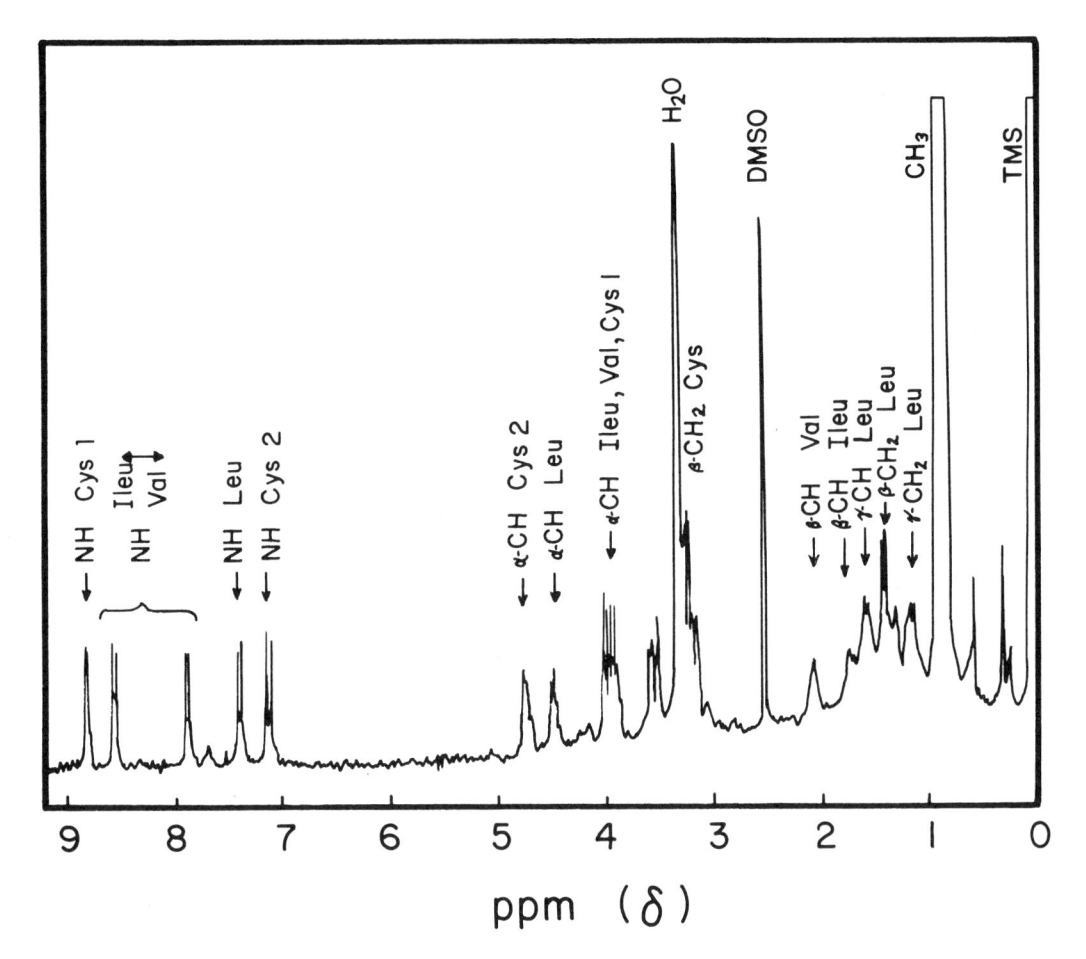

Figure 11.11b Proton spectrum of malformin.

trum. Since the aromatic groups are in different environments within the intact protein, chemical shifts for the protons of each residue are different. As denaturation occurs, the environment changes and eventually becomes equivalent. By comparing the NMR information with x-ray structure, it is possible to identify which aromatic environments change first, and indeed, the sequence

Table 11.1 NH—CH Rotation Angles for Malformin (Deduced from NMR Data)

	ϕ
D-cys (1)	+60°
D-cys (2)	+120°
D-leu	+120°
L-val	+60°
L-ile	−80°
S-S bridge (calc.)	(+60°)

of events. Figure 11.12 shows the chemical shifts of tyrosine and tryptophan aromatic protons of deuterated staphylococcal nuclease as a function of pH.

From an analysis of the preceding type it has proven possible to construct a model for the unfolding (and refolding) of staphylococcal nuclease as shown in Figure 11.13.

As we have seen, the NMR spectrum reflects the environment of atoms whose nuclei are being stimulated. It is usually not possible to determine *ab initio* what protein environment produces a specific chemical shift and thus most methods perturb structure and attempt to relate the perturbation to the properties of the group in question. For the present purposes, we shall examine basic pancreatic trypsin inhibitor (BPTI).

This protein consists of a single polypeptide chain of 58 residues, molecular weight 6,500, containing six cysteines forming three disulfide bridges, and eight aromatic amino acids, four of which are tyrosines and four of which are phenylalanines. The two characteristics (other than size) that make this protein attractive for study are that the structure is available from single crystal x-ray crystallography and that the enzyme is stable between pH 3 and 8 over the range from 1° to 85°C, and between pH 0.5 and 12 over the range from 1° to 40°C. The chain structure of BPTI is represented in Figure 11.14.

Figure 11.12 Chemical shifts of tyrosine (Y1–7) and tryptophan (W), aromatic protons of staphylococcal nuclease as a function of pH.

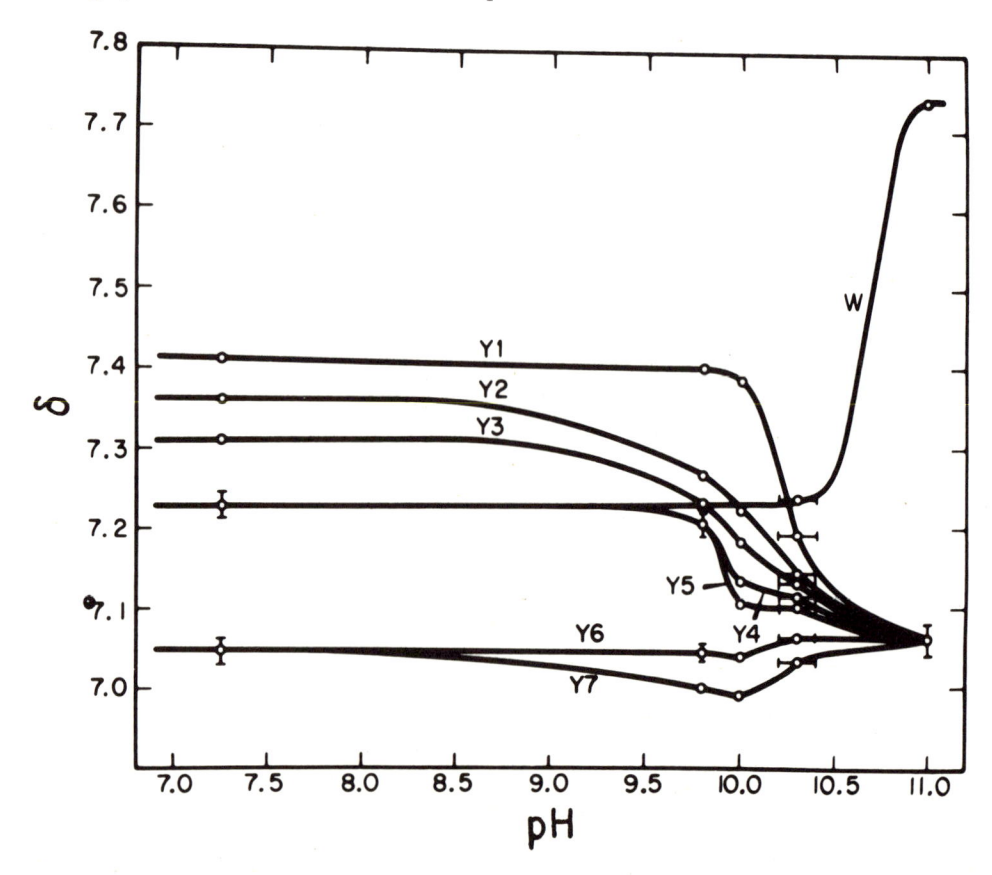

Figure 11.13 Sequence of unfolding and refolding of staphylococcal nuclease as a function of pH, inferred from NMR data. The proposed mechanism starts with the intact nuclease and follows the sequence A–E with increasing pH. Tyrosine residues Y and tryptophan W are marked. The last portion of the chain to unfold contains the tryptophan W residue.

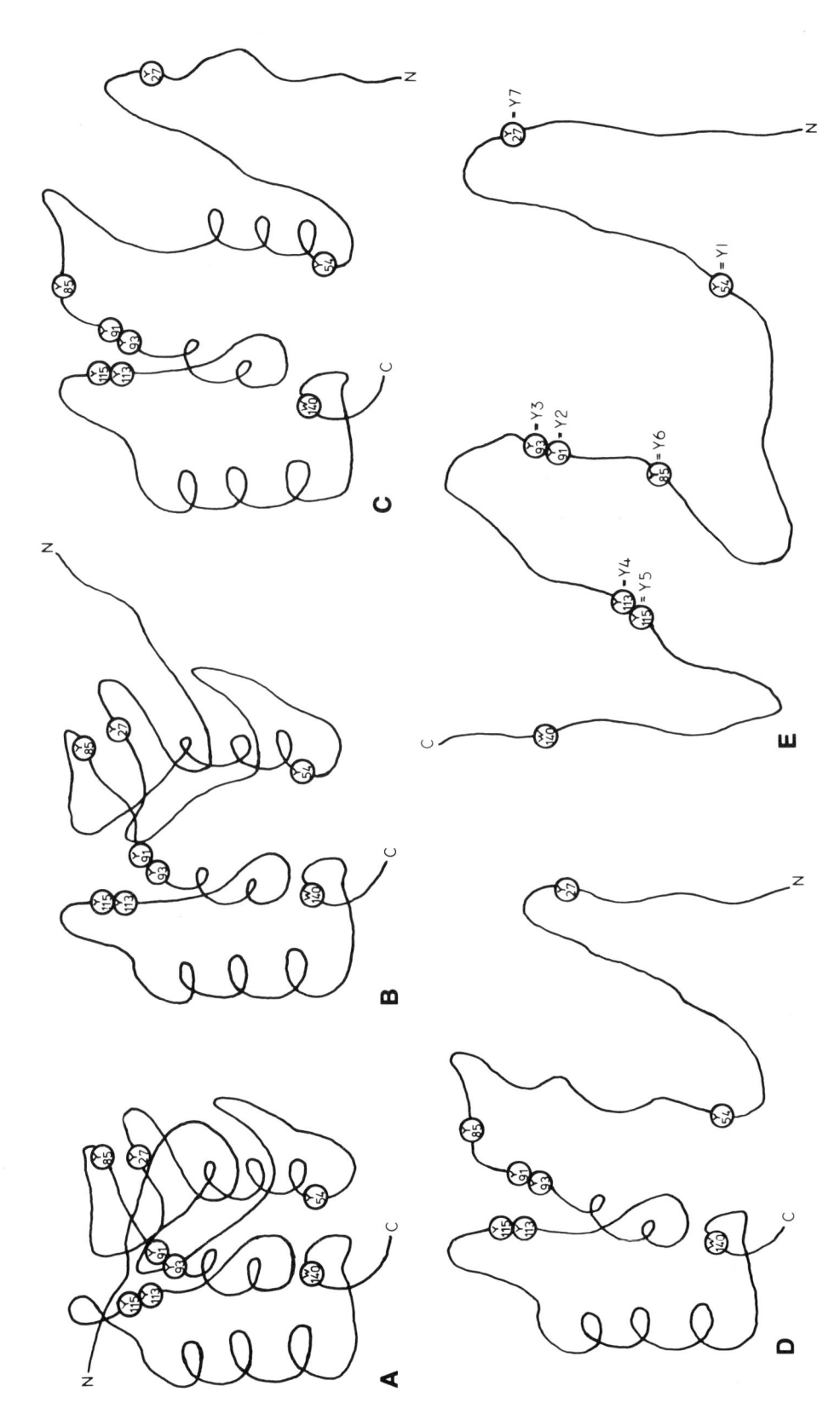

a

1   10
Arg – Pro – Asp – Phe – Cys – Leu – Glu – Pro – Pro – Tyr –
Thr – Gly – Pro – Cys – Lys – Ala – Arg – Ile – Ile – Arg –
Tyr – Phe – Tyr – Asn – Ala – Lys – Ala – Gly – Leu – Cys –
Gln – Thr – Phe – Val – Tyr – Gly – Gly – Cys – Arg – Ala –
Lys – Arg – Asn – Asn – Phe – Lys – Ser – Ala – Glu – Asp –
Cys – Met – Arg – Thr – Cys – Gly – Gly – Ala

b

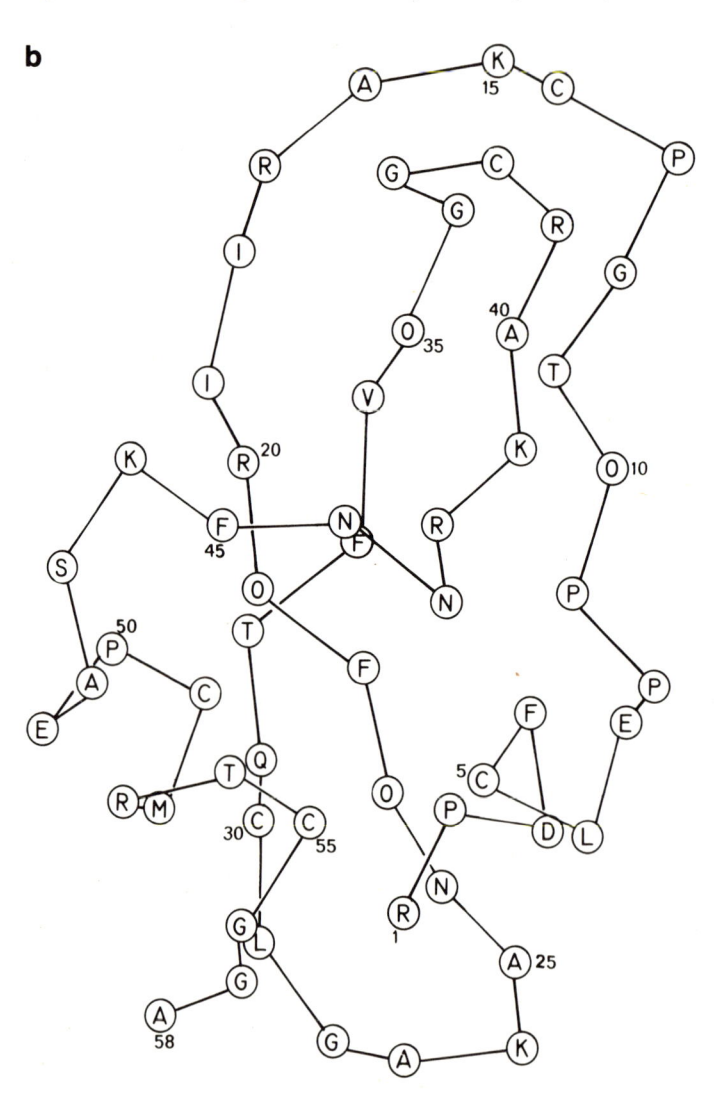

Figure 11.14 The sequence **(a)** and structure (position of backbone atoms) **(b)** for pancreatic trypsin inhibitor.

The NMR spectrum of the protein may be compared with the spectrum of its constituent amino acids (in much the same way as Raman spectra were compared in Chapter 7). In its denatured state, the protein spectrum more closely resembles that of the constituent amino acids, but in its native state considerable differences are apparent that are due to local conformational effects (see Figure 11.15). Evidently these differences occur in all regions of the spectrum and are not readily interpreted.

For this protein it is convenient to examine either the proton or the ^{13}C spectrum as a function of pH and temperature within the range that the protein maintains its native conformation. The changes of pH are likely to change the protonation state of the acidic and basic residues and perhaps tyrosine and others.

Since the proton spectrum contains resonances for around 600 protons, their resolution and interpretation are not straightforward. It is, however, possible to isolate and observe the protons in the aromatic region (6 to 8 ppm downfield from the TMS standard). In the denatured protein and theoretical (composite) spectrum based on amino acid content, there are three distinct resonance bands with the same chemical shifts. By contrast, in the native protein some of the aromatic proton chemical shifts have been moved upfield.

Consideration of a tyrosine side chain (Figure 8.16b) reveals that there are two pairs of hydrogens (2,6 and 3,5) that are related by a C_2-symmetry operation (see Appendix B) and are magnetically equivalent. In an asymmetric protein environment, these protons will only remain magnetically equivalent if rotation about the C_2 axis is fast on the NMR time scale. Analysis of the aromatic proton spectrum in the 6- to 8-ppm downfield region shows from the relative areas under the resonance bands that two tyrosines and one phenylalanine are free to rotate rapidly, the remaining aromatic protons being

Figure 11.15 Proton NMR spectra at 220 MHz of the basic pancreatic trypsin inhibitor. **(a)** Native protein at 20° C; **(b)** denatured protein at 83° C; **(c)** hypothetical spectrum computed from the amino acid spectra. In the spectra **(a)** and **(b)** the residual protons of the solvent are at −4.5 ppm.

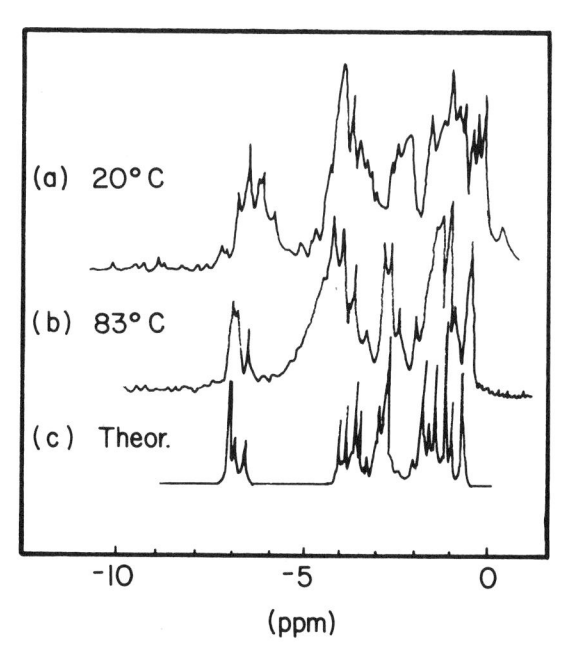

in nonequivalent environments—that is, the aromatic ring is hindered in rotation.

We have seen previously (Figure 11.12) that perturbation of the protein structure by pH also provides chemical/structural information. If a plot of chemical shift vs. pH is established for the identifiable peptide residues in the BPTI, it may be determined that the pK of one lysine (Lys[41]) and one tyrosine (Tyr[10]) are different from other lysine and tyrosine residues. An examination of the x-ray structure (Figure 11.14) shows that Lys[41] and Tyr[10] are in close proximity and probably interact through hydrogen bonding or ionic interaction, thus explaining the NMR behavior. However, in order for these two residues to interact in a manner compatible with the NMR data, there must be some structural distortion in solution from the solid-state structure. Such a result suggests that the conformation of BPTI in the solid state and solution is discernibly different.

NMR and β-Turns

One of the structural features of proteins that is difficult to detect by means other than x-ray crystallography is the β-turn (Figure 2.27). The presence of such β-turns is suspected (from conformational analysis) in other proteins and polypeptides, which do not crystallize in a form suitable for detailed x-ray crystallography. Such is true of elastin and its synthetic polypeptide analogues. These polypeptides have been examined by NMR spectroscopy and show several unique features that have been related to the β-turn. Figure 11.16 shows the primary structure of one of the elastin model polypentapeptides. The ^{13}C-NMR spectrum of this material shows five major peaks corresponding to the amide CO of the main residues. The molecular weight may be established from the ratio of the N-terminal, aldehyde ^{13}C-resonance to that of the other ^{13}C-resonances. The five $>$C=O peaks are similar to those found in α-elastin, suggesting a similar conformation. Perhaps the most interesting piece of evidence is that the Val[1] carbonyl is more shielded, that is, its chemical shift lies to the high-field side of Val[4]; and the change of resonance frequency with temperature is less than that of the other C atoms. Similarly in the proton

Figure 11.16 (a) Elastin model polypentapeptide; (b) proposed β-turn conformation based on shielding of Val[1]C$_α$ and Val[4](N)H.

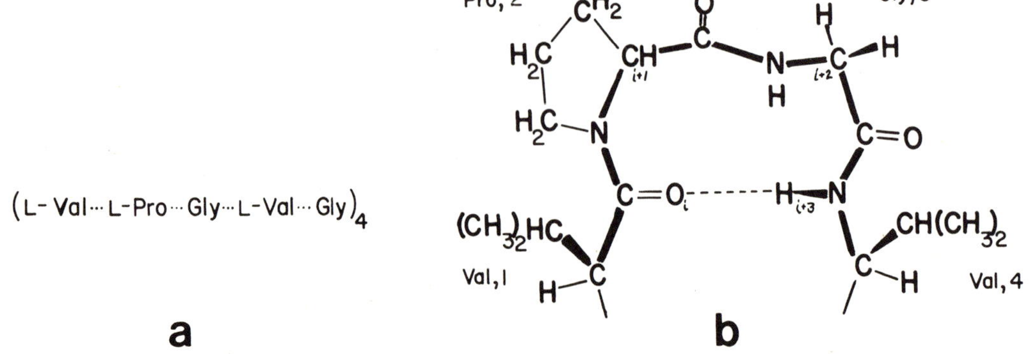

spectra, the amide hydrogen of Val[4] is more highly resistant to thermal effects. Consequently, it is assumed that a hydrogen bond exists between Val[1] and Val[4] as would be expected, uniquely, in the β-turn.

11.10 Summary

We have seen, then, that NMR spectroscopy is a tool that at present is used to explore details about specific aspects of spatial configuration, environment, and aspects of folding, rather than for providing broad conformational information. The technique seems most effective when used to understand subtle changes in structure in conjunction with models provided by x-ray crystallography.

Further Reading

Akitt, J.W. *An Introduction to N.M.R. Spectroscopy*. Halstead Press, New York, 1973.

Becker, E.D. *High Resolution N.M.R.* Academic Press, New York, 2nd ed., 1980.

Bovey, F.A. *High Resolution N.M.R.*, Academic Press. New York, 1972.

Dwek, R.A. *Nuclear Magnetic Resonance in Biochemistry*. Clarendon Press, Oxford, 1975.

James, T.L. *Nuclear Magnetic Resonance in Biochemistry, Principles and Applications*. Academic Press, New York, 1975.

Knowles, P.F. *Magnetic Resonance in Biomolecules*. Wiley-Interscience, New York, 1976.

Pople, J.A., W.E. Schneider, and H.J. Bernstein. *High Resolution N.M.R.* McGraw-Hill, New York, 1959.

Wüthrich, K. *N.M.R. in Biological Research, Peptides and Proteins*, Elsevier, New York, 1976.

III

THERMODYNAMIC AND HYDRODYNAMIC PROPERTIES OF PROTEINS

12

Proteins in Solution

Alex M. Jamieson

Department of Macromolecular Science
Case Western Reserve University
Cleveland, Ohio

The substance of this chapter is a survey of techniques and methodology that can be applied to produce information about the macromolecular structure of proteins in solution. The parameters with which we will primarily be concerned are molecular weight, degree of aggregation, molecular dimensions, and details of the degree of binding with solvent molecules (solvation), or with diffusible ions or other small molecular species present in the solution.

Techniques for molecular-weight determination are conveniently divided into *exact* or *thermodynamic* methods, and *empirical* or *hydrodynamic* methods: the former measure an absolute* molecular weight value by observing the thermodynamic properties of a protein solution at equilibrium; the latter involve a determination of the nonequilibrium transport behavior of proteins in solution and require an empirical calibration between molecular weight and hydrodynamic size. For this reason, hydrodynamic techniques are less satisfactory for accurate determination of the molecular weight of proteins than are the thermodynamic methods; however, there are important situations in which they are useful because of their convenience. There are also two types of size parameters that can be measured. The first is the *radius of gyration* (R_G); this is measured by observing the scattering of electromagnetic radiation (x-rays or visible light) from a protein solution. R_G directly characterizes the spatial distribution of mass (electron density) about the center of gravity of the molecule. The second size parameter is the *hydrodynamic radius* or Stokes radius (R_S); this is determined from a study of the hydrodynamic behavior of protein solutions. R_S is the radius of a hard inert sphere whose hydrodynamic properties are equivalent to that of the protein molecule. If the protein molecule is not spherically symmetric, both light scattering and the hydrodynamic meth-

---

*By *absolute* value one means that it is not measured relative to any other value.

ods can be used to obtain a closer representation of the geometric structure (e.g., whether it is an ellipsoid of revolution, a cylinder, or a stiff or flexible coil).

Techniques that yield absolute values of molecular weights are: total-intensity light scattering, osmotic-pressure measurement, sedimentation equilibrium, and small-angle x-ray scattering (SAXS). The more important hydrodynamic techniques are viscometry, sedimentation velocity, diffusion, gel-permeation chromatography, and gel electrophoresis.

12.1 Osmotic Pressure

In seeking a molecular-weight estimate for a protein, two types of solvent system are especially useful. The first is a solvent approximating the physiologic value for ionic strength and pH (i.e., 0.15N NaCl at pH 7.4); the second is a denaturing solvent (e.g., 6M guanidinium hydrochloride containing 0.1M mercaptoethanol). In the first solvent the protein approximates its native condition, usually a rigid globular or rodlike structure. In the second solvent the secondary and tertiary structures of a protein are destroyed by the guanidinium chloride and all disulfide linkages are disrupted by the mercaptoethanol; consequently each polypeptide chain exists as an individual random chain. A comparison of molecular weight values in the two solvents indicates whether the protein consists of subunits.

When a protein or any solute dissolves in water, the total free energy of the system (solute + solvent) decreases with respect to the separated components by an amount called the free energy of mixing, ΔG_M. This quantity is composed of contributions from the entropy of mixing ΔS_M (which predominantly reflects the fact that dispersion of the solute in the solvent increases the degree of disorder of each component), an energetic component ΔE_M (which reflects the change in the net interaction energies of the components brought about by mixing), and a contribution $P\Delta V_M$ (if there is a net volume change on mixing, ΔV_M). The magnitude of ΔG_M depends on the relative amounts of protein and solvent, that is, on the protein concentration. Each component in a solution can therefore be regarded as having lower potential energy relative to the pure state. Suppose a solution consisting of protein molecules and ions is placed in a compartment separated from a neighboring compartment containing pure solvent by a partition that is permeable to both solvent and solute molecules. Since the pure solvent has a higher chemical potential than the solvent molecules in the solution, there is a spontaneous flow of solvent into the solution compartment. The solute in the solution compartment also has a higher chemical potential than it would have in the pure solvent (i.e., infinitely dilute), so there is a simultaneous movement of solute molecules into the solvent compartment. Each flow continues until the concentration of the solute is the same in both compartments.

The result differs if the two compartments are separated by a membrane permeable to the solvent molecules and the ions but impermeable to the protein molecules. Again the solvent molecules diffuse through the membrane into the solution, but this time, no reverse flow of protein can occur. Therefore, the pressure rises in the solution compartment until it reaches a value at which it exactly opposes the thermodynamic driving force for the passage of solvent molecules into the solution. This pressure is called the *osmotic pressure* of the solution. When equilibrium is reached, diffusible small ions in the solvent are also distributed on both sides of the membrane, so that the chemical potentials of the species are identical in both compartments (if there are negligible interactions of the small ions with the protein molecules, their concentrations on either side will be equal). The condition of osmotic equilibrium is defined by the relation:

$$\mu_1 = \overline{\Delta G_1} = \left(\frac{\partial \Delta G_M}{\partial n_1} \right)_{T,P,n_2,\mu_3} = \Pi \overline{V_1}, \tag{12.1}$$

where the chemical potential of solvent μ_1 equals the excess partial molal free energy of solvent $\overline{\Delta G_1}$, which is equal to the derivative of the free energy of mixing ΔG_M with respect to the number of moles of solvent n_1 at constant temperature, pressure, and concentration of protein (component 2), and constant chemical potential of diffusible ions (component 3); Π is the osmotic pressure; and $\overline{V_1}$ is the partial molal volume of solvent defined as $\overline{V_1} = (\partial V/\partial n_1)_{T,P,n_2,\mu_3}$, where V is the total volume of the solution. The chemical potential of solvent in a solution is directly related to the molecular weight M_2 of the solute by a relation of the form:

$$\overline{\Delta G_1} = \frac{-RTV_1^\circ}{M_2} c_2 (1 + A_2 c_2 + A_3 c_2^2 \ldots), \tag{12.2}$$

where V_1° is the molal volume of solvent, A_2 and A_3 are constants known as the second and third virial coefficients, and c_2 is the protein concentration (weight/volume units). Thus, from equations (12.1) and (12.2), M_2 can be determined by measuring the osmotic pressure of a series of solutions having different concentrations of protein and extrapolating to zero concentration (i.e., $c_2 = 0$). At this limit $\overline{V_1} = V_1^\circ$ and

$$\lim_{c_2 \to 0} \frac{\Pi}{c_2} = \frac{RT}{M_2}. \tag{12.3}$$

Thus in a plot of Π/c_2 vs. c_2, the y-intercept is RT/M. (If the solution contains i proteins having molecular weights M_i, the molecular weight obtained is the number-average molecular weight (M_n), defined as

$$M_n = \frac{\sum_i n_i M_i}{\sum n_i}, \tag{12.4}$$

where n_i is the molar concentration of the i-th polypeptide).

The osmotic pressure of a protein solution is usually measured with commercially available automated membrane osmometers, an example of which is shown in Figure 12.1.

Experimentally, one uses protein solutions ranging in concentration from 1 mg/ml to 20 mg/ml, depending on molecular weight; the smaller the molecular weight, the larger the osmotic pressure for a given concentration. At high concentrations, intermolecular interactions may contribute to the osmotic

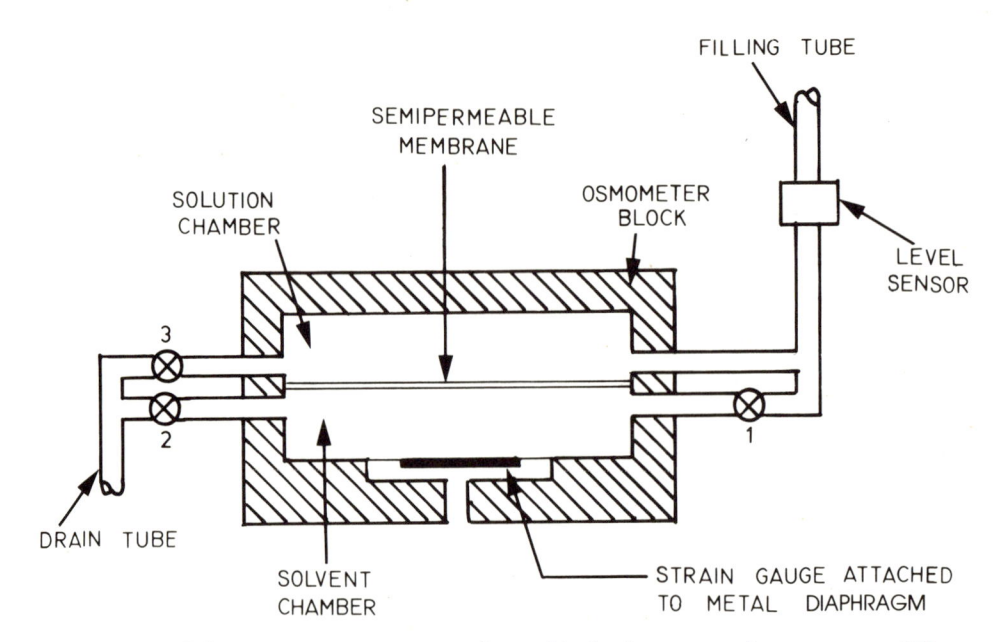

Figure 12.1 Schematic representation of one kind of automated osmometer. When determining π for a solution, the system is first filled entirely with solvent, and then valves 1 and 2 are closed. Next, solution is run into its chamber, and valve 3 is closed when the meniscus in the filling tube has fallen to its predetermined level. The tendency of solvent to pass into the solution chamber gives rise to a signal from the strain gauge, and at equilibrium it can be related to π through a calibration factor. Semipermeable membranes generally used for aqueous solutions are made from cellulose acetate (cellophane).

pressure and cause a strongly nonlinear plot of Π/c versus c. A major complication is due to electrodynamic charge–charge interactions between the proteins. These cause attractive forces between individual protein molecules due to fluctuations in the mean electric charge of the molecules and contribute a large negative term to the concentration dependence. A second problem, the Donnan effect, reflects electrostatic forces and results in an unequal distribution of small diffusible ions on the two sides of the membrane when the protein has a large excess electric charge. This results in a large positive concentration dependence. Both effects disappear rapidly on addition of salt so that osmotic pressure is most effectively measured in the presence of 0.1N NaCl.

An application of membrane osmometry is illustrated in Figures 12.2a and b. Figure 12.2a shows plots of Π/c vs. c for several native proteins in a dilute buffered salt solution. Figure 12.2b shows the corresponding plots for the denatured proteins (that is, dissolved in 6M guanidinium hydrochloride containing 0.5M mercaptoethanol). The reciprocal of the intercept provides the molecular weight of the proteins.* Comparison of the values of M_2 for each protein in the native

---

*The slope of the plots furnishes an estimate of the second virial coefficient A_2, from which one can in theory extract information regarding the charge and shape of the protein, and the protein-salt interaction. However, in practice it is rarely a profitable exercise.

and denatured form permits a determination of the number of subunits. This is summarized in Table 12.1.

When studying oligomeric proteins, there is a special problem. Such species may exist in the form of an equilibrium mixture of associated and dissociated species, the ratio depending on the strength of the protein–protein interaction. The relative amount of dissociated protein increases with decreasing concentration, resulting in a decrease in the number-average molecular weight and therefore in a strongly negative slope to the plot of Π/c vs. c. In fact, if the osmotic pressure data can be taken to sufficiently low concentrations, the y-intercept of the Π/c vs. c plot will correspond to the molecular weight of the smallest species present. In some cases analysis of the shape of the curve allows a determination of the equilibrium constants for the dissociation process. An example of osmotic pressure data of this kind is given in Figure 12.3 for various forms of ligand-bound hemoglobins in which tetramers and dimers are in equilibrium.

Osmotic pressure measurements are valuable only for proteins whose molecular weights are less than about 200,000. Above this size the value of Π becomes too small to be measured reliably unless very high concentrations are used. The necessary concentrations produce such difficulties that the method proves to be unreliable.

Figure 12.2 (a) Plots of Π c vs. c for native proteins (A) aldolase, (B) lactate dehydrogenase, (C) enolase, (D) alcohol dehydrogenase, (E) serum albumin, (F) methemoglobin, and (G) ovalbumin. (b) The same plot as (a) but using a denaturing solution of GuHCl-mercaptoethanol: (A) serum albumin, (B) ovalbumin, (C) aldolase, (D) alcohol dehydrogenase, (E) lactate dehydrogenase, (F) enolase, and (G) methemoglobin.

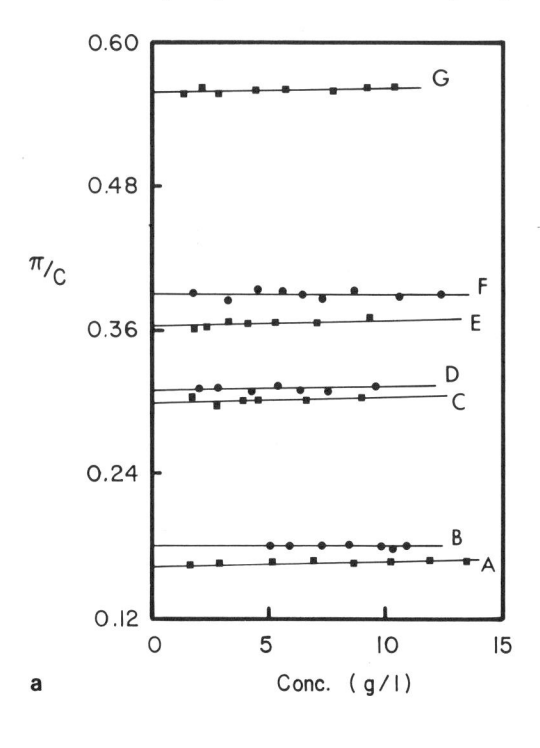

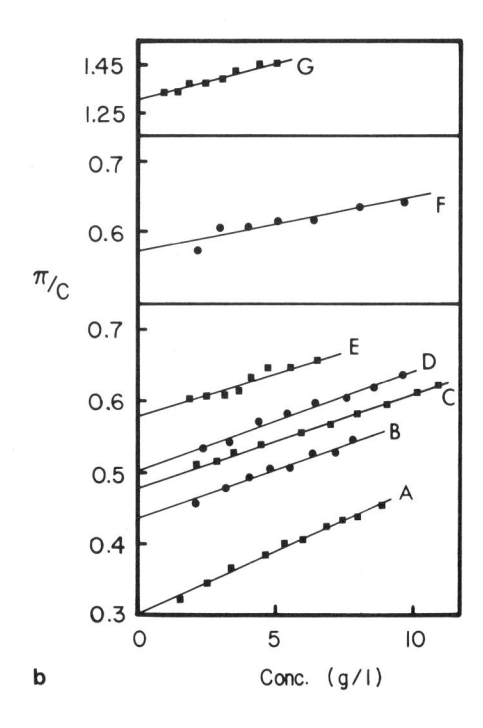

Table 12.1 The Molecular Weights of Native and Dissociated Proteins and the Number of Subunits as Determined by Osmometry*

Protein	Solvent density (g/cm³)	$RT \times 10^{-4}$ (cm 1 mole^{-1})	π/c (cm 1 g^{-1})	M_2	Subunits ± 3%
Serum albumin	1.012	2.4941	0.365 ± 0.003	68,320 ± 600	1.0
Serum albumin + G*	1.150	2.0475	0.302 ± 0.005	67,790 ± 1000	
Ovalbumin	1.012	2.4941	0.559 ± 0.003	44,620 ± 300	1.0
Ovalbumin + G	1.150	2.0475	0.440 ± 0.006	46,530 ± 600	
Alcohol dehydrogenase	1.012	2.4941	0.290 ± 0.006	86,000 ± 1750	2.1
Alcohol dehydrogenase + G	1.150	2.0475	0.502 ± 0.003	40,790 ± 300	
Enolase	1.009	2.5015	0.303 ± 0.003	82,550 ± 800	2.3
Enolase + G	1.150	2.0475	0.561 ± 0.003	36,500 ± 200	
Methemoglobin	1.008	2.5040	0.393 ± 0.007	63,720 ± 1100	4.0
Methemoglobin + G	1.150	2.0475	1.293 ± 0.006	15,840 ± 800	
Lactate dehydrogenase	1.012	2.4941	0.483 ± 0.002	136,290 ± 1400	3.8
Lactate dehydrogenase + G	1.150	2.0475	0.566 ± 0.013	36,180 ± 800	
Aldolase	1.008	2.5040	0.160 ± 0.001	156,500 ± 1000	3.7
Aldolase + G	1.150	2.0475	0.483 ± 0.003	42,400 ± 300	

*G = proteins in GuHCl-mercaptoethanol (0.5M).

Figure 12.3 Osmotic pressure data for hemoglobins in 0.4M MgCl$_2$, pH 7. ● cyanmethemoglobin; □ oxyhemoglobin.

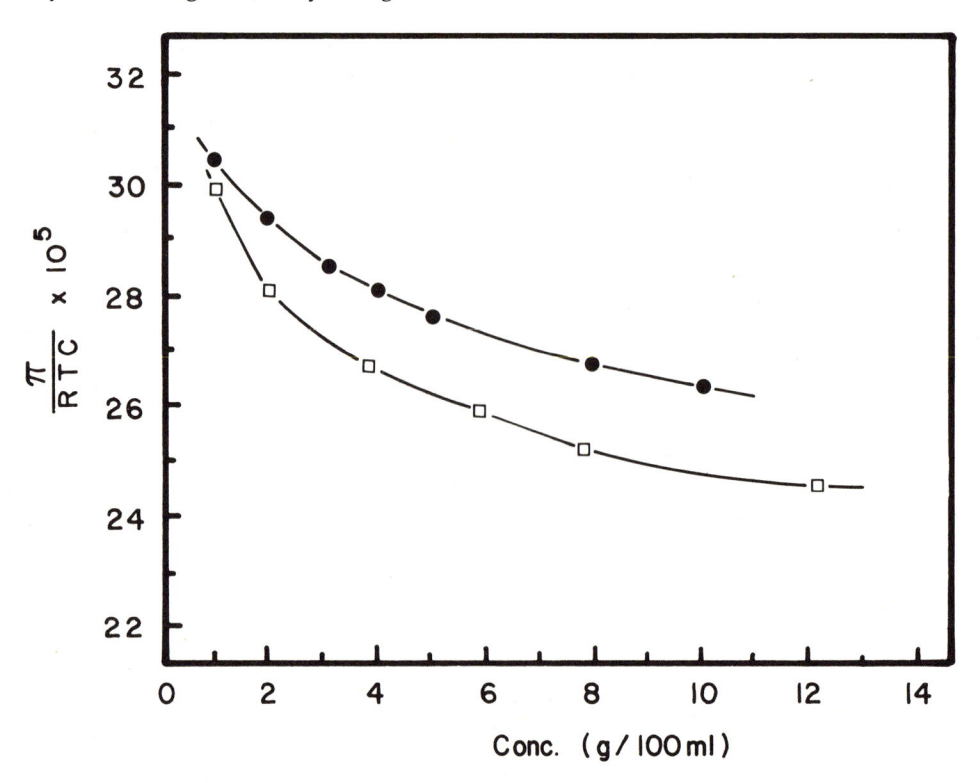

12.2 Light Scattering

An estimate of the absolute molecular weight of macromolecules in solution can also be obtained by light-scattering methods. Light passing through a solution is scattered predominantly by microscopic fluctuations in refractive index owing to localized Brownian fluctuations in the solute concentration. The intensity of the scattered light is proportional to the mean square amplitude of the concentration fluctuations, and this is inversely proportional to the derivative of the solvent chemical potential with respect to solute concentration, $(\partial \Delta \overline{G}_1/\partial c_2)_{P,T,\mu_s}$. As in the discussion of osmotic pressure, we will consider only the three-component system, water-protein-diffusible small ions; it can be shown mathematically that if the chemical potential of the ions remains constant as the protein concentration is varied, such a system can be treated as a two-component system of protein and solvent, the solvent consisting of water plus diffusible ions. Experimentally, the quantity measured in a light-scattering experiment is the ratio of intensity i_θ scattered at a certain angle θ to the intensity of the incident light I_0 falling on the solution. The components and geometry of a typical light-scattering photometer are shown in Figure 12.4a.

If the incident light is unpolarized, i_θ/I_0 is proportional to $(1 + \cos^2 \theta)$ and the reciprocal of the square of the distance r from the scattering volume to the detector. For mathematical convenience a hypothetically angle-independent quantity, the Rayleigh ratio (R_θ), is defined as:

$$R_\theta = \frac{i_\theta}{I_0} r^2 \frac{1}{1 + \cos^2 \theta}. \tag{12.5}$$

The relation between R_θ and molecular weight is given by equations (12.6)–(12.9) below:

$$\frac{K^* c_2}{\Delta R_\theta^{\mu_s}} = \frac{1}{P(\theta)} \cdot \frac{(\partial \Delta \overline{G}_1/\partial c_2)_{P,T,\mu_s}}{V_1 RT}$$

$$= \frac{1}{P(\theta)} \cdot \frac{(\partial \pi/\partial c_2)_{P,T,\mu_s}}{RT} \tag{12.6}$$

where

$$K^* = \frac{2\pi^2 \overline{n}_0^2 (\partial \overline{n}/\partial c_2)^2_{\mu_s}}{N_A \lambda_0^4} \tag{12.7}$$

and

$$\Delta R_\theta^{\mu_s} = R_\theta \text{(solution)}_{\mu_s} - R_\theta \text{(solvent)}_{\mu_s} \tag{12.8}$$

is the excess Rayleigh ratio of solution over solvent at constant chemical potential of solvent (solvent is now defined to be the sum of water and diffusible ion), $\overline{n}_0$ is the refractive index of solvent, $(\partial \overline{n}/\partial c_2)_{\mu_s}$ is the refractive index increment of the solution at constant μ_s, N_A is Avogadro's number,

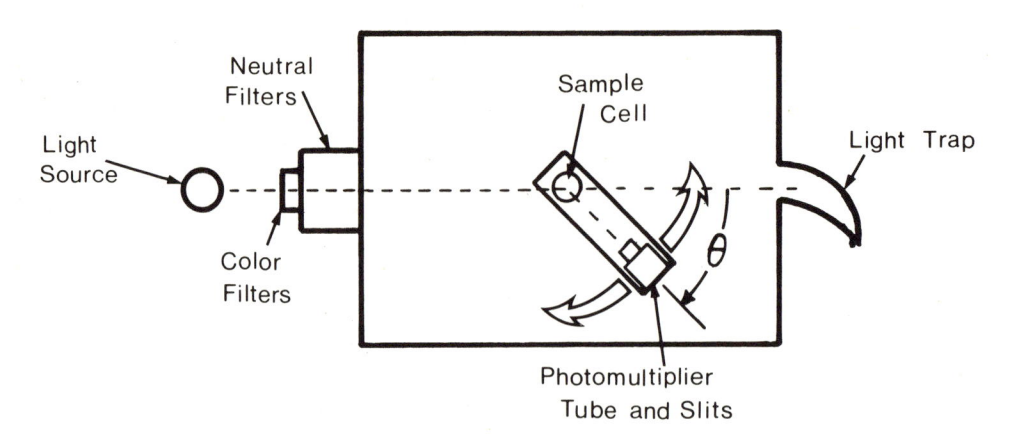

Figure 12.4a Schematic representation of a commonly used light-scattering photometer (see text for details). The photomultiplier tube is set at scattering angle θ with respect to the incident beam path, represented by a broken line. The DC current developed by the photomultiplier tube is proportional to the intensity of scattered light.

and λ_0 is the wavelength of incident light in vacuo. $P(\theta)$ is called the particle-scattering function and describes the angular variation of scattering intensity that arises from interference effects if the protein molecules are comparable in size to λ_0. For molecules whose largest dimension is smaller than $\lambda_0/20$, or 200–500 Å, $P(\theta) = 1$; for molecules larger than $\lambda_0/20$, $P(\theta)$ is not constant and can be used to provide information about macromolecular dimensions. The refractive index increment $(\partial \bar{n}/\partial c_2)_{T,P}$ with respect to protein concentration is measured using a differential refractometer (Figure 12.4b). This instrument measures the deflection of a light beam passed through a specially designed cuvette having two compartments that contain solution and solvent and that are separated by a glass partition at 45° to the incident beam path. From the deflection, the refractive index $\Delta \bar{n}$ between solution and solvent is obtained, and from the known concentration of solution, $\Delta \bar{n}/C_2$ is calculated. Since $(\partial n/\partial c_2)_{T,P}$ is linear at concentrations normally used for light-scattering experiments,

Figure 12.4b Schematic representation of a differential refractometer. The angular deflection ΔL of light passing through the specially designed cuvette is measured by an optical microscope lens mounted on a micrometer gauge. This is related to the refractive index difference $\Delta \bar{n}$ between solution and solvent by a calibration factor.

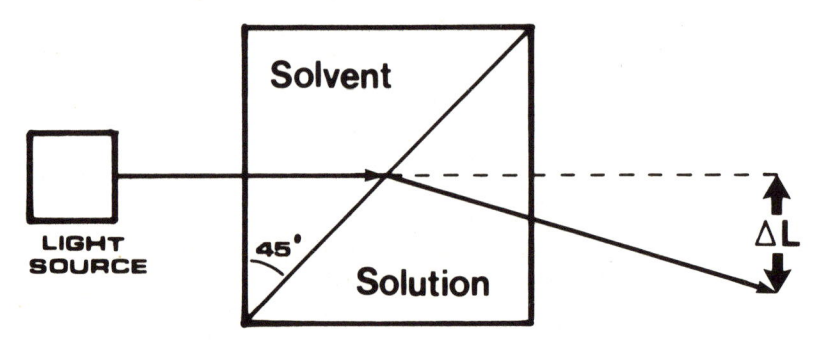

$\Delta \bar{n}/c_2$ is a direct measure of the refractive index measurement. Using equations (12.1) and (12.2), equation (12.6) becomes

$$\frac{K^*c_2}{\Delta R_\theta^{\mu_s}} = \frac{1}{P(\theta)} \cdot \frac{1}{M_w} (1 + 2A_2 c_2 + 3A_3 c_2^2 + \ldots) \qquad (12.9)$$

Thus, for molecules that satisfy the condition $P(\theta) = 1$, the molecular weight can be obtained by simply extrapolating the experimental quantities $K^*c_2/\Delta R_u^{\mu_s}$ to zero concentration; if $P(\theta)$ is not unity, then a double extrapolation, to $c_2 = 0$ and $\theta = 0$, must be performed.

For globular proteins, the most common case, $P(\theta) = 1$. A plot of $K^*c_2/\Delta R^{\mu_s}$ vs. c_2 yields the weight-average molecular weight

$$M_w = \frac{\Sigma n_i M_i^2}{\Sigma n_i M_i} = \frac{\Sigma c_i M_i}{\Sigma c_i}, \qquad (12.10)$$

from the reciprocal of the y-intercept, and the value of A_2 from the slope. The value of M_w that is obtained by the above procedure is not quite correct for the following reason. Most proteins are optically anisotropic and thus scatter a finite amount of depolarized light in addition to the isotropically scattered light to which the previous thermodynamic equations apply. This effect is expressed by the ratio ρ_u of the excess depolarized light-scattering intensity (using a horizontal polarizer in front of the detector) to the excess polarized scattered light (with a vertical polarizer before the detector), that is,

$$\rho_u = \frac{U_H}{U_V}$$

where U refers to unpolarized incident light. The true molecular weight M_w corrected for anisotropic scattering, is related to the uncorrected value M'_w obtained by extrapolation of data according to equation (12.9), through the Cabannes correcction factor

$$M'_w = M_w \left(\frac{6 + 6\rho_u}{6 - 7\rho_u} \right). \qquad (12.11)$$

Usually, the Cabannes factor is very close to unity.

Note that the relatively simple form of equation (12.9) occurs only when the chemical potential μ_s of diffusible solvent species is unaffected by variation of the concentration of protein; both $(d\bar{n}/dc_2)_{\mu_s}$ and $\Delta R_\theta^{\mu_s}$ must be evaluated when the solution and solvent are such that μ_s is constant. This can be achieved most conveniently by dialyzing the protein solution at each concentration c_2 against the solvent until osmotic equilibrium is reached; experimental light-scattering data and refractive index increments are then taken on the dialyzed solution with the dialysate as reference. Under conditions in which one of the solvent species preferentially interacts with the protein, this procedure is the only way to obtain the true anhydrous molecular weight value. If the simpler approach of maintaining constant composition of solvent is used, the measured value of molecular weight will be affected by the preferential interactions described in the next paragraph.

Native proteins in aqueous salt solutions usually exhibit specific adsorption of water (preferential hydration) rather than salt ions. For example, BSA binds 0.19 g H$_2$O per gram protein in NaCl, 0.26 g H$_2$O per gram protein in KCl, and 0.18 g H$_2$O per gram protein in 2.5 molal CsCl. In denaturing solvents, however, the protein often preferentially binds the denaturant—for example, BSA preferentially binds 0.06 g of guanidinium chloride (GuHCl) per gram of protein. In general, values for preferential binding of GuHCl to proteins are small, in the range 0–0.11 g GuHCl/gram protein so that molecular weights can be determined reliably in this solvent.

If experiments are performed at constant solvent composition, and the effects of preferential interactions are ignored, the apparent molecular weight will be too small if preferential hydration occurs (by about 20% for 0.1 g salt per gram protein). When preferential interactions are significant, it is possible, not only to obtain the true molecular weight by carrying out the dialysis technique, but also to estimate the magnitude of the preferential interaction by comparing light-scattering and differential refractometry data at constant μ_s, with the corresponding data at constant molality m_3 of the third solvent component. Thus it can be shown for the excess Rayleigh ratio at zero-angle at constant molality, $(\Delta R_0)_{m_3}$ that:

$$\frac{K^{**}c_2}{(\Delta R_0)_{m_3}} = \frac{1}{(1+D)^2}\left(\frac{1}{M_2} + 2A_2^* c_2 + \ldots\right) \tag{12.12}$$

where the constant K^{**} is the same as K^*, except that it includes $(\partial n/\partial c_2)_{T,P,m_3}$ rather than $(\partial n/\partial c_2)_{T,P,\mu_s}$; A_s^* is an apparent virial coefficient that also depends on preferential interactions and

$$D = \frac{(1 - c_3 \bar{v}_3)_{m_2}}{(1 - c_2 \bar{v}_2)_{m_3}} \frac{(\partial \bar{n}/\partial c_3)_{T,P,\mu_2}}{(\partial \bar{n}/\partial c_2)_{T,P,\mu_3}} \left(\frac{\partial g_3}{\partial g_2}\right) \tag{12.13a}$$

with

$$\left(\frac{\partial g_1}{\partial g_2}\right)_{T,\mu_1,\mu_3} = \frac{-g_1}{g_3}\left(\frac{\partial g_3}{\partial g_2}\right)_{T,\mu_1,\mu_3} \tag{12.13b}$$

$(\partial g_1/\partial g_2)$ and $(\partial g_3/\partial g_2)$ measure the preferential interactions in units of (w/w) of components 1 (water) and 3 respectively, with protein. The application of these ideas has been illustrated by the experimental study summarized in Figure 12.5, which shows Debye plots for ribonuclease A in 30% (v/v) aqueous 2-methyl-2,4-pentanediol (MPD) under conditions of constant μ_s and m_3. Under the former conditions it is found that the true weight-average molecular weight is 23,320, indicating that self-association is occurring. Using this value in equation (12.12), the value of D can be determined from the data at constant m_3, which indicates an apparent molecular weight of 19,950. For this system, $(1 + D)^2\ 23,320 = 19,950$, thus $D = -0.0751$. Since it is also found that $\bar{v}_3 = 1.073$ ml/g, $c_3 = 0.281$ g/ml, $c_2 \rightarrow 0$ at the intercept of the Debye plot, $(\partial \bar{n}/\partial c_3)_{T,P,m_2} = 0.122$ ml/g, and $(\partial n/\partial c_2)_{T,P,m_3} = 0.171$ ml/g, the preferential interaction parameters are calculated to be $(\partial g_3/\partial g_2)_{T,P,\mu_3} = -0.148$ g

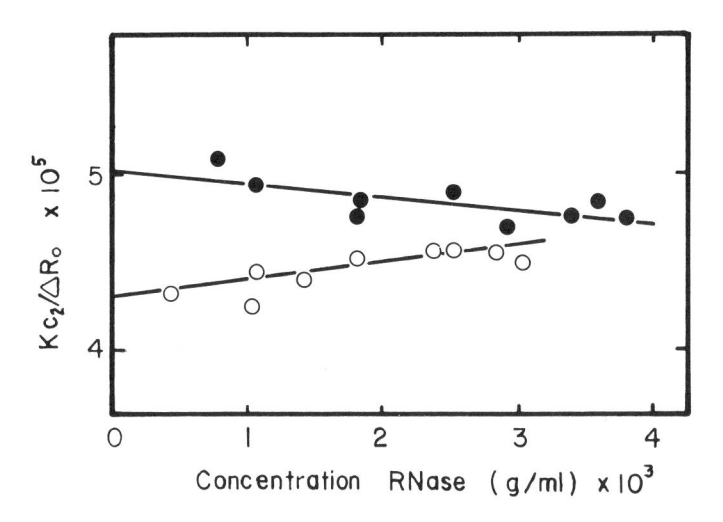

Figure 12.5 $Kc_2/\Delta R_0$ vs. concentration plots for ribonuclease A in 30% (v/v) 2-methyl-2,4-pentanediol at constant molality of component 3 (m_3) (●, 19,950) and at constant chemical potential (μ_3) (○, 23,320) (after Pittz and Timasheff, *Methods in Enzymology* XXVII, p. 255).

MPD per gram ribonuclease, and $(\partial g_1/\partial g_2)_{T,P,\mu_3} = 0.377$ g H_2O per gram ribonuclease.

12.3 Intrinsic Viscosity, Sedimentation Coefficient, and Diffusion Coefficients

Several hydrodynamic parameters from which the molecular weight of a dissolved protein may be obtained are the intrinsic viscosity $[\eta]$, the sedimentation coefficient S^0, the translational diffusion coefficient D_t^0, and the rotational diffusion coefficient D_{2r}^0, each of which is extrapolated to zero protein concentration.

The viscosity η of a fluid is a coefficient that characterizes the resistance to flow. The viscosity of macromolecular solutions is always larger than that of the solvent to a degree that is dependent on the size of the dissolved solute and its concentration. The relative viscosity is defined as $\eta_r = \eta/\eta_0$ where η and η_0 are the viscosities of the protein or polypeptide solution and solvent, respectively. This parameter can be determined in a very simple fashion, since the viscosity of a fluid is proportional to the time t for the fluid meniscus to fall, under gravity, a measured distance down a vertical capillary tube. Specifically, $\eta \propto t\rho$, where ρ is the fluid density. Thus η_r can be obtained by means of a glass capillary viscometer like the Ubbelohde viscometer shown in Figure 12.6 as the ratio

$$\eta_r = t_1\rho_2/t_2\rho_1 \tag{12.14}$$

where the subscripts 1 and 2 refer to solution and solvent, respectively. Alternatively, η_r can be measured by a rotating cylinder viscometer shown schematically in Figure 12.7. In this instrument, the protein solution is confined between two cylinders: the central one (rotor) is free to rotate, and the outside

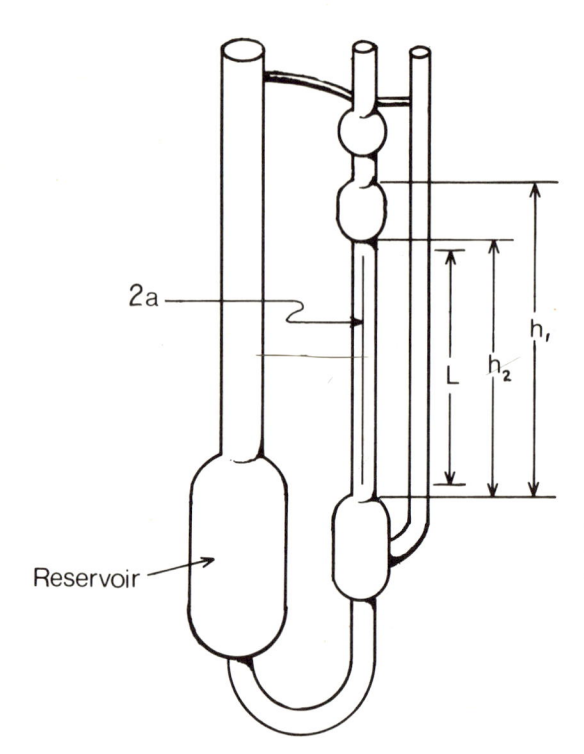

Figure 12.6 Schematic diagram of an internal-dilution capillary viscometer. The flow time is obtained by measuring the time required for solution in the center tube to drain from h_1 to h_2 by passing through the capillary section of diameter $2a$ and length L. The tube on the right always maintains the bottom of the capillary at atmospheric pressure, independent of liquid inventories above or below that point. Dilutions can be made by inserting weighed quantities of solvent directly into the reservoir bulb.

one (stator) is fixed. η_r is obtained as the ratio of the number of revolutions of the central cylinder per unit time for the solution divided by that for the solvent at the same constant torque. Alternatively, the viscosity data may be formulated in terms of the specific viscosity $\eta_{sp} = \eta_r - 1$. The intrinsic viscosity is obtained by extrapolation of experimental data to zero concentration based on either of the Huggins–Kraemer equations.

$$(\eta_{sp}/c) = [\eta] + k_1[\eta]^2 c + k_2[\eta]^3 c^2, \tag{12.15a}$$

$$\frac{\ln \eta_r}{c} = [\eta] - k_1'[\eta]^2 c - k_2'[\eta]^3 c^2. \tag{12.15b}$$

For most circumstances, since $\rho_1 \to \rho_2$ as the $c \to 0$, it is not necessary to measure the solution density and $[\eta]$ can be accurately evaluated from the ratio of flow times (i.e., $\eta_r \sim t_1/t_2$).

In the above, we have assumed that the solutions exhibit Newtonian behavior; that is, the viscosity is independent of the shear stress, defined to be the frictional force per unit area between fluid elements experiencing relative displacement in shear flow. For a Newtonian fluid, the ratio η/t is independent of capillary radius. Most globular protein solutions are in fact Newtonian. However, highly anisometric proteins, like the muscle protein myosin or high-molecular-weight synthetic polypeptides, may exhibit non–Newtonian properties. A further source of non–Newtonian behavior could be extensive intermolecular association which is sensitive to shear stress. For non–Newtonian solutions, the equations (12.15a,b) must be evaluated by dual extrapolation to zero shear stress and zero concentration. In capillary viscometry, shear stress increases as the capillary radius decreases; four-bulb

Ubbelohde viscometers can be commercially obtained (Cannon Instrument, State College, PA) that permit measurements at four shear stress values. In the rotating cylinder viscometer, shear stress is proportional to the mechanical torque applied to the rotating cylinder. For most protein solutions, including those in denaturing solvents, only the linear terms in the concentration series (12.15) are significant over the usual concentration ranges, and $[\eta]$ is obtained as the common intercept of plots of $(\eta_r - 1)/c_2$ and $(\ln \eta_r)/c_2$ vs. c_2. An example of such an analysis for BSA denatured in 6.0M GuHCl—0.1M mercaptoethanol is shown in Figure 12.8.

For spherically symmetric proteins, the intrinsic viscosity is related to the molecular weight by the equation

$$[\eta] = KM_\nu^\epsilon \tag{12.16}$$

where K and ϵ are empirically evaluated constants dependent on protein–solvent interactions. Unfortunately, most proteins in their native forms are not

Figure 12.7 Floating-rotor viscometer of Zimm and Crothers. Rotor is driven by interaction of the circulating permanent magnets with the aluminum plug. Revolution time is recorded electronically using a light beam focused by lenses on the off-center rotor mask and picked up by a photocell on the opposite side of the chamber.

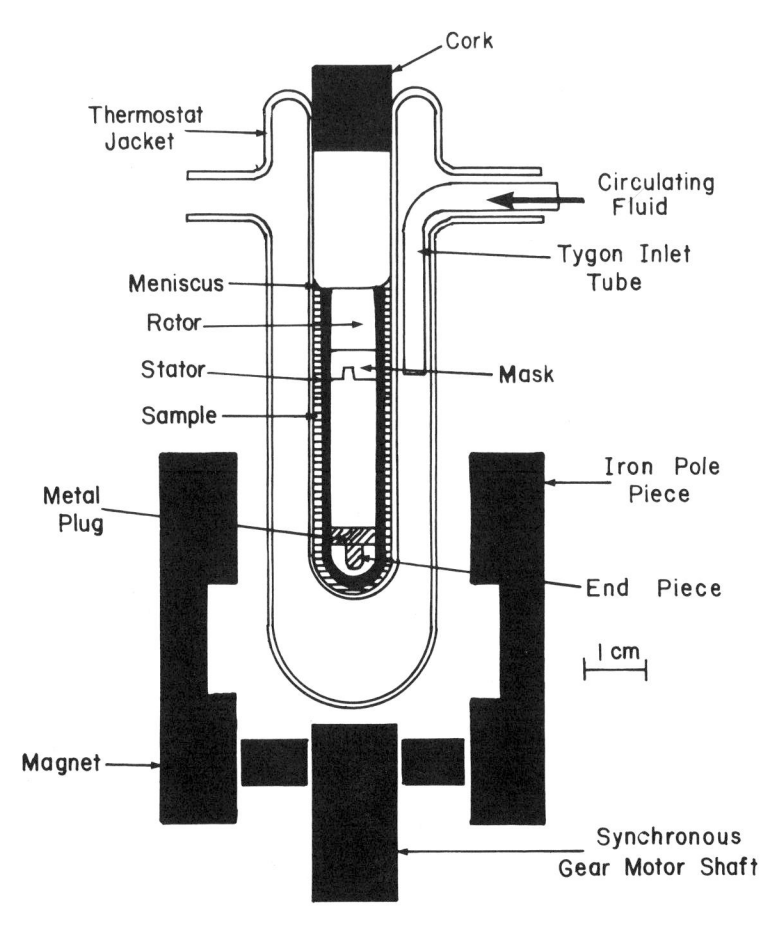

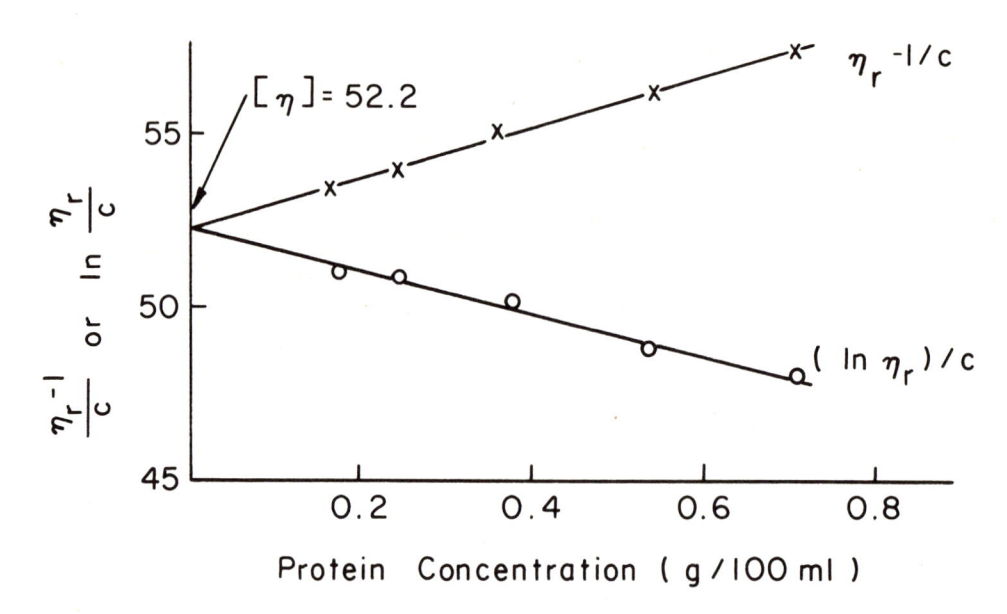

Figure 12.8 Intrinsic viscosity of BSA in 6M GuHCl—0.1M mercaptoethanol.

spherically symmetric and equation (12.16) cannot be applied to them. However as will be seen later, if the molecular weight is known, the value of $[\eta]$ provides information about the shape of rigid proteins.

For protein random coils in GuHCl—0.1M mercaptoethanol, equation (12.16) has been determined to be

$$[\eta] = 3.07 \times 10^{-2} M^{0.666}. \tag{12.17}$$

This equation is very useful in the approximate determination of the molecular weight of denatured proteins. Note that we have said approximate molecular weight. This is the case because the equation actually measures the number of amino acids per molecule rather than the molecular weight. Equation (12.17) is, however, sufficiently accurate for determining the number of subunits of a protein. Unlike proteins, synthetic polypeptides do not have a uniform molecular weight and viscosity measurements provide a viscosity-average molecular weight that lies between the weight- and number-average weights and is given by

$$M_v = (\Sigma_i c_i M_i^{1+\epsilon}/\Sigma_i c_i M_i)^{1/\epsilon} \tag{12.18}$$

i refers to the i-th species.

12.4 Sedimentation Velocity

When a linear force F is applied to a macromolecule in solution, the molecule is accelerated to a limiting velocity u, whose value is determined by a frictional coefficient f that measures the resistance imparted by the solvent to the motion of the particle—that is,

12. Solution of Proteins

$$F = fu. \tag{12.19}$$

The value of f increases with macromolecular size and can in theory be used to determine molecular weight and macromolecular dimensions. Experimentally, two kinds of force can be exerted. The first is a centrifugal force, which is described by Newton's Law:

$$F = \text{molecular mass (corrected for buoyancy)} \times \text{acceleration}$$
$$= \frac{M}{N_A}(1 - \bar{v}_2 \rho_0) \times \omega^2 r \tag{12.20}$$

where $\bar{v}_2$ is the partial specific volume of protein, ρ_o is the solvent density, ω is the angular velocity of the centrifuge, r is the radial distance from the axis of rotation. The rate of migration (sedimentation) of the protein in a centrifugal field is characterized by the sedimentation coefficient, defined as the velocity per unit field or

$$S = \frac{u}{\omega^2 r} = \frac{M(1 - \bar{v}_2 \rho_0)}{N_A f} \tag{12.21}$$

The motion in a centrifugal field corresponds to a flow of solute in the direction of centrifugal force or

$$\frac{dc}{dt} = \frac{cM\omega^2 r}{N_A f}(1 - \bar{v}_2 \rho_0). \tag{12.22}$$

The second force encountered experimentally is a thermodynamic force associated with a gradient in protein concentration. Under these conditions, a flow of solute *(translational diffusion)* occurs along the gradient and

$$\frac{dc}{dt} = \frac{-c}{N_A f} \frac{du}{dr} \tag{12.23}$$

where $(d\mu/dr)$ is the spatial gradient of solute chemical potential in the direction r. For a two-component system,

$$\left(\frac{\partial \mu}{\partial r}\right) = \frac{RT}{c}[1 + c(\partial \ln \gamma / \partial c)_{T,P}]\left(\frac{\partial c}{\partial r}\right) \tag{12.24}$$

where γ is the activity coefficient of protein. The translational diffusion coefficient of the protein D_t is defined by

$$\frac{\partial c}{\partial t} = D_t \left(\frac{\partial c}{\partial r}\right) \tag{12.25}$$

so that

$$D_t = \frac{RT}{N_A f}\left[1 + c\left(\frac{\partial \ln \gamma}{\partial c}\right)_{T,P}\right]. \tag{12.26}$$

If sedimentation and diffusion coefficient data are extrapolated to zero concentration, we obtain

$$S^\circ = M_2(1 - \bar{v}_2\rho_0)N_A f_0, \tag{12.27a}$$

$$D_t^\circ = RT/N_A f_0. \tag{12.27b}$$

The partial specific volume $\bar{v}_2 = (\partial V/\partial g)_{T,P}$ is the derivative of the total volume V with respect to g, the weight of solute in grams; ρ_0 is the solvent density and in the zero concentration limit, f_0 the frictional coefficient extrapolated to c = 0 is given by the Stokes equation:

$$f_0 = 6\pi\eta_0 R_s \tag{12.28}$$

where R_s is the Stokes hydrodynamic radius and η_0 the solvent viscosity. The limiting frictional coefficient is proportional to molecular weight:

$$f_0 = M^\nu$$

where ν is related to the coefficient ϵ of equation (12.16) by $(1 + \epsilon) = 3\nu$, which follows, as will be explained later, because $[\eta]M \propto R_s^3$ and $f_0 \propto R_s$. Since most protein solutions studied are multicomponent systems, the effect of preferential interactions with individual solvent components may be significant. These interactions can be accounted for by modification of (12.27a), replacing $\bar{v}$ with $\bar{v}'$, an apparent partial specific volume that includes both the influence of $\bar{v}$ and the effect of preferential interactions.

Both S and D_t can be measured with an analytical ultracentrifuge. The principle of this instrument is illustrated in Figure 12.9. A protein solution is placed in a sector-shaped cell whose walls are radial to avoid collision of sedimenting particles with the walls. The cell is placed in a rotor, which is rotated at very high speeds by the drive system. A photometric system is aligned so that the cell, and any other cells in counterbalanced positions, can be viewed once in every revolution of the rotor. The radial distribution of solvent in the cell can be detected in one of three ways.

The schlieren method, in which the refractive index gradient is measured from an angular deflection of light passing through the sector-shaped cell, has been outdated by the development of the more accurate Rayleigh interferometric method. This method uses a double-sector cell, one compartment of which is filled with protein solution, the other with solvent. Light passes through a slit and the compartments produce a series of interference fringes. The difference in refractive index between solution and solvent can be measured from these fringes by the simple relation

$$\Delta \bar{n} = \frac{J\lambda}{x} \tag{12.29}$$

where λ is the wavelength of light, x is the path length through the cell, and J is the fringe number. Figure 12.10b shows the fringes symmetrically located around the zero-order fringe in the region of cleared solvent and increasingly

12. Solution of Proteins   253

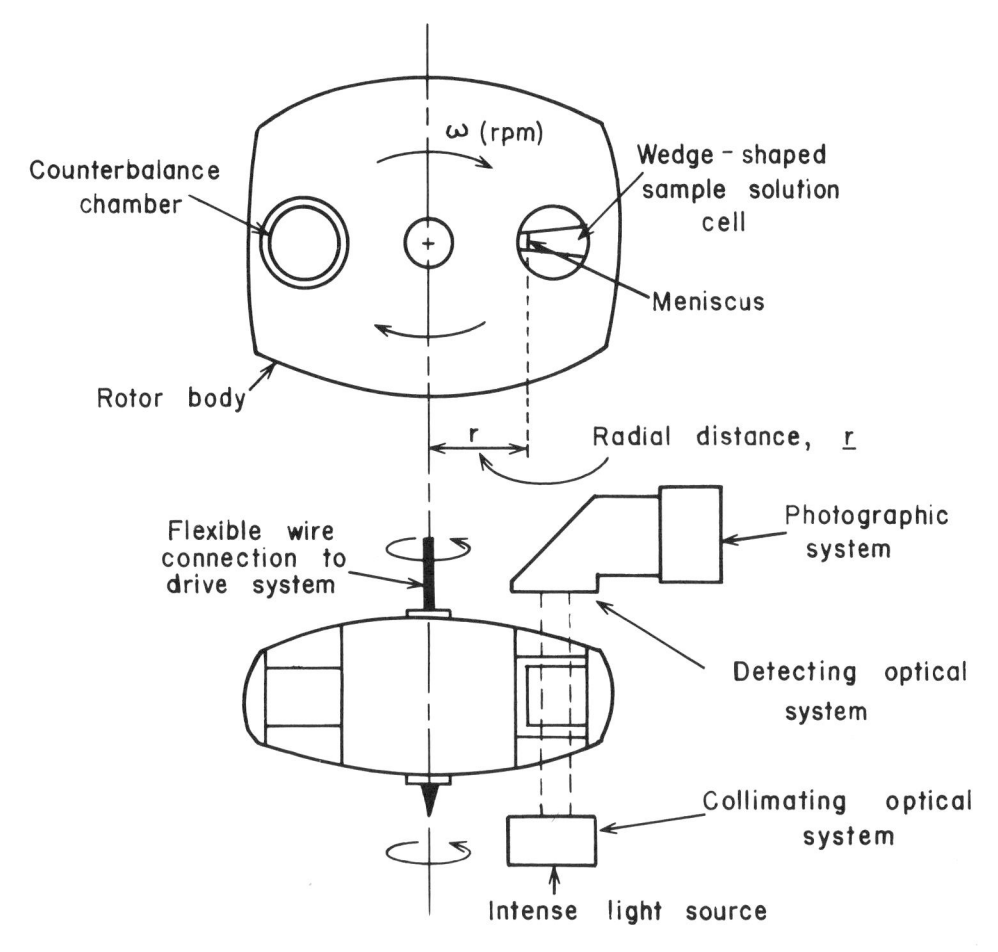

Figure 12.9 Schematic diagram of an ultracentrifuge. The rotor, which is suspended from the drive system by a flexible wire, is located such that the wedge-shaped solution cell is exposed to the schlieren optical system every revolution. When the image of a line is viewed through the sedimenting solution, it is recorded by the photographic system as a curve that gives the concentration gradient directly.

displaced at higher values of r. The Rayleigh interferometer yields the variation in refractive index $\tilde{n}$ and therefore in protein concentration c_2 as a function of r.

The third method of viewing the concentration variation involves scanning the system with light of specific wavelength. In this manner, a trace of optical density with respect to distance r in the cell is obtained that can be directly converted to concentration variation using an experimental value for the extinction coefficient of the solute. A curve of concentration vs. radial distance r as produced by this technique is shown in Figure 12.10a.

The sedimentation coefficient S is determined from the rate of migration dr/dt of the boundary by sequential observation in time of the photometric records, as shown in Figure 12.10:

$$S = \frac{dr/dt}{r\omega^2} = \frac{d\ln r/dt}{\omega^2}. \tag{12.30}$$

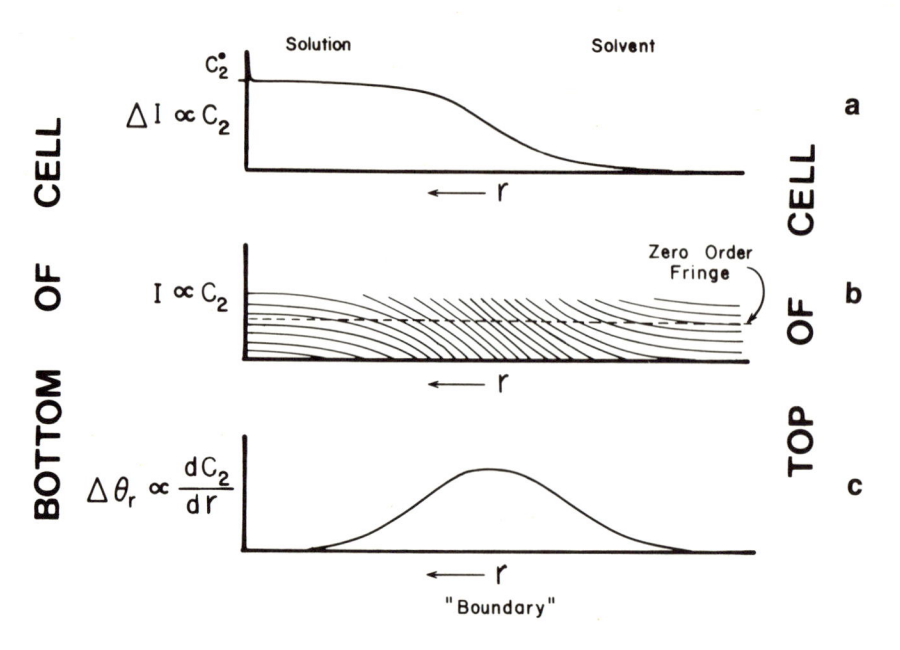

Figure 12.10 Methods for observing concentration profiles in the analytical ultracentrifuge. **(a)** Absorption optics measures optical density $\Delta I \propto c_2$ as a function of r at some specific wavelength characteristics of the solute; **(b)** Rayleigh interferometry measures fringe displacement $\propto$ refractive index $\propto c_2$ as a function of r; **(c)** schlieren optics measures radial deflection of incident light at position $r \propto (\partial \bar{n}/\partial r) \propto (\partial c_2/\partial r)$.

From equation (12.30), S can be calculated from the slope of a plot of the logarithm of the position of the boundary (i.e., maximum in concentration gradient $\partial c/\partial r$) vs. time t.

The diffusion coefficient D_t can also be determined in the ultracentrifuge from the change in width of the boundary region with time. The simplest way to understand how this is done is to consider an experiment in which a sharp boundary of protein at concentration c^o is formed in the ultracentrifuge at high speeds, then a sudden deceleration is applied and the diffusive broadening of the boundary in time is monitored. The shape of this boundary assumes a Gaussian form

$$\frac{dc}{dr} = \frac{c^o}{2\pi D_t} \exp(-r^2/D_t) \qquad (12.31)$$

In practice it is difficult to measure diffusion coefficients with an error of less than 5% in the ultracentrifuge.

12.5 Dynamic Light Scattering

More recently, a light-scattering technique that utilizes the highly monochromatic radiation from a laser source has become available. This permits measurements of D_t with very high accuracy (an error of <1%) in very short times. The method is an adaptation of the classical light-scattering method and enables analysis of

12. Solution of Proteins

the dynamics of the local spontaneous fluctuations in protein concentrations. The instrumentation involved is shown schematically in Figure 12.11. The temporal fluctuations in concentration result in a corresponding dynamic behavior of the scattered light that, in turn, produces fluctuations in the photocurrent developed when the light reaches the photodetector.

For noninteracting monodisperse proteins, the scattered intensity i(K,t) is:

$$i(K,t) = i(K,o)\exp(-D_t K^2 t) \quad (12.32)$$

where i(K,o) is the intensity of the incident light beam; K is the magnitude of the scattering vector and defines the scattering geometry;

$$K = \frac{4\pi \bar{n}}{\lambda_0} \sin \theta/2 \quad (12.33)$$

where $\bar{n}$ is refractive index of the solution, λ is wavelength of incident light, and θ is the scattering angle. The corresponding photocurrent power spectrum $P(K,\omega)$ (usually measured by a spectrum analyzer) is

$$P(K,\omega) = P(K,o) \frac{2D_t K^2}{(\omega - \omega_o)^2 + (2D_t K^2)^2} \quad (12.34)$$

where ω_o is the laser frequency in sec^{-1}. From equations (12.32), (12.33), and (12.34), D_t can be directly obtained either from the time constant of the decay function or from the half-width at half-height of the spectral distribution function. Figures 12.12 and 12.13 show, respectively, the experimental voltage spectrum of light scattered by a solution of the enzyme chymotrypsinogen, denatured in 6M GuHCl—0.1M mercaptoethanol, and a computer least-squares fit of the corresponding power spectrum to equation (12.34). The diffusion coefficient of denatured chymotrypsinogen deduced from the line width is $D_t = 3.15 \times 10^{-7}$ cm^2/sec. Integration of equations (12.32) or (12.34)

Figure 12.11 Principle for measurement of translational diffusion (D_t) by quasielastic laser spectroscopy (from Jamieson, et al.).

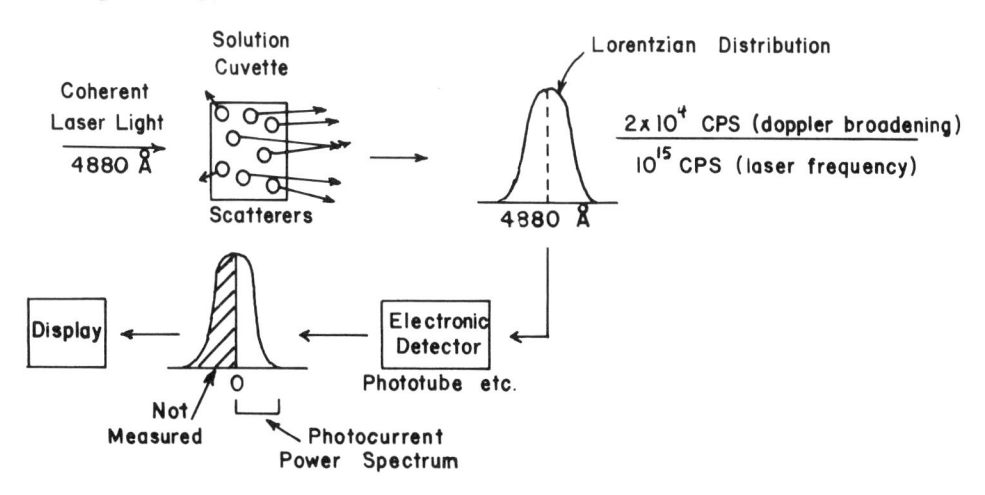

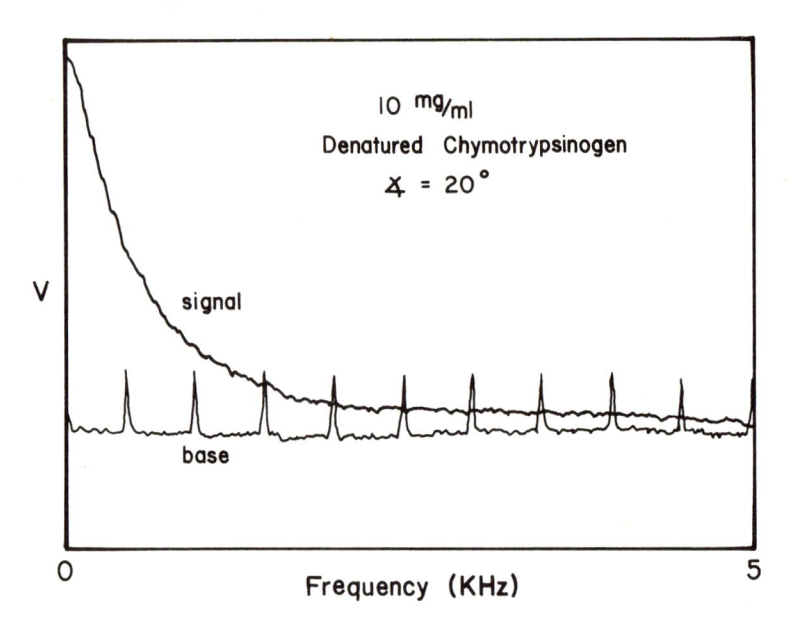

Figure 12.12 Voltage spectrum of photocurrent produced by light scattered from a solution of denatured chymotrypsinogen in 6M GuHCl—0.1M mercaptoethanol (ME McDonnell and AM Jamieson, unpublished data). Also shown is a baseline (shot-noise signal) and frequency.

Figure 12.13 Least-squares single Lorentzian fit to the power spectrum (Power P = voltage squared), using points on the voltage spectrum shown in Figure 12.12. From the half-width at half-height $\Gamma_D = 0.343$ KH, the diffusion coefficient D_t is determined to be 3.15×10^{-7} cm^2/sec using equations (12.34) and (12.35) (McDonnell and Jamieson, unpublished data, 1976).

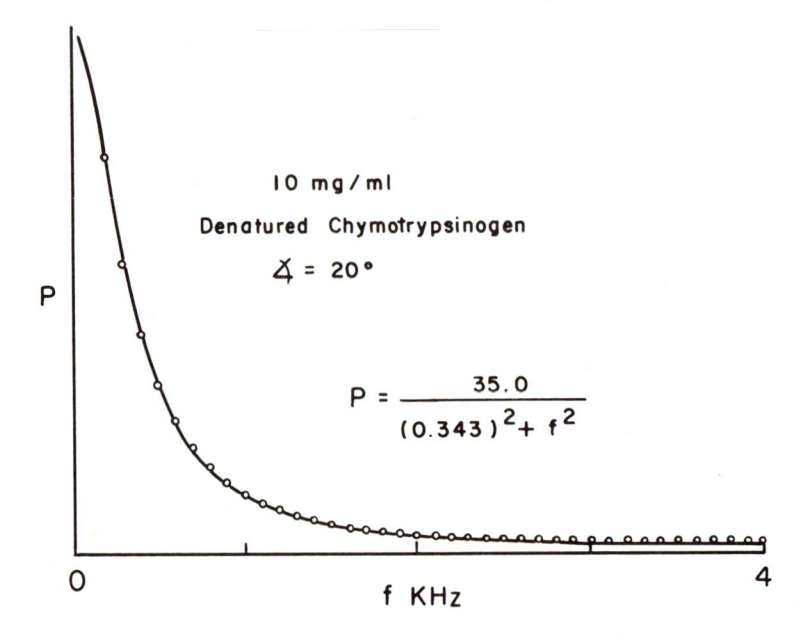

$$P = \frac{35.0}{(0.343)^2 + f^2}$$

leads to the integrated intensity i_θ, which is the parameter of interest for the conventional light-scattering experiments described earlier. With the advent of this new method, measurement of translational diffusion coefficients has become a tool for the study of macromolecules that in many cases turns out to be more convenient and precise than viscometry or sedimentation analysis.

12.6 Measurement of Partial Specific Volume

To implement molecular-weight determination by sedimentation methods, it is necessary to determine the partial specific volume $\bar{v}$. This can be calculated from the density difference $(\rho - \rho_0)$ between solution and solvent since it can be shown that

$$\bar{v} = \frac{1}{\rho_0}\left(1 - \frac{\rho - \rho_0}{c}\right). \tag{12.35}$$

High precision is required of the density measurements which must be accurate to the sixth decimal place to give 0.1% accuracy in $\bar{v}$. This can be accomplished by several alternative routes. The simplest is the technique of pycnometry (precision weighing of solution and solvent in calibrated-volume specific gravity bottles). At a specified temperature, the bottle is weighed empty and then filled with the fluid. Weighings must be accurate to 10^{-6}%. The difference furnishes the weight of a known volume of fluid from which the density is calculated. Comparison of solution and solvent permits evaluation of $\bar{v}$ from equation (12.35). More sophisticated flotation methods have been devised that involve observation of the floating equilibrium of drops of the test fluid or capillary vessels containing the fluid in a medium of known density. Flotation equilibrium occurs when the density of the medium equals the density of the test fluid.

More recently, a very convenient mechanical oscillator technique has been developed by O. Kratky and co-workers and it is now commercially available. This method involves measuring the period of oscillation of a specially designed constant volume u-tube containing the test fluid and caused to vibrate by an electromagnetic device. Since the period of oscillation is inversely proportional to the mass of the tube plus fluid, comparison of the period of oscillation of the tube filled with test fluid with the period of the same tube filled with a fluid of known density permits a sensitive determination of the density of the test fluid. Again, measurements of ρ and ρ_0 permits computation of $\bar{v}$ through equation (12.35).

In multicomponent systems, $\bar{v}$ is affected by the preferential interactions of solvent species. It can be shown that the true anhydrous molecular weight of a protein is obtained from equation (12.27a) if the true partial specific volume $\bar{v}$ is replaced by $\bar{v}'$, the apparent specific volume obtained by performing the density measurement $(\rho - \rho_0)$ between the dialyzed solution and dialysate (i.e., at constant chemical potential of the solvent species).

As for intrinsic viscosity, S^0 and D_t^0 can only be used for molecular-weight determination by means of empirical relations between the parameters and the molecular-weight relationship:

$$S^0 = K'(1 - \bar{v}'\rho_0)M^{1-\nu} \tag{12.36a}$$

$$D_t^0 = K''M^{-\nu}. \tag{12.36b}$$

As with $[\eta]$, such equations tend to be useful only for molecules of regular geometry (spheres, rods of constant cross-section coils).

For polypeptide chains denatured in GuHCl at 25°C the equation

$$D_t^0 = (9.7 \pm 0.7) \times 10^{-5} M^{-0.56 \pm 0.02} \text{ cm}^2/\text{sec} \tag{12.37}$$

describes the situation fairly well.

If both S^0 and D^0 are known, M can be calculated from the Svedberg equation:

$$\frac{S^0}{D_t^0} = \frac{M(1 - \bar{v}'\rho_0)}{RT}. \tag{12.38}$$

If $[\eta]$ is known, this can be used with either D_t^0 or S^0 in the following equations:

$$M[\eta] = \beta^3 (D^0 \eta_0 N_A/RT)^{-3} \tag{12.39}$$

and

$$M^{-2}[\eta] = \beta^3 \left[\frac{S_2^0 \eta_0 N_A}{(1 - \bar{v}'\rho_0)} \right]^{-3}. \tag{12.40}$$

If the molecules are spherically symmetric, the constant β may be assumed to be universal. As we will discuss later, if M is known equations (12.39) and (12.40) can be used to determine the shapes of aspherical proteins. The Svedberg equation does not exhibit any dependence on shape and therefore can be used to determine molecular weights of native as well as denatured protein chains.

Equation (12.39) is found to describe a wide range of native and denatured proteins. Figure 12.14 summarizes some examples of these data. All of these results can be fitted by the least-squares line shown, which corresponds to a value $\beta^3 = 2.7 \times 10^{-27}$ to an accuracy of 10%. Even the $[\eta]$ and D_t^0 values for the highly asymmetric macromolecular species such as tropomyosin and tobacco mosaic virus fall within these limits. To illustrate the usefulness of equations (12.38) and (12.39), we note that many biologic proteins are in fact copolymeric with either a lipid or a carbohydrate moiety. The molecular-weight characterization of these species is subject to technical problems in thermodynamic analysis because of the complex nature of preferential interaction of solvent species with the different chemical components of the macromolecule. Again, because of their copolymeric nature it is impossible to generate accurate empirical calibration curves for hydrodynamic parameters. An example of such a molecule is the glycoprotein ovomucoid, almost 25% of the molecular weight being saccharide units covalently linked to the protein. Interpretation of experimental data $[\eta] = 10.6$ cc/g and $D_t^0 = 6.0 \times 10^{-7}$ cm^2/sec at 25°C using equation (12.39) leads to the result M = 31,400. This compares favorably with values of 27,000 and

Figure 12.14 Graphic plot of log $[\eta]M$ vs. log D_η^o/T (see equation 12.39) for several protein/solvent systems: ○ Denatured protein in 6M guanidine hydrochloride—0.1M mercaptoethanol at 25° C; □ native BSA in water at 20° C; + native tropomyosin in water at 20° C; △ native southern bean mosaic virus in water at 20° C; × native TMV in water at 20° C.

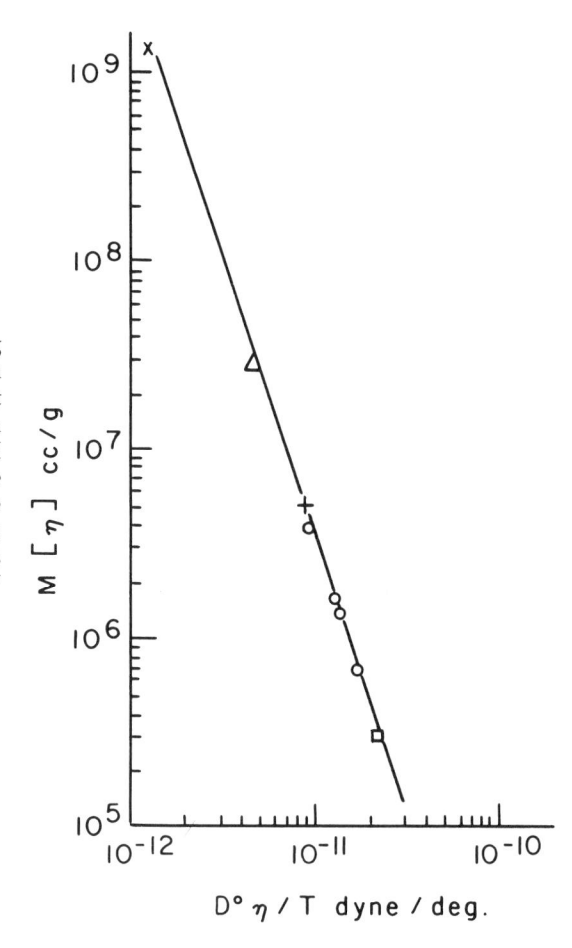

31,500 independently determined by combination of diffusion and sedimentation results.

12.7 Gel-Permeation Chromatography and Gel Electrophoresis

Two techniques that depend on the fractionation of macromolecular species in porous gel matrices provide useful methods for molecular weight determination of proteins. These are gel-permeation chromatography (GPC) and gel electrophoresis of proteins complexed to sodium dodecyl sulfate. GPC (also called exclusion chromatography) involves allowing the protein solution to pass down a column packed with a porous gel. The gel bed may be regarded as comprising two phases: the gel phase with volume V_x, and a liquid phase with volume V_o (void volume). The solvent is the mobile phase and the gel is the stationary phase in the column. If a protein solution is inserted at the top of the column, the proteins do not usually pass through the column at the same rate as the solvent, since at all points they exist in an equilibrium distribution between the gel phase and the liquid phase, depending on the amount of volume available for them in the porous gel matrix. If the fraction of available volume is K_{av}, the total available volume is $K_{av}V_x$. If the elution volume, or volume of solvent required to elute or wash down a certain protein species, is V_e, then

$$V_e = V_o + K_{av} V_x. \tag{12.41a}$$

Thus if a protein is totally excluded from the gel phase, $K_{av} = 0$ and $V_e = V_o$; it elutes in the void volume. Otherwise it elutes at a point determined by K_{av}. Since obviously the fraction of available space in the gel phase is inversely proportional to the hydrodynamic size of the protein, $K_{av} \propto R_s^{-1}$, and the proteins are eluted sequentially in order of decreasing molecular size. If a column system is designed to give highly reproducible elution characteristics, most effectively by including a constant volume pump to regulate flow of the solvent phase, then the molecular weight of unknown proteins can, in principle, be determined from V_e, provided a calibration of V_e vs. M has been established using protein standards. The instrumental arrangement of a typical analytical GPC column is shown in Figure 12.15; the detector used is usually a differential refractometer or a UV spectrophotometer. A number of different column materials can be used: crosslinked dextran gels (Sephadex); agarose gels (Sepharose); and crosslinked polyacrylamide gels (Biogel); crosslinked polystyrene gels can only be used with nonaqueous solvents. Each of these gel types is available in a wide range of pore sizes. Experimentally, data are determined as elution volumes V_e, represented by equation (12.41a), or by elution times

$$t_e = t_o + K_{av} V_x. \tag{12.41b}$$

It is best to compute the partition coefficient K_{av} using experimental estimates of V_o and V_x, since this parameter should be more reproducible in independent

Figure 12.15 General scheme of automated system for elution chromatography.

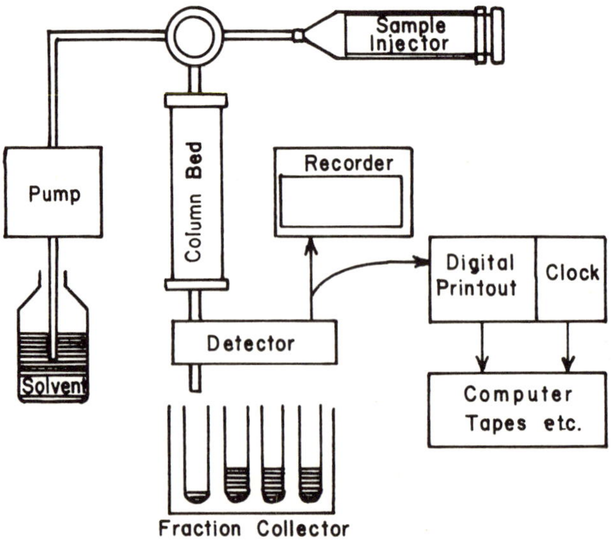

experimental studies. Most exact correlation of molecular weights with elution characteristics is found for proteins in their denatured states in 6M GuHCl—0.1M mercaptoethanol, as might be expected, since the hydrodynamic particle is a random coil under these conditions. Figure 12.16 shows a correlation of M with partition coefficient for this system. In fitting data of this kind to a mathematical function, it is unfortunate that no precise theoretical model of the exclusion phenomenon is available. Nonlinear curves of the type shown in Figure 12.16 can often be fitted over a wide range of molecular weights by a function of the form

$$K_{av} = A \log M + B. \tag{12.42}$$

A similar study of native proteins by Andrews shown in Figure 12.17 does not show such a simple relationship. Native proteins with asymmetric structures, like fibrinogen, exhibit anomalously small values of V_e or D_{av}, as do those, like ovomucoid, that contain carbohydrate.

A better correlation might be expected between the distribution coefficient and the Stokes radius R_s rather than molecular weight, since it is the former parameter that determines the exclusion properties. G. K. Ackers has proposed a linearized equation for this relationship based on a theoretical model of the exclusion phenomenon:

$$R_s = a_o + b_o \, \text{erf}^{-1}(1 - K_{av}) \tag{12.43}$$

Figure 12.16 Correlation between molecular weights and partition coefficients for proteins chromatographed on 6% agarose columns in 6M guanidine HCl—0.1M mercaptoethanol. BSA is bovine serum albumin, MDH is maleate dehydrogenase (taken from Fish, et al.).

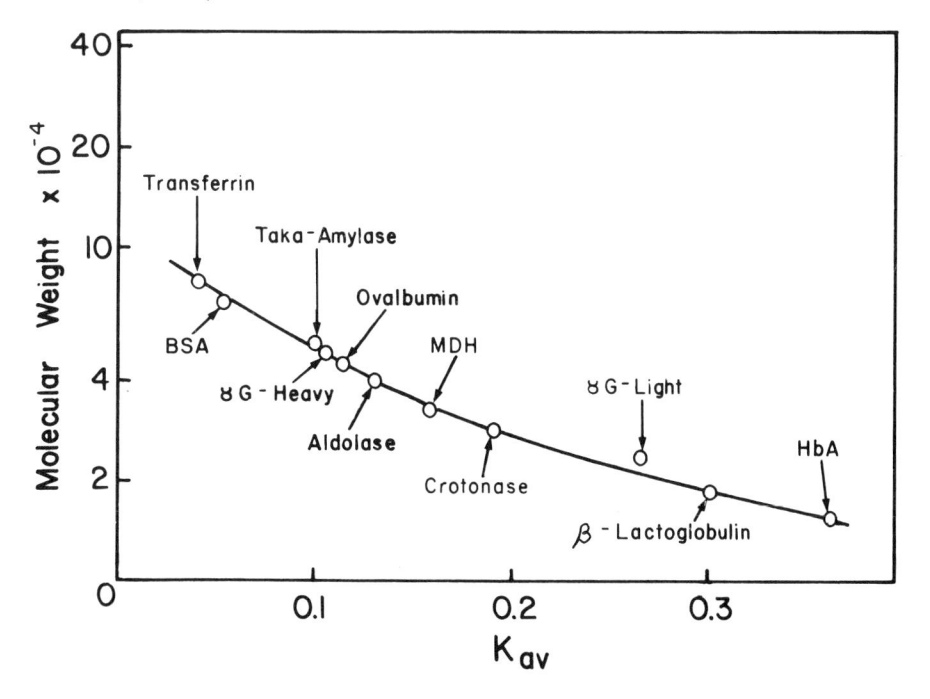

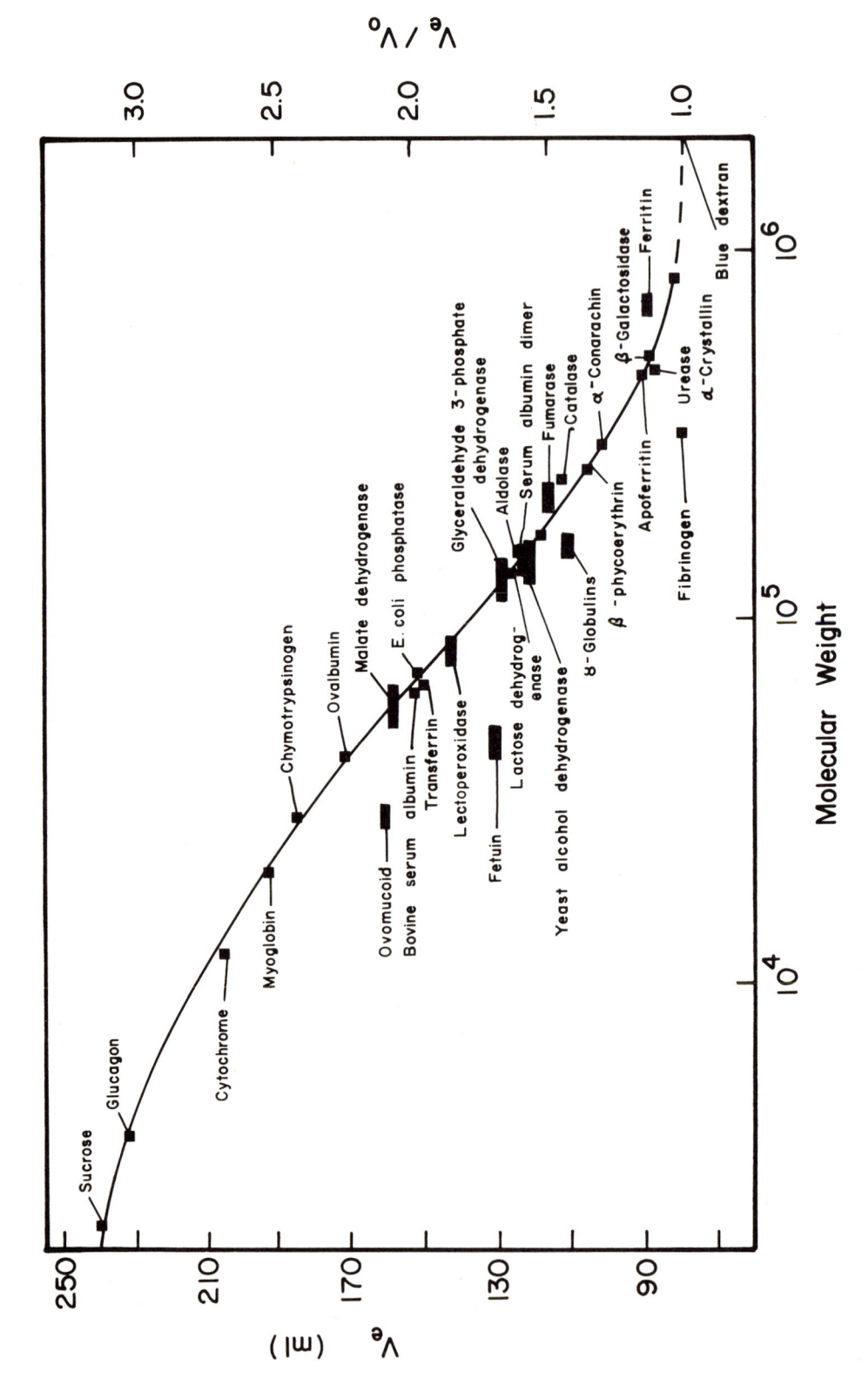

Figure 12.17 Logarithmic calibration plot for proteins of different molecular weight on Sephadex G-200 (taken from Andrews, 1965).

12. Solution of Proteins

where a_0 and b_0 are empirical constants, and $\text{erf}^{-1} x$ is the inverse of the probability integral (see equation 12.44).

$$\text{erf } x = \left(\frac{2}{\sqrt{\pi}}\right) \int_0^x \exp(-x^2) dx. \tag{12.44}$$

It may be shown that equation (12.43) can be used to fit the distribution coefficients of proteins or Sepharose 4B not only in their native forms, but also as random coils in 6M GuHCl, and as the denatured proteins complexed with sodium dodecyl sulfate. Data in all three solvents can be combined in a single calibration plot, indicating that drastic alterations in solvent medium do not significantly alter the pore-size distribution or eluting characteristics of the chromatographic columns. Once again, however, native proteins with asymmetric structures behave anomalously; their distribution coefficients this time being larger than anticipated on the basis of their Stokes radii, suggesting that end-on insertion into the gel pores may contribute significantly to increased retardation.

SDS-gel electrophoresis is a valuable technique for determining the molecular weights of denatured proteins. A protein molecule in an electric field generated by two electrodes, moves toward one of the electrodes (the direction depending on the sign of the charge of the protein). The velocity is

Figure 12.18 Comparison of the molecular weights of 37 polypeptide chains in the molecular-weight range from 11,000 to 70,000 with their electrophoretic mobilities on gels. The references to the molecular weights are given in Table 12.2.

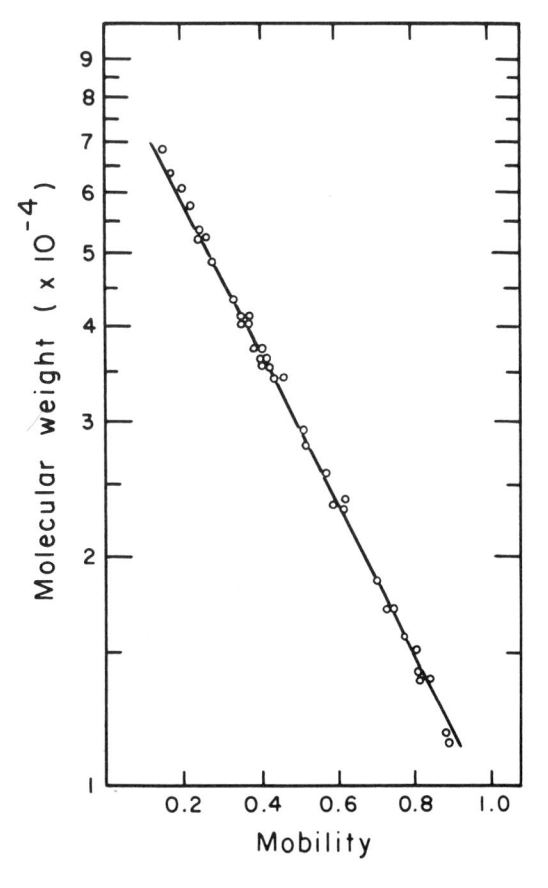

proportional to the magnitude of the charge and inversely proportional to the frictional coefficient of the particle in the fluid surrounding it. If the fluid is a molecular sieve, as is the case for a porous gel, then the frictional resistance of the macromolecules is very high so that they can be separated even if their mobilities in low-viscosity solvents are essentially identical. When protein particles are complexed with the anionic detergent SDS, the charge per unit weight is the same for all proteins (for reasons which are not well understood) and frictional control of the separation process becomes the dominant factor.

Starch and polyacrylamide gels have been successfully used for gel-electrophoresis experiments, the latter becoming more popular in recent years.

Table 12.2 Proteins Studied by SDS-Electrophoresis[a]

Protein	Mol wt of polypeptide chain
Myosin*	220,000
β-Galactosidase*	130,000
Paramyosin	100,000
Phosphorylase a*	94,000
Serum albumin	68,000
L-Amino acid oxidase	63,000
Catalase*	60,000
Pyruvate kinase*	57,000
Glutamate dehydrogenase*	53,000
Leucine amino peptidase	53,000
γ-Globulin, H chain*	50,000
Fumarase*	40,000
Ovalbumin	43,000
Alcohol dehydrogenase (liver)*	41,000
Enolase*	41,000
Aldolase*	40,000
Creatine kinase*	40,000
o-Amino acid oxidase*	37,000
Alcohol dehydrogenase (ycast)*	37,000
Glyceraldehyde phosphate dehydrogenase*	36,000
Tropomyosin*	36,000
Lactate dehydrogenase*	36,000
Pepsin	35,000
Aspartate transcarbamylase, C chain*	34,000
Carboxypeptidase A	34,600
Carbonic anhydrase	29,000
Subtilisin	27,600
γ-Globulin, L chain*	23,500
Chymotrypsinogen	25,700
Trypsin	23,300
Papain (carboxymethyl)	23,000
β-Lactoglobulin*	18,400
Myoglobin	17,200
Aspartate transcarbamylase, R chain*	17,000
Hemoglobin*	15,500
Qβ coat protein	15,000
Lysozyme	14,300
R17 coat protein	13,750
Ribonuclease	13,700
Cytochrome c	11,700
Chymotrypsin*, 2 chains	11,000 and 13,000

[a] The table lists molecular weights of the polypeptide chain taken from the literature. Proteins that under native conditions exist as oligomers are indicated by an asterisk.

Polyacrylamide gels in the concentration range 5–10% exhibit a range of pore sizes that permits the resolution of denatured proteins. They have the additional advantages over starch gels of better thermal stability, transparency, mechanical strength, and are more chemically inert.

In gel electrophoresis the relative mobility, or the distance the protein migrates relative to a standard, is measured. At the end of an electrophoresis run, the gel columns are usually stained with Coomassie blue to display the protein bands. When electrophoretic mobilities obtained in this way are plotted against the logarithm of molecular weight of standard proteins, a linear calibration curve is obtained that can be used for molecular-weight determination of unknown protein species. Figure 12.18 demonstrates that 37 different polypeptide chains ranging in molecular weight from 11,000 to 70,000 can be linearly fitted by such a procedure to an accuracy of 10%. Table 12.2 lists proteins having molecular weights up to 220,000 that have been determined by this method.

12.8 Sedimentation Equilibrium

This important technique can furnish absolute molecular weights. In this method the distribution of macromolecules is measured at equilibrium in a centrifugal field in which the sedimenting force is balanced exactly by an opposing diffusive force. The theory is the following:

For opposing diffusive and sedimenting forces, the net solute flow is derived by combining equations (12.22) and (12.23) to give

$$\left(\frac{\partial c}{\partial t}\right)_{radius} = Sc\omega r - D_t \left(\frac{\partial c}{\partial r}\right)_{time} \tag{12.45}$$

where S and D_t are defined by equations (12.21) and (12.26). Since the net flow is zero at equilibrium, we obtain

$$\frac{RT}{c}\left(\frac{\partial c}{\partial r}\right) = \frac{M(1 - \bar{v}\rho)\omega^2 r}{1 + c(\partial \ell n\gamma/\partial c)_{T,P}}. \tag{12.46}$$

Extrapolation of the left side of equation (12.46) to c = 0 permits a determination of M. It is often more convenient to use integral forms of equation (12.46) for determination of M; for example,

$$\ell nc = \frac{(1 - \bar{v}\rho)\omega^2}{2RT} \cdot M^{app} r^2 + constant \tag{12.47}$$

and $M^{app} = M/(1 + c(\partial \ln \gamma/\partial c)_{T,P})$ can be deduced from the slope of a plot of ℓnc vs. r^2. Extrapolation of several values of M^{app} to c = 0 gives M. Alternatively, by integrating equation (12.46) between the defined limits of the meniscus and the bottom of the cell, it can be shown that

$$\frac{2RT}{M^{app}\omega^2(1 - \bar{v}\rho)} = \frac{c^o}{c_b - c_m}(r_b^2 - r_m^2) \tag{12.48}$$

where c_b and c_m are the equilibrium values of c at the bottom of the cell and the meniscus respectively, r_b and r_m are the radial distances at the bottom and the meniscus respectively, and c^0 is the initial value of c everywhere in the sample. If the macromolecular solute is polydisperse, the weight-average value of M^{app} is obtained by application of equations (12.46), (12.47), or (12.48). The apparent weight-average molecular weight obtained by application of equation (12.48) corresponds to a concentration of $(c_m + c_b)/2$. In addition, point-by-point molecular weight averages at positions r can be calculated by

$$M^{app}_{\omega,r} = \frac{2RT}{(1 - \bar{v}\rho)\omega^2} \cdot \frac{d\ell nc_2}{dr^2}. \tag{12.49}$$

These values can be used to calculate M^{app}_w at the meniscus and base by extrapolation.

Experimentally, sedimentation equilibrium experiments may be performed in three ways: the low-speed, the high-speed, and the Archibald method. In the low-speed method low-rotor velocities are chosen so that the ratio of the concentrations at the base and the meniscus (c_b/c_m) is approximately 3. If the rotor speed is increased by a factor of about 3, c_b/c_m becomes $10^3–10^4$ so that the meniscus is essentially depleted of solute and $c_m = 0$. The low-speed method requires measurement of the initial concentrations c_0 in order to use equation (12.48), and one attempts to visualize the complete radial distribution of the solute. In high-speed experiments a speed is chosen so that there is zero concentration near the meniscus at a reference r_0 from which all concentrations at $r > r_0$ can be measured. Measurement of c_0 in low-speed experiments is not a trivial task. To speed up the rate of attainment of equilibrium at low speeds, small volumes of solution are studied. Under these conditions the surface-to-volume ratio is large, and loss of protein through adsorption to the cell walls may be significant. Thus c_0 cannot be assumed to be "made-up" concentration of protein added to the cell and c_0 must be measured within the cell. This can be done immediately after the rotor has attained the desired speed by appropriate calibration of the detection method (proper indexing of the interference fringes, or from calibration of the records in an absorption optics experiment). The accuracy of the low-speed method properly applied is greater because the solute concentration levels can be made higher. Conversely, however, if one has limited amounts of protein available, the high-speed technique allows one to study more dilute solutions. Number-average molecular weights can be calculated also from high-speed experiments using the equation

$$M^{app}_{n,r} = \frac{B^{-1}c_r}{\int_{r_m}^{r} c(r)d(r^2) + \frac{c_m}{BM_{n,m}}} \tag{12.50}$$

where $B = \omega^2(1 - \bar{v}_2\rho)/2RT$. The second term in the denominator is zero by virtue of meniscus depletion ($c_m = 0$).

The third method of carrying out a sedimentation equilibrium experiment is the Archibald method and is based on the fact that equation (12.46) can be applied at all stages of a sedimentation experiment, even after a few minutes, to the meniscus and also to the bottom of the cell; that is, no net flow occurs at these points. Thus M^{app} can be obtained from measurements of $(1/rc)(\partial c/\partial r)_t$ at the meniscus or bottom at time t after the start of the run. Extrapolation of M^{app} at several concentrations to $c = 0$ leads to the value of the total molecular weight M_w^t at meniscus or cell bottom at time t.

$$\frac{1}{r_m}\left[\frac{1}{c}\left(\frac{\partial c}{\partial r}\right)_t\right]_{\substack{r=r_m \\ c=0}} = \frac{M_{wm}^t(1-\bar{v})\omega^2}{RT}. \tag{12.51}$$

The value obtained at the meniscus will be different from that at the cell base. However, both can be calculated at various times and extrapolation to $t = 0$ gives M_w for the original mixture.

The methods of sedimentation equilibrium can, of course, be applied both to proteins in their native and their denatured states. As discussed earlier, because of the influence of preferential hydration or binding of small ions, in practice, the apparent partial specific volume $\bar{V}'$ must be evaluated between solutions dialyzed against the solvent system, and the dialysate. Obviously, also, the dialyzed solutions and dialysate must be used in the solution and reference compartments of centrifuge cells when using interferometric detection in order that correct concentration profiles are obtained. Examples are reported in the literature that demonstrate that, if these precautions are taken, accurate molecular weights of proteins denatured in 6M GuHCl and 0.1M mercaptoethanol can be determined, as well as molecular weights of native proteins.

As with the other techniques to which an exact thermodynamic analysis can be applied (e.g., osmometry and light scattering), the high-speed sedimentation equilibrium method can be used to investigate details of equilibrium constants that characterize the self-association behavior of oligomeric proteins. A useful analytical device for the analysis of the experimental data in such systems is to treat point-molecular-weight averages as coordinates in two-dimensional space. Assuming that only two species are present with molecular weights M_A and M_B and if the solution is thermodynamically ideal, it

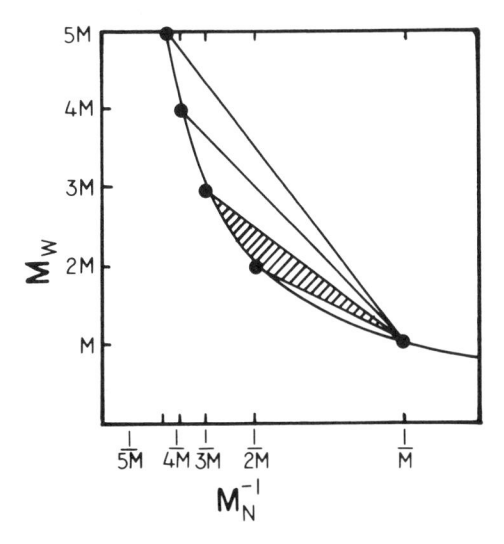

Figure 12.19 Allowed space of molecular-weight averages. The ordinate M_w represents the average molecular weight. The abscissa is the reciprocal value of the next lower average (number- or weight-average, respectively). The hyperbola is the locus of ideal, homogeneous molecular-weight averages. The lines and shaded area are discussed in the text (from DC Teller, in *Methods in Enzymology*, Vol. XXVII).

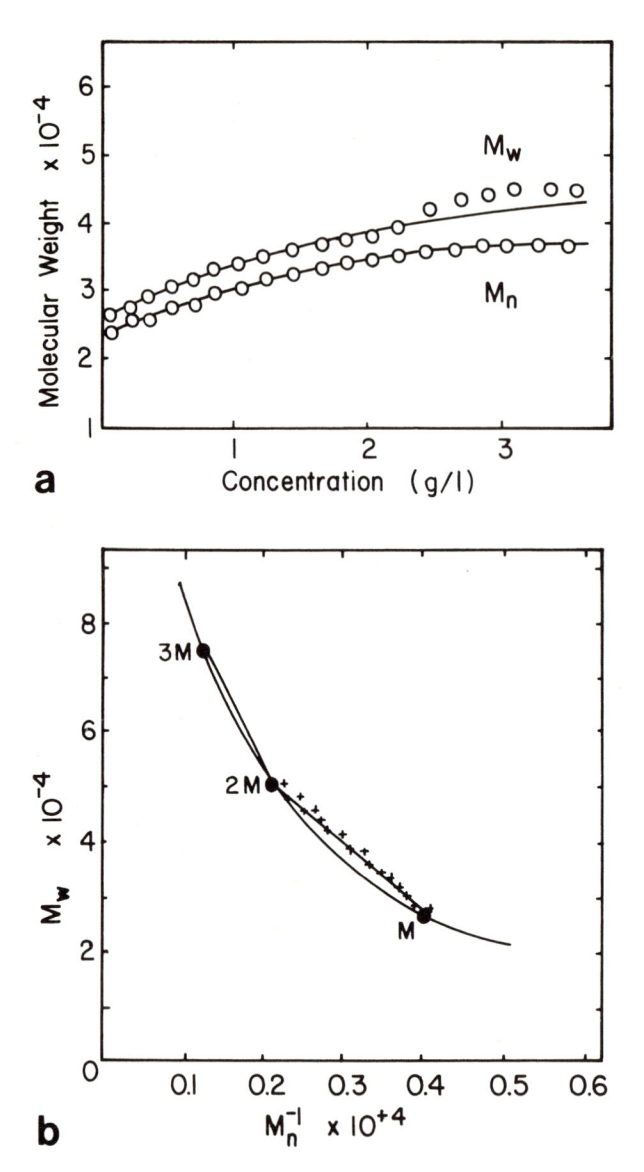

Figure 12.20 (a) Dependence of the point molecular-weight averages on α-chymotrypsin concentration in g/liter (converted from fringe number). The solid lines are theoretical fits based on a monomer–dimer association with an association constant of 2.15×10^4 liters/g. Conditions: 24,000 rpm, 25° C, 0.178M NaCl—0.01M acetate buffer at pH 4.12, initial concentration of α-chymotrypsin 0.76 g/liter (data of Aune and Timasheff). (b) Two-species plot of M_w vs. M_n^{-1} using data of (a) clearly shows presence of monomer–dimer equilibrium in solutions of α-chymotrypsin under these conditions.

can be shown that

$$M_{k,r} = -M_A M_B \left(\frac{1}{M_{k-1,r}} \right) + M_A + M_B \tag{12.52}$$

where k = 0, 1, 2 define n, w, and z-averages respectively. A two-species plot of $M_{k,r}$ vs. $M_{k-1,r}$ is then constructed as shown in Figure 12.19. If the sample is monodisperse, $M_k = M_{k-1}$, and a graph of equation (12.52) is given by the hyperbola shown in Figure 12.19. So long as the protein solution is ideal, and experimental errors are absent, then no molecular weights should appear to the left or below this curve. If two species with molecular weights $M_A = M$ and $M_B = 2M$ were present, the point averages should fall on the line connecting these points. Similarly, if a monomer–trimer equilibrium between species $M_A = M$ and $M_B = 3M$, is present, the data would fall on the line connecting M and 3M. If $M_A = 2M$, and $M_B = 3M$, the data would be along the line connecting 2M and 3M in the diagram. Thus, if a monomer–dimer–trimer system consisting of species with molecular weights M, 2M, and 3M were present, the points would all be within the shaded triangle. Departure from this stoichiometry towards higher multimers causes the data to curve vertically; thermodynamic nonideality (extreme excluded volume or charge–charge effects) moves the points on the upper part of the plot (high concentration) downward. Downward curvature in the lower part of a two-species plot is usually due to experimental error in numbering the fringes (fringe labeling) (Figure 12.10) at the lower concentrations. In most cases, however, globular proteins can be considered thermodynamically ideal for ionic strengths in the range of 0.1–0.5M, and at concentrations less than 5 mg/ml. The effects of virial coefficients are then less than the experimental errors of the measured point-by-point molecular-weight averages. Graphs of the type shown in Figure 12.19 can frequently be used to advantage to obtain an estimate of M_2, and the stoichiometry of the system by fitting linear portions of the data. These plots do not work well with only slightly associated or dissociated systems since they are dominated by experimental error and nonideality. An example of the application of such a plot to data from aqueous solutions of α-chymotrypsinogen at low pH is shown in Figure 12.20; it demonstrates that the enzyme exists as a mixture of monomer and dimer under these conditions. The molecular weight of the monomeric protein is determined by extrapolation to be 25,600 gmole^{-1}.

12.9 Small-Angle X-Ray Scattering

The theoretical basis for the method of small-angle x-ray scattering (SAXS) is essentially identical to that previously discussed for light scattering. The small-angle regime corresponds to values of the scattering angle $\theta < 5°$ (we will, for the sake of clarity, preserve our definitions of the scattering geometry as shown for the light-scattering case in Figure 12.5; other discussions of SAXS often define the angle shown to be 2θ). The small-angle x-ray scattering of macromolecular solutions generally consists of a diffuse scattering whose intensity monotonically decreases due to destructive interference in the x-rays scattered from different parts of the molecule, out to $\theta = 2°$. In the region between 2° and 5°, the internal order of macromolecules begins to affect the scattering, and secondary maxima and minima may begin to appear in the curve due to constructive interference. This produces discrete oscillations in scattered intensity which are observed at higher angles since they correspond to interatomic distances.

The fundamental equation for SAXS is identical to that characterizing the scattering of unpolarized light at angle θ

$$\frac{i_\theta}{I_o} = \frac{i(0)}{I_o} P(\theta) \left[\frac{1 + \cos^2\theta}{2}\right] \tag{12.53}$$

where i_θ, I_o now refer to scattered intensity of X-radiation at angle θ, and incident x-ray intensity, respectively. $P(\theta)$ is the particle-scattering function which describes the interference properties of the scattered x-rays, and $i(0)$ is the scattered intensity extrapolated to zero-angle. The excess x-ray scattering at zero-angle

$$\Delta i_n(0) = \frac{\Delta i(0)}{I_o} = (i(0)_{\text{solution}} - i(0)_{\text{solvent}})/I_o$$

is determined by the fluctuations of electron density $\overline{\Delta \rho_e^2}$ in a volume element δV

$$\Delta i_n(0) = \delta v \, \overline{\Delta \rho_e^2} \tag{12.54}$$

where $\overline{\Delta \rho_e^2} = \overline{\rho_e^2} - (\overline{\rho_e})^2$, the horizontal bar denotes a statistical average over the dimensions of δv, and time of experiment, and ρ_e is the electron density of the solution (number of electrons per unit volume). But $\overline{\Delta \rho_e^2}$ is related directly to the fluctuations in concentration:

$$\overline{\Delta \rho_e^2} = \left(\frac{\partial \rho_e}{\partial c_e}\right)_{T,P} \overline{\Delta c_e^2} \tag{12.55}$$

where c_e is the solute concentration expressed as the ratio of the number of electrons of solute to that of solution (essentially a weight fraction). Now evaluating $\overline{\Delta c_e^2}$ using the thermodynamic relations for concentration fluctuations leads to

$$\left(\frac{\partial \rho_e}{\partial c_e}\right)_{T,P}^2 \frac{c_e(1 - c_e)^2}{\Delta i_n(0)\rho_s^2} = \frac{1}{m_{\text{app}}} \tag{12.56}$$

where m_{app} is the apparent mass, in electrons, of the particle at the particular concentration at which scattering measurements are made, and includes the effects of thermodynamic nonideality (excluded volume, preferential interactions, etc.) summed up in the second virial coefficient A_2. At low concentrations, we may write, as for light scattering,

$$m_{\text{app}} = m - 2A_2 m^2 c_e \tag{12.57}$$

where m is the true mass of the solute. Unlike the differential refractive index used in light scattering, the electron-density increment $(\partial \rho_e/\partial c_e)_{T,P}$ is not a directly measurable quantity, but it can be calculated from the electron partial specific volume of the solute, $\bar{v}_e$. Thus, at constant T,P,

12. Solution of Proteins   271

$$(\partial \rho_e/\partial c_e)_{T,P} = \rho_e \frac{(1 - \rho_e \bar{v}_e)}{(1 - c_e)}.$$ (12.58)

At low concentrations, $\rho_e = \rho_o$, the electron density of solvent, and equation (12.56) can be rewritten:

$$m_{app} = \Delta i_n(0)(1 - \rho_o \bar{v}_e)^{-2} c_e^{-1}.$$ (12.59)

The true value m can be determined, together with the second virial coefficient A_2, by extrapolation of m_{app} to zero c_e, according to equation (12.57). The molecular weight M is related to m by

$$M = mN_a/q$$ (12.60)

where q is the number of electrons per gram of the protein calculated from its chemical composition (elemental analysis). The electron partial specific volume $\bar{v}_e$ in Å^3/electron can be converted to the partial specific volume $\bar{v}$ in ml/g by

$$\bar{v}_e = \bar{v} 10^{24}/q$$ (12.61)

In making deductions of molecular weight or A_2, it is important to remember that in multicomponent systems, the effect of preferential interactions of solvent and counterions may be significant. Under these conditions, as in the case of sedimentation equilibrium, the apparent partial specific volume $\bar{v}'$ should be used in equation (12.61) and thence (12.59). Because of the difficulties of ensuring stability of SAXS instruments over the long exposure times necessary to achieve acceptable signal-to-noise ratio, especially with respect to variations in the incident beam intensity, the absolute SAXS technique has not been applied so extensively to molecular-weight determination as most of the other techniques described above. This state of affairs may be changed with the advent of high-intensity sources and the position-sensitive detector, which have vastly decreased the exposure times required for high-precision x-ray scattering experiments. Most studies to date have involved investigation of structural features of proteins that can be based on relative intensity measurements alone and will be discussed shortly.

12.10 Comparison of Techniques for Molecular-Weight Determination

In comparing the various techniques for determining the molecular weight of proteins, we first note that exact values for the anhydrous molecular weight M can be determined only by the thermodynamically well-founded techniques of osmometry, total-intensity light scattering, sedimentation equilibrium, and SAXS; and, in addition, the combination of two hydrodynamic techniques embodied in the Svedberg and Mandelkern–Flory equations (12.38, 12.39, 12.40) respectively. If the protein is polydisperse (that is, species of different molecular weight are present), only the thermodynamic techniques provide well-defined averages M_n, M_w, or M_z; equation (12.38) provides M_w if the weight-average of S^0 and z-average of D_t^0 are used; no well-specified average can be determined by equation (12.39) or (12.40). Accuracies of 1–1.5% in estimating molecular weights by osmometry or light scattering or sedimentation equilibrium are possible if properly carried out. Comparable accuracies are possible using equations (12.38) and (12.39), or (12.40), especially with modern methods of measuring D_t^0, S_t^0, and $[\eta]$. The Flory-Mandelkern equation is less useful for native proteins because of possible particle anisometry (i.e., deviations from spherical).

The empirical methods utilizing $[\eta]$, S^0, D_t^0, GPC and SDS-gel electrophoresis have the advantage of much greater speed and convenience than the thermodynamic techniques, but have inherently lower accuracy for proteins (5–10%), and can only be applied to the denatured protein chains. Such a measurement is, however, extremely useful for characterizing the quaternary structure of oligomeric proteins. In comparing these empirical methods it seems appropriate to point out the following:

1. Measurements of $[\eta]$, D_t^0 (using dynamic light scattering), and S^0 are much more speedy than the GPC and SDS-gel electrophoresis techniques which require a complete calibration with standards before characterizing any unknown protein;
2. S^0 measurements on proteins denatured in GuHCl–mercaptoethanol and $[\eta]$ measurements require extremely accurate protein concentration values (this is also true for the thermodynamic techniques);
3. The techniques that are based on gel-separation encounter problems in analysis of proteins that contain significant amounts of nonprotein moiety such as carbohydrate or lipid;
4. The level of technical skill required for measurement of $[\eta]$, D_t^0 (by dynamic light scattering), and the gel-separation techniques is relatively low in comparison with ultracentrifuge measurements, especially sedimentation equilibrium studies, and the thermodynamic techniques;
5. Gel-separation and sedimentation-velocity experiments permit identification of individual species in multicomponent protein solutions that differ in molecular weight;
6. Separation in SDS-gel disc electrophoresis affords much higher resolution than exclusion chromatography;
7. SDS-gel electrophoresis and ultracentrifuge techniques permit studies of comparatively small quantities of protein (< 50 μg).

12.11 Structural Characterization: Radius of Gyration and Related Information

Information about the molecular dimensions of native proteins can be obtained from scattering of electromagnetic radiation. These data are inferred from studies of the angle-dependence of the relative scattered intensity that is described by the particle-scattering function $P(\theta)$ in equation (12.6) for light scattering, and equation (12.53) for SAXS. Since $P(\theta)$ measures an interference behavior at the detector of the radiation scattered from different points on the same molecule, the size of the structural features that can be discerned by this method is critically dependent on the radiation wavelength. Thus light scattering cannot provide information about particles smaller than 200 Å; SAXS, on the other hand, can distinguish features as small as 2 Å.

Debye derived an explicit mathematical relation for $P(\theta)$

$$P(\theta) = \frac{1}{\sigma^2} \sum_i^\sigma \sum_j^\sigma \left[\frac{\sin Kr_{ij}}{Kr_{ij}} \right] \quad (12.61)$$

based on the model that the scattering particle is composed of a finite number σ of scattering elements, the sum is taken over all pairs i,j of such elements separated by distances r_{ij}. The parameter $K = (4\pi/\lambda)\sin \theta/2$. Later, Guinier showed that this equation could be used to extract a characteristic geometric parameter from $P(\theta)$ that is independent of any assumptions regarding the shape of a particle. This is the radius of gyration R_g. Expanding (12.61), he obtained:

$$P(\theta) = 1 - \frac{K^2 R_g^2}{3} + \ldots \quad (12.62a)$$

$$\sim \exp\left(-\frac{K^2 R_g^2}{3}\right). \quad (12.62b)$$

In light scattering, most of the experimental scattering envelope (variation of scattering intensity with angle θ) usually falls within the approximation (12.62a). Inserting (12.62a) in (12.9), it is apparent, using the approximation

$$\frac{1}{P(\theta)} \sim 1 + \frac{K^2 R_g^2}{3} \quad (12.63)$$

that:

$$\frac{K^* c_2}{\Delta R_\theta} \sim \left(\frac{1}{M_w} + 2A_2 c_2 + \ldots \right)\left(1 + \frac{16\pi^2}{3\lambda^2} R_g^2 \sin^2 \frac{\theta}{2} \ldots \right). \quad (12.64)$$

The best way to utilize equation (12.64) is the Zimm plot or $K^*c_2/\Delta R_\theta$ vs. $(\sin^2 \theta/2 + kc_2)$ where k is an arbitrary constant chosen to permit a suitable graphic display. An example of such a plot is shown in Figure 12.21 for a sample of poly-L-proline in propionic acid. The weight-average molecular weight is obtained from the reciprocal of the intercept for $\theta = 0$ and $c = 0$. The radius of gyration is obtained from the slope of the data at $c_2 = 0$ and constant θ, which obey the linear relation

$$\lim_{c=0} \frac{K^* c_2}{\Delta R_\theta} = \frac{1}{M_w}\left(1 + \frac{16\pi^2}{3\lambda^2} R_g \sin^2 \frac{\theta}{2} + \ldots \right). \quad (12.65)$$

Because of the smaller wavelength, the scattering envelope of SAXS includes the higher terms in the expansion (12.62a) even at the smallest angles, necessitating use of the exponential approximation (12.62b). Utilizing (12.62b) in conjunction with (12.53), it is easy to show

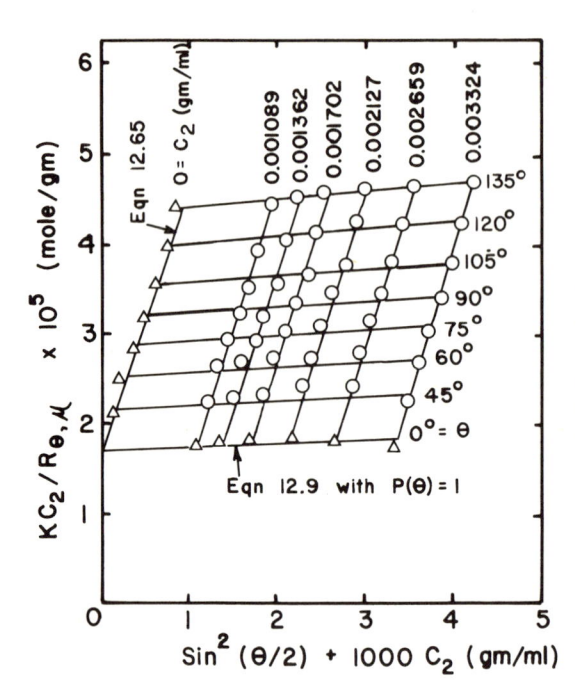

Figure 12.21 Reciprocal intensity plot of light-scattering data of poly-L-proline in propionic acid. The stock solution and subsequent dilutions are put in the photometer, and the angular dependence of scattered light is measured in each case. ○ Actual data; △ points derived by extrapolation.

$$\Delta i_n(\theta) = \Delta i_n(0) \exp\left(-\frac{K^2 R_g^2}{3}\right). \tag{12.66}$$

Thus R_g^2 can be determined from a linear plot of log $i_n(\theta)$ vs. K^2, the slope of which is $(16\pi^2/3\lambda^2)R_g^2$.

The radius of gyration can be a very useful parameter for probing the structure of macromolecular species. For solid spheres of radius R, it can be proven that

$$R_g^2 = \frac{3}{5} R^2. \tag{12.67}$$

For a long thin rod of length L,

$$R_g^2 = L^2/12. \tag{12.68}$$

A flexible random coil model consisting of σ dimensionless points (each having the mass of one peptide unit), separated by a distance ℓ, has a value R_g given by

$$R_g^2 = \frac{1}{6}\sigma\alpha^2\ell^2 \tag{12.69}$$

where α is a constant that describes the effect of polymer–solvent interactions on chain expansion ($\alpha \geq 1$). As an example of the application of these ideas, we cite the example of the muscle protein myosin. Figure 12.22 shows results obtained for the absolute reciprocal scattering envelope $1/M_2 P(\theta)^{-1}$ as a function

12. Solution of Proteins

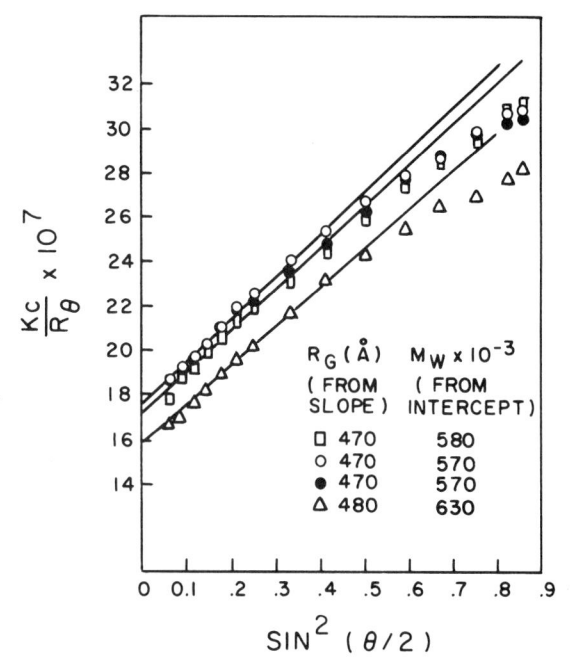

Figure 12.22 Reciprocal scattering envelope of myosin as a function of pH at ionic strength 0.6M: □, 0.1M phosphate, pH 7.8; ○, 0.1M borate, pH 8.5; ●, 0.6M KCl, pH 6.3; △, 0.1M borate pH 9.0 (after Holtzer and Lowey).

	R_G (Å) (FROM SLOPE)	$M_W \times 10^{-3}$ (FROM INTERCEPT)
□	470	580
○	470	570
●	470	570
△	480	630

of pH. These show that the root-mean-square value $(R_g^2)^{1/2}$ and molecular weight M_w are essentially constant. Table 12.3 shows a comparison of possible structural models for this protein using the known molecular weight and partial specific volume $\bar{v} = 0.728$ ml/g.

Obviously, the observed value $R_g = 470$ Å indicates that the model of a cylinder 1620 Å × 26 Å is most plausible in this case, though one cannot exclude the possibility of a longer, thinner rod. An experimental determination of R_g by itself does not therefore completely define a structural model, but does, however, greatly narrow the range of choices.

In order to characterize the macromolecular dimensions in more detail, one must either have recourse to independent methods for determining structural parameters, such as the hydrodynamic techniques that will be described later, or one must study the details of $P(\theta)^{-1}$ at higher angles and compare them with a theoretical computation of $P(\theta)^{-1}$ for a particular structural model. Figure 12.23 shows theoretical curves for $P(\theta)^{-1}$ for models of a sphere, disc, random chain, and a rigid thin rod; between the curves for a rigid rod and flexible chain lies a continuum of curves, one of which is shown as a hatched line, that

Table 12.3 Calculated Radius of Gyration for Models of Myosin

Model type	R_g
Solid sphere	$R_g = 41$ Å
Flexible coil*	
Poor solvent $\alpha = 1.00$	$R_g = 234$ Å
Good solvent $\alpha = 1.47$	$R_g = 343$ Å
Cylinder (1,620 Å × 26 Å)	$R_g = 470$ Å

*For the flexible coil model a random peptide chain with n = 4,555 and e = 85 Å has been used.

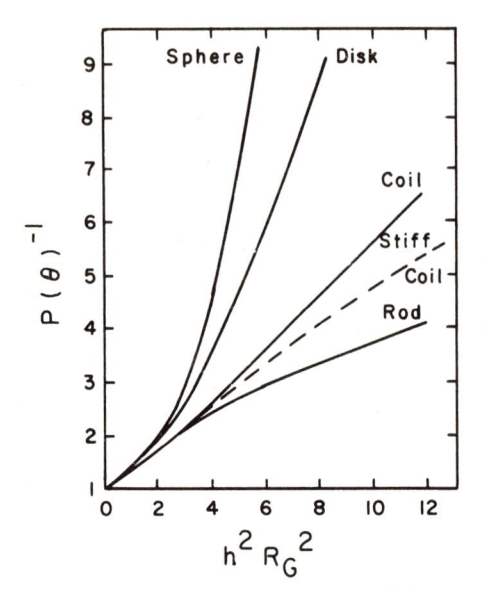

Figure 12.23 The reciprocal particle-scattering factor $P^{-1}(\theta)$ for a sphere, disc, two forms of coil and rod.

represents coils with varying degrees of stiffness, calculated on the basis of a wormlike coil model. Depending on the precision of experimental data, the fitting of these theoretical curves to experimental $P(\theta)^{-1}$ values may permit a more accurate characterization of structure. One of the problems in such analyses is illustrated by Figure 12.24 which compares $P(\theta)^{-1}$ data obtained by light scattering for tobacco mosaic virus (TMV) with theoretical curves for a rigid thin rod of length L. The data fit the curve for L = 3200 Å at small angles and the curve for L = 2900 Å at large angles. This can only be the result of a slight polydispersity of the TMV since deviations from rodlike behavior would

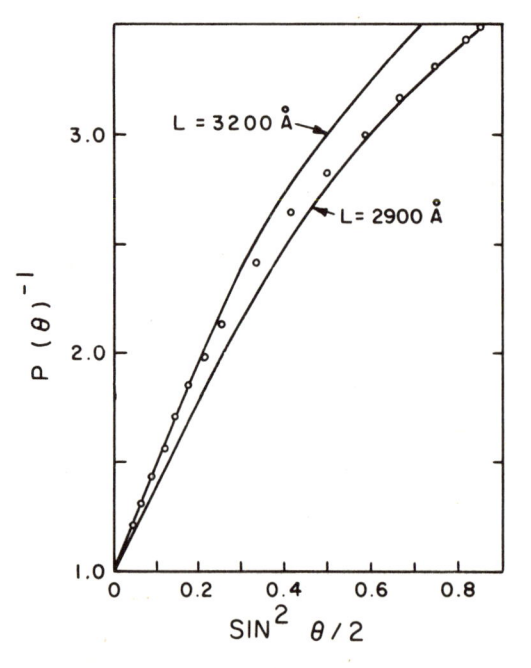

Figure 12.24 Reciprocal particle-scattering factor of TMV solution. ○ Experimental points, ——— theoretical curves (from Boedtker and Simmons).

cause the high-angle points to lie above rather than below the theoretical curve, as evident in Figure 12.23. It may also be shown theoretically that $P(\theta)^{-1}$ for a polydisperse sample of particles always lies below the curve for a monodisperse system of equal length. The value of L = 3200 Å agrees well with the electron microscope (EM) length of 3020 Å when one realizes that $(R_g^2)^{\frac{1}{2}}$ for a rod measured by light scattering represents a $[z(z+1)]^{\frac{1}{2}}$-average, but the number-average value is calculated in EM studies. Utilizing the known molecular weight and density or specific volume as was described for myosin above, it turns out that the data predict a single TMV virus can be represented by a cylindrical model of length L = 3000 Å, diameter 149 Å.

The problem of polydispersity severely hinders efforts to apply $(R_g^2)^{\frac{1}{2}}$ measurements to deduce helical conformational parameters of synthetic polypeptides. In principle, polypeptides in helicogenic solvents might be expected to resemble rigid rods, and it should be possible to determine the conformational nature of the helix from the length of the molecule, measured by estimation of $(R_g^2)^{\frac{1}{2}}$. From equation (12.68), the peptide repeat h of the helix can be calculated according to

$$h = 12L^2 M_0/M \tag{12.70}$$

where M_o is the molecular weight of the monomeric residue. Thus, in 1956, P. Doty and colleagues first reached the conclusion, based on light-scattering analysis and viscometry (which will be discussed later) that poly-γ-benzyl-L-glutamate (PBLG) exists as an α-helix in certain solvents, such as dimethyl formamide (DMF) and a chloroform-formamide mixture (CF), with a value h = 1.5Å. The correctness of this conclusion has since been established by x-ray diffraction analysis.

However, in the interim, a number of other studies, using light scattering or SAXS, summarized in Figure 12.25, suggested that different h-values, and therefore different helices, might exist in these solvents. It is also obvious from

Figure 12.25 Comparison of peptide repeat values h for PBLG in various helocogenic solvents. ○ Dimethyl formamide; ● chloroform-formamide; × ethylene dichloride.

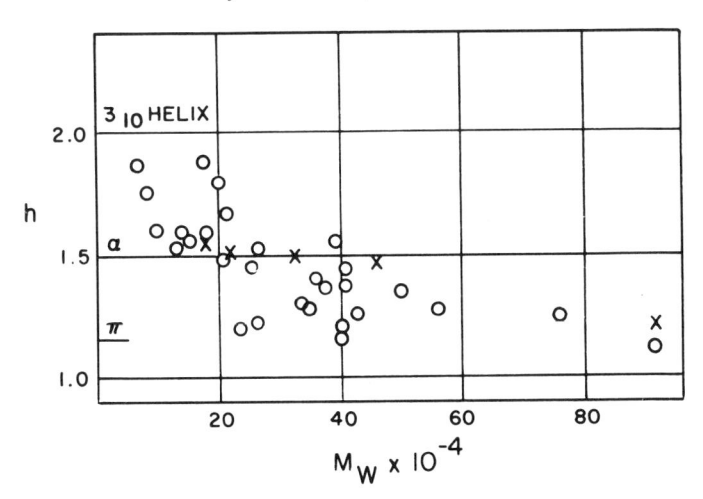

Figure 12.25 that h is molecular-weight dependent, and decreases as M_w increases. There are many reasons for these discrepancies: the polydispersities of the sample may alter the h-values; there are considerable difficulties in experimentally determining R_g of low-molecular-weight polymers; increasing molecular weight introduces the possibility of chain flexibility, so that at higher values of M_2 the end-to-end distance becomes smaller than the rigid-rod value.

In the theory of light scattering at large scattering angles and large values of M_2, one may write a linear relation of the form

$$\left(\frac{K^*c}{R_\theta}\right)_{c=0} = \frac{2}{M_n \pi^2} + \left(\frac{4L_n}{M_n \lambda}\right) \sin \theta/2 \qquad (12.71)$$

where λ is the wavelength of incident light and L_n and M_n are the number-average length and molecular weight respectively. A similar equation may be written for the weight-average parameters and therefore equation (12.71) is independent of molecular-weight distribution. An analysis of this kind for a PBLG sample having $M_n = 162{,}000$ and $M_w = 410{,}000$ illustrated in Figure 12.26 leads to values h = 0.85 Å in dimethyl formamide and h = 1.15 Å in chloroform formamide. Such results indicate the molecular flexibility at high values of M_2 since the true conformation is certainly α-helical.

Experimental studies of the change in polypeptide dimensions during the conformational transition from a random chain to helix can be observed by light-scattering measurements of R_g. Figure 12.27 shows the variation in $\langle R_g^2 \rangle$ for PBLG in mixtures of dichloracetic acid–heptane. As the temperature is

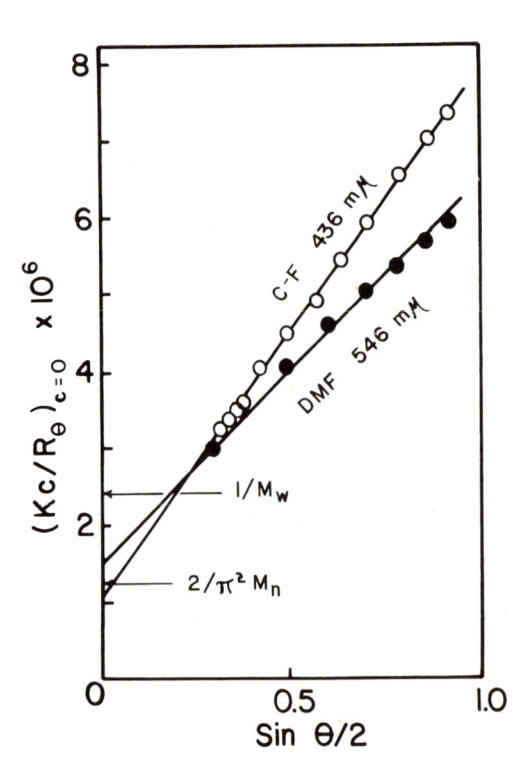

Figure 12.26 Light-scattering data for PBLG in two different solvents plotted in terms of equation (12.71).

12. Solution of Proteins 279

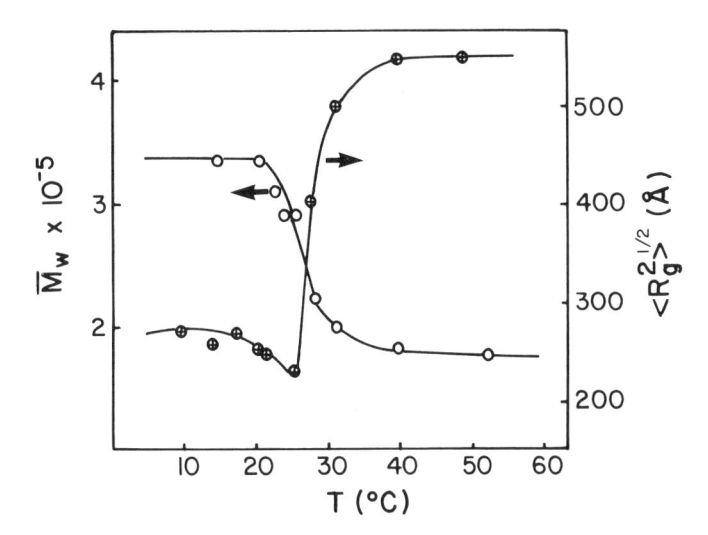

Figure 12.27 Temperature-dependent helix-random chain transition in poly-γ-benzyl-L-glutamate followed by variation in radius of gyration —⊕— and by changes in the apparent molecular weight resulting from variations in the preferential adsorption with structure —○—.

raised, a rapid cooperative transition from random to helix* occurs between 25° C and 35° C corresponding to a sudden increase in $(R_g^2)^{1/2}$. The transition is evidently preceded by a slight contraction of the chain dimensions due to formation of a small number of short helical segments prior to the appearance of the fully extended helical form. Also shown in Figure 12.27 is a decrease in the preferential adsorption of dichloracetic acid molecules that accompanies the transition and causes a discrete decrease in light-scattering intensity (and, therefore, the apparent molecular weight).

A variety of structural parameters (in addition to R_g) can be determined by the SAXS technique using the Guinier plot shown in Figure 12.28a. For an isotropic particle of homogeneous electron density, the surface areas of the particle can be determined from a Soulé-Porod plot as shown in Figure 12.28b. If the intercept is A, then

$$s = 16\pi^2 A(\Delta\rho)^{-2} \tag{12.72}$$

where $\Delta\rho$ is the excess electron density. The slope of the Soulé-Porod plot provides a parameter δ^* which reflects internal structure of the particle. The hydrated volume of the particle V is calculated from:

$$V = \frac{i_n(0)}{\int_0^\infty 2\pi\kappa j_n^*(\kappa)d\kappa} \tag{12.73}$$

---

*Most texts use the terminology helix-coil transitions. We prefer the use of random for denatured proteins or polypeptides, reserving the term coil for the charged state of such poly(amino acids) as poly(glutamic acid) and propolylysine at neutral pH.

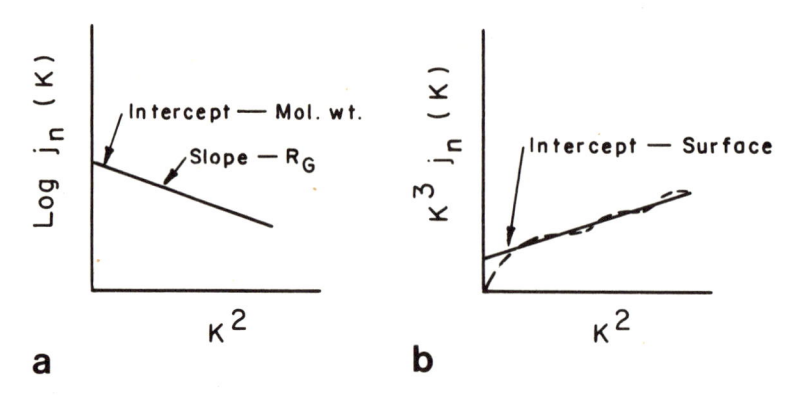

Figure 12.28 Types of plots used in small-angle x-ray scattering. (a) Guinier plot; (b) Soulé-Porod plot.

where

$$j_n^*(\kappa) = j_n(\kappa) - \delta^* \tag{12.74}$$

and $j_n(\kappa)$ is the normalized experimental "smeared" scattering intensity. Table 12.4 shows examples of structural information determined for three globular proteins using the kind of SAXS analysis outlined above. The R_g values of the proteins are all approximately the same, but V and the ratio S/V are significantly different. From these values, it can be concluded that lysozyme and lactalbumin have very similar structures, but ribonuclease is significantly different. Calculation of the axial ratio a/b, assuming a prolate ellipsoid model, can be accomplished from the quantities $(3V/4\pi R_g^3)$ and $(R_g S/V)$. Obviously, ribonuclease has more anisometric structure than the other two proteins. For comparison, we point out that the axial ratio for lysozyme measured by x-ray diffraction from single crystals is 1.5. The larger values for a/b calculated from $(R_g S/V)$ are a direct result of the fact that proteins cannot be modeled as smooth ellipsoids, but have topologically complex surfaces consisting of bumps and holes or clefts. Thus the surface area S is larger than predicted for a smooth prolate ellipsoid.

At higher protein concentrations, maxima and minima become apparent in the scattering patterns at higher angles. The positions and magnitudes of these extrema are determined by the protein structure and can be used as an additional means of characterization. As a cautionary note, the protein concentration in such studies should be low enough (< 100 mg/ml) to avoid the possibility of interparticle interference effects.

In Figure 12.29 we show comparison of the normalized SAXS from β-lactoglobulin solutions with theoretically predicted curves for various models. Under the conditions of the experiment β-lactoglobulin exists as a dimer. None of the geometric models fits the experimental data closely, though a parallelepiped is a more accurate model than two spheres. Again the reason for the discrepancy is that proteins cannot be exactly modeled as smoothly isotropic geometric structures.

12. Solution of Proteins

Table 12.4 Structural Parameters

Parameter	Ribonuclease	Lysozyme	α-Lactalbumin
R_g, Å	14.8	14.3	14.5
M	12,700	13,600	13,500
V, Å^3	22,000	24,200	25,100
S/V, Å^{-1}	0.29	0.25	0.24
H, g_{H_2O}/g_{Prot}	0.46	0.33	0.37
a/b from $\left(\dfrac{3V}{4\pi R_g^3}\right)^*$	1.87	1.42	1.43
a/b from $\left(R_g \dfrac{S}{V}\right)^*$	3.70	2.92	2.82

*Assuming a prolate ellipsoid of revolution.

More recently, with the advent of stable higher power generators and position-sensitive detectors, it has become feasible to attempt more difficult characterization problems such as characterizing the internal structure of certain internally heterogeneous proteins. The system in which most success has been achieved in this direction is that of the plasma lipoproteins of human

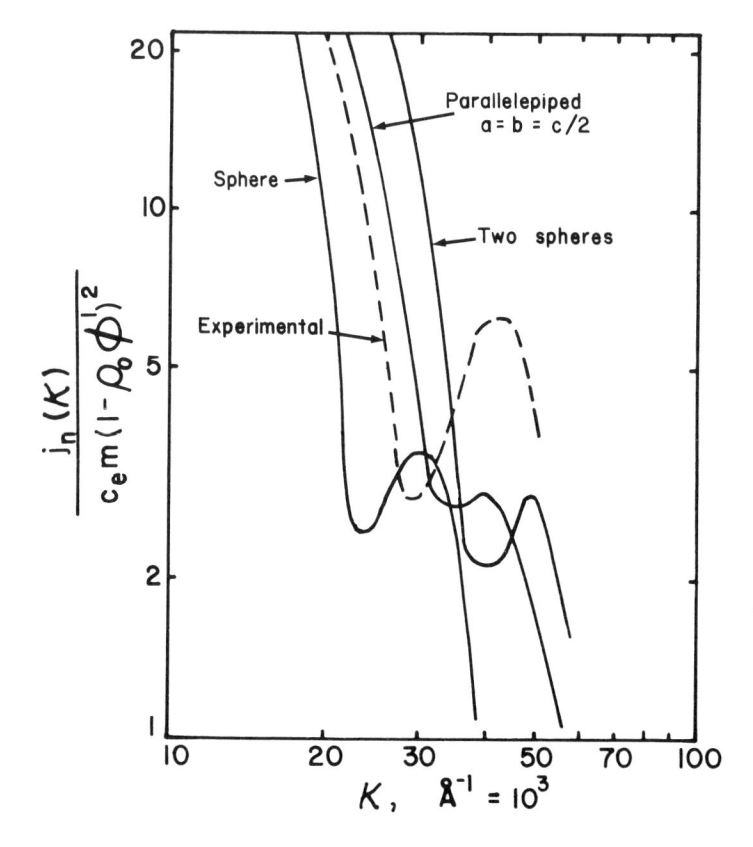

Figure 12.29 Normalized scattering of β-lactoglobulin A and B in the higher angle range in 0.1M sodium acetate at pH 5.7. ---- The experimental curve; ——— the convoluted theoretical curves calculated for various models.

serum. These species are chemically complex particles containing protein, phospholipid, cholesterol esters, free cholesterol, and triglycerides. They exist in a range of molecular weights but have been classified on the basis of their sedimentation behavior in a sucrose density gradient as high-density lipoprotein (HDL), low-density lipoprotein (LDL), and very-low-density lipoprotein (VLDL). (The lipoproteins play a central role in lipid metabolism as the major vehicle for cholesterol transport.) Subfractions of each of the major classes can be defined in terms of a particular range of sedimentation coefficients, or by immunochemistry. SAXS analysis shows them to be essentially spherically symmetric objects (to within 5% radial symmetry), which simplifies the task of interpretation.

Because of the significant contributions of internal scattering of the SAXS patterns, the apparent radius of gyration R_g^a obtained from a Guinier plot is influenced by the internal structure to an extent determined by the electron-density contrast $\Delta\rho_e$ between solvent and the protein particle. The true value R_g must be obtained by extrapolation in solvents of varying electron density to infinite contrast.

$$R_g^2(\rho_e^0) = R_g^2 + \frac{\text{const}}{\Delta\rho_e}. \tag{12.75}$$

SAXS studies of this type using aqueous solutions of NaCl or sucrose of varying concentration show that both HDL and LDL are pseudospherical structures, the former having smaller dimensions R_g = 54 Å, the latter having R_g = 90 Å; these correspond to spheres of radii 70 Å and 115 Å, respectively.

At lower electron-density contrast, the scattering from internal structure is dominant and has been interpreted as indicating a spherically symmetric layered structure. The conclusion is that the HDL particle comprises an outer shell of protein and phospholipid head groups with high electron density, which is almost 15 Å wide, centered at a radius of 63 Å; the interior low electron-density core extends out to about 55 Å and contains the phospholipid carbon chains, cholesterol, cholesterol esters, and the triglycerides. The bigger LDL particle is structurally more complex. The radial electron-density distribution indicates three maxima centered around 30-, 65-, and 105-Å radii, and a low electron-density core. Structural interpretation of this distribution indicates an outer shell of protein and phospholipid head groups and an internally ordered arrangement of the cholesterol esters, probably in concentric layers.

12.12 Structural Characterization: Hydrodynamic Volume and Related Information

As we have discussed, the hydrodynamic quantities that are experimentally accessible are intrinsic viscosity $[\eta]$, translational and rotational diffusion coefficients at zero concentration D_t^0 and D_r, and sedimentation coefficient, also extrapolated to $C_2 = 0$, S^0. These parameters can be used to obtain structural information about the macromolecules in the form of hydrodynamic dimensions. Measurement of the intrinsic viscosity defines the hydrodynamic volume V_h of the macromolecule through the Einstein-Simha relation:

12. Solution of Proteins

$$[\eta] = \frac{\nu N_A V_h}{M} \tag{12.76}$$

where ν is a geometric constant that depends on the deviation of the particle from spherical symmetry. The hydrodynamic volume includes all solvent species that behave as part of the molecule in hydrodynamic experiments. V_h can be defined in terms of the partial specific of volumes of the various components

$$V_h = \frac{M}{N_A}\left(\bar{v}_2 + \sum_i \delta_i V_i^0\right) \tag{12.77}$$

where $\bar{v}_2$ is the partial specific volume of the biopolymer, V_i^0 is the partial specific volume of solvent component i, and δ_i is the solvation factor equal to the weight of species i bound to a particular weight of the macromolecule.

For proteins in dilute salt solutions, as we have seen, $\Sigma \delta_i \bar{V}_i^0$ is negligible for all components other than water, unless strong specific interactions occur with salt species. Thus,

$$V_h = \frac{M}{N_A}(\bar{v}_2 + \delta_1 V_1^0) \tag{12.78}$$

The magnitude of the hydrodynamic radius for a protein of specified density increases with departure from spherical symmetry in much the same manner that R_g increases. This effect is described by the Simha constant ν in equation (12.76) which, evaluated for ellipsoids of revolution, takes the form:

$$\text{prolate ellipsoids } \nu = \frac{a^2/b^2}{15(\ln(2a/b) - 3/2)} \quad \text{(semi-axes a, b, b)} \tag{12.79}$$

$$\text{oblate ellipsoids } \nu = \frac{16\, a/b}{15 \tan^{-1} a/b} \quad \text{(semi-axes a, a, b)} \tag{12.80}$$

Equations (12.79) and (12.80) are special cases [(a/b) > 10] of Simha's more general theory, displayed graphically in Figure 12.30.

Figure 12.30 Simha's factor for the viscosity increment of ellipsoids for relatively low values of the axial ratio, a/b.

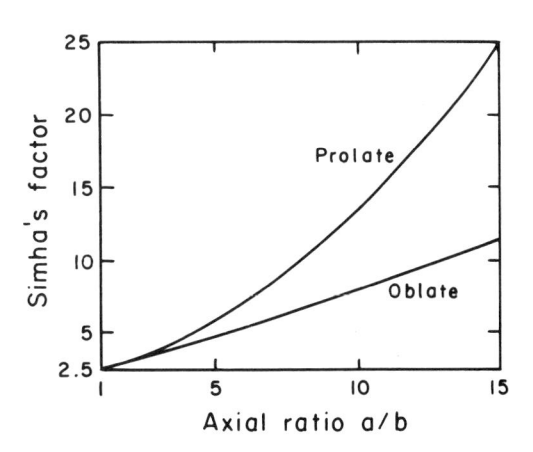

The parameters S^0 or D_t^0 can also furnish an estimate of macromolecular dimensions in the form of the Stokes radius R_s. The frictional coefficient at zero protein concentration f_0 defines the Stokes radius through the equation

$$f_0 = 6\pi \eta_0 R_s. \tag{12.81}$$

For rigid protein molecules, one might expect that the hydrodynamic volume and the Stokes radius are related by $V_h = \frac{4}{3}\pi R_s^3$. This is true only for impermeable rigid protein structures; if the protein exhibits a degree of flexibility or porosity, it is predicted that the effective size, measured in a viscosity experiment (where local shear fields are exerted on the molecule), will differ from that in a diffusion or sedimentation experiment, where such fields are absent. However, one may always write

$$M[\eta] \propto V_h \propto (R_s')^3 \text{ and } f_0 \propto R_s = k R_s'$$

where k is a constant for a molecular species of particular physical type (flexibility, porosity, shape) and R_s' denotes the Stokes radius measured from $[\eta]$, that is, different from R_s, obtained through f_0. Thus, one will always expect the molecular-weight dependence of the hydrodynamic parameters of this class of particle to obey equations (12.16) and (12.36) where $a + 1 = 3\gamma$, since both R_s and R_s' will depend on M in the same way, that is, $R_s = kR_s' \propto M_2^\gamma$

The effect of particle geometry or frictional coefficient can be conveniently expressed in terms of the ratio f_0/f_0^s where f_0^s is the frictional coefficient of a sphere of the same volume. For prolate ellipsoids,

$$\frac{f_0}{f_0^s} = \frac{f}{6\pi\eta_0 R_s} = \frac{(1 - b^2/a^2)^{1/2}}{(b/a)^{2/3} \ln \frac{1 + (1 - b^2/a^2)^{1/2}}{b/a}} \tag{12.82}$$

and for oblate ellipsoids,

$$\frac{f_0}{f_0^s} = \frac{[(a^2/b^2) - 1]^{1/2}}{(a/b)^{2/3} \tan^{-1}[(a^2/b^2) - 1]^{1/2}}. \tag{12.83}$$

The Stokes radius R_s in equations (12.82) and (12.83) can be related to the dimensions of the equivalent ellipsoid:

$$\frac{4}{3}\pi R_s^3 = \frac{4}{3}\pi ab^2 \text{ (prolate ellipsoid) or } \frac{4}{3}\pi a^2 b \text{ (oblate ellipsoid)}.$$

A graphic representation of these functions is shown in Figure 12.31 for low values of the axial ratio a/b.

In the application of equations (12.78)–(12.83) and Figures 12.30 and 12.31, it is evident that there are essentially two unknowns, if M and $\bar{v}_2$ are characterized. These are the axial ratio a/b, and the solvation parameter δ_1;

$$f_0 = 6\pi\eta_0 \frac{f_0}{f_0^s} \left[\frac{3M(\bar{v}_2 + \delta_1 V_1^0)}{4\pi N_A} \right]^{1/3}. \tag{12.84}$$

12. Solution of Proteins

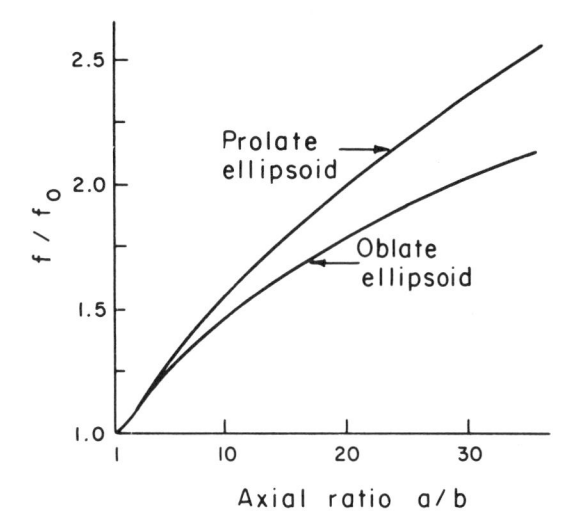

Figure 12.31 Perrin's factor for the frictional coefficient of ellipsoids for relatively low values of the axial ratio, a/b.

Thus, in principle, in order to elucidate both a/b and δ_1, two independent hydrodynamic experiments must be performed. It may be shown that by combining equations (12.76), (12.77), and (12.84) the parameter β in equations (12.39) and (12.40) depends solely on the axial ratio, that is,

$$\beta = \psi \, \nu \, f_0/f_0^s \qquad (12.85)$$

where ψ is a numerical constant. The parameter β is shown graphically for prolate and oblate ellipsoids in Figure 12.32.

Unfortunately, for oblate ellipsoids, β does not change significantly with axial ratio, and varies only slowly for prolate ellipsoids for a/b < 10. In this region, cumulative errors in the parameters D_t^0 [η] and M may become too large

Figure 12.32 Plot of β versus axial ratio. If β can be calculated, it is, in principle, possible to obtain a value of a/b for ellipsoidal particles that would behave equivalently.

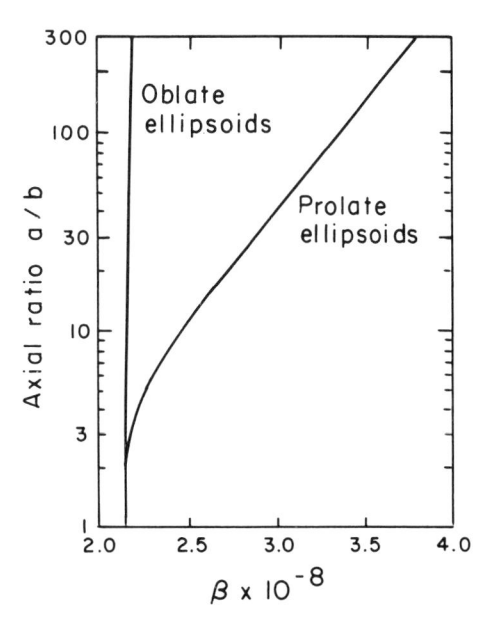

even to distinguish spheres from prolate ellipsoids. In principle, having established a/b, one can then proceed to measure δ_1. For completeness, we note that in discussing the radius of gyration, we included an expression for particles that are rod-shaped, having length L and diameter d. One might consider that such a particle is equivalent to a prolate ellipsoid of the same length (2a = L) and having the same volume. If this were true, the relationship between axial ratio of the two objects would be

$$a/b = (2/3)^{1/2} L/d. \tag{12.86}$$

Before considering applications of these ideas to structural characterization of proteins, we observe that it is possible to determine experimentally a fourth important hydrodynamic quantity, the rotational diffusion coefficient D_r^0, where the superscript again refers to an extrapolation to zero solute concentration. The rotational diffusion coefficient characterizes the time taken for a molecule to rotate in solution during its spontaneous Brownian tumbling motions. For a molecule having several axes of symmetry, one may expect that there are different rotational diffusion coefficients associated with rotation about each axis. Thus for an ellipsoid of revolution there will be two rotational diffusion coefficients, one for the major axis and one for the minor axis. The time scale of these motions will be more separated, as the ratio a/b increases. D_r^0 can be measured by a variety of instrumental methods including flow dichroism, transient electric birefringence, fluorescence depolarization of fluorescence decay, dielectric relaxation, and depolarized dynamic light scattering.

12.13 Flow Dichroism

The most widely used classical technique has been flow dichroism, which is based on the fact that the shearing force produced in a flowing liquid can be used to orient asymmetric molecules. This orientation is opposed by the rotational diffusion of the molecules. Thus the dependence of the degree of orientation of the particles on the rate of shear can be used to calculate D_r. The degree of orientation can be estimated by using the fact that a solution of oriented molecules exhibits birefringence, that is, the refractive index of the solution varies depending on the orientation of the plane of polarization of the incident light relative to the orientation of the molecular species in the solution. If the refractive index ($\bar{n}_2$) corresponding to the plane of polarization parallel to the axis of the oriented molecules (optical axis) is greater than that perpendicular to this axis ($\bar{n}_1$), then the birefringence is positive. If $\bar{n}_2 < \bar{n}_1$, it is negative. The birefringence of a material is determined experimentally by observing the transmission of polarized light through the sample using crossed polarizers. The first polarizer defines the plane of polarization of incident light; and the second, at right angles to the first, defines the polarization of transmitted light that reaches the detector. Some light will always reach the detector if the sample is birefringent, because a portion of the incident light will be transmitted with its plane of polarization perpendicular to the incident plane, unless the incident plane is parallel or perpendicular to the optical axis. At this position, no light passes (complete extinction). By rotation of the crossed

12. Solution of Proteins

polarizers, it is thus possible to determine the position of the optical axis. The apparatus for flow dichroism consists of two concentric cylinders, one of which is rotated, the other fixed (Figure 12.33). The solution is confined in the space between. A uniform gradient is set up in the fluid; this gradient produces orientation of anisometric particles. Polarized light passes vertically through the solution; the transmitted light then falls on a second polarizer whose plane of polarization is perpendicular to that of the first polarization. Complete extinction occurs at four positions. This extinction pattern is called *the cross of isocline*. The extinction angle χ locates the axis of orientation of the molecules, and determination of this angle as a function of the rate of shear rate $\dot{\gamma}$ yields D_r by means of the relation

$$\chi = 45° - \frac{1}{12}\frac{\dot{\gamma}}{D_r} + \text{const.} \left(\frac{3\dot{\gamma}}{D_r}\right)^3 + \ldots \tag{12.87}$$

Figure 12.33 Experimental manifestation of flow birefringence when all solute particles have the same orientation angle ϕ_1 with respect to the flow lines. The upper diagram shows the cross of isocline, which points to the four locations where the optic axes of solute particles are exactly parallel to the analyzer or polarizer plane. The lower diagram shows the observable result in terms of transmission of light through the annular space between the two cylinders of the apparatus. It is to be noted that the extinction angle χ, which is the angle between the cross of isocline and the polarization planes, is here equal to the angle ϕ_1.

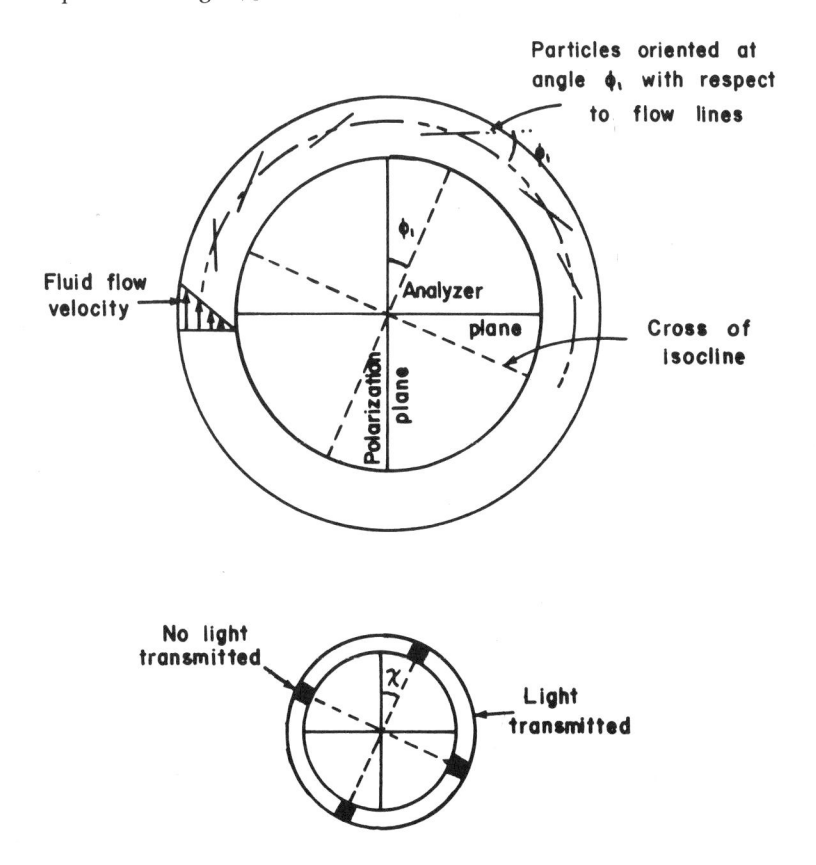

Equation (12.87) shows that χ approaches 45° as $\dot{\gamma} \to 0$, and that a linear relation between χ and $\dot{\gamma}$ should be found for sufficiently low values of $\dot{\gamma}$ from which D_r can be calculated.

12.14 Transient Electric Birefringence

More recently, transient electric birefringence has been used to determine rotational diffusion coefficients in some protein solutions. This method depends on the fact that, in an electric field, anisometric macromolecules in solution are oriented because such molecules have a net electric dipole moment that will tend to align with the direction of the applied field. When the field is applied, the solution becomes birefringent. If the field is then turned off, the birefringence decays to zero, and the rate of the decay is determined by the rotational diffusion coefficient of the molecules. Figure 12.34 shows experimental data for solutions of BSA. An exponential relaxation process is observed for the BSA birefringence whose characteristic time τ_n is related to D_{rb} by

$$\tau_n = \frac{1}{6} D_{rb}. \tag{12.88}$$

The subscript b refers to the fact that rotation of the major axis about the minor axis was observed in the experiment. Voltage pulses of 1–6 μsec duration and amplitudes up to 5,000 volts are applied to the BSA molecules in solution.

12.15 Dielectric Relaxation

The dielectric dispersion technique involves measuring the dielectric constant ϵ of a protein solution as a function of the frequency of an applied oscillating electric field. The dielectric constant of a solution containing electric dipoles is frequency dependent because its magnitude depends on the ability of the dipoles to fluctuate in orientation in response to the oscillations of the applied field. At low frequencies, the dielectric constant is relatively high because the dipoles can follow the field and at all times there is a Boltzmann distribution of dipolar orientations with respect to the applied field. At higher frequencies when the field oscillates too quickly for the dipoles to follow, there is a random

Figure 12.34 Birefringence signals from deionized BSA solution, 15 g/liter, 25° C.

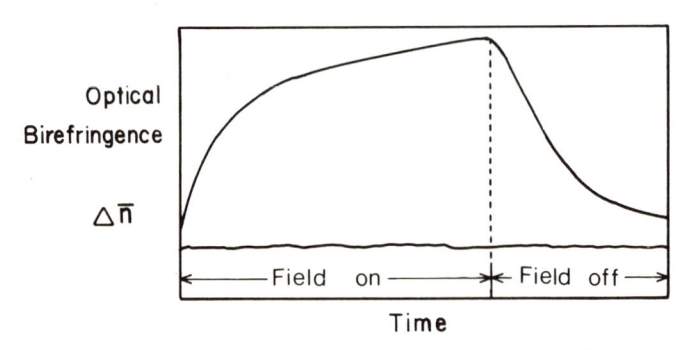

distribution of orientations and a lower value for the measured dielectric constant. At some intermediate frequency, there will be a sudden decrease in the magnitude of D. This is illustrated in Figure 12.35; the dispersion for protein molecules occurs at a much lower frequency than that of the solvent because large molecules cannot reorient as fast as small molecules. The dispersion frequency, as might be intuitively guessed, is determined by the relaxation time for rotational diffusion and permits one to calculate D_r.

The technique of fluorescence depolarization is outlined in Chapter 10. Depolarized dynamic light scattering has been less widely applied to measurements of rotational diffusion of proteins than other methods and will not be detailed here.

Usually, only the smaller rotational diffusion coefficient D_{rb} is detected by experiments of the type explained so far. For dielectric dispersion and transient birefringence studies of BSA, however, the higher diffusion coefficient corresponding to rotation around the major axis a has been observed.

The parameter D_{rb} contains structural information about the macromolecule that can be characterized in terms of an equivalent ellipsoid of revolution as was discussed for (η) and f^0 earlier. A rotational frictional coefficient f_r^0 can be defined by

$$D_r^0 = \frac{RT}{N_A f_r^0} \tag{12.89}$$

and it can be shown for a prolate ellipsoid of revolution that

$$f_r^0 = \frac{16\pi\eta a^3}{3}\left[\frac{1-(b^2/a^2)^2}{(2-b^2/a^2)G(b/a)-1)}\right] \tag{12.90}$$

Figure 12.35 Schematic diagram illustrating the data obtained in measurement of the dielectric properties of protein solutions (concentration c in grams per ml).

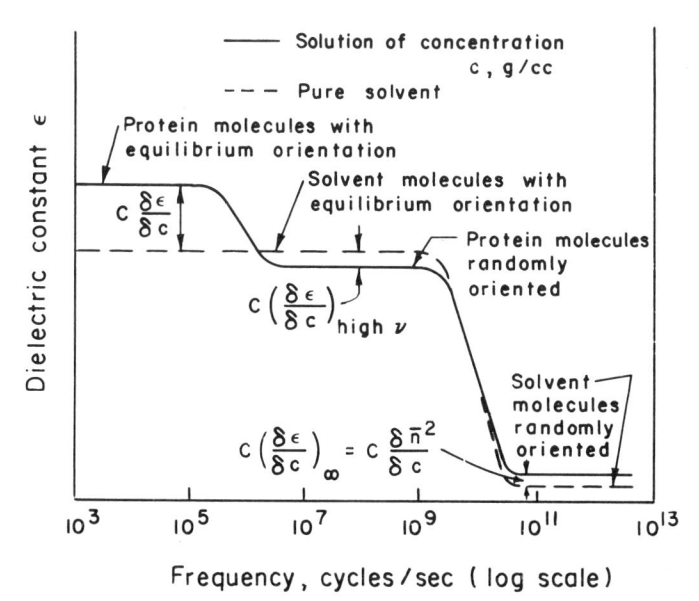

where

$$G(b/a) = \left[\frac{1 + (1 - b^2/a^2)^{1/2}}{b/a}\right] \bigg/ (1 - b^2/a^2)^{1/2} \quad \text{(prolate)} \tag{12.91}$$

for an oblate ellipsoid, eq. (12.90) applies with

$$G(b/a) = \frac{\tan^{-1}(a^2/b^2 - 1)^{1/2}}{(a^2/b^2 - 1)^{1/2}} \quad \text{(oblate)}. \tag{12.92}$$

Using this format, equations (12.82) and (12.83) can be rewritten concisely:

$$f_0 = \frac{6\pi\eta a}{G(b/a)}. \tag{12.93}$$

The utility of rotational diffusion experiments for protein characterization is that they are much more sensitive to shape than are translational diffusion of $[\eta]$ measurements. We will emphasize the advantage of combining D_r^0 with D_t^0 or with S or with $[\eta]$. A geometric parameter δ_b may be defined as follows:

$$\nu f_{rb}^0/f_r^s = 6\eta_0 D_{rb}^0 [\eta] M/RT = \delta_b. \tag{12.94}$$

The sensitivity of δ_b to particle anisotropy may be calculated and is shown in Figure 12.36. In the preceding equation (12.94), f_r^s is the rotational diffusion coefficient of a sphere of equivalent hydrodynamic volume. A similar relation could be proposed for f_{ra}^0, but is not often used. The advantage of equation (12.94) over (12.85) is that δ_b is more sensitive to particle geometry as illustrated by comparing Figure 12.36 with Figure 12.32.

Application of equations (12.76) and (12.84) to proteins and polypeptides is limited by the fact that one cannot unequivocally assign a value to a/b and, independently, to δ_1, the solvation parameter (see equation 12.78). One may assume the protein is unsolvated, $\delta_1 = 0$, and calculate a maximum asymmetry $(a/b)_{max}$ from Figures 12.30 or 12.31 or, alternatively, assume the protein is spherically symmetrical $(a/b) = 1$, and calculate a maximum amount of solvation $(\delta_1)_{max}$ from the experimental data. The true picture inevitably will lie somewhere between the two extremes.

Results of an analysis of this kind, applied to various proteins in their native configurations, are summarized in Table 12.5. From this table it is obvious just from the solvation data that there are two distinctly different types of protein. The first five proteins are moderately solvated and/or slightly anisometric and are globular proteins; the other four are highly solvated (i.e., flexible coils) or extremely anisometric (rigid rods). To distinguish a flexible coil from a rigid rod one must examine the radius of gyration R_g. From the hydrodynamic data it is possible to calculate an expected value for R_g for each alternative model. For example, for a random chain it may be shown that

$$[\eta] = \frac{10\pi N_a}{3M} \xi^3 R_g^3 \tag{12.95}$$

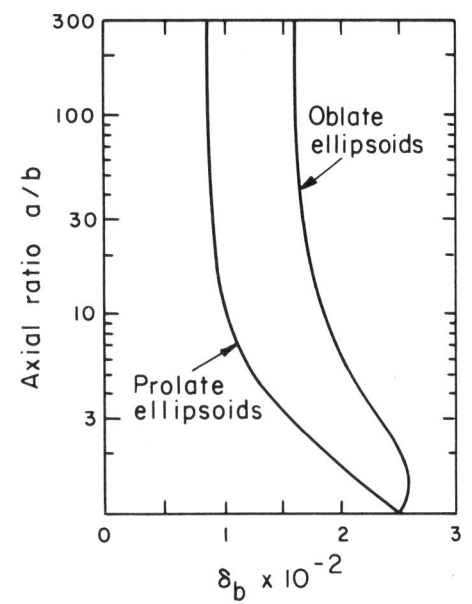

Figure 12.36 Plot of δ_b vs. axial ratio. This plot is applicable to flow birefringence or other measurements of D_{rb}.

and

$$D_t^0 = RT/N_A f = RT/6\pi N_A \eta \xi_f R_g \qquad (12.96)$$

where $\xi = 0.875$ and $\xi_f = 0.66$ are constants predicted by the theory. On the other hand, one can compute the length of the equivalent rigid rod $L = 2a$ from the maximum value of a/b shown in Table 12.5 together with the knowledge that $V_h = \frac{4}{3}\pi ab^2$. Then R_g for the rigid rod can be calculated using equation (12.68). Comparison of these alternative values of R_g obtained from the hydrodynamic data with experimental results for R_g can be used to determine the "correct" structure. There is some uncertainty associated with the theoretical constants ξ and ξ_f but this is small in comparison with the large dimensional differences between random-chain and rigid rods noted earlier (Table 12.3).

Table 12.5 Molecular Dimensions of Proteins from Intrinsic Viscosities

	Maximum solvation ($\nu = 2.5$)		Maximum asymmetry ($\delta = 0$)	
	δ_1 grams/gram	R_g, Å	ν	a/b, Prolate ellipsoid
Ribonuclease	0.59	19.3	4.5	3.9
β-Lactoglobulin	0.61	26.6	4.5	3.9
Serum albumin	0.75	33.7	5.0	4.4
Hemoglobin Catalase	0.69	34	4.8	4.1
Tropomyosin	20	91	70	29
Fibrinogen	10.1	112	38	20
Collagen	460	400	1,660	175
Myosin	86	257	298	68

It is important to note that there is no significant difference between globular proteins in the calculated axial ratio and, indeed, one cannot absolutely exclude the possibility that each protein is in fact a solvated sphere. It might seem that the dilemma would be simply resolved if one could assign δ_1 the value corresponding to the hydration number of the protein, measured by, say, differential refractometry or light scattering. In this case one can certainly determine a value for a/b, but there is no a priori reason to believe that the ellipsoidal structure determined by this procedure even remotely resembles the true protein structure because a protein is unlikely, in reality, to be rigid or impermeable or to have a smooth surface.

Equation (12.85) circumvents the question of solvation and permits one to define the dimensions of an equivalent prolate ellipsoid by determining the constant β (from Figure 12.32, it is obvious that all oblate ellipsoids will be indistinguishable from a sphere). However, it can realistically be applied only to highly anisometric structures (a/b > 10) even in this case. Further, it appears that for at least some globular proteins (e.g., BSA) the experimental value of β is significantly less than the value predicted for a hard sphere! This unexpected result has recently been interpreted theoretically in terms of a degree of porosity in the protein structure. The individual hydrodynamic parameters are more sensitive to anisometry in molecular structure than is a ratio of any pair, and therefore an analysis like that embodied in Table 12.5 may be more fruitful.

With regard to the structural analysis of globular proteins, measurement of the rotational diffusion coefficient turns out to be a most useful probe when combined with another independent hydrodynamic measurement. With the modern development of convenient and accurate methods for determining diffusion coefficients D_t^0 and D_r^0, which have been described previously, implementation of equations (12.90)–(12.93) becomes a rather attractive route. Analytical and numerical methods for inverting these equations to define a/b have been developed. As an example, for BSA, D_t^0 at 20°C in water has been determined to be 6.1×10^{-7} cm²/sec by dynamic light scattering, and D_{rb}^0 (20°C, water) = 1.93×10^6 sec^{-1} by transient birefringence or dielectric dispersion. These lead to dimensions of the equivalent prolate ellipsoid of 2a = 140 Å and 2b = 40 Å in agreement with earlier experiments in which comparison of D_{rb}^0 and S^0 with $[\eta]$ was utilized.

In studies of more highly anisometric proteins or polypeptides, the parameters D_t^0, S^0, and $[\eta]$ have provided some significant insights. In such cases these hydrodynamic quantities can provide a useful index of particle anisometry. It appears possible to study the conformational properties of synthetic polypeptides by measuring D_t^0 or $[\eta]$ as a function of molecular weight for fractionated samples of narrow polydispersity. In this analysis, the Stokes radius R_s or hydrodynamic volume is utilized in much the same fashion as was described earlier for the radius of gyration. The axial length 2a of the polymer is determined by comparing D_2^0 and $[\eta]$ and utilizing the Mandelkern–Scheraga theory (Figure 12.32) in conjunction with the equation $V_h = (4/3)\pi ab^2$ for a cylindrical rod. The rise per residue h is then computed by dividing 2a by the degree of polymerization of the sample, $h = 2aM_0/M$. Results of such an analysis for several molecular-weight fractions of PBLG in an α-helix directing solvent are shown in Figure 12.37. Once again, as observed in the analogous

Figure 12.37 Rise per residue h calculated from hydrodynamic parameters for poly-γ-benzyl-L-glutamate sample as a function of molecular weight: unfractionated samples.

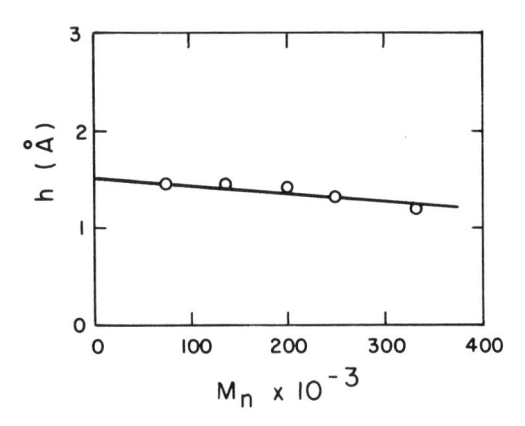

studies using R_g, values of h at higher molecular weights (M > 200,000) are significantly influenced by chain flexibility leading to erroneously low values of h. Extrapolation of h values for samples of narrow polydispersity to M = 0 leads to a value of h = 1.5 Å, very close to the true value for α-helical pitch measured by x-ray diffraction.

Hydrodynamic parameters can also be used to monitor changes in the conformational structure of proteins and polypeptides. Figure 12.38 represents data for the pH-driven denaturation of BSA. The sigmoidal behavior of D_t^0 and S^0 as the pH is reduced indicates that there is an increase in the frictional coefficient. The ratio S^0/D_t^0 remains unchanged, however, showing that the molecular weight is unchanged (equation 12.35). The transition corresponds to a swelling or unfolding of individual BSA molecules.

Finally, we note that it is possible, using D_t or S measurements, to obtain information about the self-association behavior of oligomeric proteins similar to that derived by thermodynamic techniques. In the absence of aggregation

Figure 12.38 Sedimentation and diffusion coefficients of serum albumin as a function of pH.

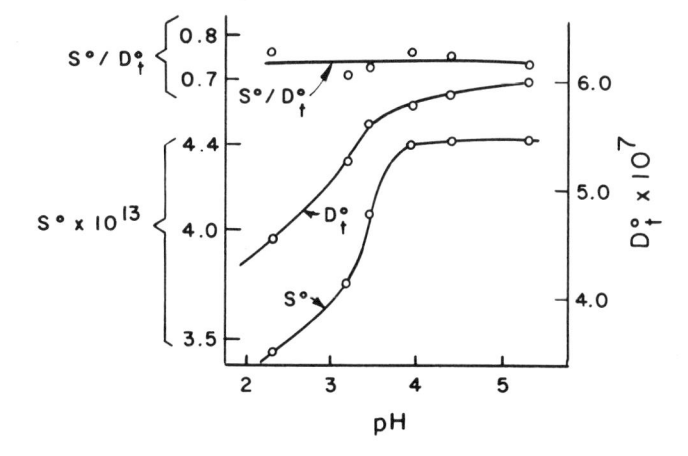

behavior, the concentration dependence of the translational diffusion coefficient of a globular protein is described by

$$D_t = \frac{kT}{f^0} [1 + (2A_2 - k_f) C_2] \qquad (12.97)$$

where A_2 is the osmotic second virial coefficient (equation 12.2) and k_f characterizes the concentration dependence of the translational frictional coefficient f

$$f = f^0(1 + k_f C_2 + \cdots). \qquad (12.98)$$

In the presence of sufficient supporting electrolyte to minimize long-range electrostatic or electrodynamic interactions, the result of equation (12.97) is that the concentration dependence of D_t is very weak and may be slightly positive or negative, depending on the relative magnitude of A_2 and k_f. Likewise, the concentration dependence of S is of the form

$$S = S^0 (1 - k_f C_2) \qquad (12.99)$$

and is also small and negative.

If reversible self-association of the protein occurs, however, dramatic, often nonlinear, behavior of the concentration dependence of D_t or S will be observed corresponding to the transition from predominantly multimer at high concentrations to the monomeric species at low concentrations. Analysis of the concentration dependence of D_t or S can therefore be used to evaluate the equilibrium constants for the self-association using either experimental values for A_2 and k_f of the monomer and multimer, or theoretically predicted values based on a hard- or soft-sphere model.

Such an analysis has been applied to the concentration dependence of the sedimentation constant of aqueous solutions of β-lactoglobulin at pH = 1.6. Under these conditions the protein exists as an equilibrium mixture of monom-

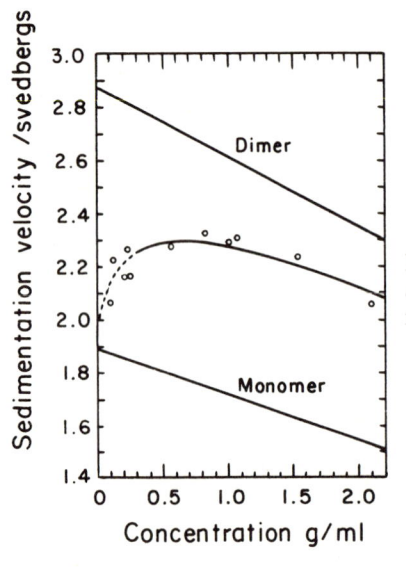

Figure 12.39 Sedimentation of β-lactoglobulin at pH 1.6. Limiting values for the velocities of monomer and dimer are also shown.

12. Solution of Proteins

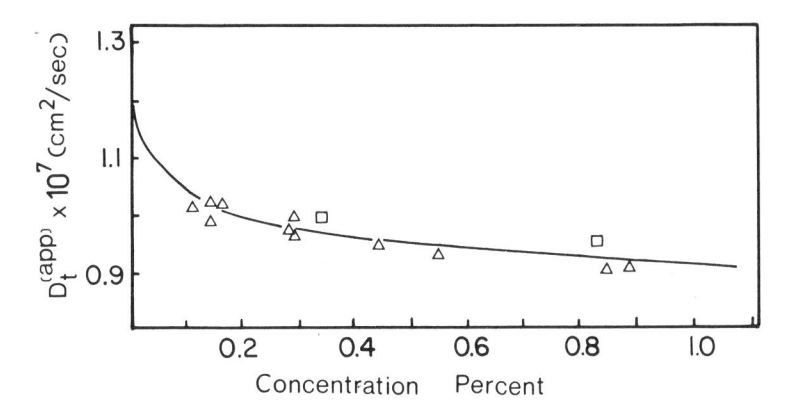

Figure 12.40 Apparent diffusion constants of myosin as a function of concentration. △, in 0.5M KCl, 0.01M EDTA, 0.2M PO$_4$ (pH 7.3); □ in 0.5M KCl, 0.01M EDTA, 0.01M Tris, 0.2M SO$_4$ (pH 7.3). (Data of Herbert and Carlson).

ers and dimers. Theoretical analysis of these data, shown in Figure 12.39, leads to a value of the association constant (monomer ⇌ dimer) $K_{12} = 4.34 \pm 0.58$ dl/g. The limiting sedimentation coefficients of monomer and dimer are $S_1^0 = 1.89S$ and $S_2^0 = 2.87S$, respectively, and the value of k_f (see equation 12.99) is assumed the same for monomer and dimer, $k_f = 0.092 \pm 0.008$ ml/g.

Diffusion data, measured by dynamic light-scattering techniques from a solution of the muscle protein myosin, may be similarly treated in terms of a monomer–dimer equilibrium as shown in Figure 12.40. The analysis indicates that the equilibrium constant K_{12} decreases from 10.6 ml/g in 0.2M phosphate or 0.2M sulfate to 1.30 ml/g in 0.5M phosphate or 0.5M sulfate. The diffusion data for myosin in 2.0M KCl at pH 8.5 in the absence of sulfate or phosphate show similar concentration dependence to that of myosin in 0.5M KCl + 0.2M phosphate or 0.2M sulfate, indicating that neither higher ionic strength nor higher pH is responsible for the observed changes in K_{12} from the diffusion constants at zero concentration. Thus the association equilibrium appears to be specifically influence by phosphate and sulfate ions. For monomer and dimer, $D_{t1}^0 = 1.24 \times 10^{-7}$ cm^2/sec and $D_{t2}^0 = 0.84 \times 10^{-7}$ cm^2/sec. Lengths for monomer and dimer correspond to 1481 Å and 2121 Å, respectively.

12.16 Concluding Remarks on Structural Characterization

We have covered a number of highly useful instrumental methods by means of which structural information about protein species can be derived in solution. In summarizing the salient features of this discussion we select a number of specific points. First, techniques abound by which accurate molecular weights of native proteins and their denatured counterparts can be conveniently determined. These provide a powerful tool for a priori investigation of oligomeric structure. Second, thermodynamic and hydrodynamic approaches exist that can be used fruitfully to study the association equilibria of oligomeric proteins; however, only relatively simple equilibria are amenable to accurate analysis. Third, R_g and R_s can be used to distinguish the larger structural features of protein moieties (coil, sphere, rod) but are less useful for studying the geometry of globular proteins. Fourth, R_g and R_s can be used to study the conformational nature of helical polypeptides but polydispersity must be controlled and extrapolation to samples having M < 100,000 must be carried out. Fifth, rotational diffusion experiments provide, by themselves or in conjunction with other size measurements, an important route for studying geometry of globular protein species (modest anisotropy). Sixth, the development of advanced characterization technology (dynamic light-scattering methods, improvements in SAXS instrumentation, split-beam photoelectric scanners in ultracentrifuge analysis) now permits one to observe very subtle changes in the conformational properties of proteins even though the dimensional changes accompanying such transitions are exceedingly small (< 1%) and, in addition, allows the nature of internal structure of complex proteins in solution to be probed.

Further Reading

Berne, B.J. and R. Pecora, *Dynamic Light Scattering with Applications to Chemistry, Biology, and Physics*. Wiley, New York, 1976.

Cantor, C.R. and P.R. Schimmel. *Biophysical Chemistry*, Vol. 2, Chapters 10–12. W.H. Freeman and Co., San Francisco, 1980.

Chu, B. Laser Light Scattering. Academic Press, New York, 1974.

Hirs, C.H.W. and S. Timasheff, eds. *Methods in Enzymology XXVII*, Enzyme Structure Parts C–H. Academic Press, New York, 1973–1979. (Especially Vol. 26, Sections I and III; Vol. 27, Sections I and II; Vol. 28, Sections I and II; Vol. 61, Sections I and II.)

van Holde, K.E. *Physical Biochemistry*. Prentice-Hall, Englewood Cliffs, NJ, 1971.

Morawetz, H. *Macromolecules in Solution*, 2nd Ed. Wiley-Interscience, New York, 1975.

Tanford, C.H. *Physical Chemistry of Macromolecules*. Wiley, New York, 1961. (Particularly Chapters 3–7).

IV

PROTEIN SYSTEMS

13

Collagen and Connective Tissue

Collagen is the most abundant protein of connective tissue, in which it plays a major structural role. Commercially, collagen is important in industries dealing with leather, gelatin, some membranes, hair treatment, and so on. As with such fibrous proteins as α-keratin (in hair, wool, etc.) and muscle, different hierarchies or levels of organization may be observed. In rat tail tendon, several such levels are evident, as shown in Figure 13.1.

Because the main function of collagen is to act as a mechanically strong fiber (for example in tendons), nature has undertaken to intertwine molecules into an intricate array that possesses the necessary load-bearing features. The path that we shall follow in this chapter is that spelled out in other chapters where an understanding of primary sequence is followed through conformation, molecular aggregation and assembly, and structural function.

Before proceeding with this analysis it is necessary to point out that although collagens from different sources have structural similarities, they also have structural differences. Even from one species (for example, man) it is now becoming clear that there are several different collagens, differing only slightly in primary structure, but sufficiently so that their organization and function are different. Although most work has been carried out with water-soluble, extracted "tropocollagen," this material is known to be preceded in cellular production by a "procollagen" molecule. The procollagen molecule contains three chains that are crosslinked together at one end. Subsequent cleavage into two fragments produces the tropocollagen molecule that will be the basis of our further examination (see Figure 13.2).

13.1 Tropocollagen

Early examination of tropocollagens from various sources, including calfskin and rat tails, indicated that the molecular weight of approximately 300,000 decreased to 100,000 on heating. From this type of evidence it has become

TENDON HIERARCHY

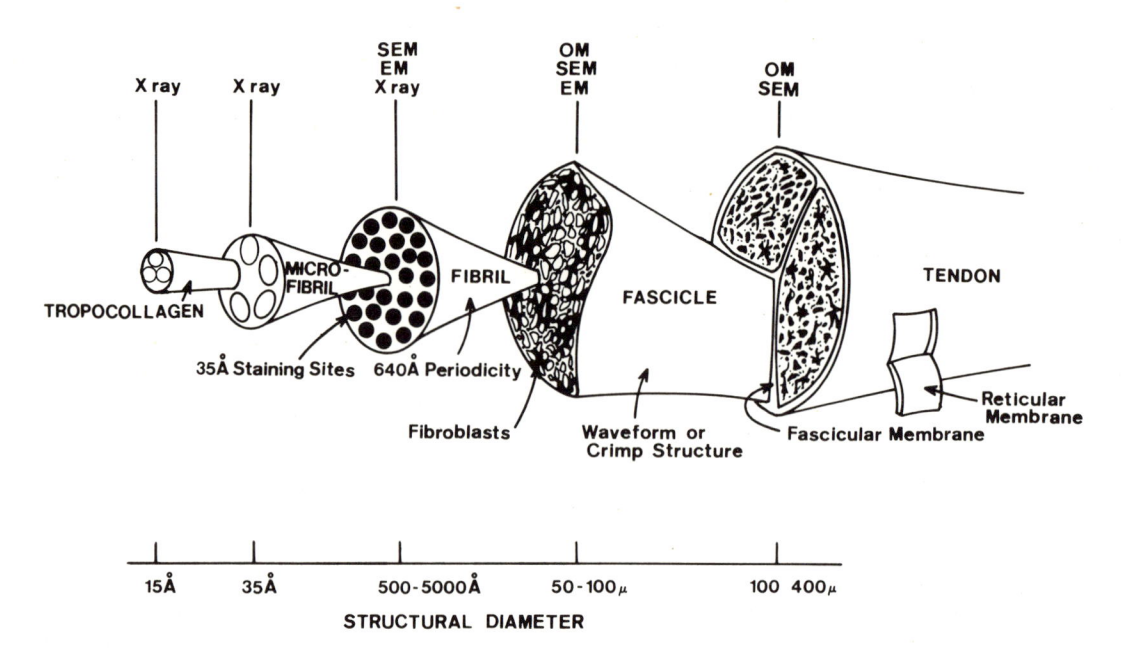

Figure 13.1 Levels of organization of collagen in adult rat tail tendon.

clear that tropocollagen contains three (α-) polypeptide chains. (The nomenclature of α-chains should not be confused with the α-helix.) The most abundant tropocollagen consists of three α-chains, of which two are identical in amino acid content and sequence (α_1-chains) and one is slightly different (α_2-chain). At least four types of human collagen are now known that vary in a relatively minor way. These are classified as follows:

Type 1. Skin, bone, tendon—two α_1- and one α_2-chains

Type 2. Cartilage—three α_1-chains—differ slightly from type 1, α_1

Type 3. Arteries, some skin, uterus—three α_1-chains—differ from type 1 and 2, α_1. Contain an S—S bridge in the helical C-terminal region

Type 4. Basement membrane—extended nonhelical regions.—S—S— bridge in nonhelical region

Many forms of collagen are insoluble due to intramolecular (e.g., α_{12}) and intermolecular (e.g., β_{11}) crosslinks. The β_{11} entity consists of two α_1-chains from separate molecules in which one or more crosslinks have formed. Similarly, β_{12} represents a covalently linked combination of an α_1- and α_2-chain and γ_{112} represents three chains, two α_1 and one α_2, which may come from the same or different tropocollagen molecules, covalently linked, often by condensation between lysyl ϵ-amino groups. For present purposes, we shall concentrate on uncrossed tropocollagen containing two α_1- and one α_2-chains, unless otherwise specified.

13. Collagen and Connective Tissue

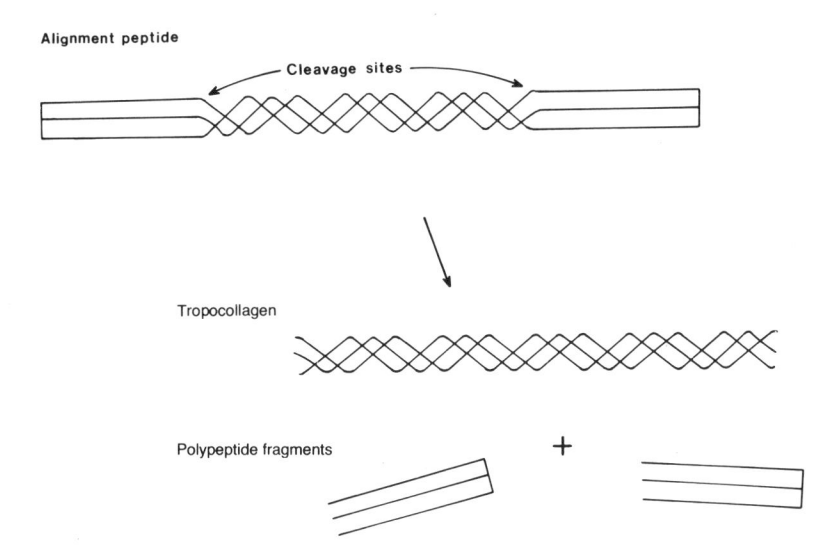

Figure 13.2 Schematic diagram of the cleavage of procollagen to form tropocollagen.

We can now employ a standard scheme or procedure pertaining to structure and function. This scheme will be used repeatedly with both fibrous and globular proteins. The scheme is:

1. Amino acid analysis and sequence;
2. Conformational determination and physical characterization;
3. Electron microscopy;
4. Relation of various levels of structure;
5. Relation between structure and function.

In some cases the methodology does not produce useful answers; in others, slight variations in approach make certain answers much more appropriate and useful. For collagen, which has been very extensively studied, all of the above questions have pertinent answers.

Amino Acid Analysis and Sequence

It was evident from early analysis that collagens from various sources possessed similar (but not identical) amino acid analyses in which two unusual amino acids, hydroxyproline and hydroxylysine, are found. In general, it can be said that tropocollagen contains approximately 33% glycine and 20–25% proline and hydroxyproline. Alanine is the next major component at 10–12% of the total amino acid content.

(With this information alone and knowing that collagen is a fibrous protein of uniform conformation, one could guess at the secondary structure. Since proline is not capable of taking part in α-helix or uniform β-sheet structures, the polyglycine II (polyproline II) conformation is most likely. However, a β-structure with proline in the β-turn is not ruled out.)

The most definitive aspects of the primary structure have come in the past few years when a composite structure for the α_1-chain has been de-

Table 13.1 Composite Sequence of the α-1 Chain of Tropocollagen

pGlu-Met-Ser-Tyr-Gly-Tyr-Asp-Glu-Lys-Ser-Ala-Gly-Val-Ser-Pro-

1	Gly-Pro-Met-	Gly-Pro-Ser-	Gly-Pro-Arg-	Gly-Leu-Hyp-	Gly-Pro-Hyp-	Gly-Ala-Hyp-	Gly-Pro-Hyp-	Gly-Phe-Gln-	Gly-Pro-Hyp-
28	Gly-Glu-Hyp-	Gly-Glu-Hyp-	Gly-Ala-Ser-	Gly-Pro-Met-	Gly-Pro-Arg-	Gly-Pro-Hyp-	Gly-Pro-Hyp-	Gly-Lys-Asn-	Gly-Asp-Asp-
55	Gly-Glu-Ala-	Gly-Lys-Pro-	Gly-Arg-Hyp-	Gly-Gln-Arg-	Gly-Pro-Hyp-	Gly-Pro-Gln-	Gly-Ala-Arg-	Gly-Leu-Hyp-	Gly-Thr-Ala-
82	Gly-Leu-Hyp-	Gly-Met-Hyl-	Gly-His-Arg-	Gly-Phe-Ser-		Gly-Leu-Asp-	Gly-Ala-Lys-	Gly-Asn-Thr-	Gly-Pro-Lys-
109	Gly-Glu-Hyp-	Gly-Ser-Hyp-	Gly-Glx-Asx-	Gly-Ala-Hyp-		Gly-Gln-Met-	Gly-Pro-Arg-	Gly-Leu-Hyp-	Gly-Arg-Hyp-
136	Gly-Pro-Hyp-	Gly-Ser-Ala-	Gly-Ala-Arg-	Gly-Asp-Asp-		Gly-Ala-Val-	Gly-Ala-Ala-	Gly-Pro-Hyp-	Gly-Pro-Thr-
163	Gly-Pro-Hyp-	Gly-Phe-Hyp-	Gly-Ala-Ala-	Gly-Ala-Lys-		Gly-Pro-Gln-	Gly-Pro-Ala-	Gly-Ala-Arg-	Gly-Ser-Glu-	Gly-Pro-Gln-
190	Gly-Val-Arg-	Gly-Glu-Hyp-	Gly-Pro-Hyp-	Gly-Pro-Ala-		Gly-Ala-Ala-	Gly-Pro-Ala-	Gly-Asn-Hyp-	Gly-Ala-Asp-	Gly-Gln-Hyp-
217	Gly-Ala-Lys-	Gly-Ala-Asn-	Gly-Ala-Hyp-	Gly-Ile-Ala-		Gly-Ala-Hyp-	Gly-Phe-Hyp-	Gly-Ala-Arg-	Gly-Pro-Ser-	Gly-Pro-Gln-
244	Gly-Pro-Ser-	Gly-Ala-Hyp-	Gly-Pro-Lys-	Gly-Asn-Ser-		Gly-Glu-Hyp-	Gly-Ala-Hyp-	Gly-Asn-Lys-	Gly-Asp-Thr-	Gly-Ala-Lys-
271	Gly-Glu-Hyp-	Gly-Pro-Ala-	Gly-Val-Gln-	Gly-Pro-Hyp-		Gly-Pro-Ala-	Gly-Glu-Glu-	Gly-Lys-Arg-	Gly-Ala-Arg-	Gly-Glu-Hyp-
298	Gly-Pro-Ser-	Gly-Leu-Hyp-	Gly-Pro-Hyp-	Gly-Glu-Arg-		Gly-Gly-Hyp-	Gly-Ser-Arg-	Gly-Phe-Hyp-	Gly-Ala-Asp-	Gly-Val-Ala-
325	Gly-Pro-Lys-	Gly-Pro-Ala-	Gly-Glu-Arg-	Gly-Ser-Hyp-		Gly-Pro-Ala-	Gly-Pro-Lys-	Gly-Ser-Hyp-	Gly-Glu-Ala-	Gly-Arg-Hyp-
352	Gly-Glu-Ala-	Gly-Leu-Hyp-	Gly-Ala-Lys-	Gly-Leu-Thr-		Gly-Ser-Hyp-	Gly-Ser-Hyp-	Gly-Pro-Asp-	Gly-Lys-Thr-	Gly-Pro-Hyp-
379	Gly-Pro-Ala-	Gly-Gln-Asp-	Gly-Arg-Hyp-	Gly-Pro-Ala-		Gly-Pro-Hyp-	Gly-Pro-Ala-	Gly-Ala-Arg-	Gly-Gln-Ala-	Gly-Val-Met-	Gly-Phe-Hyp-
406	Gly-Pro-Lys-	Gly-Ala-Ala-	Gly-Glu-Hyp-	Gly-Lys-Ala-		Gly-Glu-Arg-	Gly-Val-Hyp-	Gly-Pro-Hyp-	Gly-Ala-Val-	Gly-Pro-Ala-
433	Gly-Lys-Asp-	Gly-Glu-Ala-	Gly-Ala-Gln-	Gly-Pro-Hyp-		Gly-Pro-Ala-	Gly-Glu-Arg-	Gly-Glu-Gln-	Gly-Pro-Ala-
460	Gly-Ser-Hyp-	Gly-Phe-Gln-	Gly-Leu-Hyp-	Gly-Pro-Ala-		Gly-Pro-Hyp-	Gly-Glu-Ala-	Gly-Lys-Hyp-	Gly-Glu-Gln-	Gly-Val-Hyp-
487	Gly-Asp-Leu-	Gly-Ala-Hyp-	Gly-Pro-Ser-	Gly-Ala-Arg-		Gly-Phe-Hyp-	Gly-Glu-Arg-	Gly-Val-Glu-	Gly-Pro-Hyp-

514	Gly-Pro-Ala-	Gly-Pro-Arg-	Gly-Ala-Asn-	Gly-Ala-Hyp-	Gly-Asn-Asp-	Gly-Ala-Lys-	Gly-Asp-Ala-	Gly-Ala-Hyp-	Gly-Ala-Hyp-
541	Gly-Ser-Gln-	Gly-Ala-Hyp-	Gly-Leu-Gln-	Gly-Met-Hyp-	Gly-Glu-Arg-	Gly-Ala-Ala-	Gly-Leu-Hyp-	Gly-Pro-Lys-	Gly-Asp-Arg-
568	Gly-Asp-Ala-	Gly-Pro-Lys-	Gly-Ala-Asp-	Gly-Ala-Pro-	Gly-Lys-Asp-	Gly-Val-Arg-	Gly-Leu-Thr-	Gly-Pro-Ile-	Gly-Pro-Hyp-
595	Gly-Pro-Ala-	Gly-Ala-Hyp-	Gly-Asp-Lys-	Gly-Glu-Ala-	Gly-Pro-Ser-	Gly-Pro-Ala-	Gly-Thr-Arg-	Gly-Ala-Hyp-	Gly-Asp-Arg-
622	Gly-Glu-Hyp-	Gly-Pro-Hyp-	Gly-Pro-Ala-	Gly-Phe-Ala-	Gly-Pro-Hyp-	Gly-Ala-Asp-	Gly-Gln-Hyp-	Gly-Ala-Lys-	Gly-Glu-Hyp-
649	Gly-Asp-Ala-	Gly-Ala-Lys-	Gly-Asp-Ala-	Gly-Pro-Hyp-	Gly-Pro-Ala-	Gly-Pro-Ala-	Gly-Pro-Hyp-	Gly-Pro-Ile-	Gly-Asn-Val-
676	Gly-Ala-Hyp-	Gly-Pro-Hyl-	Gly-Ala-Arg-	Gly-Ser-Ala-	Gly-Pro-Hyp-	Gly-Ala-Thr-	Gly-Phe-Hyp-	Gly-Ala-Ala-	Gly-Arg-Val-
703	Gly-Pro-Hyp-	Gly-Pro-Ser-	Gly-Asn-Ala-	Gly-Pro-Hyp-	Gly-Pro-Hyp-	Gly-Pro-Ala-	Gly-Lys-Glu-	Gly-Ser-Lys-	Gly-Pro-Arg-
730	Gly-Glu-Thr-	Gly-Pro-Ala-	Gly-Arg-Hyp-	Gly-Glu-Val-	Gly-Pro-Hyp-	Gly-Pro-Hyp-	Gly-Pro-Ala-	Gly-Glu-Lys-	Gly-Ala-Hyp-
757	Gly-Ala-Asp-	Gly-Pro-Ala-	Gly-Ala-Hyp-	Gly-Thr-Pro-	Gly-Pro-Gln-	Gly-Ile-Ala-	Gly-Gln-Arg-	Gly-Val-Val-	Gly-Leu-Hyp-
784	Gly-Gln-Arg-	Gly-Glu-Arg-	Gly-Phe-Hyp-	Gly-Leu-Hyp-	Gly-Pro-Ser-	Gly-Glu-Hyp-	Gly-Lys-Gln-	Gly-Pro-Ser-	Gly-Ala-Ser-
811	Gly-Glu-Arg-	Gly-Pro-Hyp-	Gly-Pro-Met-	Gly-Pro-Hyp-	Gly-Leu-Ala-	Gly-Pro-Hyp-	Gly-Glu-Ser-	Gly-Arg-Glu-	Gly-Ala-Hyp-
838	Gly-Ala-Glu-	Gly-Ser-Hyp-	Gly-Arg-Asp-	Gly-Ser-Hyp-	Gly-Ala-Lys-	Gly-Asp-Arg-	Gly-Glu-Thr-	Gly-Glu-Thr-	Gly-Ala-Hyp-
865	Gly-Pro-Hyp-	Gly-Ala-Hyp-	Gly-Ala-Hyp-	Gly-Pro-Val-	Gly-Pro-Ala-	Gly-Lys-Ser-	Gly-Asp-Arg-	Gly-Glu-Thr-	Gly-Pro-Ala-
892	Gly-Pro-Ile-	Gly-Pro-Val-	Gly-Pro-Ala-	Gly-Ala-Arg-	Gly-Pro-Ala-	Gly-Pro-Gln-	Gly-Pro-Arg-	Gly-Asx-Hyl-	Gly-Glx-Thr-
919	Gly-Glx-Glx-	Gly-Asx-Arg-	Gly-Ile-Hyl-	Gly-His-Arg-	Gly-Phe-Ser-	Gly-Leu-Gln-	Gly-Pro-Hyp-	Gly-Pro-Hyp-	Gly-Ser-Hyp-
946	Gly-Glu-Gln-	Gly-Pro-Ser-	Gly-Ala-Ser-	Gly-Pro-Ala-	Gly-Pro-Arg-	Gly-Pro-Hyp-	Gly-Pro-Ala-	Gly-Ser-Hyp-	Gly-Lys-Asp-
973	Gly-Leu-Asn-	Gly-Leu-Hyp-	Gly-Pro-Ile-	Gly-Hyp-Hyp-	Gly-Pro-Arg-	Gly-Arg-Thr-	Gly-Asp-Ala-	Gly-Pro-Ala-	Gly-Pro-Hyp-
1000	Gly-Pro-Hyp-	Gly-Pro-Hyp-	Gly-Pro-Hyp-	Gly-Pro-Pro-					

Ser-Gly-Gly-Tyr-Asp-Leu-Ser-Phe-Leu-Pro-Gln-Pro-Pro-Gln-Glx-Lys-Ala-His-Asp-Gly-Gly-Arg-Tyr-Tyr

termined. (The presented sequence is made up from sequences obtained from collagens from different sources, but is believed to be similar for skin collagens from all higher mammals.) The sequence is given in Table 13.1.

It is noteworthy that, except for the initial 16 and final 25 residues, the sequence has glycine in every third position and a typical sequence would be —Gly Pro Hypro—. The N- and C-terminal residues, not in the triplet sequence, are often referred to as "telopeptides" and are the remnants of the chains of procollagen which presumably act to align residues from adjacent chains.

There are many other notable aspects of the sequence. For example, hydroxyproline only occurs in the "third position" if we number the triplet from glycine at the N-terminus; that is, $\overset{1}{\text{Gly}}\ \overset{2}{\text{X}}\ \overset{3}{\text{Hypro}}$. Hydroxyproline is in fact produced enzymatically after the chain sequence has been put together at the ribosome. The substrate configuration apparently limits hydroxylation to the third position. Other aspects include Phe appearing in only the second position (probably for steric reasons) and Hylys appearing only in the third position, again resulting from enzymatic action. In addition, depending upon the origin of the collagen, one or more saccharide or disaccharide units are linked to hydroxylysine residues. In particular, a disaccharide unit is normally attached to Hylys[103] (abbreviated Hyl).

There are, however, many interesting questions that may be asked concerning the primary sequence of collagen, such as why is it necessary to have glycine occupy every third residue position?; what role do the various residues play in organizing chains and molecules?; what specificity is there in the collagen sequence and why is specificity necessary if collagen plays a rather nonspecific, load-bearing role?

The answers to these questions emerge as we study the physical aspects of collagen structure.

Physical Characterization

Physical characterization methods may be conveniently grouped into solid-state methods (x-ray diffraction, electron microscopy, infrared spectroscopy) and solution methods (hydrodynamics, circular dichroism spectroscopy). These two groups are not entirely separate; for example, infrared spectroscopy can be used on solutions under certain circumstances and CD spectroscopy may be used on films. Not one of the methods is comprehensive in its scope, but together they have proven to represent very powerful methodology that generally provides much of the required information.

Although the sequence in which physical methods are applied is somewhat arbitrary, it is usual for some solution measurements to become available first, because they are generally part of the biochemical preparation scheme. (In some cases, intact native samples may be used in structural characterization—e.g., hair, tendon, etc.—but ultimately the role of each component is sought.) For collagen, light-scattering measurements show the tropocollagen molecule to be about 3,000 Å in length.

An electron micrograph of native fibrils (Figure 5.8) has already been presented and shows a fibrillar width of about 1,000 Å (the width varies depending on age and source) and a repeat pattern along the length of the fiber of 640 Å. The main questions to be addressed are, then, how do 3,000-Å long molecules fit together to form fibrils, what is their conformation, and what are the forces controlling self-assembly?

X-ray diffraction

X-ray fiber diffraction patterns of collagen have been studied for many years. Unstretched dried rat tail tendon gives a fairly diffuse diffraction pattern (Figure 13.3a); stretching the tendon 8–10% gives a much more detailed pattern (Figure 13.3b). The diffraction lines are indexed (in Å) in Figure 13.4 and the layer lines indicated.

According to the procedure spelled out in Chapter 4, if we look first for meridional reflections we find that the first lies on the 10th layer line and indeed the predicted diffraction from a 10_3 helix has several features in common with Figure 13.3. Using equation (4.27) (i.e., $\ell = Nm + tn$) we expect strong meridional reflections (Bessel function order n = 0) on the $\ell = 10$, 20 . . . lines and strong off-meridionals (n = 1) on the 3rd, 7th, 13th, 17th, etc.

The peptide repeat calculated from the first meridional is ~2.9 Å for unstretched collagen. Comparison between calculation and experiment shows that the predicted off-meridionals on layer lines 13 and 17 are smeared onto the meridional in the experimental case (Figure 13.4). Even if the agreement had been identical we would have been faced with an apparently new conformation, a 10_3 helix with d = 2.9 Å. In fact, the sequence —GlyProX— gives us the clue that a polyglycine II (polyproline II) helix should be the basic chain conformation with a 3_1 left-hand helix of d = 3.1 Å. Since it is known that three chains form the tropocollagen molecule, it is reasonable to expect the chains to be coiled together, thus presenting us with a somewhat more complex situation to analyze.

Figure 13.3 X-ray fiber patterns from **(a)** unstretched, **(b)** 8% stretched rat tail tendon (after Cowan et al.).

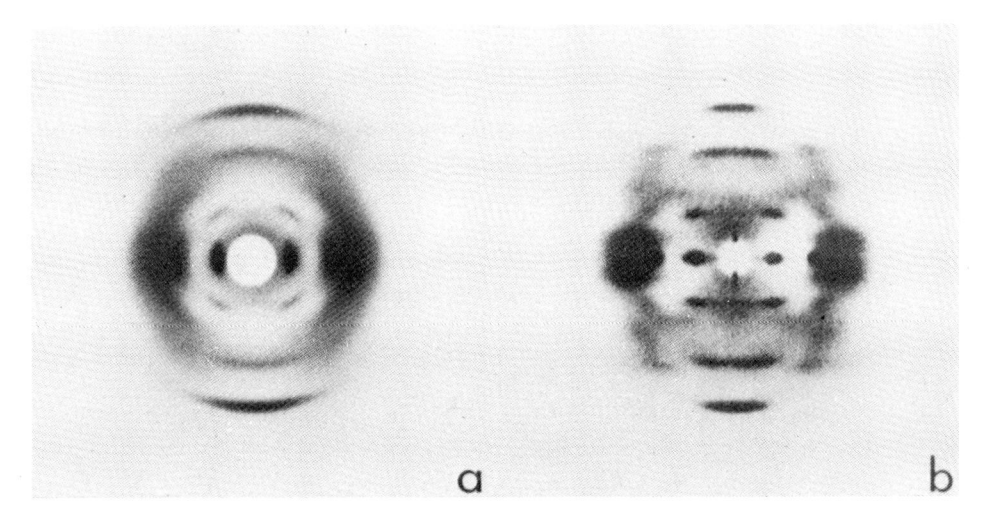

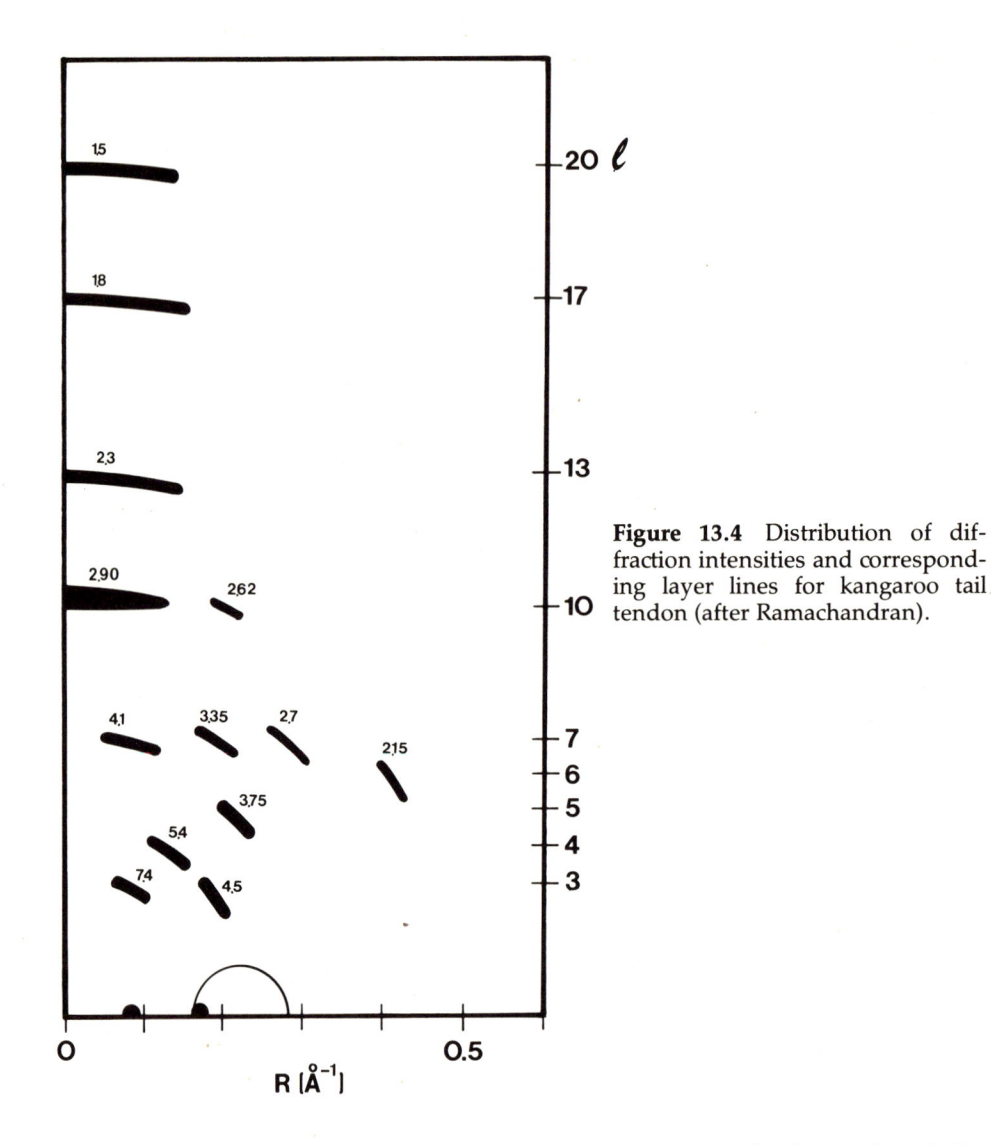

Figure 13.4 Distribution of diffraction intensities and corresponding layer lines for kangaroo tail tendon (after Ramachandran).

Under these circumstances, it is appropriate to use selection rules of the type outlined in equation (4.30). By comparison with the α-helix coiled coil where there is an 18_1 screw axis of the major helix and there are 7 residues in 2 turns of the minor helix, and $18 \times 7 = 126$ residues/turn of the major helix: for collagen we have the x-ray fiber-diffraction pattern revealing a 10_3 helix. If there are three residues/turn of the minor helix, then we have 30 residues/major helix repeat. The repeat distance would be 30×2.9 Å = 87 Å; however, since there are three turns in this distance, we have a pseudorepeat of 87 Å/3 = 29 Å as shown in Figure 13.5.

Circular dichroism/optical rotary dispersion

If one uses the Moffitt–Yang equation to calculate b_0 (equation 9.6) and hence the percentage of the α-helix in tropocollagen, the answer is close to zero. The CD curves, on the other hand, confirm, as expected, that the

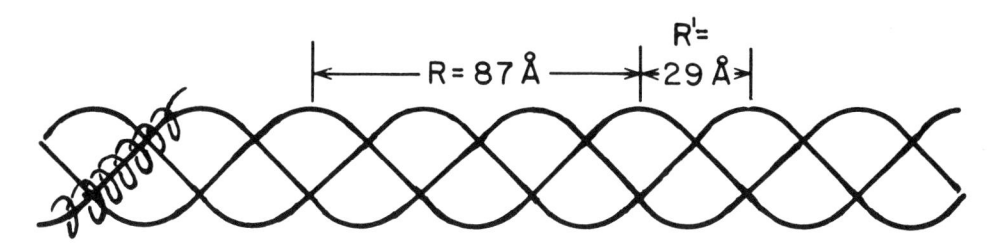

Figure 13.5 Schematic diagram of the arrangement of the three chains of tropocollagen showing the superhelix repeat R and pseudorepeat R'.

collagen spectrum resembles that of poly-L-proline or the 3_1 left-hand helix. Figure 13.6 shows CD spectra of collagen below 37° C and in the denatured (gelatin) form. These spectra may be compared with the spectrum of polyproline in Figure 9.11. Since tropocollagen is presumably in the triple helical coiled coil state and poly-L-proline is in a single helix, one might expect subtle differences in the two CD spectra and indeed the presence of the positive band $[\theta]_{221} = \sim 6{,}000$ deg. cm²/decimole in the collagen spectrum is taken as an indication of supercoiling. Since the ellipticity is proportional to the concentration of a certain structure, the proportion of triple helix may be followed by the appearance or disappearance of the $[\theta]_{221}$ band.

The similarity in band position and shape between the curves confirms that the individual chains of the triple helix are in the left-hand configuration.

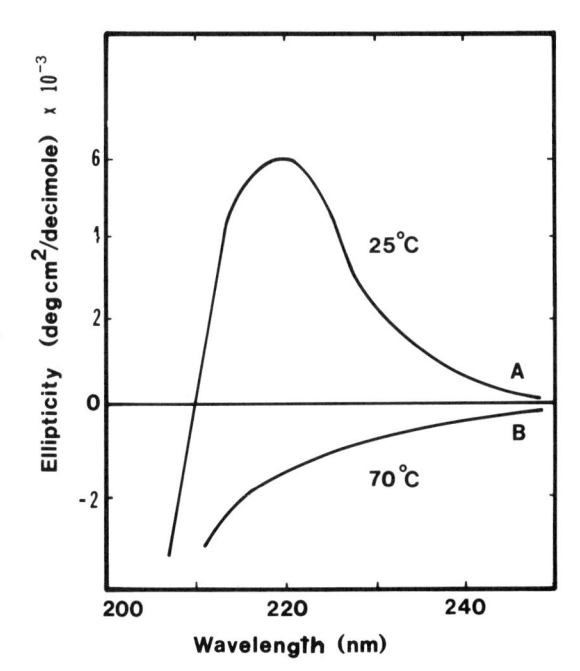

Figure 13.6 Comparison of collagen CD spectrum (A) with that of denatured collagen (gelatin) (B).

Polypeptide models

If we wish to inquire more deeply into the relation between sequence and secondary structure, it is instructive to examine the properties of synthetic polypeptide analogues. A table of solid-state structures for polypeptides is given in Table 13.2. Although many other polytri- and polyhexapeptides have been studied, Table 13.2 provides us with many of the salient details. We note first of all, that it is necessary to have glycine every third residue for a triple helix to form. A simple way of visualizing this effect is to look down the end of a group of three helices, each of which is a 3_1 helix with a glycine every third residue. Figure 13.7 shows such an array.

The small glycine residues lying on the "inside" of the triple helix allow a close steric interaction between chains and the development of hydrogen bonds between the peptide groups of adjacent chains.* An additional "twist" of the parallel helices would take them into the coiled coil configuration.

In general, the presence of proline in the second (and third) position strengthens the basic conformation and causes supercoiling.

Polypeptide models with ionic groups in them (e.g., lysine) show that in aqueous solution, the presence of charge is a disruptive influence and the triple helix is much more stable when the charge is neutralized.

Infrared spectroscopy

Infrared and/or Raman spectroscopy are often useful for rapid identification of chain conformation, particularly if the fibrous protein is α-helix or β-sheet. Since collagen is neither, infrared spectroscopy has been most useful in identifying more subtle aspects of structure. With oriented specimens and infrared dichroism it is, for example, possible to identify the orientation of structural water molecules. Again, based on model polypeptide studies with (Gly Pro Pro)$_n$ it has been suggested that there are water molecules bridging adjacent chains. One of the most interesting examples of the application of infrared spectroscopy to collagen is the examination of the amide A band. In air-dried tendon collagen, the amide A band occurs at about 3325 cm^{-1}, which

Table 13.2 Structure of Synthetic Polytripeptides

Polypeptide	Structure*
1 2 3	
(Ala Pro Pro)$_n$	Polyproline II, no triple helix
(Gly Pro Pro)$_n$	Triple helix
(Gly Pro Ala)$_n$	Triple helix
(Gly Ala Pro)$_n$	Triple helix or pPII
(Gly Ala Ala)$_n$	α-helix or β-sheet
(Gly Gly Ala)$_n$	β-sheet or polyglycine II (pPII)

*Based on high-molecular weight material.

---

*Although it would appear from Figure 13.7 that glycine would hydrogen-bond to glycine in the adjacent chain, x-ray evidence suggests that the —NH of glycine in one-chain hydrogen bonds to the $>$C=O in the residue next to glycine in the adjacent chain.

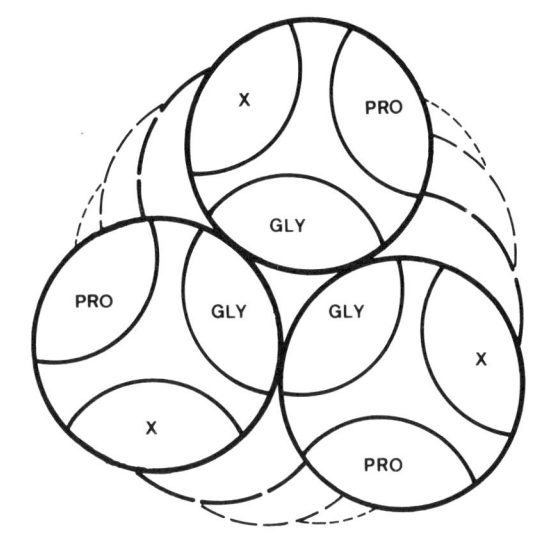

Figure 13.7 Schematic view of the three collagen helices "end-on," showing relative location of residues in the tripeptide sequence.

is considerably higher than for other known conformations. As shown in Chapter 6, hydrogen-bond lengths may be deduced from a combination of the various vibration assignments and in this case the (interchain) hydrogen-bond length is calculated to be 2.94 Å, which is somewhat shorter than the normal value. Since the hydrogen-bond length reflects the steric arrangement of supercoiled chains and we have already seen that the degree of supercoiling is affected by primary sequence, we would expect the amide A position to be different for triple-helical synthetic polypeptides with different primary structure (which it is) and also to change under conditions affecting supercoiling in collagen. Increases in humidity cause the amide A band to decrease in frequency, presumably as a result of changes in fundamental structure that involve hydration.

In review then, the primary molecular unit of collagen fibers consists of three chains, each in a slightly distorted left-hand polyproline II conformation, wound together into a rope or coiled coil. The supercoil probably has a right-hand twist (the twist sense of various levels of organization seems to alternate from one level to the next). The chains are believed to be more or less in phase and have a tripeptide repeat, glycine occurring every third residue, except in small "tail" regions. Forces holding the three chains together are predominantly hydrogen bonds and hydrophobic interactions.

Higher orders of structure—the protofibril, microfibril, and so on—are not easily accessible to any one particular technique, but structural details have been revealed mainly by a combination of x-ray diffraction and electron microscopic measurements.

13.2 Fibrillar Structure

Electron Microscopy

Electron microscopy may be performed either on native fibrils (usually 1–4,000 Å in diameter) or on reconstituted material, that is, fibrils produced by precipitating tropocollagen molecules under specific conditions. The latter method has proven more useful and is the one that will be examined here. As outlined in detail in Chapter 5 on electron microscopy, various morphologies may be examined by uranyl acetate (UA) or phosphotungstic acid (PTA) staining.

Normal "native-type" collagen fibrils show a striated staining pattern in which a *directional* repeat of about 640 Å is evident (Figures 5.8, 5.9). A number of minor bands are apparent and from the directional property of the staining pattern we may conclude that the molecules are parallel to the fibrillar axis (confirmed by x-ray diffraction and infrared spectroscopy) and that the molecules are in *parallel* array. The first and most basic question is, where are the molecules located in the fibrils? Since the molecules are ~ 3,000 Å long they must clearly be overlapped in a manner to produce a 640-Å repeat (small-angle x-ray diffraction of hydrated molecules shows the repeat to be 668 Å; some shrinkage occurs in the vacuum of an electron microscope). One definitive piece of evidence comes from "negative" staining, which effectively fills in any voids in the structure. Figure 5.9 shows that in collagen fibrils voids occur, suggesting that molecules are not arranged end to end, but have a space corresponding to approximately 400 Å at the ends. This observation led to the suggestion that molecules are staggered such that the repeat of 668 Å is approximately one-fifth of 3000 + 400 Å, that is, the molecules are arranged as depicted schematically in Figure 5.9. One notices that the representation is two-dimensional and much attention has been directed recently to a consideration of the three-dimensional array. The current view, based on x-ray diffraction data, on model building, and to some extent on electron microscopy, is that the molecules grow around a central (void) axis such that the sixth molecule repeats the first with an axial translation of 3,400 Å, as shown in Figure 13.8. However, the pentamer is then imagined to be twisted such that a

Figure 13.8 Schematic arrangement of collagen molecules in which pentamers are twisted into a fourfold "protofibril."

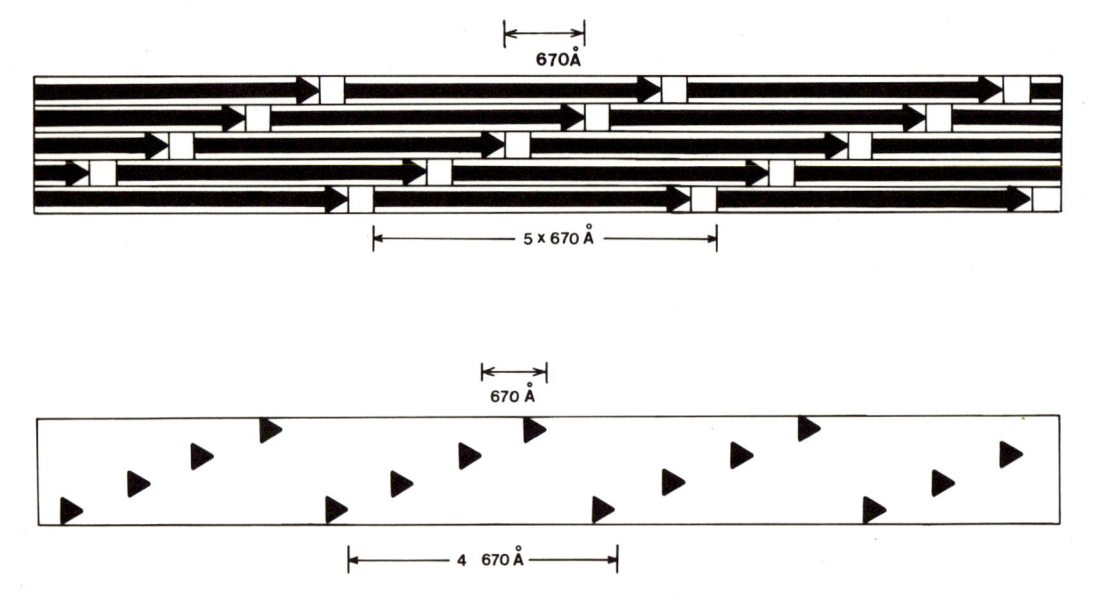

fourfold axis is produced. This twisting into a fourfold arrangement is also shown in Figure 13.8. Presumably if our tentative rule of alternating twist is observed, the smallest fibril or "protofibril" would have a left-hand twist. The forces holding the molecules together are now fairly clear. These forces are next explored in some detail.

Protofibrils and Forces Dictating Structure

The major interaction force causing aggregation of tropocollagen molecules into protofibrils is ionic; the negative aspartate or glutamate groups are (predominantly) paired with arginyl or lysyl groups in adjacent molecules. It is noteworthy that since there are three chains in each molecule but only two direct neighbor molecules, some charges are not balanced by this procedure. In addition, hydrophobic interactions help maintain molecular phasing. Figure 13.9 shows that interaction maxima occur as the molecules are staggered by a phasing of D (= 670 Å) or multiples thereof. (Ionic interactions occurring between molecules are assumed to interact over a range of ±2 residues; hydrophobic interactions are only for adjacent residues.) By such means, values of 233 ± 2 residues have been calculated for the residue stagger between molecules. Thus the long period of 233 × 2.87 Å = 668 Å is accounted for. Other computations show that intermolecular ionic repulsions also tend to localize molecular arrangements.

It is also interesting that saccharide moieties abut the triple helical region of adjacent molecules and also seem to play a role in molecular phasing (Figure 13.10).

Figure 13.9 Summed electrostatic and hydrophobic interactions as two tropocollagen molecules are moved with repeat to each other. Interaction maxima can be seen at multiples of D (= 670 Å).

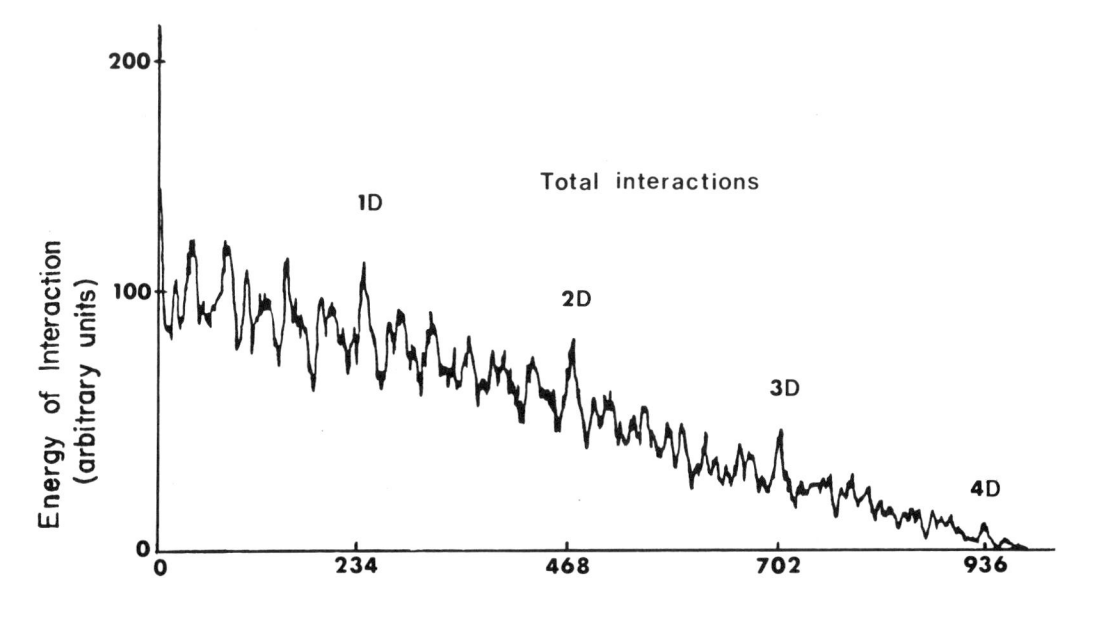

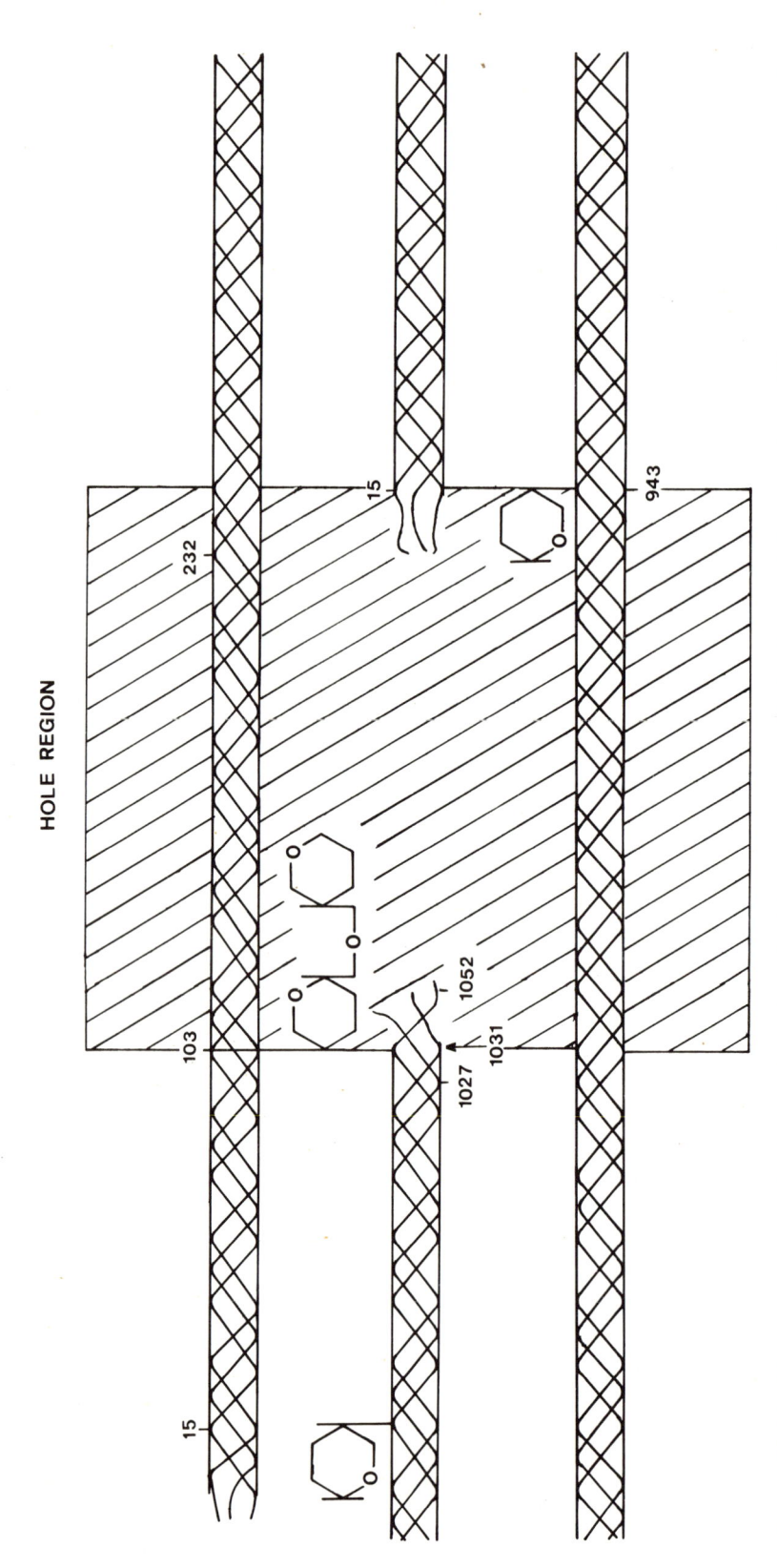

Figure 13.10 The arrangement of telopeptide and saccharide units in assembled collagen fibrils assuming a 232-residue displacement. (Depending on the source of collagen, the sugar unit at residue 943 may be missing.)

Microfibrils/Fibrils

The arrangement of protofibrils into higher orders of structure, such as in tendon, was originally thought to be in hexagonal or spiral array; however, evidence now seems to be growing that there is a tetragonal array. This deduction would be in accord with the fourfold axis of the protofibril since symmetry would be maintained. Although it is not at all clear what forces hold adjacent protofibrils in register, it must presumably be long-range ionic interactions. The fact that protofibrils are in register is evident from the fact that electron micrographs show fibrils to have a uniform staining pattern that could only arise from the uniform packing of protofibrils.

As mentioned in Chapter 5, the observation that PTA- and UA-stained fibrils give essentially identical patterns indicates that basic and acidic groups are clustered in the same regions and must, therefore, be involved in the assembly of fibrils in accord with calculations.

Fibers

The assembly of fibrils into fibers involves the introduction of a new entity, a matrix material consisting mainly of proteoglycan. Figure 13.11 shows a cross-section of fibrils embedded in a proteoglycan matrix from rat tail tendon. There is no obvious correlation between the molecular stagger in various fibrils (unless they are in very close proximity); however, the fibrils are generally of fairly uniform thickness (this changes with age) and distribution.

It used to be thought that the matrix was amorphous and acted essentially as a transport medium for ions, water, and so on. However, it is now clear that the matrix interacts with collagen fibrils in a fairly specific manner. Changes in the composition of proteoglycans with age may well be related to the changing properties of tissue, including mechanical properties and moisture content. The proteoglycan consists of protein chains linked at one end to a polysaccharide,

Figure 13.11 Electron micrograph of collagen fibrils in a cross-section of a young rat tail (mag. 99,000×). The matrix surrounding the fibrils contains proteoglycan molecules.

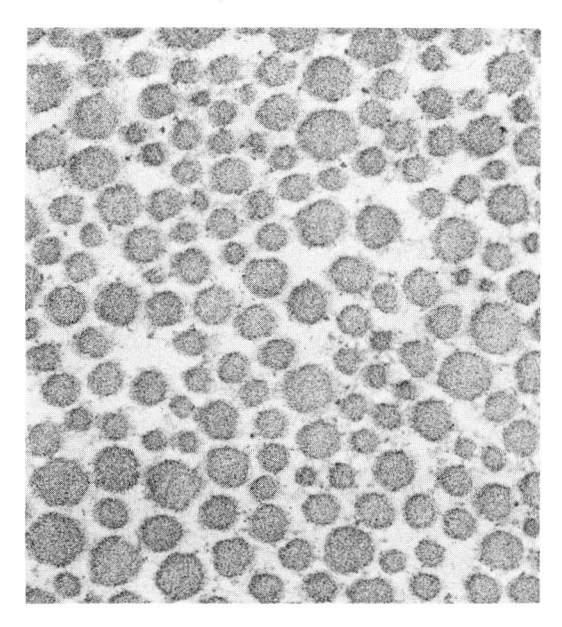

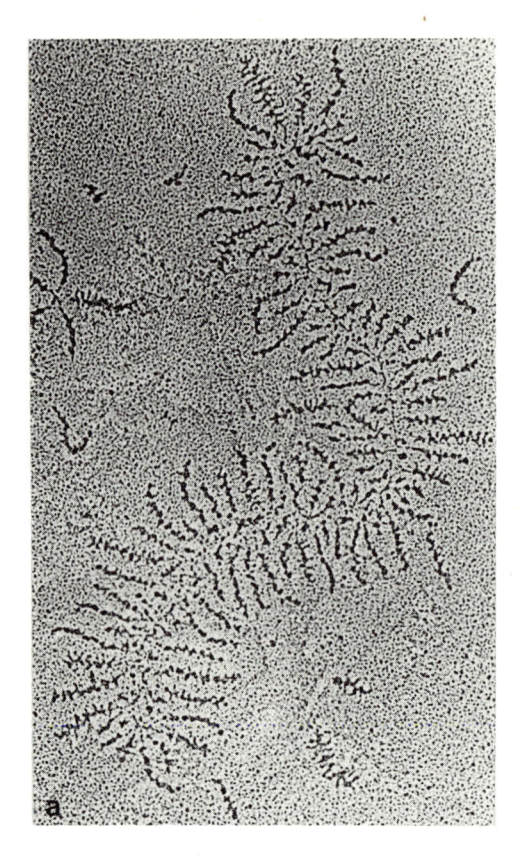

Figure 13.12a Electron micrograph of bovine nasal cartilage proteoglycan (mag. 350,000×).

hyaluronic acid. Other polysaccharides, chondroitin sulfate and keratan sulfate, are linked to the protein covalently (through serine groups) in a "bottle brush"–type structure. Figures 13.12a and b show experimental and schematic diagrams of the proteoglycan structure.

It is probably the chondroitin and keratan sulfate chains that bind to the collagen fibrils, the interaction being between the sulfate group of the polysaccharide and the lysyl or arginyl groups on the collagen fibrils.

In local regions then, collagen fibrils are generally assembled into fibrils embedded in a highly hydrated proteoglycan matrix. It must not, however, be assumed that connective tissue contains collagen fibrils organized only in parallel array. At yet higher levels of organization, the fibrils have higher orders of twist and crimp. The factors organizing collagen in gross tissue are highly complex. The collagen molecules are precipitated in a matrix after originating from fibroblasts; the molecular fibrillar and higher order organization must originate from a diffusion/precipitation sequence in which cell motion stress, rate of collagen production, and enzyme processes all play a part.

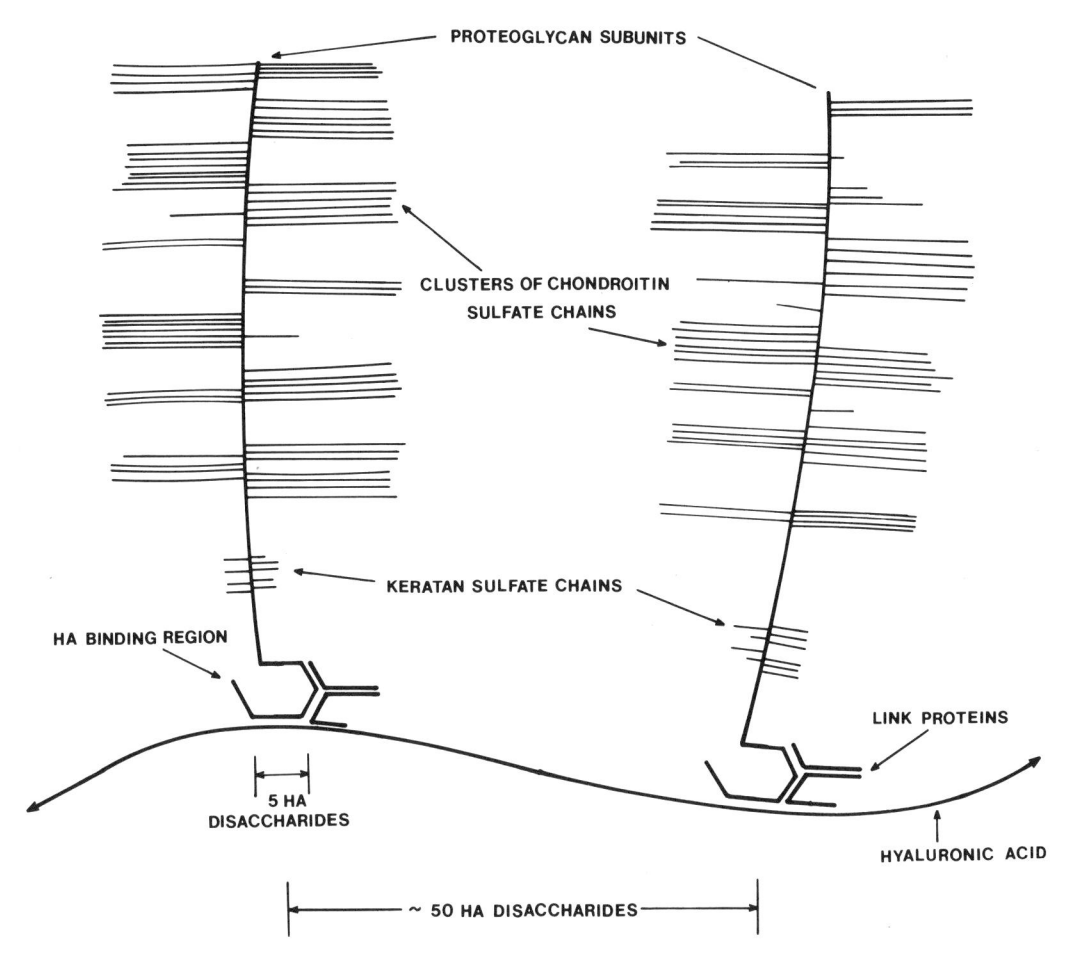

Figure 13.12b Schematic representation of proteoglycon aggregate.

Further Reading

Bornstein, P. and W. Traub. The Chemistry and Biology of Collagen. In *The Proteins*, H. Neurath and R.L. Hill, eds., Academic Press, New York, 3rd ed., Volume 4, Chapter 3, 1979.

Fraser, R.D.B. and T.P. MacRae. *Conformation in Fibrous Proteins*. Academic Press, New York, 1973, Chapter 14.

Ramachandran, G.N., ed. *Treatise on Collagen*, Vols. 1 and 2. Academic Press, New York, 1967, 1968.

Ramachandran, G.N. and A.H. Reddi. *Biochemistry of Collagen*. Plenum Press, New York, 1976.

Traub, W. and K.A. Piez. The Chemistry and Structure of Collagen. *Advances in Protein Chemistry* 25:243–352, 1971.

14

Molecular Structure and Function of Muscle

14.1 Muscle Structure

Muscle is the most complicated fibrous protein system we shall examine, because it combines the mechanical characteristics of the fibrous protein with chemical activity and complex function generally associated with globular proteins. Muscle is, in fact, composed of a variety of different proteins, each of which forms a part of the structural and/or contraction function. As with collagenous connective tissue, muscle shows various levels of organization, shown schematically in Figure 14.1. Gross muscle is connected, as a functioning unit, by collagenous tendon at either end. The muscle fibers, each lying within a muscle cell, measure 50–100 μ and the myofibril is 1–2 μ in diameter. End on, the myofibril can be seen to contain two distinctly different types of fibril, known as the thick (150 Å) and thin (50–70 Å) filament. The interdigitation of these filaments, as well as their control by chemical stimuli, represents the basis of the molecular mechanism of muscle function.

In thin sections of muscle (the section direction is parallel to the myofibril axis), electron microscopy reveals a highly banded structure (Figure 14.2). These bands have historically been labeled alphabetically, as is shown in Figure 14.3.

The repeat unit in the myofibril is known as a sarcomere, which in resting (rabbit) muscle is approximately 2.3 μ in length. The relationship between the thick and thin filaments is suggested by electron microscope observations of thin sections, as shown in Figure 14.4. Thick filaments extend through the length of the "A" band and are overlapped partially with the thin filaments in a hexagonal array. There are three main protein components in muscle: actin and tropomyosin in the thin filaments, and myosin in the thick.

14. Molecular Structure and Function of Muscle

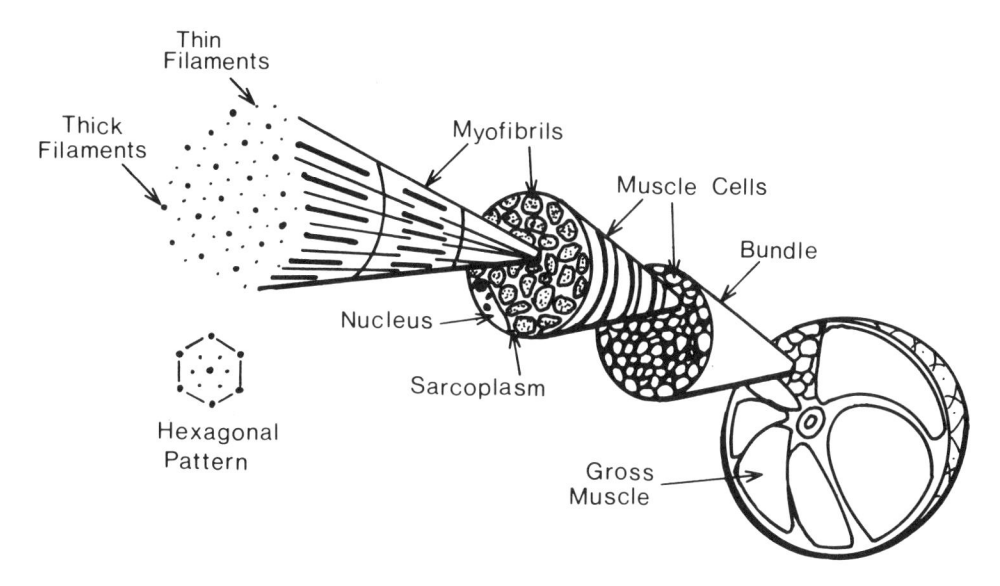

Figure 14.1 Hierarchy of structure in muscle fibers. Primary units are the thick and thin filaments.

14.2 The Thick Filaments

Myosin is the principal protein of the thick filaments. Individual myosin molecules appear as long rods with a double globular head (Figure 14.5). The molecule can be fragmented by enzymatic digestion into three or more fragments. Trypsin treatment produces two main fragments of molecular weight, ~150,000 and ~340,000, termed light meromyosin and heavy meromyosin, respectively. Further proteolysis causes fragmentation of heavy meromyosin (HMM) into subfragments S1 and S2. Further, there are four additional light chains ($M_w \sim 20,000$) found to be in the head region, which are not covalently bound.

Myosin Structure

If we employ our standard approach to elucidating molecular structure, it would follow the procedure:

1. Preparation of pure material;
2. Amino acid analysis;

Figure 14.2 Electron micrograph of a section of muscle showing characteristic banding pattern.

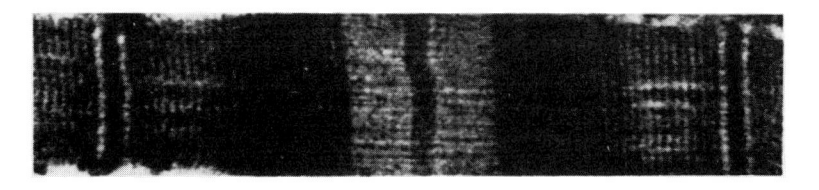

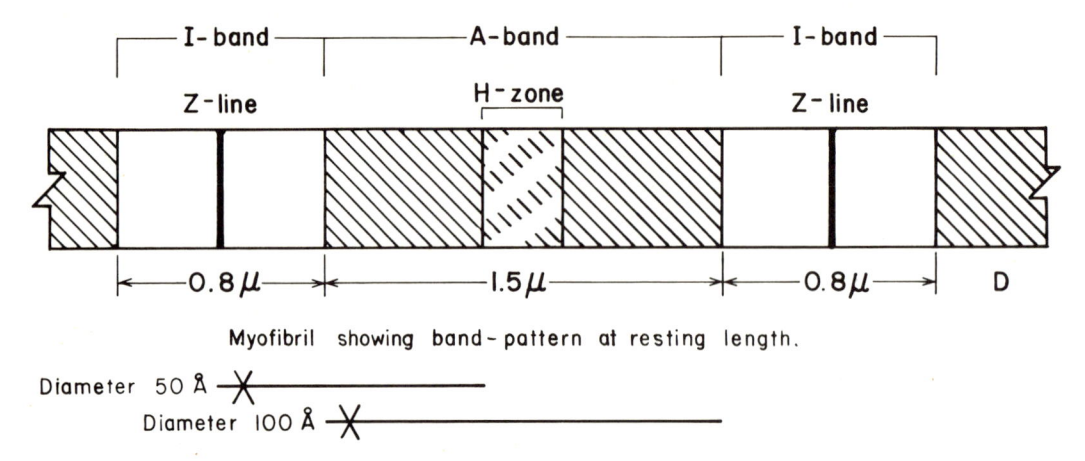

Figure 14.3 Schematic diagram of muscle structure with band dimensions.

3. Sequencing;
4. Conformational determination and physical characteristics;
5. Electron microscopy on (a) molecular material; (b) reconstituted material;
6. Relation of tertiary and quaternary structure to native morphology;
7. Identification of relation between structural arrangement and function.

In the case of myosin, we have already "cheated" somewhat by using electron microscopy ahead of schedule, but this has proved useful in identifying various aspects of the preparation. For example, papain digestion of myosin essentially removes the globular head (subfragment S1) so that we may examine the subfragments separately. We now proceed with the "detective" work.

Figure 14.4 Arrangement of thick and thin filaments within a myofibril (after Cohen).

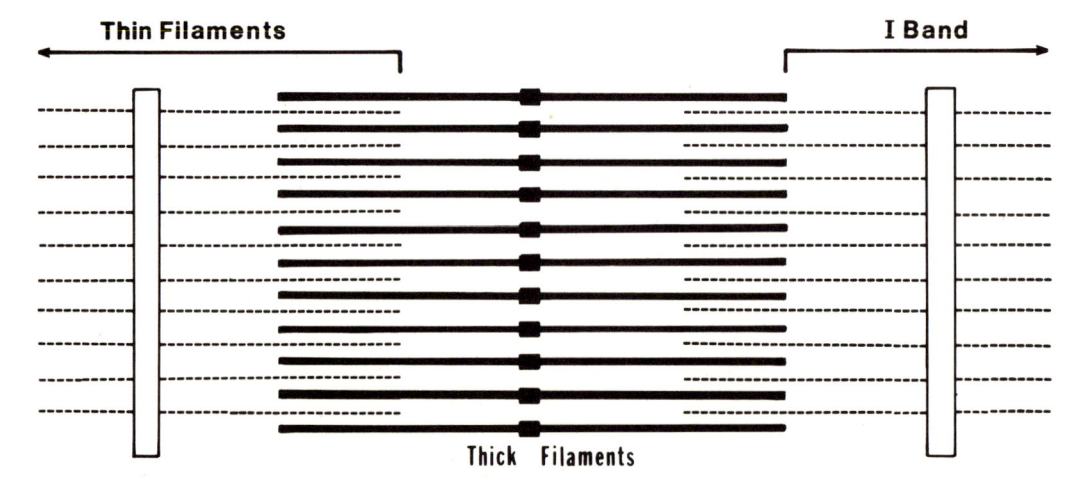

14. Molecular Structure and Function of Muscle

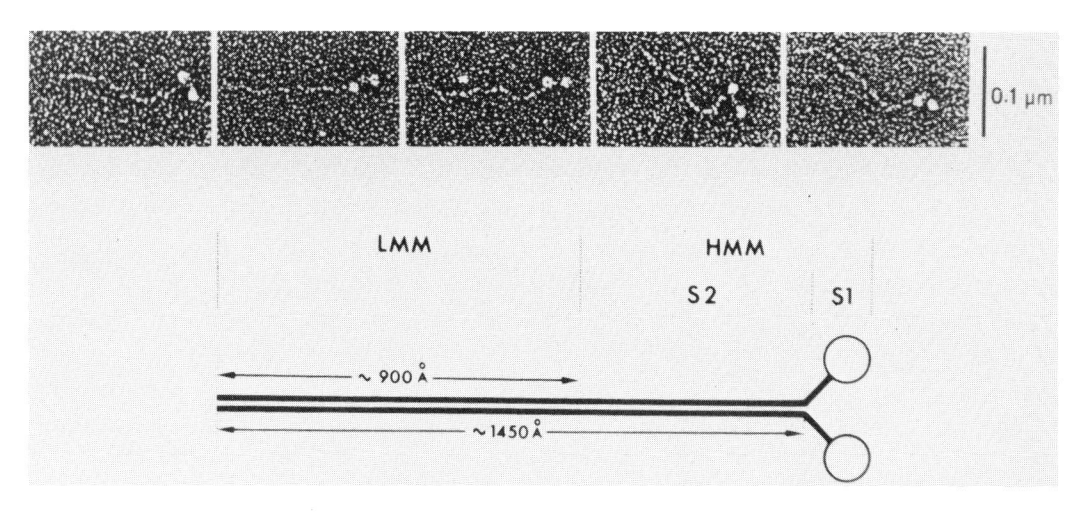

Figure 14.5 Electron micrographs and schematic representation of the myosin molecule showing fragment-cleavage positions. The molecule can be seen to consist of a thin tail with two globular heads (after Lowey et al.).

Amino acid analysis

As we have seen from the silk structures, and from collagen, there are often clues in the amino acid analysis, particularly for fibrous proteins, which give us some idea of structure (Table 14.1). For example, proline is rarely found in α-helix or β-sheet regions of a protein (see Chapters 2 and 3); cystine is far more abundant in globular than fibrous proteins; glycine and serine are rarely found to any extent in α-helical structures; and large amounts of leucine are rarely found in β-sheet structures.

Table 14.1 Amino Acid Analysis of Myosin and Fragments

Amino acid	Myosin	HMM Subfragment 1	Papain rod
Alanine	9.0	7.8	9.7
Arginine	5.0	3.8	6.4
Aspartic acid	9.9	9.4	10.1
Glutamic acid	18.2	13.0	25.0
Glycine	4.6	6.8	2.3
Half-cystine	1.0	1.2	0.5
Histidine	1.9	2.0	1.7
Isoleucine	4.9	5.9	4.1
Leucine	9.4	8.3	11.3
Lysine	10.7	9.2	12.2
Methionine	2.7	3.1	2.6
Phenylalanine	3.4	5.8	0.8
Proline	2.5	4.1	0
Serine	4.5	4.5	4.5
Threonine	5.1	5.4	4.3
Tyrosine	2.3	3.8	0.7
Valine	5.0	6.1	3.8

In the present case the amino acid analysis of the "papain rod" section of the meromyosin molecule is remarkable because of its very high content of polar amino acids (~54%), with acid residues outnumbering basic residues in the ratio of approximately 2:1. This feature, as it turns out, is characteristic of all known α-helix coiled coils (α-keratin; tropomyosin). Furthermore, the absence of proline and low half-cystine content suggests the concept of a "fibrous"-type protein. On the other hand, there is no particularly outstanding feature of the amino acid content of HMM (S1), suggesting (as we already suspected) that it is representative of a globular protein having chemical function.

The detailed sequence of myosin is not known at this time; however, the similarity of the amino acid analysis of the "papain rod" portion to tropomyosin, which we shall examine shortly, suggests there may be sequence similarities. The high incidence of α-helix directing residues suggests a high α-helical content for this segment.

Conformation

ORD/CD measurements indicate that the rod portion of myosin is 90%, or more, α-helical in aqueous solution at neutral pH. Furthermore, denaturation of the rod produces two molecules of essentially equal molecular weight in random conformation. This evidence is very similar to experimental findings for tropomyosin and seems to be indicative of a two-strand α-helix coiled coil. HMM (S1 and S2) is approximately 40% α-helical, of which most probably lies in the S2 region, leaving virtually no α-helix in the globular head region.

Infrared dichroic measurements are also consistent with the assignment of near 100% α-helix for the rod fragment (LMM) of myosin.

Light meromyosin (LMM) may be cast in films suitable for x-ray fiber diffraction and under these conditions, diffraction typical of an α-helix coiled coil (Chapter 3) with a meridional spacing of 1.5 Å is observed. Low-angle x-ray diffraction patterns show that the rods are arranged in periodic array with a repeat of 430 Å.

Electron microscopy

The main objective of electron microscopy is to provide information pertaining to the assembly of molecules. In the present case it is possible to examine the reconstituted rod fragments of myosin, or assemblies of intact molecules. Because of the rod shape of the LMM fragment, it provides particularly uniform packing. It is possible to obtain at least two different morphologies of reconstituted LMM, both of which appear to be intimately related to the way in which myosin forms heavy filaments in muscle.

Figure 14.6 shows an electron micrograph of a bipolar segment formed by the rod portion of myosin when precipitated with calcium ions. By the principles espoused in Chapter 4, the molecules must be antiparallel; however, it is not known whether the head-to-head (tail-to-tail) arrangement brings N- or C-termini together.

Tactoids may also be formed from LMM with a continuing 430-Å repeat

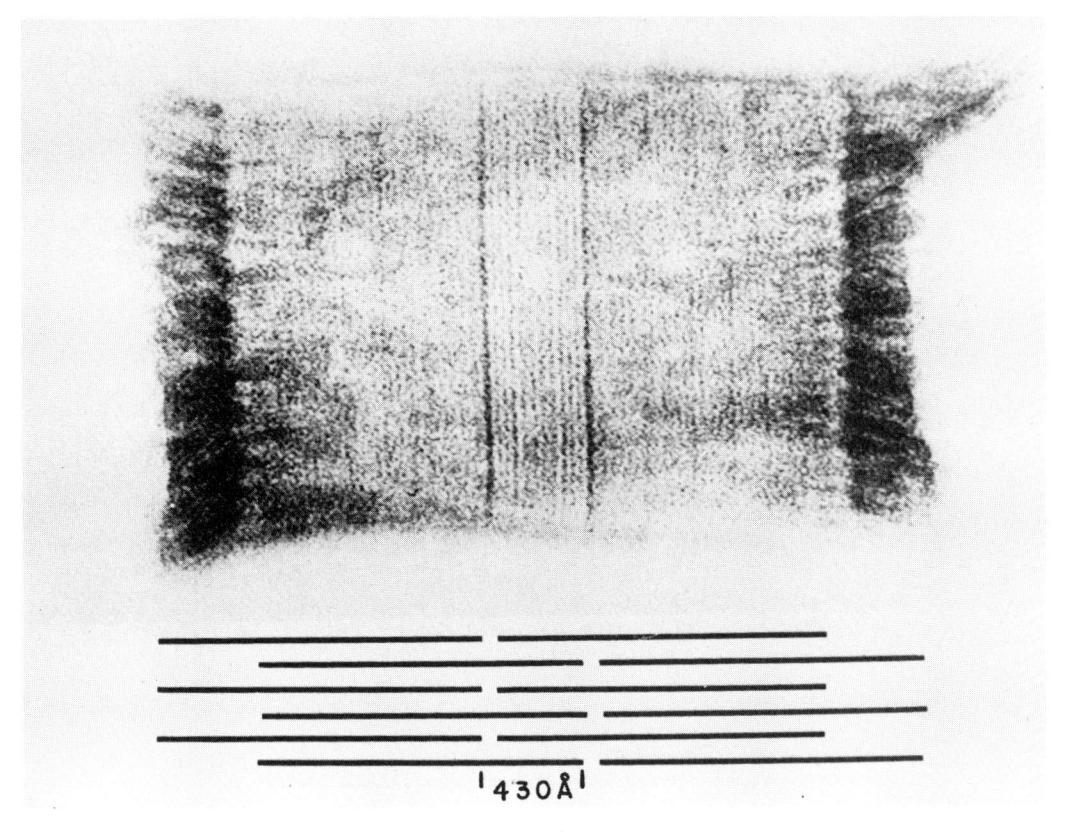

Figure 14.6 Electron micrograph of the reconstituted rod portion of myosin stained with UA, showing a 430-Å block. Beneath the micrograph is a schematic model of the molecular arrangement.

and negative staining suggests the arrangement shown in Figure 14.7. It is tempting to speculate that the arrowhead end of the rod is where the globular S2, S1 fragments would have been. Such an argument is consistent with the proposed arrangement of intact myosin in the heavy filament and also in reconstituted myosin.

Figure 14.7 Reconstituted myosin-rod segments form tactoids; models of the probable structure are shown.

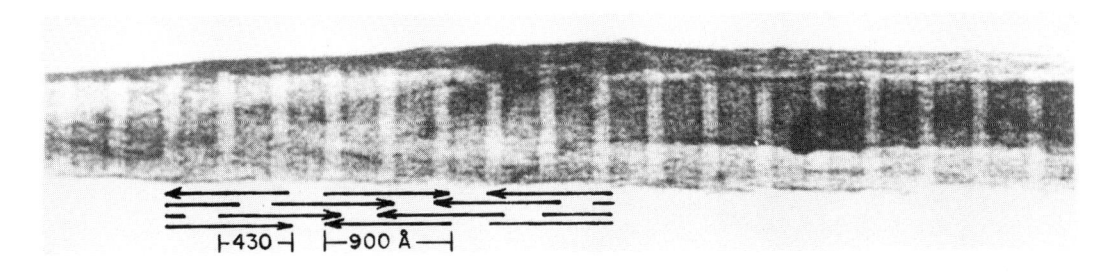

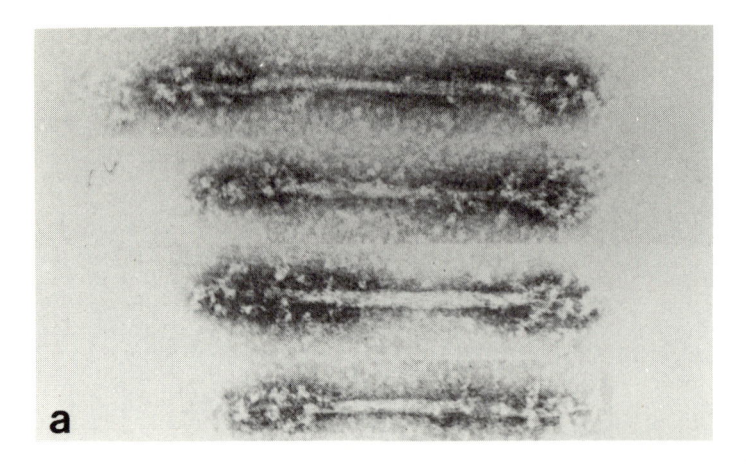

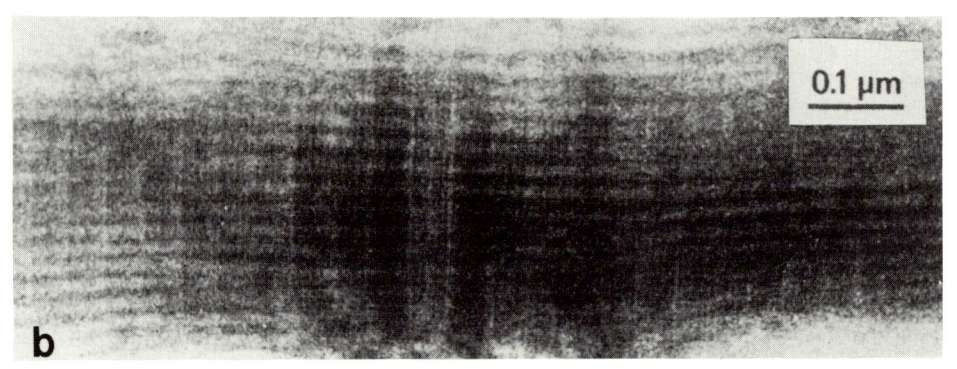

Figure 14.8 Electron micrographs of assemblies of myosin molecules.

Relation of structure to morphology

Figures 14.8a and b show bundles of reconstituted myosin molecules in which a tail-to-tail arrangement in the central portion of the aggregate is occurring.

In Figure 14.8b the large assembly of myosin molecules reproduces the M band (about 495 Å wide) and the B zones (about 395 Å), with a remaining periodicity of 420 Å in the rest of the aggregate.

X-ray diffraction of living, resting muscle gives a 429-Å low-angle period, which is believed to originate from the stagger of the myosin chains.

The detailed structure of the heavy filaments is not known at this time, but it is generally accepted that the structure must arise from an arrangement of molecules such as that shown in Figures 14.9a, b, and c. In order for the reconstituted myosin aggregate to correlate with the micrograph of Figure 14.8b, the tails of the two central molecules of Figure 14.9a must overlap. In the native structure, that interaction may be mediated by "M"-protein. However, the diameter of the core is 150–200 Å and it appears to contain finer filaments, 20–30 Å in diameter, which probably correspond to the coiled coil of single molecules. Evidently, a large number of the primary filaments would fit into the larger thick-filament core and there has

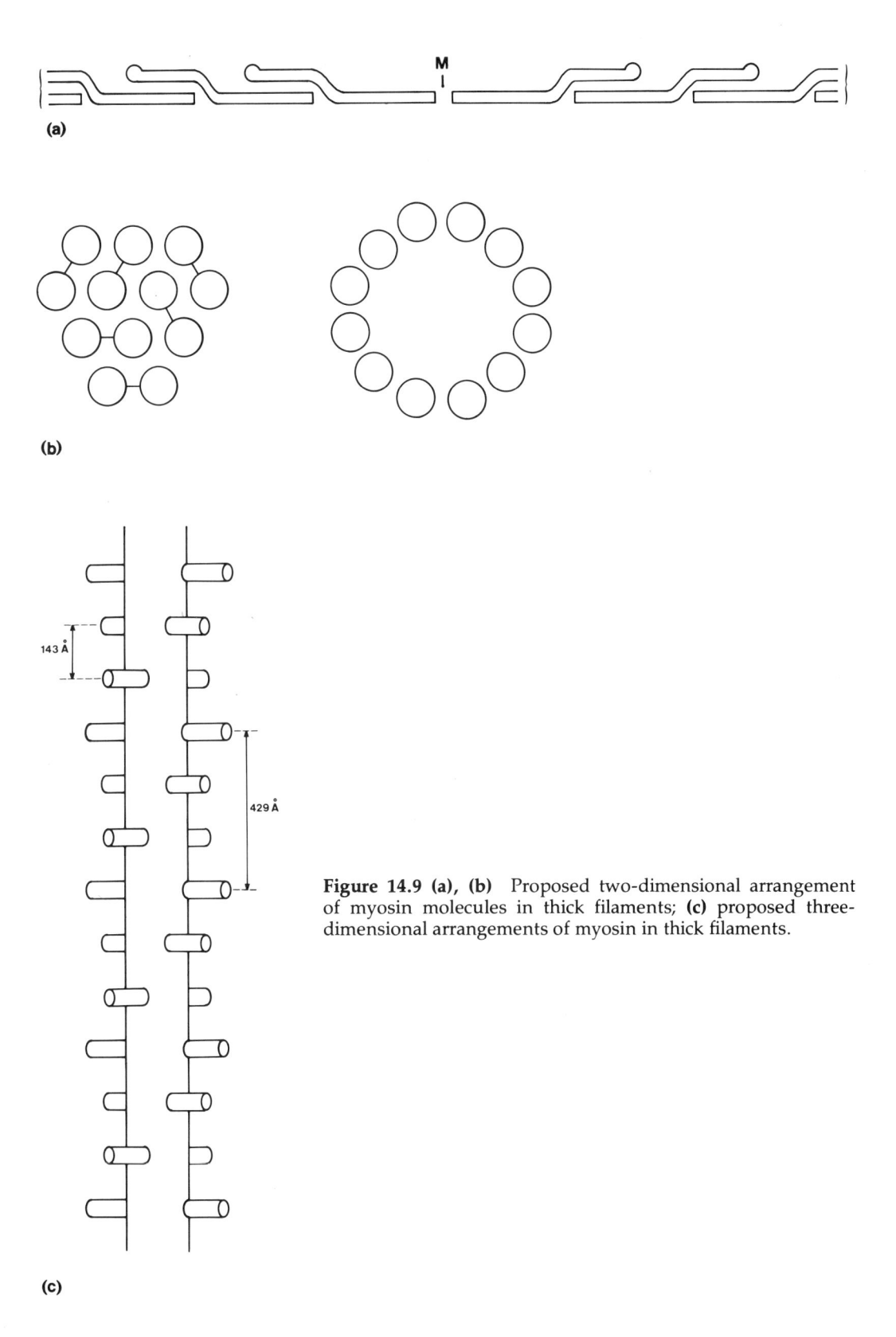

Figure 14.9 **(a), (b)** Proposed two-dimensional arrangement of myosin molecules in thick filaments; **(c)** proposed three-dimensional arrangements of myosin in thick filaments.

been some suggestion that it may be hollow or contain other protein. In any case, the arrangement of molecules in Figure 14.9 is consistent with the LMM tactoid stagger of 430 Å in Figure 14.7.

Relation between structure and function

The heavy filaments are composed primarily of myosin, which, as we have seen, contains a rodlike α-helix coiled coil tail and a globular head. Since the fibrous tail is likely to be the inert "structural" part of the molecule, we may expect the response to chemical stimulus to lie in the head region. It is thought that the head region binds to actin of the thin filament. The mode of binding, response to adenosine triphosphate (ATP) and calcium flux will be discussed in the following sections.

The arrangement of globular heads shown in Figure 14.9b is such that an end-on view displays a hexagonal array and brings the myosin heads into line with the hexagonally displaced thin filaments shown in Figure 14.1.

14.3 Thin Filaments

Two sets of thin filaments (one at each end of the thick filament) 70–80 Å in diameter are located within each sarcomere (see Figure 14.4). The thin filaments are composed of two major protein components, actin and tropomyosin, and a minor (but important) component, troponin. Arrays of actin filaments in one sarcomere are linked to those in the adjacent sarcomere by the Z-line, containing the protein α-actinin in such a manner that the filaments on one side of the Z-line are antiparallel to those on the other side.

Actin

Actin is a most unusual protein in that it has a primarily globular fundamental structure (G-actin), which acts as a monomer. In the presence of ATP in acidic solution, or upon addition of 0.1M KCl, the G-actin polymerizes into a beadlike filament (F-actin). Thus actin does not fit into the category of either globular or fibrous protein, but is a combination of both. The reason that nature turned to such a curious and unique structure appears to be that it needed a structural filamentous entity that carried out detailed and specific chemical functions, a process with which fibrous proteins are not normally involved.

Following our procedure of amino acid analysis, sequence, characterization, visualization, and assessment of relationship of structure to function we have:

Amino acid analysis and sequence

As might be expected for a globular protein, G-actin does not have any particular amino acid in excess. In fact there is considerable similarity with the analysis for the myosin head fragment (HMM, S1).

14. Molecular Structure and Function of Muscle

The complete amino acid sequence of actin from rabbit skeletal muscle was recently identified. The actin polypeptide chain is composed of 374 residues, including one residue of the unusual amino acid N-methyl histidine. The sequence is presented in Table 14.2.

The sequence contains several interesting features. There are some highly charged regions, notably near the N-terminus, where four of the first five residues are negatively charged, and the first 25 residues carry a net negative charge of 7 (at neutral pH). Also noteworthy are free sulfhydryls at residue positions 10, 217, 256, 284, and 373. N-methyl histidine is an unusual amino acid found only in actin and myosin; its biologic function is obscure.

Table 14.2 Amino Acid Sequence of Actin

1								10	191								
Ac- D	E	T	E	D	T	A	L	V	C	K	I	L	T	E	R	G	Y	S	F
11								20	201								
D	D	G	S	G	L	V	K	A	G	V	T	T	A	F	R	E	I	V	R
21								30	211								
F	A	G	D	D	A	P	R	A	V	D	I	K	Q	K	L	C	Y	V	A
31									221								
F	P	S	I	V	G	R	P	R	G	L	D	F	E	N	E	M	A	T	A
41									231								
Q	G	V	M	V	G	M	G	Q	G	A	S	S	S	L	E	K	S	Y	E
51									241								
D	S	Y	V	G	D	A	Q	S	K	L	P	D	G	E	V	I	T	I	G
61									251								
K	R	G	I	L	T	L	K	Y	P	N	E	R	F	R	C	P	E	T	L
71									261								
I	E	HMe	W	G	I	I	T	N	D	F	Q	P	S	F	I	G	M	E	S
81									271								
D	M	E	K	I	W	H	H	T	F	A	G	I	H	E	T	T	Y	N	S
91									281								
Y	N	E	L	R	V	A	P	E	E	I	M	K	C	D	I	D	I	R	K
101									291								
H	P	T	L	L	T	E	A	P	L	D	L	Y	A	N	N	V	M	S	G
111									301								
N	P	K	A	N	R	E	K	M	T	G	T	T	M	Y	P	G	T	A	D
121									311								
Q	I	M	F	E	T	F	N	V	P	R	M	Q	K	E	I	T	A	L	A
131									321								
A	M	Y	V	A	I	Q	A	V	L	P	S	T	M	K	I	K	I	I	A
141									331								
S	L	Y	A	S	G	R	T	T	G	P	P	E	R	K	Y	S	V	W	I
151									341								
I	V	L	D	S	G	D	G	Y	A	G	G	S	I	L	A	S	L	S	T
161									351								
H	N	V	P	I	Y	E	G	Y	A	F	Q	Q	M	W	I	T	K	Q	E
171									361								
L	P	H	A	I	M	R	L	D	L	Y	D	E	A	G	P	S	I	V	H
181									371								
A	G	R	D	L	T	D	Y	L	M	R	K	C	F						

Physical characterization

G-actin has a molecular weight of 41,785 and is an essentially spherical molecule. It contains 26–29% α-helix by CD/ORD and estimates of 26% for β-sheet and 48% for irregular structure have been calculated for the remainder of the structure. On polymerization of G-actin to form F-actin (fibrous actin), there is no discernible change in conformation.

X-ray diffraction of F-actin fibers produces a pattern that can be indexed by the helical diffraction formula (equation 4.27)

$$\ell = 13m - 6n$$

where ℓ is the layer line, m an integer, and n the order of Bessel function. Thus there are 13 units in 6 turns of a helix. Unlike other fibrous proteins, however, the basic repeat is not that of a peptide unit but is rather the *actin monomer*. The meridional spacing reveals the actin repeat to be 54 Å (actually 26.9 Å for a two-strand helix), which is taken as the diameter of the globular actin monomer.

A second arrangement of actin subunits forms with 28 subunits in 13 turns. The first arrangement is found at low calcium levels ($< 10^{-6}$M), the second at high calcium levels ($Ca^{2+} > 10^{-5}$M). It has been speculated that the arrangement may actually change during muscle contraction. How and why actin monomers form a helical arrangement is revealed by electron microscopy.

Electron microscopy

High-resolution electron microscopy of the F-actin filaments reveals beadlike ropes that are twisted into a two-strand helix. Figure 14.10 shows such a micrograph in which the small spheres have a diameter of approximately 55 Å (a spherical molecule of molecular weight 47,000 would be expected to have a diameter of approximately 50 Å). The micrograph has been interpreted as indicating 13 subunits in each repeat twist (for each helix), the pitch being 700 Å and the diameter of the strands being ~ 80 Å, that is, the same as the thin filament in muscle.

Figure 14.10 Electron micrograph of F-actin negatively stained with UA.

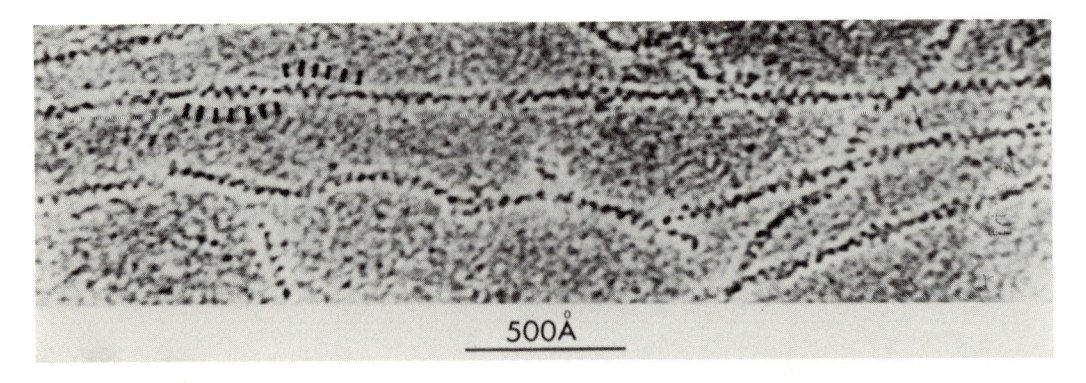

14. Molecular Structure and Function of Muscle

A schematic reconstruction of the F-actin filament showing the arrangement of globular subunits is shown in Figure 14.11a. At first sight, it would appear that since there are 13 units in one complete helical "turn" of the twisted rope, the terminology should be a 13_1 double helix. This terminology does not match the 13_6 description arrived at by x-ray diffraction. However, the arrangement is best regarded, for diffraction analysis, as a single helix in which the dotted spheres are the true repeat units (~ 350 Å) and each sphere is drawn on an imaginary helix (Figure 14.11b). In this manner, it can be seen that each primary unit lies on the helix and is rotated ~ 165°. Thus there are really 2.17 G-units per turn, as is shown in the radial projection (Figure 14.11b), and the 13_6 nomenclature is appropriate.

Electron-imaging reconstruction (see Chapter 5) has also been carried out on the F-actin helix, which gives us a clearer idea of the detailed structure. A reproduction of these data is presented in Figure 14.12a.

The nature of the site involved in actin aggregation is of some interest. Since each globular unit is at a fixed orientation with respect to each unit before and

Figure 14.11 Representation of the actin helix. **(a)** Actin molecules drawn on an imaginary 13_6 helix; **(b)** the two-dimensional representation of (a).

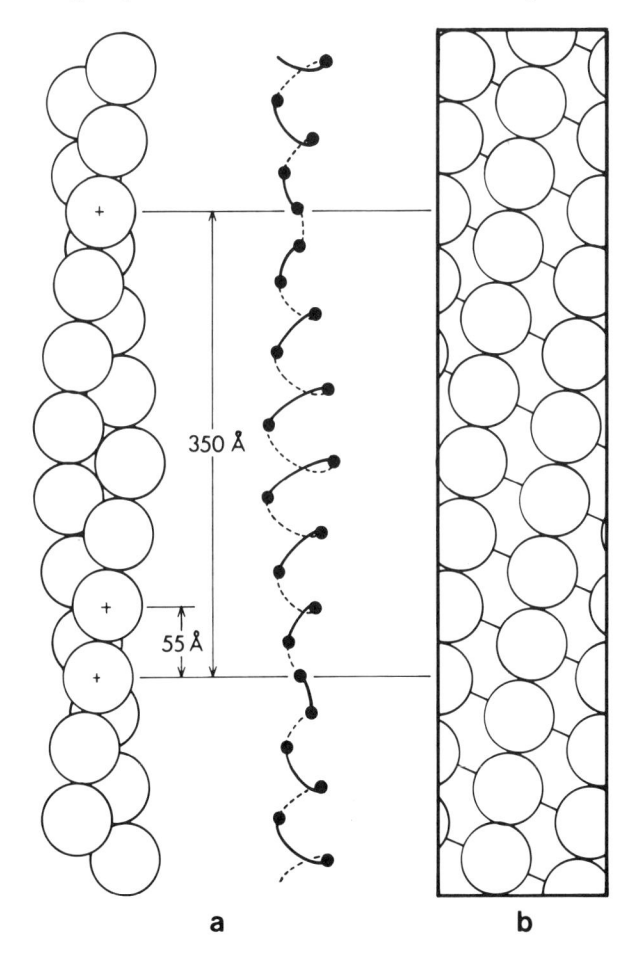

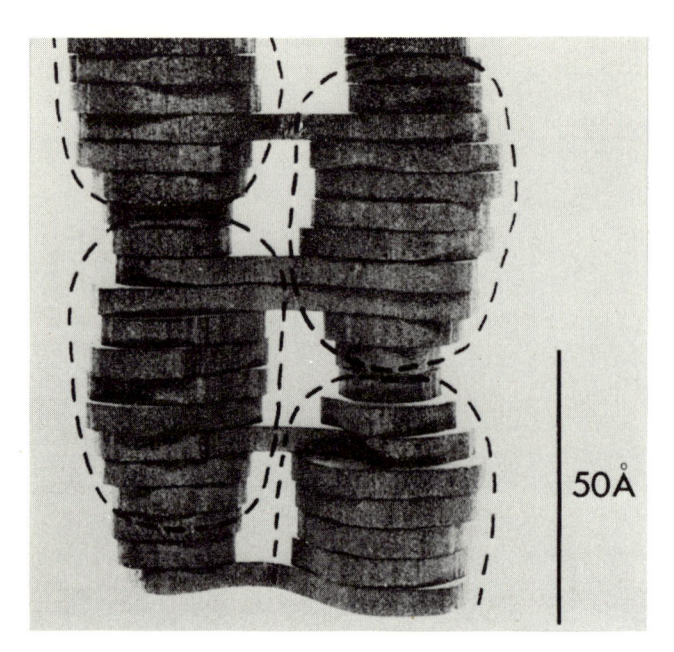

Figure 14.12 Electron-image reconstruction of a section of the actin helix.

after it in the chain, there must be a constant rotation of the binding site through each globular entity. The exact nature of the binding site is, of course, uncertain. However, crystallographic studies of actin currently underway may reveal this detail.

Assessment of function

F-actin serves as a fibrous matrix that (a) supports the other protein components of the thin muscle filament (following sections); and (b) provides specific activity towards the thick filament. The x-ray diffraction data, along with electron microscopic studies of HMM (S1) binding to F-actin suggest that the monomeric units all have a well-defined orientation to the fiber axis, that is, the exterior portion of the actin monomer presented to the outside (and inside) of the "bead" helix is the same along its length. Thus interactions between F-actin and other components are likely to involve the same forces and factors along its length. We have already seen that the thick myosin filaments also have a repeat structure and the nature of the interaction between these two and the other thin-filament proteins clearly holds the ultimate clue to the mechanism of muscle function at the molecular level.

Tropomyosin is a double-stranded coiled coil, approximately 422 Å long, found intertwined as a continuous filament in the actin grooves, one tropomyosin molecule on each side of the actin double helix, in a ratio of one tropomyosin molecule to every seven actin monomers. Located at regular 380 to 385 Å intervals along the thin filament is the troponin complex. It comprises three globular subunits: (1) troponin-T (TN-T, molecular weight 38,000), which is rich in lysine and glutamic acid and binds to tropomyosin; (2) troponin-C

(TN-C, molecular weight 18,000), rich in aspartic and glutamic acids. It binds Ca^{2+} ions and can bind to TN-I or TN-T but not to actin or tropomyosin; (3) troponin-I (TN-I, molecular weight 22,000), again rich in lysine and glutamate. It can bind to actin, but not tropomyosin and it apparently inhibits the binding of actin to myosin. Figure 14.13 shows the proposed relationship between the components of the thin filament.

Tropomyosin

Tropomyosin is a fibrous protein that is particularly susceptible to analysis by methodology described in this text. Rabbit tropomyosin has a molecular weight of ~ 66,000, and denatures to yield two similar chains of essentially 33,000 molecular weight. Recently, the complete amino acid sequence has become available so that the structure at various levels can be traced to the primary sequence—an objective rarely attainable but highly desirable.

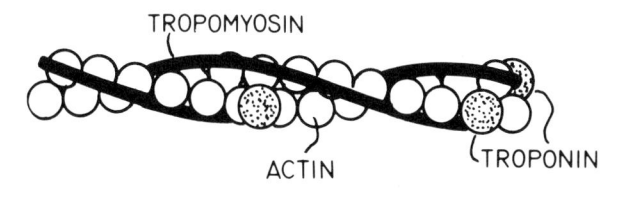

Amino acids and sequence

The sequence of one chain of rabbit tropomyosin is presented in Table 14.3. Unlike collagen, no specific amino acid repeat is evident; nevertheless, from the conformational principles spelled out in Chapters 2 and 3, some type of repeat is expected for a fibrous protein. In fact, it appears that the hydrophobic nature of residues occurs in a repeat sequence placed at positions, 1, 4, 8, 11, and so on, along the chain. These residues are marked with squares and circles that clearly line up in sets of seven. We may then regard tropomyosin as a polyheptapeptide in which the *properties* of the residues are periodic. (There are three exceptions to the hydrophobic sequence, residues 15, 29, and 218 being hydrophilic.) It is noticeable that, as with the rod portion of myosin, there is a high concentration of acidic and basic residues (~ 54%) in the approximate ratio of 2:1.

Physical characterization

Tropomyosin is found to be more than 80% α-helical in content by CD/ORD measurements at neutral pH; however, at higher pH's there is a decrease in α-helix content that appears to be associated with a denaturing process. Figure 14.14 shows the CD spectra for tropomyosin at various pH's. In explaining the structure of tropomyosin it is necessary to account for this phenomenon.

Conformational analysis of the type proposed by Chou and Fasman (Chapter 3) would predict ~ 90% α-helix based on the sequence (note that the Chou–

Figure 14.13 Assembly of protein components in thin-muscle filament (after Cohen).

Table 14.3 Amino Acid Sequence of Rabbit Tropomyosin

	1 AcMet	2 -Asp	3 -Ala-	4 Ile	5 -Lys	6 -Lys	7 -Lys-	8 Met	9 -Gln	10 -Met-	11 Leu	12 -Lys	13 -Leu
14 Asp-	15 Lys	16 -Glu	17 -Asn-	18 Ala	19 -Leu	20 -Asp	21 -Arg-	22 Ala	23 -Glu	24 -Glu-	25 Ala	26 -Glu	27 -Ala
28 Asp-	29 Lys	30 -Lys	31 -Ala-	32 Ala	33 -Glu	34 -Asp	35 -Arg-	36 Ser	37 -Lys	38 -Gln-	39 Leu	40 -Glu	41 -Asp
42 Glu-	43 Leu	44 -Val	45 -Ser-	46 Leu	47 -Gln	48 -Lys	49 -Lys-	50 Leu	51 -Lys	52 -Gly-	53 Thr	54 -Glu	55 -Asp
56 Glu-	57 Leu	58 -Asp	59 -Lys-	60 Tyr	61 -Ser	62 -Glu	63 -Ala-	64 Leu	65 -Lys	66 -Asp-	67 Ala	68 -Gln	69 -Glu
70 Lys-	71 Leu	72 -Glu	73 -Leu-	74 Ala	75 -Glu	76 -Lys	77 -Lys-	78 Ala	79 -Thr	80 -Asp-	81 Ala	82 -Glu	83 -Ala
84 Asp-	85 Val	86 -Ala	87 -Ser-	88 Leu	89 -Asn	90 -Arg	91 -Arg-	92 Ile	93 -Gln	94 -Leu-	95 Val	96 -Glu	97 -Glu
98 Glu-	99 Leu	100 -Asp	101 -Arg-	102 Ala	103 -Gln	104 -Glu	105 -Arg-	106 Leu	107 -Ala	108 -Thr-	109 Ala	110 -Leu	111 -Gln
112 Lys-	113 Leu	114 -Glu	115 -Glu-	116 Ala	117 -Glu	118 -Lys	119 -Ala-	120 Ala	121 -Asp	122 -Glu-	123 Ser	124 -Glu	125 -Arg
126 Gly-	127 Met	128 -Lys	129 -Val-	130 Ile	131 -Glu	132 -Ser	133 -Arg-	134 Ala	135 -Gln	136 -Lys-	137 Asp	138 -Glu	139 -Glu
140 Lys-	141 Met	142 -Glu	143 -Ile-	144 Gln	145 -Glu	146 -Ile	147 -Gln-	148 Leu	149 -Lys	150 -Glu-	151 Ala	152 -Lys	153 -His
154 Ile	155 Ala	156 -Glu	157 -Asp-	158 Ala	159 -Asp	160 -Arg	161 -Lys-	162 Tyr	163 -Glu	164 -Glu-	165 Val	166 -Ala	167 -Arg
168 Lys-	169 Leu	170 -Val	171 -Ile-	172 Ile	173 -Glu	174 -Ser	175 -Asp-	176 Leu	177 -Glu	178 -Arg-	179 Ala	180 -Glu	181 -Glu
182 Arg-	183 Ala	184 -Glu	185 -Leu-	186 Ser	187 -Glu	188 -Gly	189 -Lys-	190 Cys	191 -Ala	192 -Glu-	193 Leu	194 -Glu	195 -Glu
196 Glu-	197 Leu	198 -Lys	199 -Thr-	200 Val	201 -Thr	202 -Asn	203 -Asn-	204 Leu	205 -Lys	206 -Ser-	207 Leu	208 -Glu	209 -Ala
210 Gln-	211 Ala	212 -Glu	213 -Lys-	214 Tyr	215 -Ser	216 -Gln	217 -Lys-	218 Glu	219 -Asp	220 -Lys-	221 Tyr	222 -Glu	223 -Glu
224 Glu-	225 Ile	226 -Lys	227 -Val-	228 Leu	229 -Ser	230 -Asp	231 -Lys-	232 Leu	233 -Lys	234 -Glu-	235 Ala	236 -Glu	237 -Thr
238 Arg-	239 Ala	240 -Glu	241 -Phe-	242 Ala	243 -Glu	244 -Arg	245 -Ser-	246 Val	247 -Thr	248 -Lys-	249 Leu	250 -Glu	251 -Lys
252 Ser-	253 Ile	254 -Asp	255 -Asp-	256 Leu	257 -Glu	258 -Asp	259 -Glu-	260 Leu	261 -Tyr	262 -Ala-	263 Gln	264 -Lys	265 -Leu
266 Lys-	267 Tyr	268 -Lys	269 -Ala-	270 Ile	271 -Ser	272 -Glu	273 -Glu-	274 Leu	275 -Asp	276 -His-	277 Ala	278 -Leu	279 -Asn
280 Asp	281 -Met	282 -Thr-	283 Ser-	284 Ile									

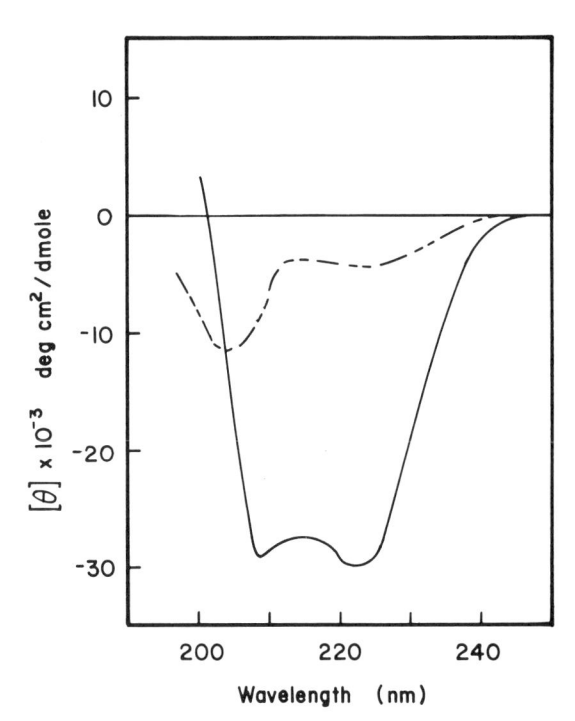

Figure 14.14 CD curves of native (———) and denatured (----) tropomyosin.

Fasman method is not expected to be particularly accurate for fibrous proteins) and the Raman and infrared spectra are in accord with a very high proportion of α-helix.

X-ray diffraction of oriented films reveals the typical α-helix coil pattern (Chapter 4), having a meridional arc at 5.15 Å and a peptide repeat of 1.49 Å.

The physical evidence thus supports the concept that tropomyosin is a two-chain α-helix coiled coil. Whether it is 100% α-helical or has some defect structure is not entirely clear, but the analysis of tropomyosin tactoids (fibers of tropomyosin precipitated with divalent cations) suggests that there is probably little, if any, deviation from a 100% α-helical rod.

There are, however, some interesting questions associated with the structure of the molecule, for example:

1. How can a molecule have such a high density of ionic groups, particularly negatively charged carboxyl groups, without undergoing disruptive charge interaction—such as poly(glutamic acid)—and ceasing to be α-helical?
2. What is the role of the hydrophobic groups?
3. What criteria are necessary to form an α-helix coiled coil?
4. Are the two chains in the molecule parallel or antiparallel?
5. Are the two chains in the molecule staggered or in phase?

Some of these questions are easily answered, others are not. The fact that one can form asymmetrically banded tactoids (fibrous crystals) with lead ions shows that the chains must be parallel. (All combinations of antiparallel chains or molecules produce symmetrical staining patterns). Also, the arrangement of hydrophobic residues in a heptad repeat is expected because of the 7.2 residues per two-turn repeat of the α-helix. In fact, if two adjacent α-helices were to interact preferentially, one would expect complementary residues in positions 1, 4.6, 8.2, 11.8, and so on, compared with the 1, 4, 8, 11, and so on, found experimentally. A slight twist of the two helices (see Chapter 4) brings the desired sequence 1, 4.5, 8, 11.5, 15, very close (in fact, identical if the nearest lower integral number is taken) to that observed. This situation is shown schematically in Figure 14.15.

One concludes, therefore, that the hydrophobic residues lie on the "inside" of the coiled coil, away from any aqueous environment. The ionic groups Glu, Asp, Arg, Lys, on the other hand, probably lie on the outside of the coiled coil. It would seem that the forces holding the two chains together are predominantly London/van der Waals dispersive forces; however, the ionizable groups are sufficiently long that groups of opposite charge could wrap around to cause ionic stabilization. That such a stabilizing feature does indeed occur is revealed by the denaturation behavior exhibited in the CD spectra of Figure 14.4, which seem to indicate that deprotonation of lysyl or arginyl groups at high pH causes a collapse of the structure.

Another point revealed by CD spectroscopy is that if the chains are thermally denatured and separated, they take on a random conformation, indicating that in all probability the sequence of the single chain alone does not dictate the formation of α-helix, but rather the complementary interaction with an adjacent chain induces the helix in both.

Efforts to produce synthetic sequential polyheptapeptides that take on a double-helix coiled coil have included $(Ala_2GluAla_3Glu)_n$, $(Ala_2LysAla_3Glu)_n$, $(AlaLeuGluAla_2LysGlu)_n$, and $(LysLeuGluSerLeuGluSer)_n$, of which only the last two seem to show evidence of α-helix coiled coil structure. Since these are

Figure 14.15 Two α-helices twisted slightly such that hydrophobic bonding is optimized between residues 1, 4, 8, and so on. Formation of the two-strand α-helix coiled coil is stabilized by ionic interchain forces as indicated for a representative tropomyosin sequence.

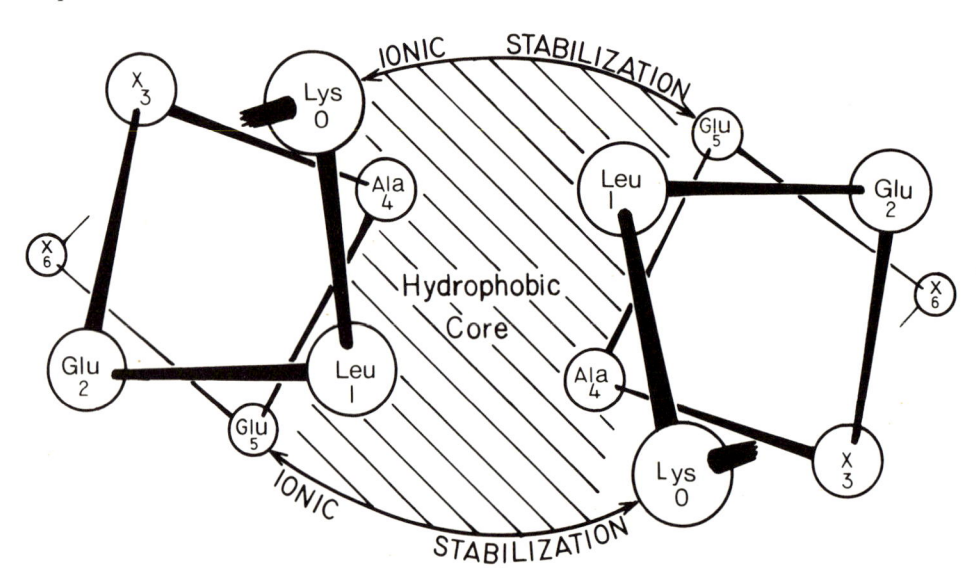

14. Molecular Structure and Function of Muscle

all polyheptapeptides with the 1, 4, 8, 11 hydrophobic sequence, it seems again that the subtler aspects of intramolecular ionic stabilization are required.

Finally, although it was originally thought that the two chains of tropomyosin were staggered to optimize hydrophobic interactions and steric requirements, it is now known from chemical studies and computer simulation of tactoids that there is no chain stagger within the molecule.

Troponin

As mentioned previously and demonstrated in Figure 14.13, a minor protein component (troponin) complexes with other muscle elements and is believed to play an important role in muscle function.

Troponin-T

Troponin-T (TN-T) interacts with tropomyosin and troponin C. The main objectives in studying TN-T have been to identify its mode of binding and its structure, and potentially, its function. The methods used are similar to our standard approach.

Amino acid analysis and sequence. TN-T has a single polypeptide chain of 259 amino acids and a molecular weight of 30,503. The sequence is given (in single-letter code) in Table 14.4.

Table 14.4 Amino Acid Sequence of Troponin-T

	1									131							
Ac	S	N	E	E	V	E	H	V	E	E	E	D	A	K	R	R	A	E	E
	11									141							
	E	A	E	E	E	A	P	S	P	A	D	L	K	K	K	K	A	L	S	S
	21									151							
	E	V	H	E	P	A	P	E	H	V	M	G	A	N	Y	S	S	Y	L	A
	31									161							
	V	P	E	E	V	H	E	E	E	K	K	A	D	Q	K	R	G	K	K	Q
	41									171							
	P	R	K	L	T	A	P	K	I	P	T	A	R	E	M	K	K	K	I	L
	51									181							
	E	G	E	K	V	D	F	D	D	I	A	E	R	R	K	P	L	N	I	D
	61									191							
	Q	K	K	R	Q	N	K	D	L	M	H	L	S	D	E	K	L	R	D	K
	71									201							
	E	L	Q	A	L	I	D	S	H	F	A	K	E	L	W	D	T	L	Y	Q
	81									211							
	E	A	R	K	K	E	E	E	E	L	L	E	T	D	K	F	E	F	G	E
	91									221							
	V	A	L	K	E	R	I	E	K	R	K	L	K	R	Q	K	Y	D	I	M
	101									231							
	R	A	E	R	A	E	Q	Q	R	I	N	V	R	A	R	V	E	M	L	A
	111									241							
	R	A	E	K	E	R	E	R	Q	N	K	F	S	K	K	A	G	T	T	A
	121									251							
	R	L	A	E	E	K	A	R	R	E	K	G	K	V	G	G	R	W	K	

The most remarkable feature of the sequence of this protein is its high content of charged residues. Out of a total of 259 residues, 61 are aspartates and glutamates, 64 are arginines and lysines (and 6 are histidines). Thus approximately 50% of the amino acids of TN-T are charged at neutral pH. In addition, inspection reveals that the distribution of these residues is not random. The N-terminal region (residues 1–39) is highly acidic (18 Asp + Glu); the C-terminus (residues 221–259) is highly basic. It is noticeable that frequent

Figure 14.16 Schematic representation of the secondary structure of TN-T, as predicted by the method of Chou and Fasman.

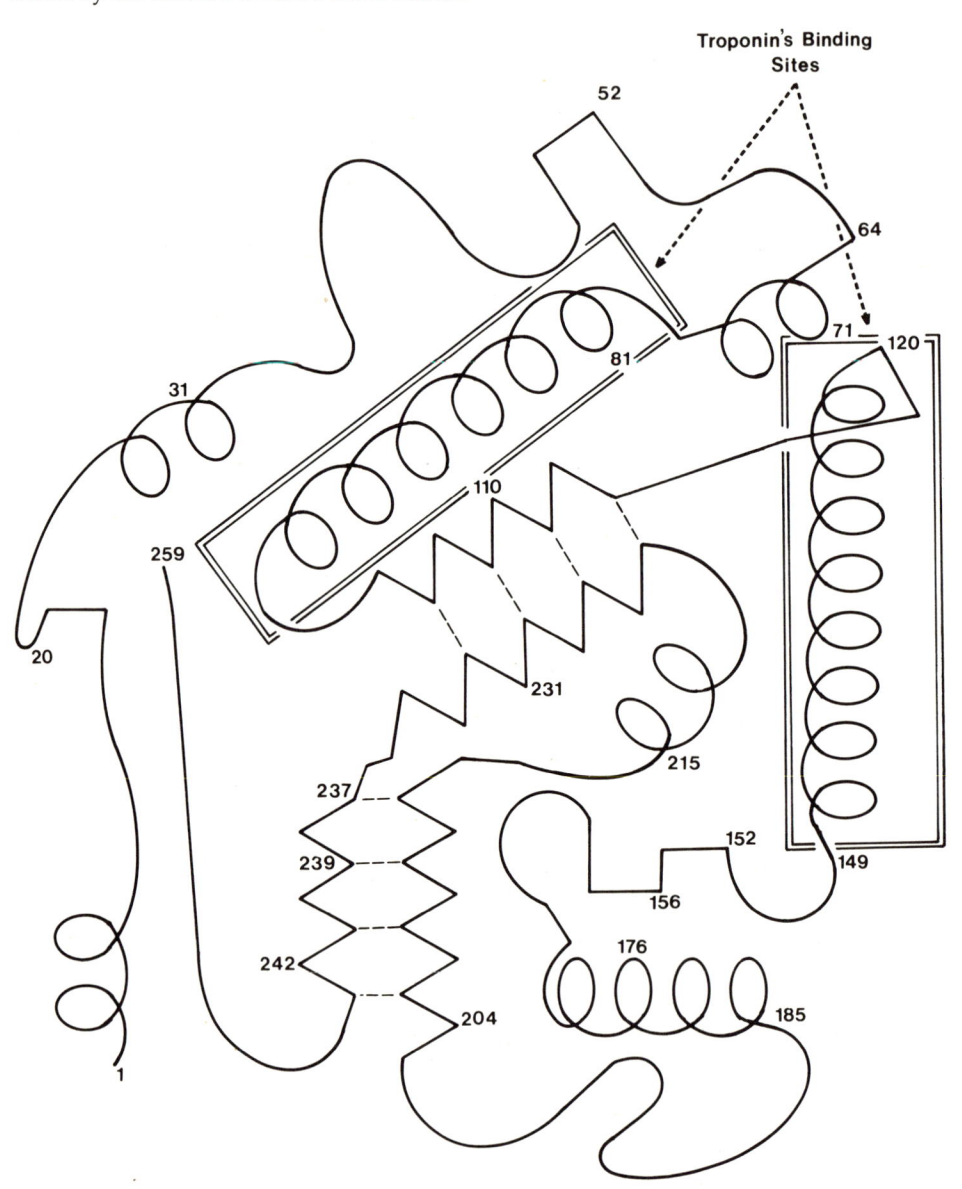

clusters of three or four charged residues occur. Also, there are no clusters of hydrophobic residues. These observations alone suggest that ionic interactions with tropomyosin, troponin C, and ions are likely to be important.

Secondary structure. One of the most interesting approaches to the conformation of TN-T has been the application of the Chou–Fasman method (Chapter 3) to conformation prediction. Such calculations suggest that the protein has ~ 37% α-helix, ~ 10% β-sheet, and seven β-turns. A schematic diagram is shown in Figure 14.16.

Structure and function. Although the detailed structure of TN-T is unknown, it is tempting to speculate that the two charged α-helical regions associate with the α-helices of tropomyosin. Since the other two troponin moieties—TN-C and TN-I—are highly acidic and basic, respectively, it may be that the troponin complex is formed by association of TN-C with the basic TN-T, C-terminal region; and TN-I with the acidic TN-T, N-terminal region. Although such speculation is highly tentative, the procedure spelled out above is an interesting exercise in forming functional hypotheses based on primary-sequence data.

Troponin-I

Troponin-I (TN-I) is a component of troponin that inhibits certain enzymatic activity. Its relation to actin and TN-T binding has already been mentioned. Using the same approach to structure and function as demonstrated previously for TN-T, we have:

Amino acid sequence. The sequence of TN-I has recently become available and shows it to be a single polypeptide chain of 179 residues. The sequence is shown in Table 14.5.

Table 14.5 Amino Acid Sequence of Troponin-I

1										91									
X	G	A	E	E	K	R	N	R	A	I	E	L	E	D	M	N	Q	K	L	F
11										101									
T	A	R	R	Q	H	L	K	S	V	D	L	R	G	K	F	K	R	P	P
21										111									
M	L	Q	I	A	A	T	E	L	E	L	R	R	R	V	R	M	S	A	D
31										121									
K	E	E	G	R	R	E	A	E	K	A	M	L	K	A	L	L	G	S	K
41										131									
Q	N	Y	L	A	E	H	C	P	P	H	K	V	C	M	D	L	R	A	N
51										141									
L	S	L	P	G	S	M	A	E	V	L	K	Q	V	K	K	E	D	T	E
61										151									
Q	E	L	C	L	Q	L	H	A	K	K	E	R	D	V	G	D	W	R	K
71										161									
I	D	A	A	E	E	E	K	V	D	N	I	E	E	K	S	G	M	E	G
81										171									
M	E	I	K	V	Q	K	S	S	K	R	K	K	M	F	E	S	E	S	

Table 14.6 Troponin-C Sequence of Rabbit Skeletal Muscle*

	1										81								
X	(Q	D	Q	T)	A	E	A	R	S	Y	Q	M	K	E	D	A	K	G	K S
	11										91								
	L	S	E	E	M	I	A	E	F	K	E	E	E	L	A	E	C	F	R I
	21										101								
	A	A	F	D	M	F	D	A	D	G	F	D	R	N	A	D	G	Y	I D
	31										111								
	G	G	D	I	S	V	K	E	L	G	A	E	E	L	A	E	I	F	R A
	41										121								
	T	V	M	R	M	L	G	Q	T	P	S	G	E	H	V	T	D	E	E I
	51										131								
	T	K	E	E	L	D	A	I	I	E	E	S	L	M	K	D	G	D	K D
	61										141								
	E	(V	D	E	D	G	S	G	T)	I	N	D	G	R	I	D	F	D	E F
	71										151								
	D	F	E	E	F	L	V	M	V	R	L	K	M	M	E	G	V	Q	

*The order of residues in parentheses is tentative.

Figure 14.17 Schematic diagram of the (predicted) secondary structure of TN-I, showing probable sites of interaction with TN-C and actin.

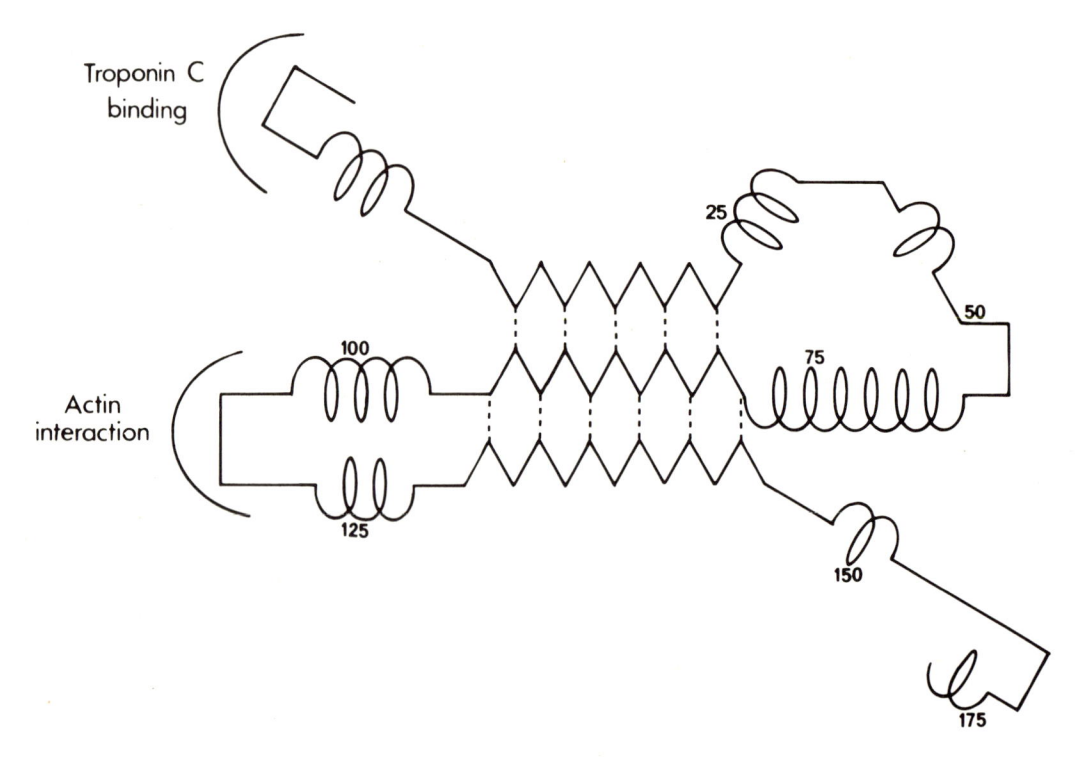

14. Molecular Structure and Function of Muscle

Secondary structure and proposed binding sites. As with TN-T, TN-I has been subjected to a Chou–Fasman predictive conformational analysis. The results predict 45% α-helix, 9% β-sheet, and four β-turns. The detailed structure is shown in Figure 14.17.

It is suggested that TN-I binds to actin through the basic region (residues 102–135) and to TN-C through the shorter basic N-terminal region (residues 5–27).

Troponin-C

Troponin C binds calcium ions reversibly and also binds in the calcium salt form to TN-I. It is also thought to bind, as noted previously, to TN-T.

Amino acid sequence. There are 158 residues in the primary sequence, as shown in Table 14.6.

Secondary structure. Conformational analysis (Ptitsyn approach, see Chapter 3) predicts 49–56% α-helix and 3–9% β-sheet, which is remarkably similar to TN-I. Regions of α-helix are predicted at positions 13–29, 39–48, 54–65, 72–82(?), 91–99, 111–121, 128–136, and 147–155; β-bends are predicted between the α-helical sections. Experimental data (CD/ORD) show 52% α-helix and ~ 10% β-sheet, respectively. No detailed suggestions for binding sites have yet been presented.

14.4 Muscle Contraction

Putting the above information for all muscle components together, it is concluded that as muscle contraction occurs, the arrays of thin filaments within the sarcomere slide towards one another. The I-bands and H-zone decrease in length (see Figures 14.3 and 14.18). Presumably contraction is accomplished by changes in d-interaction between actin in the thin filaments and myosin in the thick filament by a type of "rowing" action.

During contraction, the position of tropomyosin changes within the actin helix. In relaxed muscle ($Ca^{2+} < 10^{-6}$ M), tropomyosin is found at the edge of

Figure 14.18 Model of changes in muscle sarcomere with contraction, showing interrelation of thick and thin filaments.

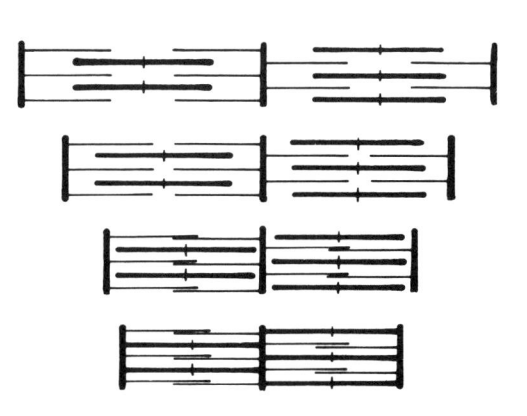

the actin groove, close to the point of attachment of the myosin globular head (S1, HMM) and may physically block actin–myosin attachment. With addition of calcium ($Ca^{2+} > 10^{-5}$ M), tropomyosin movement is mediated through calcium binding to TN-C, which induces conformational changes in the protein and in its sister protein TN-I. The "rolling" of the tropomyosin in the actin groove is propagated axially along the head-to-tail tropomyosin units in a cooperative effect. At this time the details of the actin/tropomyosin, actin/myosin, troponin/tropomyosin, and so on, interactions are not known; however, as the sequence and structure of these proteins become more fully elucidated, it is expected that many of these problems will be resolved.

Further Reading

Fraser, R.D.B. and T.P. MacRae. *Conformation in Fibrous Proteins*. Academic Press, New York, 1973, Chapter 15.

Harrington, W.F. Contractile proteins of muscle. In *The Proteins*, H. Neurath and R.C. Hill, eds., 3rd ed., Vol. 4, Chapter 3, 1979.

Laki, K. *Contractile Proteins and Muscles*. Marcel Dekker, New York, 1971.

Mannherz, H.G. and R.S. Goody. Proteins of contractile systems. *Annual Reviews of Biochemistry* 45:427–465, 1975.

15

Proteins of the Blood

In previous chapters we have examined the relation between various levels of protein structure and function in connective tissue and muscle. In the former, the major protein (collagen) is fibrous; that is, it has the property of self-assembly and its primary function is mechanical, providing structural stability to tissue. Muscle, as we have seen, is more complex, having several protein components ranging from the fibrous tropomyosin through proteins having both structural and chemical properties—for example, myosin—to globular proteins, which regulate the contractile mechanism (troponins). Following this procedure of examining systems of proteins rather than classes, we turn now to the blood system, in which there are a large number of proteins that are predominantly globular and have chemical function. The blood system is very complex in that a large number of proteins has been isolated, some of which are understood in some detail, such as hemoglobin, others of which have not even been obtained in pure form. Similarly the function of some blood proteins is well known and that of others is entirely unknown.

We shall seek to identify the relation between sequence, conformation, tertiary structure, and function where possible. Although much of the functionality of blood is carried out through cellular species, we shall, for present purposes, ignore such structures and consider only the proteins, even though in some cases they function in tandem with cells.

It is convenient to divide the blood proteins into two classes: (1) those that are involved in the blood-clotting process and are enzymes or their precursors (zymogens); and (2) nonclotting proteins—such as the globulins responsible for immune response, proteins involved in carrying oxygen (hemoglobin), albumins, complement, and so on. It is also logical to start with the protein about which most is known, that is, hemoglobin.

15.1 Hemoglobin

When examining the structural features of globular proteins, the same type of systematic procedure may be applied as that used for fibrous proteins: that is, purification, identification of molecular weight, primary sequence, conformation by physical methodology, tertiary structure of crystallizable material by x-ray diffraction and electron microscopy, and analysis of structure/function relationships. Since so much is known about hemoglobin, we will start first with composition and sequence.

Hemoglobin is the principal oxygen-transporting substance in vertebrate blood and is present in high concentrations in red blood cells. It contains four similar polypeptide chains, consisting of two α-chains and two β-chains (not to be confused with α-helix and β-sheet conformation). Each chain acts essentially as a "box" surrounding a *heme* group, the formula for which is given in Figure 15.1. As can be seen, an iron atom is chelated by the porphyrin ring system and is the binding site for oxygen.

For fibrous proteins we followed the investigative sequence of examining primary sequence, physical methodology and conformation, tertiary structure, and function. Because globular proteins generally have a much more specific chemical function than do fibrous proteins, it is often possible to follow the role of defects in the primary sequence which cause changes or defects in the function of the protein. We shall, therefore, complete the cycle by referring back from structure and function to changes induced by defects in primary sequence. The human α-chains have a molecular weight of 15,526 and contain 141 amino acid residues; the β-chains are slightly larger with a molecular weight of 15,867 and contain 146 residues.

Primary Structure

The sequence of human α- and β-chains is given in Table 15.1. It is noticeable that when the amino acid sequence is arranged as in Table 15.1, there are many similarities (63 duplications), and at the same time many

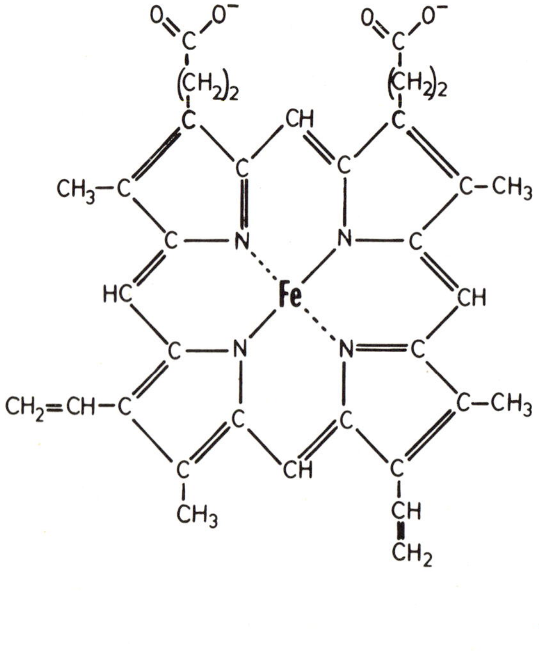

Figure 15.1 The heme group of hemoglobin.

Table 15.1 The Sequence of Human Hemoglobin α- and β-Chains

		5					10					15					20					25					30
α 1	V	*	L	S	P	A	D	K	T	N	V	K	A	A	W	G	K	V	G	A	H	A	G	E	Y	G	A	E	A	L
β 1	V	·	H	L	T	P	E	E	K	S	A	V	T	A	L	W	G	K	V	N	·	*	V	D	E	V	G	G	E	A	L

		35					40					45					50					55					60
α 31	E	R	M	F	L	S	F	P	T	T	K	T	Y	F	P	H	F	·	D	L	S	H	·	*	·	*	G	S	A
β 31	G	R	L	L	V	V	Y	P	W	T	Q	R	F	F	E	S	F	G	D	L	S	T	P	D	A	V	M	G	N	P

		65					70					75					80					85					90
α 61	Q	V	K	G	H	G	K	K	V	A	D	A	L	T	N	A	V	A	H	V	D	D	M	P	N	A	L	S	A	L
β 61	K	V	K	A	H	G	K	K	V	L	G	A	F	S	D	G	L	A	·	A	L	D	N	L	K	G	T	F	A	T	L

		95					100					205					110					115					120
α 91	S	D	L	H	A	H	K	L	R	V	D	P	V	N	F	K	L	L	S	H	C	L	L	V	T	L	A	A	H	L
β 91	S	E	L	H	C	D	K	L	H	V	D	P	E	N	F	R	L	L	G	N	V	L	V	C	V	L	A	H	H	F

		125					130					135					140					145	
α 121	P	A	E	F	T	P	A	V	H	A	S	L	D	K	F	L	A	S	V	S	T	V	L	T	S	K	Y	R
β 121	G	K	E	F	T	P	P	V	Q	A	A	Y	Q	K	V	V	A	G	V	A	N	A	L	A	H	K	Y	H

differences between α- and β-chains. Since the chains are known to have very similar structure and function, it could be argued that nature can accommodate wide variability in sequence without wide alteration in tertiary structure. Indeed, the sequences of a large number of vertebrate hemoglobins are known and as many as 50 substitutions in the α-chain are not uncommon (e.g., carp). Whereas the detailed structure of these hemoglobins is often not known, it is reasonable to suppose that the conformation and higher order structures are very similar from species to species. The reason for this permitted variation would seem to be that most of the sequence leads only to the formation of the protective "container" for the heme group and substitutions that do not occur in the region of the heme group do not drastically affect the chemical function of the protein. However, as we shall see later, certain substitutions cause quite drastic changes in function, some of which may be readily understood based on a detailed study of the three-dimensional structure.

Secondary Structure

It is perhaps inappropriate to apply the Chou–Fasman conformational approach to hemoglobin, since this is one of the proteins from which the empirical rules were derived. Nevertheless, if the Chou–Fasman method is applied to hemoglobin it suggests that the molecule is mainly α-helical with a number of β-turns. Similarly, ORD/CD measurements, infrared/Raman spectra, and so on, support the concept of a highly α-helical molecule.

We shall now presuppose the information obtained from x-ray crystallography in the following section and note that hemoglobin consists of a number of α-helices folded around the heme group. The helical regions are:

A Residues 4–19

B Residues 22–36

C Residues 37–43

D Residues 52–58

E Residues 59–78

F Residues 87–95

G Residues 101–119

H Residues 126–148

In Table 15.1 the nonhelical regions have been outlined. As can be seen, the D-helix is completely missing in the α-chain. Otherwise 41 of the 63 identical residues are in α-helical regions, from which we can conclude that substitutions are independent of the conformation of the region in question. That is not to say, however, that substitutions are independent of the conformational needs of the structure. Amino acid residues in the α-helical regions are replaced by other α-helix formers in general.

Attention should be focused on four residues: F(Phe)[44], H(His)[65], L(Leu)[90], and H(His)[94]. These four residues are involved with binding the heme group (the two histidines being coordinated directly to the iron atom) and are crucial to the performance of the protein. Not surprisingly, these residues are conserved in α- and β-chains and indeed throughout the vertebrate hemoglobins.

Tertiary Structure

One of the most remarkable advances in protein chemistry came about as a result of the solution of the x-ray diffraction patterns of myoglobin and hemoglobin. These studies reveal the details of the subunit structure, conformation, and atomic geometry. Figure 15.2 shows one view of the four chains along with a schematic representation of the heme groups. As may be seen, there is little contact between the two α- or the two β-chains; the main interactions are between α- and β-chains.

The question may well be put: what holds the four chains together? This question is perhaps best answered by considering the difference in sequence between hemoglobin and its sister protein myoglobin, which has a very similar structure but is monomeric. It is the hydrophilic groups, Glu, Asp, Arg, and Lys, that are normally considered to render proteins water soluble. Comparison of the frequency of these residues in human myoglobin (152 residues in length) with α- and β-chains of human hemoglobin (141 and 146 residues) shows that the hydrophilic content of hemoglobin, particularly lysine, is much reduced (Table 15.2). Examination of the sequence indicates that lysyl groups are replaced by hydrophobic residues, some of which are involved in interchain interaction. Thus the main force binding hemoglobin subunits is hydrophobic or London/van der Waals interaction.

Detailed x-ray diffraction studies have shown that in the oxygen-carrying ("oxy") form, the heme group is approximately planar and the heme-linked N of the histidine residue (94 α in Table 15.1) is about 2.0 Å from the plane of the porphyrin ring. (The oxy form is known as the "r" or "relaxed" form.) On the other hand, in the deoxy form ("tense" or "t" form), the iron lies out of the heme plane by 0.75 Å and the histidyl N is about 2.9 Å from the plane.

Structure, Function, and Sequence

With the deoxygenation of oxyhemoglobin and the subsequent binding and removal of carbon dioxide, the relative positions of the α- and β-chains change, almost as though the molecule itself were breathing. Thus the behavior of the molecule cannot simply be treated as a series of heme groups, surrounded by an inert sea of amino acid residues. Evidently the functioning of the molecule must be dependent to a large degree on overall conformation and chain arrangement. Very many abnormal hemoglobins have now been studied and the residues marked with a dot (Ẋ in Table 15.1) are known to undergo single-residue deletions in various detected abnormalities. Others will undoubtedly become known and yet others will go undetected because they do not produce noticeable physiologic abnormalities. It is interesting to note that the majority of defects detected in α-chains also occur at equivalent conformational positions in the β-chain.

Figure 15.2 Relation between α- and β-chains in hemoglobin.

Table 15.2 Comparison of Hydrophilic Groups in Myoglobin and Hemoglobin (Human)

	Myoglobin	Hemoglobin α	Hemoglobin β
Glu	10	4	8
Asp	7	8	7
Lys	20	11	11
Arg	2	3	3
Total	39	26	29

All four of the heme-interacting "constant" residues are known to undergo replacement in genetically defective hemoglobin. Although the structural effect of most replacements is unknown, some structural and chemical effects have been studied in some detail. For example, "sickle-cell" hemoglobin has a Val in place of Glu at position 6 in both β-chains. The increased hydrophobicity causes a fibrous precipitate to form, which modifies the shape of the red blood cell and reduces oxygen affinity.

The crystal structures of some of the abnormal hemoglobins are known in detail and provide clues to the malfunction in certain blood disorders. In the mutant form known as deoxyhemoglobin Yakima, Asp (101 β) is replaced by His. In normal deoxyhemoglobin, this aspartic acid, which lies at the $\alpha_1 \beta_2$ interface, forms a crucial hydrogen bond with an adjacent Tyr in the α_1-subunit. The replacement results in the removal of the hydrogen bond and the bulky imidazole ring constrains the structure towards the oxy form. These constraints lead to a vastly diminished heme-heme interaction and increased oxygen affinity.

A summary of some other abnormal hemoglobins and their physiologic properties and structures is presented in Table 15.3.

Table 15.3 Physiologic Properties of Abnormal Hemoglobins (Structures Investigated by X-ray Diffraction)

Mutant	Position*	From	To	Oxygen affinity	Clinical symptoms†	X-ray data
Ranier	147 β	Tyr	Cys	Very high	Polycythemia	3.5 Å Difference Fourier
Kansas	104 β	Asn	Thr	Very low	Cyanosis	5.5 Å Difference Fourier
Richmond	104 β	Asn	Lys	Normal	Normal	4.0 Å Difference Fourier
Chesapeake	99 α	Arg	Leu	High	Polycythemia	5.5 Å Difference Fourier
J. Captown	99 α	Arg	Gln	Slightly high	Normal	5.5 Å Difference Fourier
M. Iwate	94 α	His	Tyr	Low	Cyanosis	5.5 Å Difference Fourier
M. Hyde Park	94 β	His	Tyr	Normal	Methemoglobinemia	3.5 Å Difference Fourier
Hiroshima	148 β	His	Asp	High	Polycythemia	3.5 Å Difference Fourier

*Positions are defined in Table 15.1.
†Polycythemia—excess of red cells; cyanosis—blue color of skin due to deoxyhemoglobin in capillaries.

Source: T.L. Blundell in *Amino Acids, Peptides and Proteins,* Vol. 4, The Chemical Society, London, 1972, p. 210.

It now appears that substitution of amino acids on the outside of hemoglobin does not generally lead to changes in tertiary structure, whereas changes of residues in positions that affect subunit interaction or heme interaction may have considerable conformational effect and changed oxygen affinity.

A comparison of the detailed structural changes induced in a deoxyhemoglobin by a single-residue change in the Ranier mutant is shown in Figure 15.3.

Figure 15.3 Comparison of the structural effect of chain substitution in the Rainier mutant of hemoglobin.

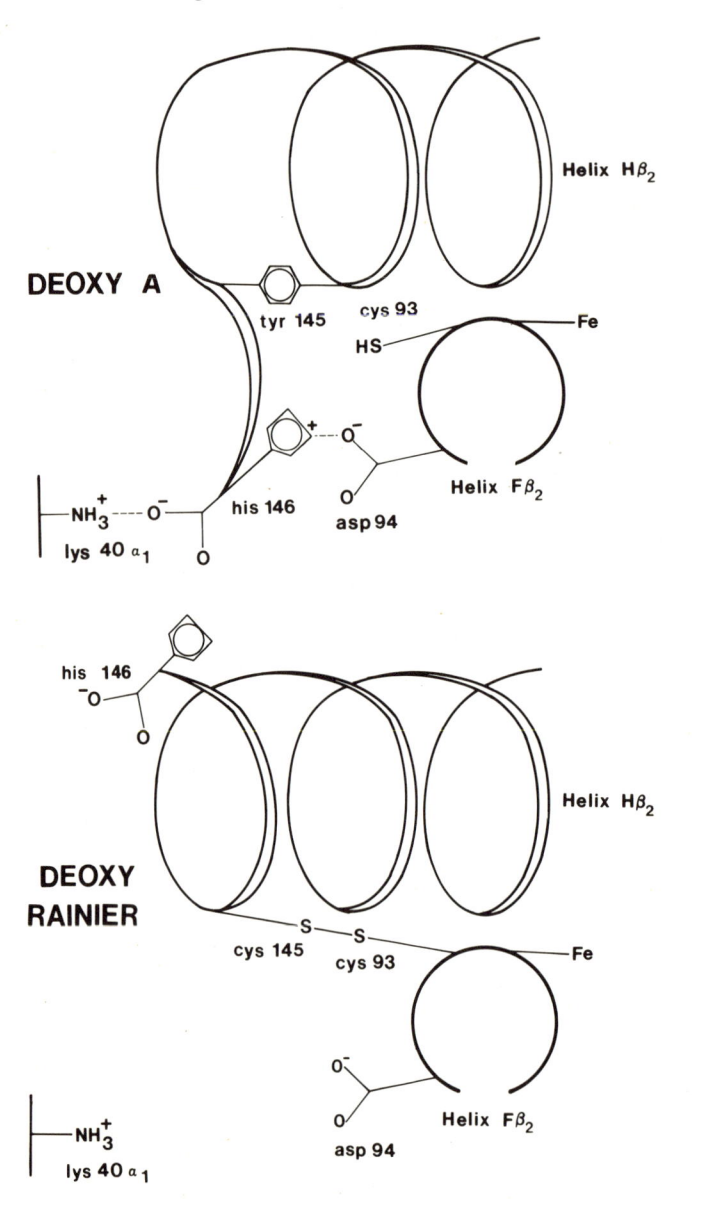

15.2 Immunoglobulins

When an individual is challenged by a foreign body (protein, polysaccharide, etc.), known as an antigen, the body responds by producing an immunoglobulin (antibody) that combines with the antigen, normally forming an insoluble complex.

Once the immune response has been provoked, an individual has increased resistance to the antigen subsequently. The blood serum of normal individuals contains many different immunoglobulins because of the many responses that have been provoked. Hence, it is virtually impossible to isolate complete immunoglobulins with uniform residue sequence. It is the versatility of the variable peptide sequence that allows the human system to respond to large numbers of different antigens. There are five major classes of globulins found in human serum.

All of the immunoglobulins contain two identical "light" chains and two identical "heavy" chains, differentiated by their molecular weight. Each chain has a variable amino acid region at the N-terminus (the variable region) and a constant region of the C-terminus. There are five different types of heavy chains, which have different molecular weight and different sequence from each other in the C-terminal region. These five different types are the basis of the five classes of immunoglobulins. Table 15.4 presents these classes and their properties.

To repeat, then, the designation of heavy chains γ_1, and so on, is based on the sequence and molecular weight of the heavy chains. The number of subclasses refers to the different disulfide-linkage systems. Figure 15.4 shows the four subclass systems for the γ-globulins or IgG. In addition to the subclasses based on heavy chains, there are two types of light chain, λ and κ, again based on the C-terminal sequence.

The four subclasses vary in the interheavy chain —S—S— bridges. In 2,3,4 the bridging to the light chains is similar but the number of inter-"heavy-chain" bridges varies between two and five. The position of the cysteine that binds the light chain in $\gamma 1$ is quite different.

The variable portion of the polypeptide chains (shaded in Figure 15.4) is about 120 residues long and represents part of the antigen-binding site.

A schematic diagram of the total IgG1 molecule (i.e., $\gamma 1$ molecule) is shown in Figure 15.5. V_H, V_L refer to the variable regions of the heavy and light chains. C_H and C_L refer to the constant regions. The oligosaccharide moiety binding site is indicated by $\boxed{CHO}$. The enzyme, papain, cleaves the heavy chain in the

Table 15.4 Properties of Human Immunoglobulins

Class	IgG	IgA	IgM	IgD	IgE
Heavy-chain class	γ	α	μ	δ	ϵ
No. of subclasses	4	2	2	1	1
% of serum globulins	70–80	15–20	5–10	< 1	< 1
Usual aggregate*	Monomer	Dimer	Pentamer	Monomer	Monomer
Aggregate mol. wt.	150,000	370,000	900,000	185,000	200,000
Carbohydrate (%)	2.9	8	12	12	11
Mol. wt. of heavy-chain protein	50,000	50,000	59,000	63,000	58,500

*Monomer = 2 heavy + 2 light chains.

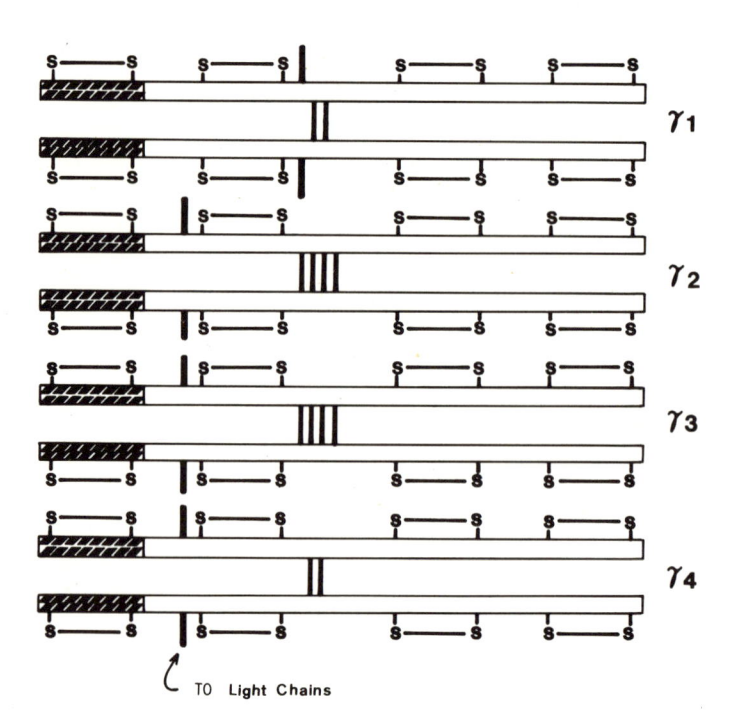

Figure 15.4 The four subclasses of human γ-globulin heavy chains showing variable region (shaded) and inter- and intrachain sulfur bridges.

position shown, producing three fragments. Two of the fragments are identical and contain the two light chains. These fragments are called F_{ab} (fragment antigen binding) peptides and are usually noncrystallizable. The other fragment formed from the C-terminal ends of the two heavy chains is called the F_c crystallizable fragment and its function is to interact with cell membranes (and complement). The variable portion of the chains is known to have at least 100 different amino acid substitution sites, with an average of approximately four possibilities per site. However, the main variations occur in the so-called hypervariable loops, namely three segments each in the V_L and V_H chain consisting of 5 to 10 residues each. Although it is not known whether more than one combination is effective against one antigen, clearly the human body has a defense mechanism adaptable to a fantastic number of foreign molecular invaders. (The mechanism by which the cell system is triggered to respond to a specific antigen is beyond the scope of this text.)

Structure of IgG

Deduction of the structure of immunoglobulins is a completely different type of problem from that encountered with hemoglobin and myoglobin. First, it is difficult to obtain material of uniform purity (sequence); then it is necessary to crystallize the intact material. Detailed x-ray analysis is difficult for such a large molecule and for identification of the antigen-binding mechanism it would be necessary to co-crystallize antigen and the appropriate F_{ab} fragment. These are the sorts of problems typically encountered with all but a limited

15. Proteins of the Blood

number of globular proteins (although the chain variability adds an additional constraint in the present case). Most emphasis has been placed on more indirect methods of deducing structure. Some of the details follow.

Primary structure

It would have been extremely difficult to identify the sequence of the variable region of any immunoglobulin but for the fact that multiple myeloma patients produce an excess of uniform sequence light (L) chains, known as Bence–Jones protein. A number of uniform heavy chains of all classes have also been obtained from myeloma patients. The protein is excreted in the urine and may be collected and purified. Since each patient produces predominantly one sequence, but generally produces a different sequence from other patients, it has been possible to identify many substitution sites. It has also proven possible to obtain the complete sequence of an IgG from a multiple myeloma patient, as shown in Table 15.5. The Table is arranged such that the amino acid sequence in the light and heavy chains of the variable region may be compared.

The boxes indicate positions where only one amino acid residue is known at this time. As can be seen, the constant positions in the V_H chain are repro-

Figure 15.5 A schematic diagram of an IgG_1 molecule showing variable (shaded) and fixed regions of heavy and light chains.

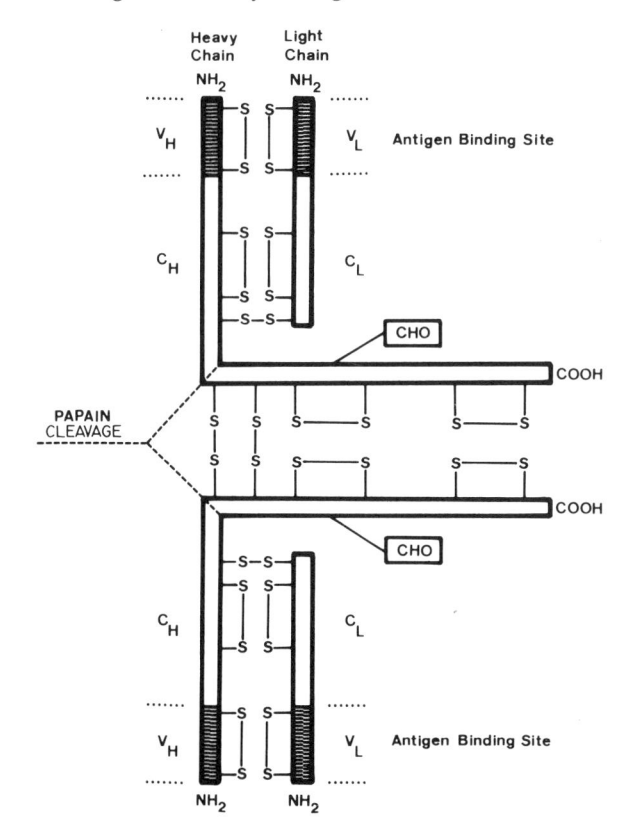

Table 15.5 Amino Acid Sequence of the Variable Region of Human (Eu) IgG

```
                      1                    10                       20              30
Variable light chain  V_L  *  D   I Q M T Q S P S S L S A S V G D R V T I T C R A S Q *
Variable heavy chain  V_H  *  E   V Q L V Q S G † A E V K K P G S S V K V S C K A S G G T
                                         40                    50                    60                    70
V_L  * * * * * K I N T W L A W Y Q Q K P G K A P K L L M Y K A S S † L E S G V P S R F
V_H  F S * * † † R S A I I W V R Q A P G Q G L E W M G G I V P M F G P P N Y A Q K F
                      80                    90                   100                  110
V_L  I G S G S G T E F T † † † † † L T I S S L Q P D D F A T Y Y C Q Q † Y N S
V_H  Q G † R V T I T A D E S T N T B T Y M E L S S L R S E D T A F Y F C A G G Y G I
     71
                     111               120
V_L  * * D S K M F G Q G T R V E V K G
V_H  † † Y S P E E Y N G L V T
```

*Indicates that in other IgG proteins there may be a residue present.
†Represents an arbitrary space to optimize homology.

15. Proteins of the Blood

duced in the V_L chain; their significance is unknown. In general, relatively few substitutions at each position are known, although up to 13 different residues have been found in position 113 and several others are known with 6 or more.

The first and most obvious feature of the sequence is that very few residues in this region are hydrophilic. Rather than proceeding to a discussion of the remainder of the protein, we shall examine the conformational implications of the sequence of the variable portions.

Conformational analysis

Apart from the large number of hydrophobic residues present in the variable chains, there is evidently a high concentration of residues usually associated with β-sheet structures, that is, Val, Ser, Gly. In addition, there appears to be a distinct periodicity in the properties of the residues in the light chain. For example, ionic residues occur at positions 18,19; 46; 57; 69; 78; 97,98; 113,115; and so on. Residue 35 (=29) is commonly Arginine. Thus there is an apparent separation of 10½, 11, 11, 11, 9, 11½, 17½, residues, respectively. There are only four other charged residues in the main body plus those that tie at the ends. This periodicity is highly reminiscent of that found in the cross-β structure of chrysopa silk, where the hydrophilic residues are believed to lie in the folds. Thus a tentative folded structure for the L chain might be as indicated in Figure 15.6.

Figure 15.6 Schematic diagram of the predicted antiparallel β-sheet arrangement for the light chains of Bence–Jones protein. Ionic residues lie on the exterior of the protein segment.

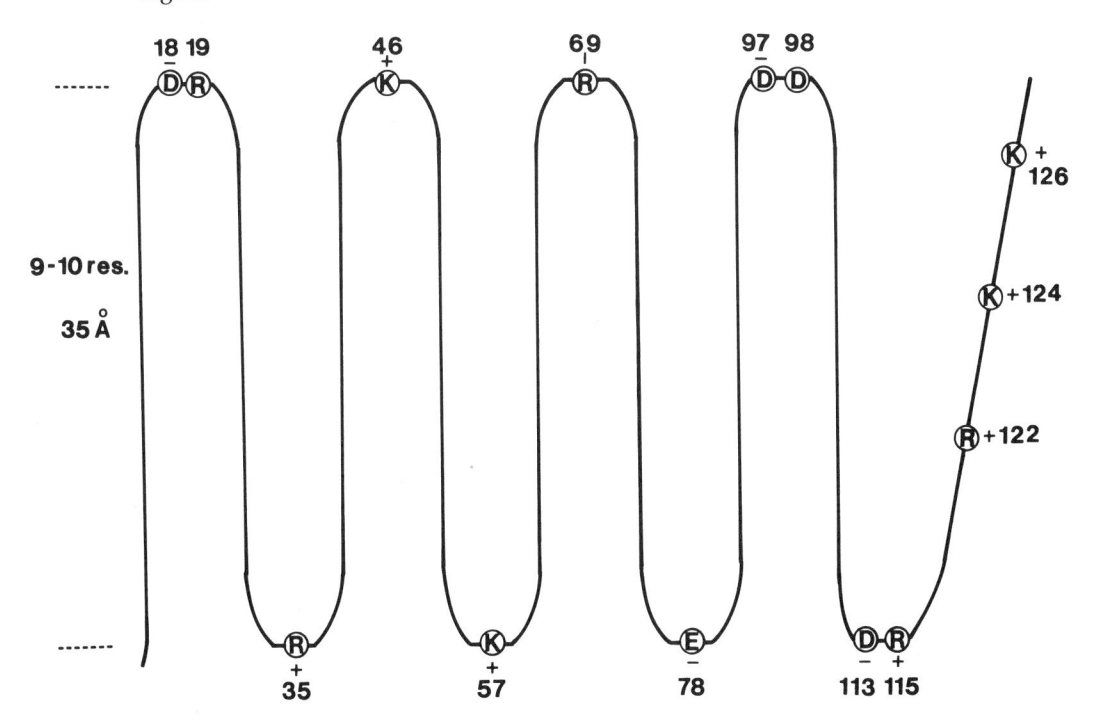

The hydrophilic periodicity in the heavy chain is not as marked as the light chain, but it too seems to have a periodicity in which 7–8 residues intervene. Similarly the constant regions, though rich in β-directing residues, do not seem to show a very regular hydrophilic sequence.

Spectroscopic measurements (ORD/CD and infrared) have confirmed that Bence–Jones proteins (light chains) are highly β-sheet in structure.

Tertiary structure

Despite the great difficulties involved in crystallizing F_{ab} peptides and bound antigen, some progress has been made in this direction and x-ray diffraction data have been resolved at low resolution. Perhaps the most detailed information is available for a λ-type Bence–Jones protein (immunoglobulin light chain). Figure 15.7 shows a schematic drawing based on the x-ray diffraction structural analysis of such a protein.

The figure shows, as expected, an antiparallel β-sheet arrangement. The bends are more complicated than the simple prediction of Figure 15.6, but the structures are in good qualitative accord. (Actually the sequence of Table 15.4 and the prediction of Figure 15.6 are for a κ light chain but the homology is similar to the λ-chain of Figure 15.6.) Using the same numbering scheme as that of Figure 15.5, the observed fold residues lie in the neighborhood of 14, 26, 43, 53, 60, 68, 82, and 96 compared with 17½, 29, 39, 50, 61, 70, 81½, and 97 predicted. We see that the sulfur bridges (heavy line in Figure 15.7) are in essentially equivalent positions on each side of the molecule and that the variable portion of the molecule is separated by a short stretch of residues from the constant region. The equivalency of the S—S bridges and their importance in the structure accounts for the C—C homology in the sequence at positions 24 and 104 of Table 15.4.

Figure 15.7 Schematic diagram of the arrangement of an immunoglobulin light chain from a λ-type Bence–Jones protein, based on x-ray diffraction data.

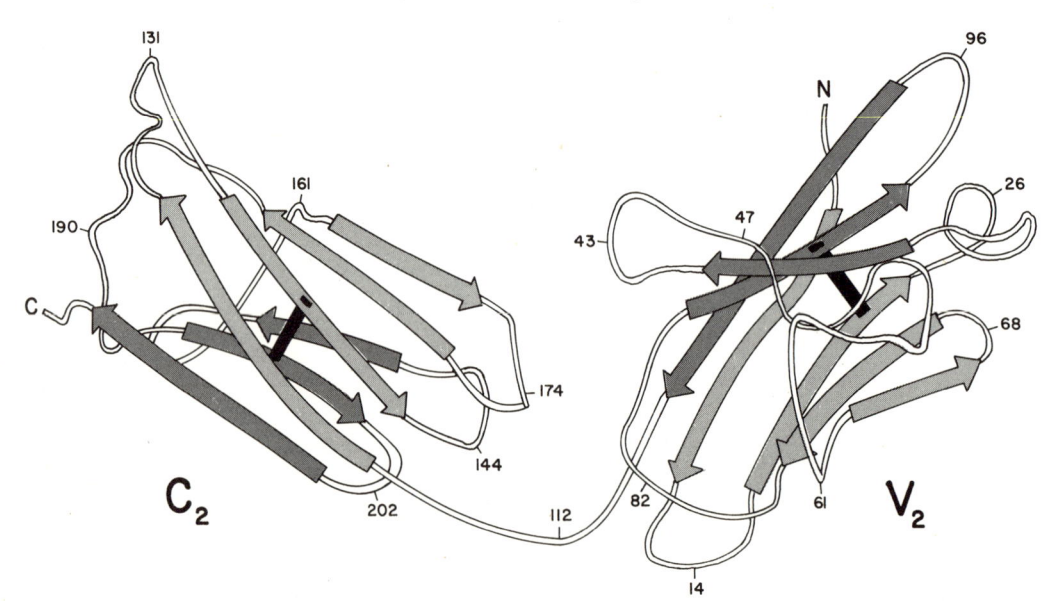

Functional Domains

It appears from our preliminary description of immunoglobulins that certain regions of the molecule have specific functions. Such regions, or domains, possess the characteristics of a complete globular protein and are called functional domains. These functional domains may consist of one or more structural domains; thus IgG may be represented as a series of structural and functional domains. The heavy- and light-chain variable regions represent two structural domains combining to act as a functional domain (antigen-binding site). A single structural domain in the region of the bound saccharide acts as a functional domain in binding complement, and so on.

Binding Sites

Quite recently it has proven possible to crystallize a mouse myeloma F_{ab} fragment of an immunoglobulin with phosphoryl choline and to examine the structure by x-ray crystallography. By this means information has been provided concerning the nature of the antigen-binding site.

The site of binding of the hapten (chemical group on the antigen molecule) is in a large wedge-shaped cavity at the N-terminal end of the molecule, that is, in the hypervariable region. The cavity is approximately 12 Å and 15 Å wide at the mouth; it is 20 Å long and is lined with hypervariable residues. A two-dimensional model of the structure is shown in Figure 15.8.

Figure 15.8 Schematic representation of the polypeptide-chain arrangement at the binding site of phosphoryl choline * to a fragment of mouse myeloma immunoglobulin McPC 603.

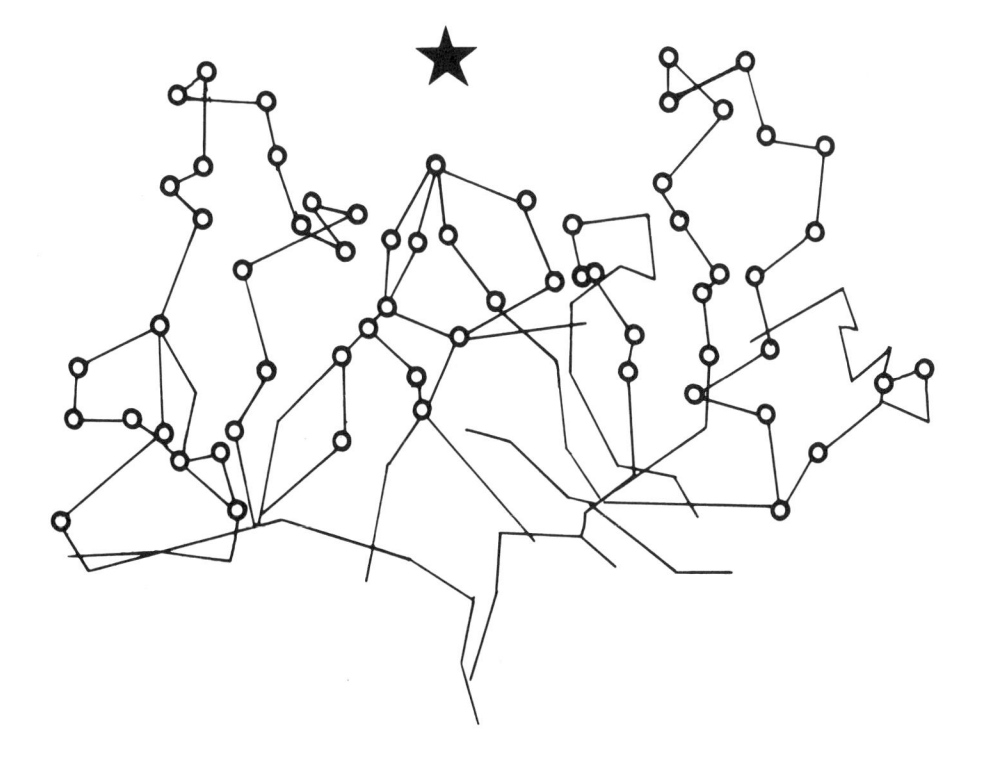

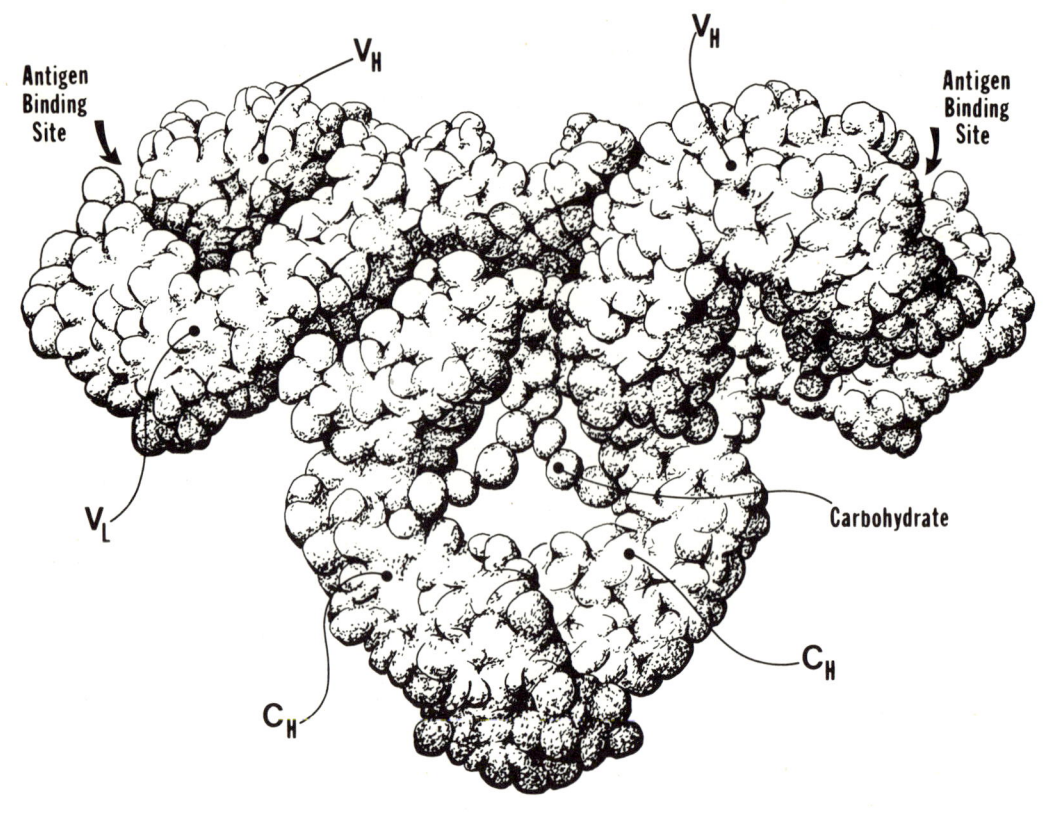

Figure 15.9 The antibody IgG molecule shown at low resolution, illustrating the various domains.

The phosphorylcholine molecule is much smaller than the cavity and is bound asymmetrically such that it is closer to the V_H-chain than the V_L. Although the details of binding are not entirely clear, it seems that the molecule of phosphorylcholine is held in position by a combination of ionic and hydrogen bonding. The phosphate group appears to be hydrogen-bonded to the tyrosyl (33 V_H) hydroxyl and ionically bound to Arg (52 V_H) and Lys (54 V_H). The choline moiety, which is buried more deeply in the cavity, appears to interact both with the V_L- and V_H-chains, the negatively charged Glu(35 V_H) being close to the positively charged nitrogen of choline. In addition, residues 102–103 of the heavy chain and 91–94 of the light chain are close to the choline and may interact by London/van der Waals forces.

Low resolution x-ray diffraction data have recently revealed the overall shape of the IgG molecule (known as Dob protein). A representation is shown in Figure 15.9.

15.3 Serum Albumin

Albumin is the most abundant protein component of human serum, yet remarkably little is known about the details of its function. Until recently little was also known of its structure. It is known that human serum albumin (HSA)

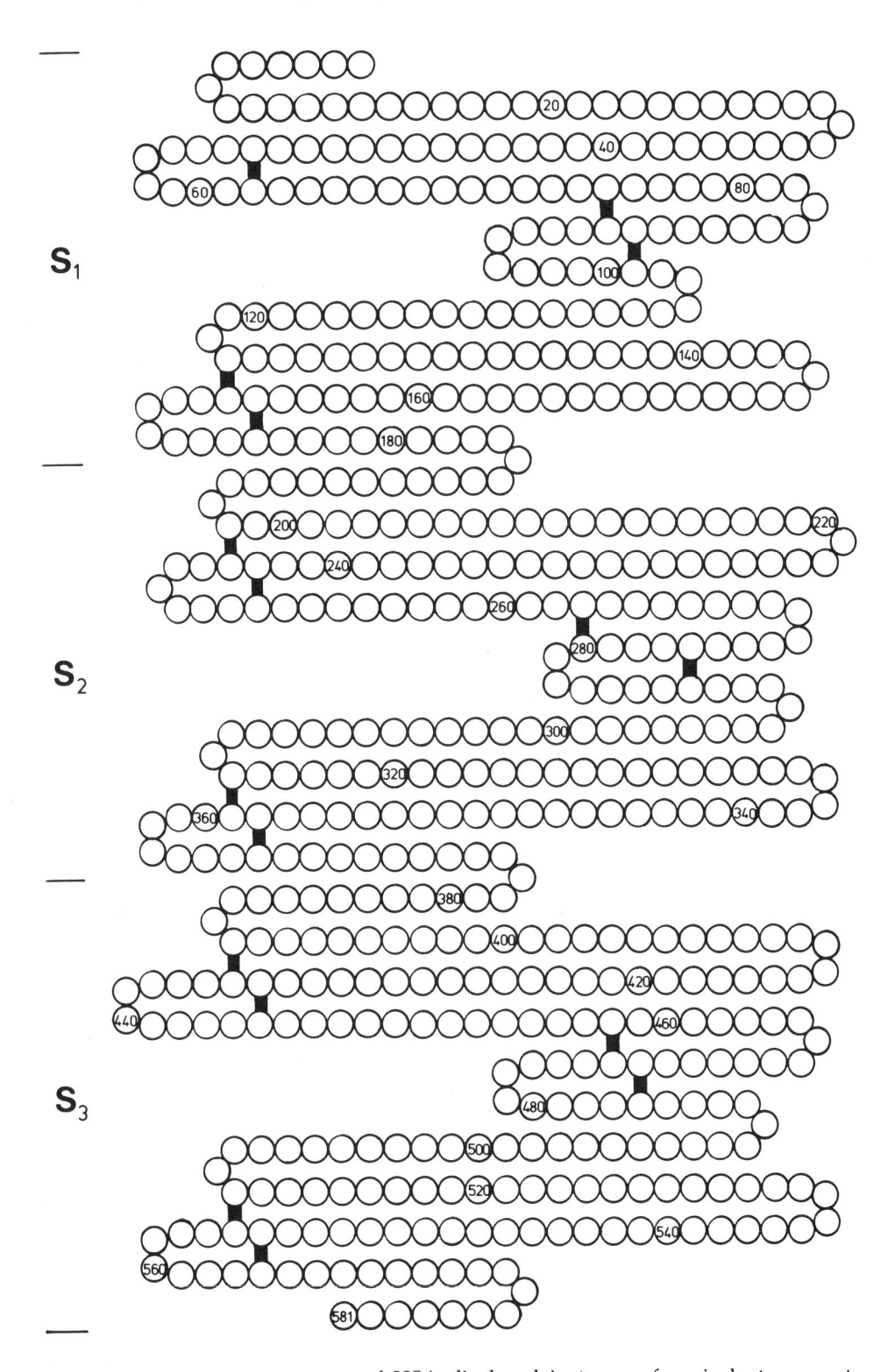

Figure 15.10 Primary sequence of HSA displayed in terms of equivalent segments (after Behrens et al.).

binds fatty acids (lipids), bilirubin, amino acids, hormones (particularly steroids), dyes, ions, and so on. It also plays a mediating role between surfaces (native or artificial) and cellular components. None of the preceding functions seems to warrant a specific sequence and conformation, yet such exists. The conformation is pH-sensitive and it may be that albumin acts as a release carrier for a series of different important compounds.

Perhaps the most interesting and unusual aspect of HSA is the fact that most of the cystines occur as doublets, that is, —Cys—Cys— (total of seven times), and are essentially equally spaced, suggesting a domain structure.

Primary Sequence

The amino acid sequence of HSA is shown in Figure 15.10, arranged such that the sequence appears to be repeated in three equivalent segments S1, S2, and S3. The segments often have repeats of hydrophilic residues in positions 1, 4, 7, 11, 14, 18, typical of sequences expected for α-helices in which there is one side exposed to solvent and another to a hydrophobic milieu.

Secondary Structure

As suggested by the repeat arrangements of hydrophilic residues and predicted by a Chou–Fasman conformational analysis, the HSA molecule is substantially α-helical (CD/ORD), that is, 50% or more. It has been suggested that the helical regions form a cylindrical molecule, perhaps with a highly ionically interactive C-terminus where some 11 ionic residues are found in a sequence of 19 residues. Such a suggestion would account for the anisotropic shape of HSA deduced from light-scattering and other solution measurements. If the cylinder has a hydrophobic interior it may be particularly appropriate to the transport of lipids and other apolar materials.

15.4 Blood-Clotting Proteins

The blood-clotting proteins are comprised mainly of a number of enzymes and their inactive precursors. In choosing enzymes for the purpose of textbook presentation, in terms of structure and function it would perhaps be logical to choose those about which most is known, for example, lysozyme and ribonuclease. To do so would, however, overlook many of the problems that confront the biologic research scientist, namely that most important systems are highly complex, the proteins are rarely crystallizable for x-ray diffraction analysis, and it is not necessarily clear, even when x-ray data are available, how the protein functions.

We shall thus choose a system that is only partially understood—namely, the blood-clotting system—because it poses many of the questions that the physical methodology of the preceding sections seeks to answer. The point must be made, however, that physical science is completely limited in its ability to make structural deductions unless material in pure form is available. Many of the blood-clotting proteins are extremely scarce and difficult to purify and thus the progress made to date has been achieved only at the expenditure of much time, effort, and dedication.

15. Proteins of the Blood

Table 15.6 Properties of Some of the Well-Defined Blood-Coagulation Factors

Roman numeral	Common name	Mol. wt.
Factor I	Fibrinogen	340,000
Factor II	Prothrombin	68,700
Factor III	Tissue Factor	220,000
Factor IV	Calcium Ions	
Factor V	Proaccelerin	300,000
Factor VII	Proconvertin	63,000
Factor VIII	Antihemophilic Factor	1.1 million
Factor IX	Christmas Factor	55,000
Factor X	Stuart Factor	55,000
Factor XI	Plasma Thromboplastin Antecedent	160,000
Factor XII	Hageman Factor	90,000
Factor XIII	Fibrin-Stabilizing Factor	α 39,000
	Thrombin	β, γ 28,000

The blood-clotting process involves a cascade or amplification series, in which a number of proteins are involved. These proteins are often referred to by Roman numerals, although some are known by common names. The blood-clotting protein that is most abundant in human blood serum is factor I or fibrinogen. Thirteen factors have been fairly well characterized; names and molecular weights are listed in Table 15.6. Thrombin, which is produced from prothrombin, is also included.

Their mode of action involves a complicated feedback process but may be simplified as shown in Figure 15.11.

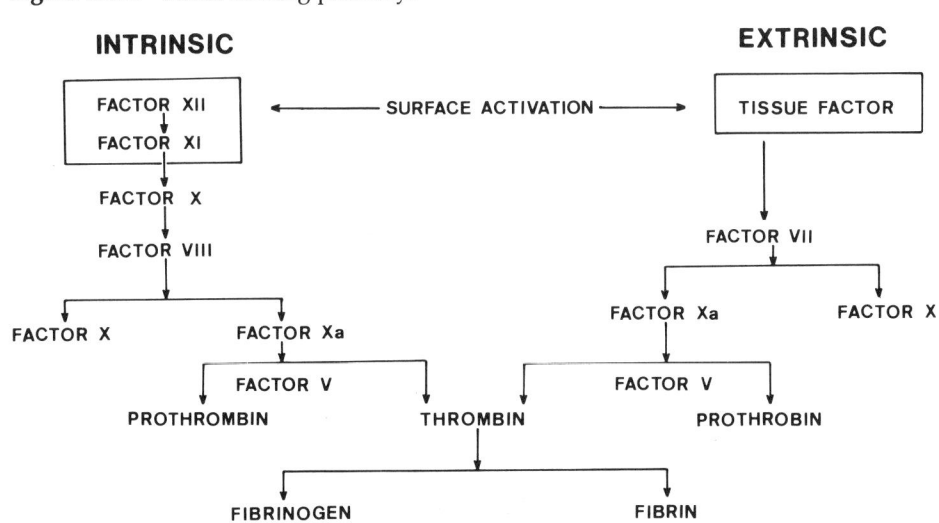

Figure 15.11 Blood-clotting pathway.

The intrinsic pathway is so named because the various proteins involved are all found in serum (although some may be more commonly found in the boundary layer of cells, particularly blood platelets). The extrinsic pathway depends, for initiation, upon one or more components drawn from sources external to serum, that is, tissue.

It is believed that contact of factor XII with collagen at an injury site initiates the coagulation process in vivo (in the body) by forming the activated factor XIIa. Activated XII (XIIa) then activates factor XI, converting it to an enzyme, factor XIa. In turn, factor XIa activates IX in the presence of calcium ions. Thus a series of stepwise reactions is triggered, leading to the formation of thrombin and a fibrin clot. This series of reactions has considerable potential for amplification since a few molecules of XIIa can activate hundreds of molecules of factor XI, which in turn will activate thousands of molecules of factor IX.

The question before us is what role does structure play in the blood-clotting proteins. Since so little is known of the detailed structure of these proteins we must address somewhat more limited questions, such as what type of proteins are these, what role does the primary structure play, is there a conformational change on activation, and how are fibrin (blood clot) fibers formed?

Primary Sequence and Function

Partial amino acid sequences of several of the blood factors in their active and inactive forms have been obtained. A comparison of aspects of sequence between factors X, IX, and II (prothrombin) indicates that each is activated by hydrolytic enzyme cleavage and that each involves serine at the active site. Sequential information and homology are presented in Table 15.7.

Table 15.7 Sequential Information for Blood Factors and Related Proteins

N-terminal sequences of proenzymes

Prothrombin	A N K G F L Ga Ga - V R K G N L · · ·
Factor IX	T N S G K L E E F V R - G N L · · ·
Factor X	A N S - F L E E - V K Q G N L · · ·

N-terminal sequences of the heavy chain of active blood factors compared with similar regions in related zymogens

Thrombin	I V E G Q D A E V G L S P W Q
Factor IXa	V V G G E D A E R G E F P W Q
Factor Xa	I V G G R D C A E G E C P W Q
Trypsinogen	I V G G Y T C G A N T V P Y Q
Chymotrypsinogen (B)	I V N G E D A V P G S W P W Q

Sequence of the active site

Thrombin	D A C E G D S G G P F V M K
Factor IXa	D S C Q G D S G G P H V T Z
Factor Xa	D A C Q G D S G G P H V T R
Trypsin (Bovine)	D S C Q G D S G G P V V - -
Chymotrypsin (Bovine)	S S C M G D S G G P L V - -

Secondary and Tertiary Structure of Blood Factors

As noted previously, not all of the blood factors are available in highly purified form and thus little structural work has been performed overall. Circular dichroism studies of factors XII, XI, X, IX, V, II, IIa, I, and Ia have been carried out. Later we shall consider the structure and function of the last two, namely fibrinogen and fibrin. The other factors show relatively low ellipticities and therefore relatively little ordered (α-helix, β-sheet) structure. The structure of a sister enzyme α-chymotrypsin is known in some detail. Indeed, based on sequence homology between α-chymotrypsin and the β-chain of thrombin, it has been suggested that the tertiary structure is very similar. Figure 15.12 shows a predicted structure for thrombin produced by comparing the insertions and deletions from the chymotrypsin sequence and predicting the conformation of the insertion by the Chou–Fasman method of Chapter 3.

Figure 15.12 Tertiary (predicted) structure of the β-chain of thrombin showing the major regions of structural difference from chymotrypsin.

Fibrinogen/Fibrin

Fibrinogen is the second most abundant protein found in human plasma and because of its abundance has been purified and examined extensively. Despite this extensive study, there is still controversy over the structure and details of the aggregation and polymerization into fibrin, the fibrous component of blood clots.

Fibrinogen has a molecular weight of ~340,000 and is composed of three types of peptide chains designated α (A), β (B), and γ. The amino ends of the three peptide chains are linked together by cystine bridges. The fibrinogen molecule is believed to be a dimer. Molecular weights of the α-, β-, and γ-chains are 63,500, 56,000, and 47,000, respectively.

Primary structure

The primary structures of the α-, β-, and γ-chains of fibrinogen have only recently been determined. Several features emerge from the residue sequence that appear to be key to the higher order structure and function of the molecule. In the N-terminal region the chains do not appear to have distinctive features indicative of secondary or tertiary structure until a sequence of CysProxxCys is encountered that acts to connect the three chains in a disulfide "knot." The knot structure is shown in Figure 15.13.

From the region of the disulfide knot proceeding toward the C-terminus of the chains there are regions of predominantly α-helix forming residues that alternate in hydrophilic and hydrophobic residues reminiscent of the heptad sequence of tropomyosin, except that the repeats are less distinct. Finally, in the C-terminal region is another disulfide knot followed by a nondistinctive sequence.

Higher Order Structure of Fibrinogen

Conformational analysis based on the primary structure suggests that 30–40% of the molecule should be α-helical and that the α-helical region may be in the form of a three-strand coiled coil. Circular dichroism studies of fibrinogen in solution show ~33% α-helix content, and x-ray diffraction studies of dried films indicate the presence of α-helix in a coiled coil form.

Direct visualization of fibrinogen by electron microscopy reveals an apparent trinodular structure, as indicated in Figure 15.14.

Although micrographs of fibrinogen have been available for many years, the relation between conformation and structure and even the location of N- and C-termini of the chains has not been clear. The most recent model suggests that the N-termini lie in the middle of the molecule, as indicated in Figure 15.15.

Fibrinogen/Fibrin Conversion

One of the final stages in the formation of a blood clot is the polymerization of fibrinogen into the fibrous protein fibrin. This process is achieved by the cleavage of peptides from the N-terminal region of α- and β-chains by the enzyme thrombin. Figure 15.16 shows the site of attack, namely the —Arg—Gly— dipeptide. In the presence of thrombin and calcium ions, the fibrinogen

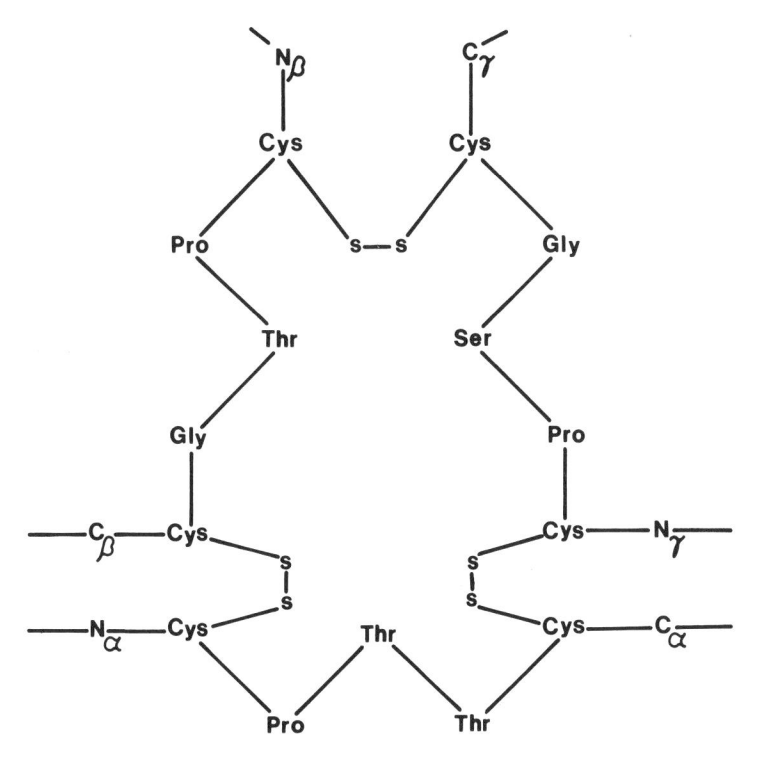

Figure 15.13 The disulfide "knot" of fibrinogen.

Figure 15.14 Electron micrograph of fibrinogen showing molecular structure and schematic model of the molecule.

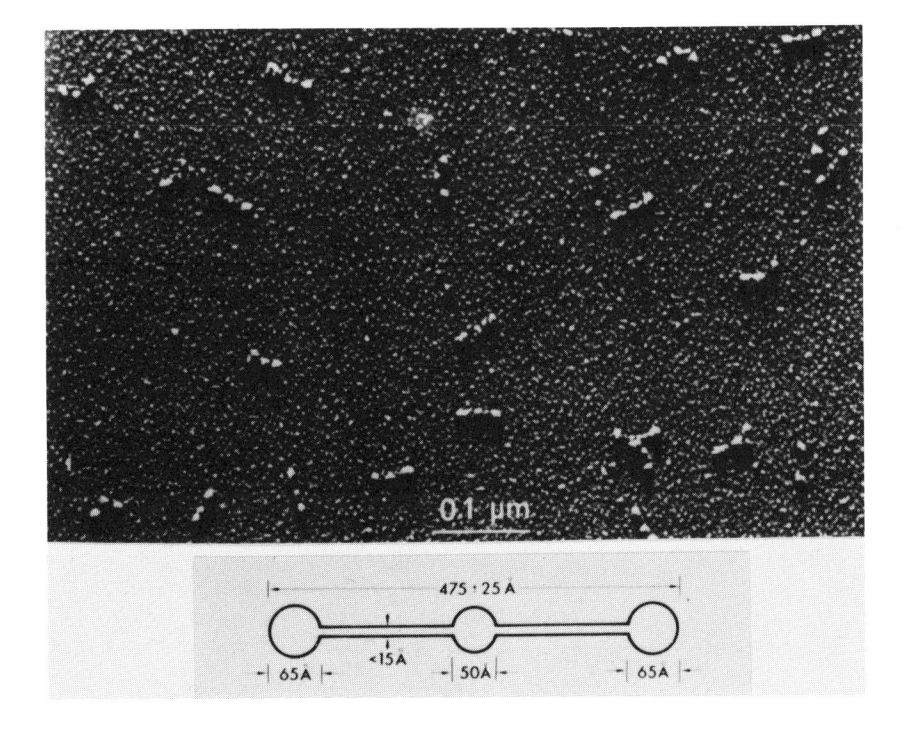

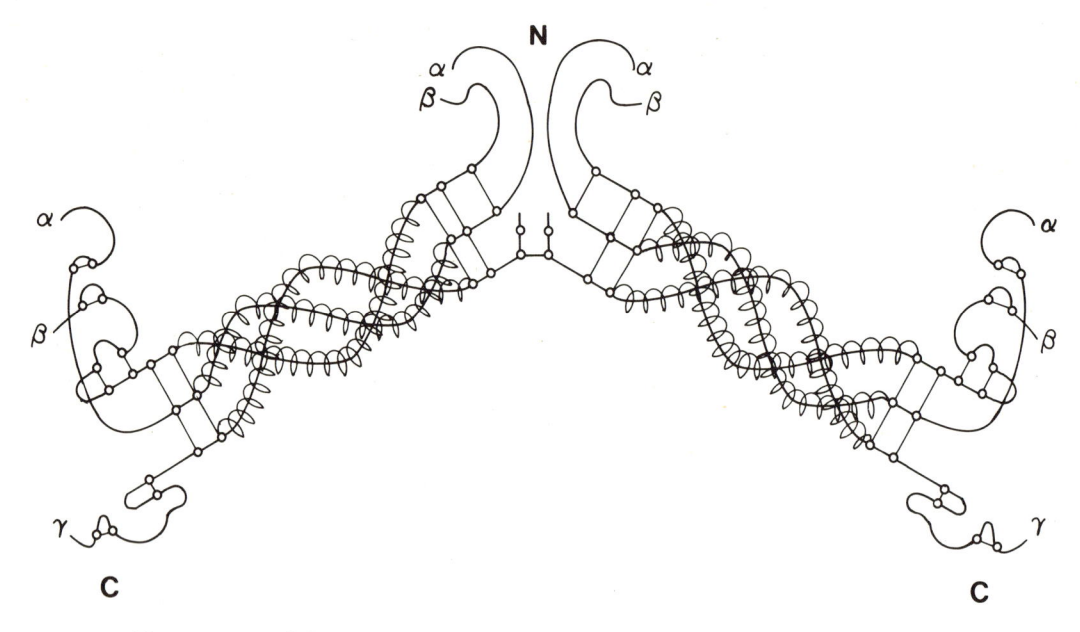

Figure 15.15 Schematic model of fibrinogen showing the location of N- and C-termini, cystine residues, and disulfide "knots." The triple-strand α-helix coiled coil lies between the knots and the major nodules lie in the C-terminal region.

Figure 15.16 Schematic diagram showing the cleavage of two molecules of peptide A and two of peptide B from bovine fibrinogen by thrombin prior to the polymerization-process.

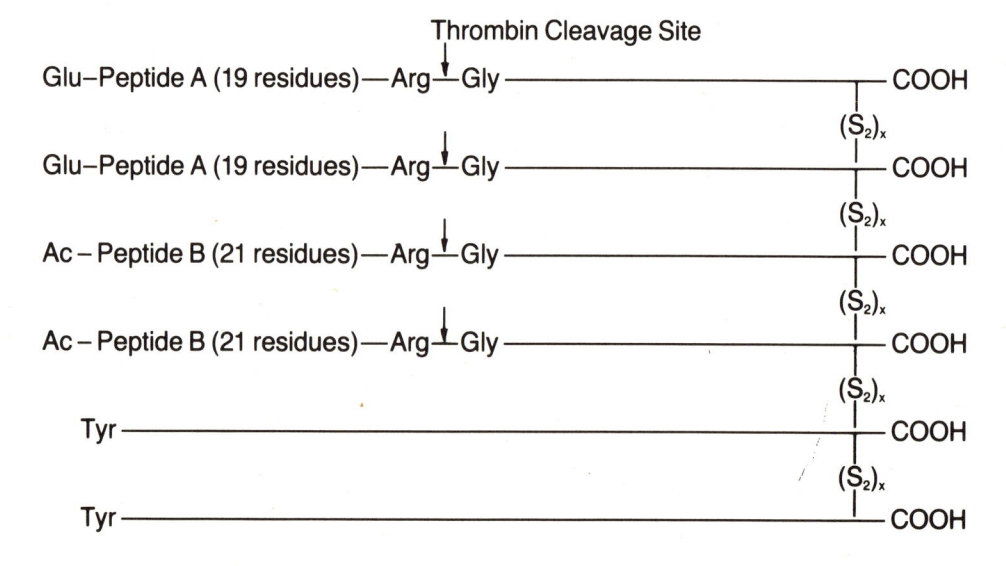

```
                          Thrombin Cleavage Site
                                    ↓
Glu–Peptide A (19 residues)—Arg—Gly——————————————COOH
                                                 |
                                               $(S_2)_x$
Glu–Peptide A (19 residues)—Arg—Gly——————————————COOH
                                                 |
                                               $(S_2)_x$
Ac–Peptide B (21 residues)—Arg—Gly———————————————COOH
                                                 |
                                               $(S_2)_x$
Ac–Peptide B (21 residues)—Arg—Gly———————————————COOH
                                                 |
                                               $(S_2)_x$
Tyr——————————————————————————————————————————————COOH
                                                 |
                                               $(S_2)_x$
Tyr——————————————————————————————————————————————COOH
```

Figure 15.17 A fibrin microcrystal.

polymerizes first into a soluble fibrin molecule, which then aggregates and becomes insoluble. In the "in vivo" process the fibrin aggregates are crosslinked by another enzyme (fibrin-stabilizing factor).

Under certain circumstances, fibrin can be produced as fibrous microcrystallites, which produce a characteristic staining pattern when viewed in the electron microscope. Such a fibrin microcrystal is shown in Figure 15.17. The repeat banding structure is 450 Å, which is comparable with the molecular length of the fibrinogen molecule, as noted in Figure 15.14. In the form of fibrin found in the normal blood-clotting process, PTA staining reveals a repeat

Figure 15.18 Blood cells (primarily erythrocytes) enmeshed with filamentous fibrin in part of a blood clot (mag. ~ 20,000×).

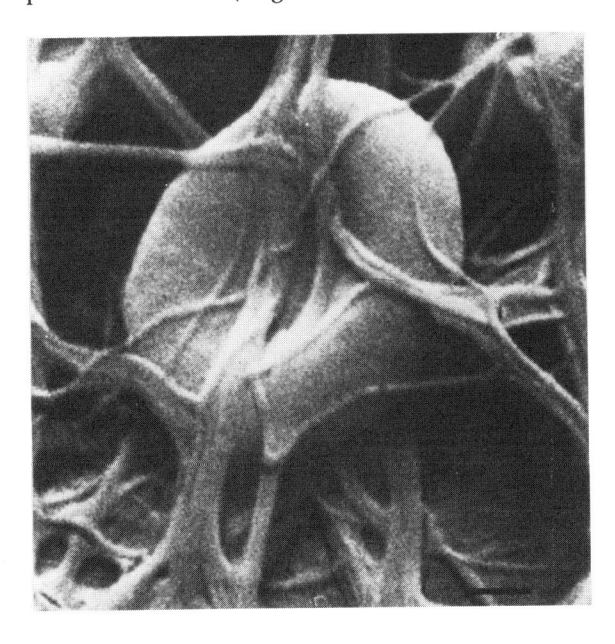

banding structure of 230–250 Å, suggesting that the molecules overlap in a half-staggered array. Usually, in clots, fibrin is found as a thin filamentous material that entraps blood cells to act as a coherent plug. Figure 15.18 shows clotted cells with some associated fibrin.

15.5 Summary

In the blood system we have seen that some protein components are abundant and are well understood, for example, hemoglobin. Others, such as fibrin, are abundant, much studied but still not fully evaluated from a structural point of view. The globular components of blood are often only poorly understood, from the abundant albumin to the rare blood-clotting factors. Hence the proteins of the blood offer many remaining challenges in understanding their chemistry, structure, and function. Some aspects will be resolved by methodology spelled out in this text; others will require new techniques and approaches. In any event there remains much fundamental research to be carried out in understanding proteins and their role in biology and medicine.

Further Reading

Doolittle, R.F. Structural aspects of the fibrinogen to fibrin conversion. *Advances in Protein Chemistry* 27:1–110, 1973.

Kabat, E.A. The structural basis of antibody complementarity. *Advances in Protein Chemistry* 32:1–76, 1978.

Perutz, M.F. Regulation of oxygen affinity of hemoglobin: Influence of structure of the globin on the heme iron. *Annual Reviews of Biochemistry* 48:327–386, 1979.

Peters, T. Serum albumin. In *The Plastic Proteins*, F.W. Putnam, ed., Academic Press, New York, 2nd ed., Vol. 1, Chapter 3, 1975.

Appendix A
Fourier Transform Analysis

Since the pioneering application in 1822 of the Fourier integral theorem to the characterization of thermal diffusion (J. Fourier, *Theorie Analytique de la Chaleur*), this mathematical transform technique has become an indispensable tool for spectral analysis of fluctuating signals. The basic notion of the technique is that any arbitrary continuous function f(x) can be represented as a particular sum of sine and cosine terms

$$f(x) = \sum_{y=0}^{\infty} (a_y \cos yx + b_y \sin yx) \tag{A.1}$$

or, equivalently,

$$f(x) = \sum_{-\infty}^{\infty} g_y \exp(iyx) \tag{A.2}$$

where y is an integer, a_y, b_y, g_y are constants. The coefficient g_y is defined by the integral

$$g_y = \frac{1}{2\pi} \int_{-\pi}^{\pi} f(x) e^{-iyx} dx. \tag{A.3}$$

In more general terms, the variables x and y may be assigned continuous values, and one may define

$$f(x) = \int_{-\infty}^{\infty} g(y) e^{ixy} dy. \tag{A.4}$$

It can be shown, conversely, that

$$g(y) = \frac{1}{2\pi} \int_{-\infty}^{\infty} f(x) e^{-ixy} dx. \tag{A.5}$$

Thus the representation of equations (A.2) and (A.3) is a limiting form of equations (A.4) and (A.5), with g(y) taken as a periodic function having period 2π.

An example of a Fourier-transform pair is shown in Figure A.1, which represents a cosine function in the x-variable and its twin, a double-valued Dirac delta function in the y-variable. This is shown to underline the idea that a Fourier-transform operation can be used to analyze the harmonic components of a fluctuating function. Any arbitrary function that is a

Figure A.1 Example of Fourier transform pair:

(1) $f(n) = \cos 2\pi y_0 n$;
(2) $g(y) = \dfrac{1}{4\pi} [\delta(y - y_0) + \delta(y + y_0)]$.

For more detail, see *The Measurement of Power Spectra*, R. B. Blackman and J. W. Tukey, Dover Publishing Inc., New York, 1958.

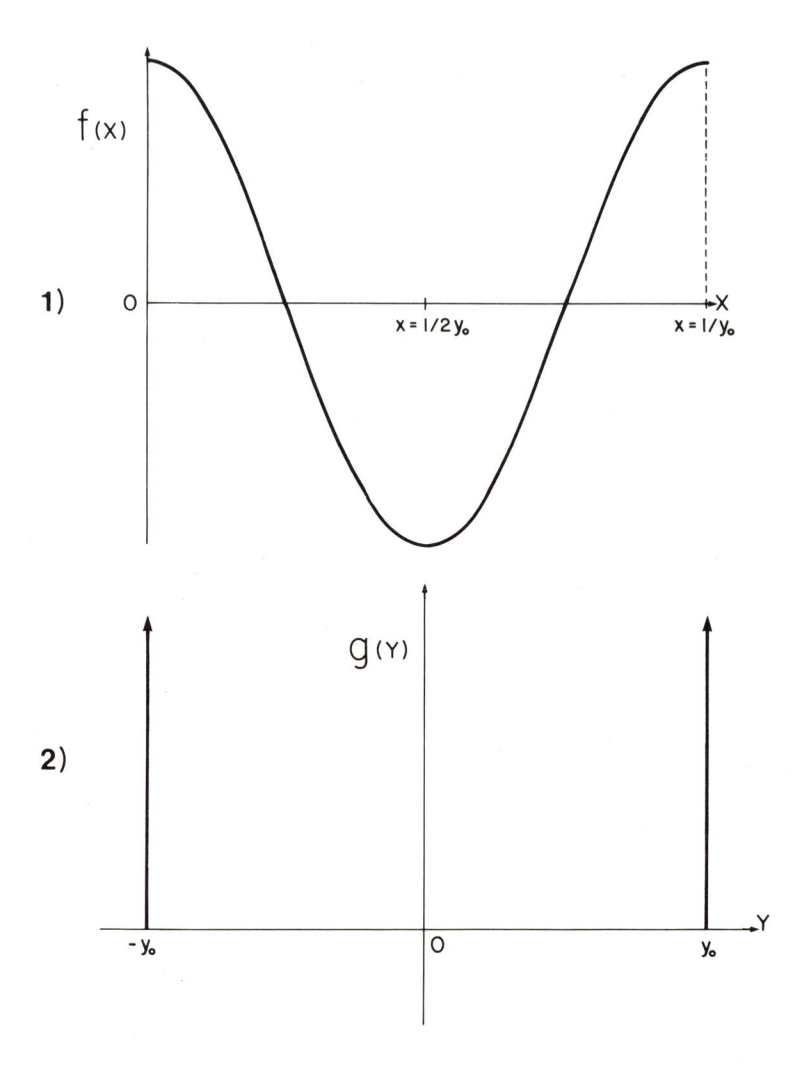

superposition of oscillating components can be transformed into a sum of delta functions, each weighted by an appropriate amplitude.

In the spectroscopic analysis of biologic macromolecules, the Fourier transform–variable pairs of interest are *time*, t (sec), *frequency*, ω (sec⁻¹), *wavelength*, λ (cm), and *wave vector*, q (cm⁻¹), respectively. As a general statement, it may be noted that most spectroscopic techniques require analysis of the frequency distribution of electromagnetic radiation sensed at a detector after interaction with a biopolymer sample, and comparison with the intrinsic frequency distribution of the radiation source. Based on the Fourier-transform theorem, it should now be apparent that the spectroscopic information can be *equivalently* described in either of two forms: in frequency, as the spectral distribution function S(ω), or in time as the temporal correlation function C(t). These are related by

$$S(\omega) = \frac{1}{2\pi} \int_{-\infty}^{-\infty} C(t) e^{-i\omega t} dt \tag{A.6}$$

and

$$C(t) = \frac{1}{2\pi} \int_{-\infty}^{\infty} S(\omega) e^{i\omega t} dt. \tag{A.7}$$

Conventional spectroscopic techniques have employed swept-filter modes of generating the spectral information in the form of S(ω). This approach involves a fundamental limitation in signal-to-noise performance based on the fact that the spectral "filter" has a finite band-width and therefore detects only a small portion of the total energy content of the signal at any instant. In ultraviolet, infrared, or Raman spectroscopy, the "filter" is an optical monochromator system that utilizes prisms or diffraction gratings to spatially "spread" or "disperse" the radiation as a function of wavelength (inverse frequency) prior to observation at the detector. The band-width or range of wavelengths (frequencies) permitted to pass to the detector is then determined by the width of a slit interposed between detector and monochromator. In NMR spectroscopy, the signal is obtained in the form of an oscillating voltage that is amplified and recorded as a function of frequency. The band-width of the spectrometer is determined by the band-widths of the radiofrequency transmitter ued to excite the sample and the amplifiers used to boost the resonance signal. Again only a small portion of the total radiative energy at any one instant is detected.

More recently it has become possible to generate S(ω) with vastly improved signal-to-noise ratios (S/N) by interfacing instrumentation that will generate and detect the correlation function C(t) with an on-line minicomputer system. The C(t) data are digitized and rapidly transferred and stored in the computer and a fast Fourier-transform operation is numerically performed (equation A-6) to obtain the S(ω). A dramatic increase in S/N is achieved because all of the energy in the detected signal at any instant is sampled in the C(t).

These operations are the basis for the techniques of Fourier-transform infrared (FTIR) and Fourier-transform NMR spectroscopic instrumentation. In

FTIR spectrometers, the incident infrared radiation is passed through not a monochromator, but an oscillating interferometer, which imposes a temporal modulation on the incident source spectrum at each wavelength. The signal that reaches the detector after passing through the sample consequently contains the spectral information as C(t). In FT NMR, a strong radiofrequency pulse is rapidly applied to the sample to saturate the spin population. The decay in time of the resulting NMR signal at the detector contains all the spectral information, that is, C(t). In each case rapid storage and averaging of the C(t) in digital form, followed by Fourier transform to S(ω), is accomplished by on-line minicomputers.

Appendix B
Symmetry and Interpretation of Structure

Symmetry considerations are fundamental to many areas of structure determination, including x-ray diffraction and most spectroscopic techniques. In fact, most of the selection rules for spectroscopy cannot be truly appreciated without a knowledge of the application of symmetry considerations and group theory. Since these topics are in themselves subject matter for an entire book and a detailed approach is far beyond the scope of the current text, we will consider only in a very simplified manner the implications of group theory and symmetry.

Symmetry Elements

If we wish to consider a small molecule or, for the case of biologic macromolecules one of the repeat units, then we can confine our attention to five types of symmetry elements, which are: (1) the center of symmetry; (2) the identity; (3) the rotation axis; (4) the mirror plane; and (5) the rotation-reflection axis. Most of these elements are self-explanatory and will be considered in terms of specific examples shortly. Each element is given a symbol: these are i, E, C, σ, and S. For the purposes of x-ray diffraction analysis, there are additional symmetry parameters to be considered, and in the spectroscopic analysis of helices we need to consider line symmetry as well as group symmetry; but for the present purposes we will consider the implications of the preceding five symmetry elements.

Example

Let us first examine some symmetry properties of the water molecule. As shown in Figure B-1, the water molecule has three of the five symmetry elements previously listed. The axis through the oxygen atom is a rotation axis (C). Rotation of 180° about this axis produces an indistinguishable con-

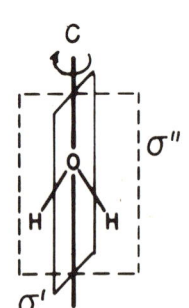

Figure B.1 The twofold rotation axis C and the two mirror planes σ' and σ'' of the water molecule.

figuration. Since the operation of two rotations about this axis brings the molecule back to its original position, this is called a C_2 axis. There are also two mirror planes indicated as σ' and σ'', which are drawn through the C_2 axis and are in the plane of the two hydrogen atoms or are 90° to that plane. Obviously, reflection of the atoms through these planes produces a mirror image. The other symmetry element present in the water molecule is the identity element E. Such an element exists for all molecules because it represents an operation that does not change the position of the atoms or molecule in any way.

Because of the combination of a C_2 axis and mirror planes that are drawn through it, the symmetry of the water molecule in shorthand notation is said to be of type C_{2v}.

Evidently, the groups that we wish to deal with in proteins are generally of low symmetry. For example, the peptide group has only a mirror plane in the plane of the peptide bond (σ) and an identity (E). In shorthand this is known as a C_s group and is non-axial.

Perhaps the most symmetrical group found in proteins is the benzene ring. The benzene molecule has a sixfold rotation axis perpendicular to the ring (C_6), indicating that each rotation of 360°/6 = 60° brings the molecule to an identical configuration. Apart from the sixfold axis, there are two threefold axes (C_3), seven twofold axes of three different types (C_2, $3C_2'$, $3C_2''$), a center of inversion (i), two threefold rotation-reflection axes (S_3), two sixfold rotation-reflection axes (S_6), and six mirror planes (σ). In a sense, then, the symmetry of the molecule represents fully its spatial characteristics. The shorthand nomenclature for the symmetry elements of benzene is D_{6h}.

Character Tables

In nearly all books on symmetry and group theory, one finds character tables. Many students who do not have a background in group theory or matrix algebra are totally intimidated by such tables. Rather than consider how they arise, we shall examine a typical table and see how it can be applied to a structural problem. Let us return to a consideration of the water molecule which, in symmetry nomenclature, is C_{2v}. If we look up the C_{2v} character table we find:

C_{2v}	E	C_2	σ'	σ''		x^2, y^2, z^2
A_1	1	1	1	1	z	
A_2	1	1	−1	−1	R_z	x, y
B_1	1	−1	1	−1	x, R_y	xz
B_2	1	−1	−1	1	y, R_x	yz

Obviously the E, C_2, σ', and σ'' are the symmetry elements that are appropriate for water and C_{2v} is the shorthand nomenclature for this combination of symmetry elements. (Similarly, if we had found all of the symmetry elements for benzene we would have found these listed in a D_{6h} character table.)

The A's and B's listed in the column under C_{2v} refer to symmetric and antisymmetric operations with respect to the rotation axis C_2, respectively. We shall consider the meaning of the numbers shortly, but they represent the number of operations necessary to produce a given configuration. Finally, the columns with x, y, z, and R refer to operations that have symmetry with respect to the x, y, z axes or a rotation about one of these axes.

Application of Symmetry Operations and Construction of Matrices

If we wish to apply symmetry theory to, say, infrared spectroscopy, then we need to be concerned with the relative motion of atoms. Let us consider first the application of some symmetry operations to the simplest molecule, H_2. If we assign a set of axes x_1, y_1, z_1 and x_2, y_2, z_2 to the hydrogen molecule shown in Figure B-2 and then perform a rotation of the molecule about the C_2 axis, we obtain the new axes x_1', y_1', z_1', and x_2', y_2', z_2'. We may now construct a table to show how the C_2 operation has changed these axes as follows:

	x_1	y_1	z_1	x_2	y_2	z_2
x_1'	0	0	0	−1	0	0
y_1'	0	0	0	0	−1	0
z_1'	0	0	0	0	0	1
x_2'	−1	0	0	0	0	0
y_2'	0	−1	0	0	0	0
z_2'	0	0	1	0	0	0

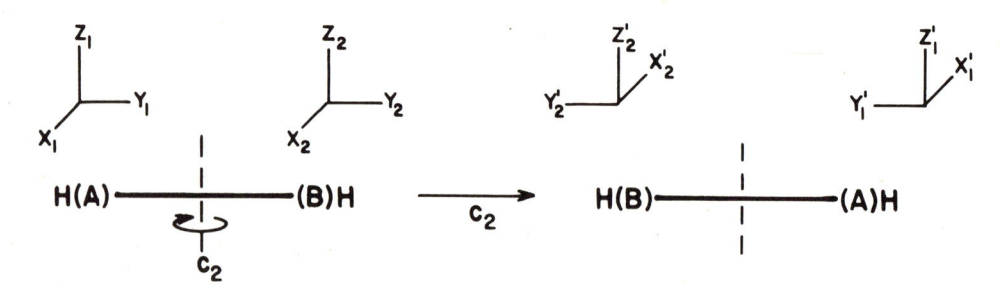

Figure B.2 Application of a C_2 rotation to a hydrogen molecule.

Across the top we have the original axes (row) and in the vertical column, the new axes. We now make the assumption that the C_2 operation either leaves the direction of the axis the same (+1), or changes it to the opposite direction (−1). In addition, we can assign a zero (0) if two parameters are unrelated to an operation. Thus if we examine the first row, x_1', we see that the C_2 operation produced $-x_2$ from x_1' and is assigned -1. However, there is no relation between x_1' and x_1, z_1, y_2 and z_2 so they are assigned 0. Similarly for the other rows; we note, however, that the z's are not changed in direction and are thus assigned $+1$'s. We are thus able to complete the matrix for a C_2 operation on the two hydrogen atoms A and B. It is readily possible to complete other matrices corresponding to other symmetry operations such as E and σ.

Now there are many properties of matrices we should be aware of, but two that will be mentioned here are: (1) multiplication of the matrix by the letter representation in the first row is equivalent to the production of the letter representation in the first column, that is,

$$\begin{vmatrix} x_1' \\ y_1' \\ \cdot \\ \cdot \\ \cdot \end{vmatrix} = M \begin{vmatrix} x_1 \\ y_1 \\ \cdot \\ \cdot \\ \cdot \end{vmatrix}$$

and (2) the sum of the diagonal numbers in the matrix is known as the trace and is written χ. For our six-dimensional matrix $\chi = 0$, that is, $\chi(C_2) = 0$. If we had chosen the identity operation E we would have found $\chi(E) = 6$, and so on. Applying an identical procedure for water we would find

	E	C_2	σ_v''	σ_v'	
χ_T	9	−1	+3	+1	(α)

(The trace $\chi(E)$ always has the same numerical value as the dimension of the matrix.)

Application to Vibrational Spectroscopy

For the water molecule, there are three atoms that may move in any one of three directions (x, y, z), producing $3 \times 3 = 9$ combinations of atomic movement. For a molecule containing N atoms there are 3N combinations. When each of the atoms is moving in, say, the x direction simultaneously, translation of the molecule occurs. Therefore there are three effective translation modes (x, y, z). Similarly, three combinations of atomic motion (for a nonlinear molecule) give rise to rotational motion, so that there are said to be $3N - 6$ vibrational modes in which the atoms are traveling in a non-cooperative fashion. For the water molecule there are then $9 - 6 = 3$ vibrational modes. In terms of our symmetry nomenclature we say that there are irreducible symmetry representations (Γ) and that

$$\Gamma_{motion} = \Gamma_{rotation} + \Gamma_{translation} + \Gamma_{vibration}.$$

To find out what these irreducible representations are and how to assign them, we must use the character table and our calculations for the values of the matrix traces (χ), and then apply what is known as a decomposition formula, which is

$$a_i = \frac{1}{h} \sum g \chi_i(R) \chi_T(R)$$

where a_i is the number of irreducible representations of the i-th type, h is the order of the group, g is the number of elements in a given class, χ_i is obtained from character tables, and χ_T is our calculated trace value. To see how this works, let us apply it to the water molecule. In the C_{2v} character table we have one E operation, one C_2, one σ' and one σ'', for a total of four discrete operations. The order of the group h is thus equal to four.

Writing out the decomposition formula for the A_1 mode (in the character table) we have

$$a_{A_1} = \frac{1}{4} [g\chi_{A_1}(E)\chi_T(E) + g\chi_{A_1}(C_2)\chi_T(C_2) + g\chi_{A_1}(\sigma_v')\chi_T(\sigma_v)$$
$$+ g\chi_{A_1}(\sigma_v'')\chi_T(\sigma_v'')]$$
$$= \frac{1}{4}[(1 \times 1 \times 9) + (1 \times 1 \times -1) + (1 \times 1 \times 1) + (1 \times 1 \times 3)] = 3$$

↑ there is only one E element
↑ only one C_2
↑ one σ_v'
↑ one σ''

The number of elements g is one in each case (remember this was not so for benzene). The second number (1) in the series of three numbers multiplied together is obtained from the character table from the row for A_1. The third number is the calculated matrix trace.

Similarly, for the four rows of the character table we obtain $a_{A_2} = 1$, $a_{B_1} = 3$, $a_{B_2} = 2$, so that the irreducible representation for the motion of molecules in water would be

$$\Gamma_{motion} = 3A_1 + A_2 + 3B_1 + 2B_2$$

We notice that there are nine irreducible representations corresponding to the nine (3N) motions of atoms in the water molecule. We now have to sort out which is which. Since we know that translation of the molecule involves motion in the x, y, or z direction, we examine the character table and find that A_1 transforms as z, B_1 as x, and B_2 as y, and thus the irreducible representation for translation could be written

$$\Gamma_{trans} = A_1 + B_1 + B_2$$

Similarly, rotation from the character table involves R_x, R_y, and R_z so

$$\Gamma_{rot} = A_2 + B_1 + B_2.$$

Therefore

$$\Gamma_{vib} = 2A_1 + B_1$$

Since the A vibrations are symmetrical to the C_2 axis of the molecule, they correspond to modes of the type

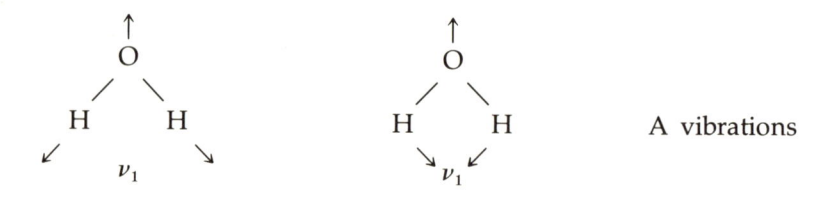

A vibrations

whereas B vibrations are antisymmetric:

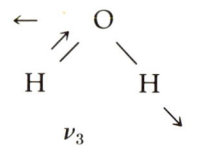

However, the fact that there are three vibrations does not mean they are all infrared or Raman active. We have seen (Chapters 6 and 7) that infrared-active vibrations involve a change in dipole moment and Raman lines, a change in polarizability. The dipole moment of a molecule is affected by the same symmetry operations as x, y, z. Thus if the normal mode is of the same representation as x, y, z it will be infrared active. Looking again at the C_{2v} character table, we see that A_1, B_1, and B_2 modes are infrared active.

Raman activity requires terms in x^2, y^2, z^2, xy, yz, xz, that is, A_1, A_2, B_1, and B_2. Therefore all three vibrations are infrared and Raman active. As a general rule, when there is no center of symmetry the infrared and Raman bands (lines) are the same.

In a similar manner, the infrared and Raman bands of all biologic elements may be established.

Overtones and Combinations

It might be supposed from the preceding discussion that there will be only three bands predicted for the infrared or Raman spectrum of water. This, however, is not true, since overtone and combination bands are allowed. The selection rules are: *If the character product of combination or overtone bands is the same as the character of the allowed fundamentals, then the overtone is allowed.*

Let us again perform the simple calculation for the water molecule. Suppose we wish to know whether there will be an allowed overtone of the A_1 band. We again refer to the character table for C_{2v} and ask whether $\chi_{A_1} \times \chi_{A_1} = \chi_{A_1}^2$ is allowed.

	E	C_2	σ_v'	σ_v''
χ_{A_1}	1	1	1	1
χ_{A_1}	1	1	1	1
$\chi_{A_1}^2$	1	1	1	1

Note that this is not an addition table but that the numbers below the line are the product of the two numbers above the same column. Since $\chi_{A_1}^2$ has the same character as χ_{A_1}, it is allowed. Similarly A_2, which is not an allowed fundamental, has an allowed overtone because

	E	C_2	σ_v'	σ_v''
χ_{A_2}	1	1	-1	-1
χ_{A_2}	1	1	-1	-1
	1	1	1	1

That is, $\chi_{A_2}^2$ has the same character as χ_{A_1} and is allowed. By the same methodology, allowed and unallowed combination bands may be calculated.

Helices and Line Symmetry

We note that poly(amino acid) helices have symmetry that is not of the same type as that discussed in the aforementioned examples. A polyproline II helix, for example, has a threefold *screw* axis rather than a C_3 axis. This means that the symmetry depends upon a combination of rotation and translation. Without going into detail, the vibrational analysis of such a helix, which is said to have *line* symmetry, may be treated in a similar manner to the previous examples showing *group* symmetry. Thus if we look in the character tables for a C_3 symmetry we find that there are two characters, A and E. For less symmetrical helices we always find the appearance of A and E characters and these are the origin of A and E assignments made in Chapters 6 and 7.

Symmetry and Electronic Spectroscopy

In the formulation of selection rules for electronic spectroscopy, the same types of symmetry considerations are observed except this time it is necessary to consider the orbital symmetry and the result of symmetry operations instead of atomic coordinates. As an example, let us consider a molecule of formaldehyde (which has certain similarities with the carbonyl group of the peptide bond). It is convenient to use formaldehyde as an example because it also has C_{2v} symmetry and we can use the same character table as before.

Let us consider the molecular orbitals of formaldehyde in the carbonyl bond. We may consider the sharing of two electrons from carbon and oxygen as forming a σ-bond (not to be confused with σ-mirror planes) and one electron each from carbon and oxygen in a π-bond. In Figure B-3 the ground σ- and π-bonded states are shown, along with the excited σ^* and π^* states. In addition, the oxygen p-electron nonbonded states are shown. By convention the atoms are chosen to be in the yz plane. If the operations E, C_2, $\sigma_v'(xz)$, and $\sigma_v''(yz)$ are performed on the π-orbital, the result is $+1$, -1, $+1$, -1; that is, a -1 operation changes the sign of the wave function, while $+1$ leaves the sign unchanged. This result is the same as that listed for B_1 in the C_{2v} character table and the orbital is said to be of the symmetry species b_1. Similarly, if the n_a, n_b, π^*, σ, and σ^* orbitals of formaldehyde are subjected to the above symmetry operations it can be shown that the orbitals belong to the irreducible representations (or transform as) a_1, b_2, b_1, a_1, and a_1 respectively.

If we wish to examine the symmetry species of a state, it is found by taking the product of the odd electron state symmetry. For example, the n → π^* transition is b_2 → b_1.

Figure B.3 Representation of the molecular orbitals of the C=O bond of formaldehyde. The dotted and solid lines represent portions of the wave function of opposite sign (dashed line = negative).

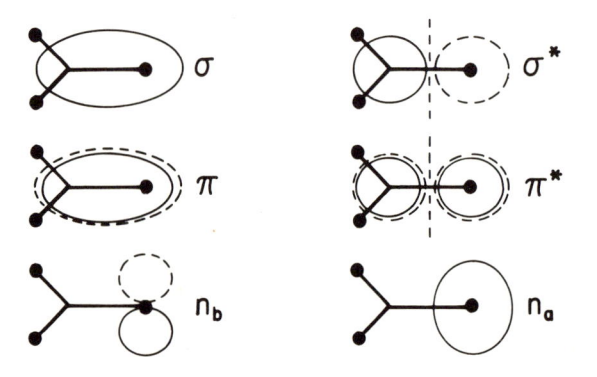

	E	C_2	$\sigma_v'(xz)$	$\sigma_v''(yz)$	
$b_1 \times b_2$	1×1	-1×-1	1×-1	$-1 \times +1$	
Result	$+1$	$+1$	-1	-1	$= A_2$

The resulting irreducible representation is A_2 and the excited state is described as A_2. The transition is often written $^1A_2(n,\pi^*) \leftarrow {}^1A_1$ where the excited state is written first and the spin multiplicity is included.

Allowed Transitions

In general it may be stated that if there is no charge displacement induced by incident light during a transition, then that transition will be forbidden. Charge displacement is represented by the transition moment integral (Chapter 8), which can be broken down into various components $\int \psi_g \hat{M}_{x,y,z} \psi_e dv$, where ψ_g and ψ_e are the wave functions for ground and excited states, $\hat{M}$ is a vector operator, and dv a volume element. Thus, for an allowed transition, the integral has to be nonzero in one or more of the x, y, and z directions.

Although this integral is complicated, a simple approach is made possible by recognizing that $\hat{M}_x$ transforms as the x vector in the appropriate character table; similarly with $\hat{M}_y$ and $\hat{M}_z$. Consequently, for formaldehyde we note from the C_{2v} character table that A_1, B_1, and B_2 transform as z, x, and y. Since we have seen that the $n - \pi^*$ transition involves an excited state of A_2, the transition is *not* allowed. Conversely, the $\pi - \pi^*$ transition involves an excited A_1 state and is therefore allowed.

In addition, transitions between states of different multiplicity are forbidden.

There are many additional complexities that enter into the prediction of allowed and forbidden states, particularly those involving simultaneous changes in vibrational energy (vibronic states). However, the role of symmetry is again central to the derivation of these rules, which follow similar principles.

Symmetry in Crystallography

Whereas with small molecules or parts of larger molecules we deal with point groups, there are more symmetry operations involved with crystal arrangements that define a *space group*. Two new symmetry elements that must be introduced are the screw axis and the glide plane. The screw axis has already been discussed in terms of the properties of helices. For example, if one has a 3_1 helix in which all residues are equivalent, then a rotation of 120° followed by a translation of ⅓ of the helix repeat brings residues into an equivalent position. Thus the helix has a threefold screw axis. Similarly in the unit cell, for an n-fold screw axis, rotation of $2\pi/n$ followed by translation of $1/n$ of the operation brings the structure into an equivalent configuration. The glide operation involves reflection in a plane followed by translation of half a unit cell in some direction in the same plane.

A third aspect of space group symmetry not found with point groups is involved with centered lattices. Lattices may be face centered (abbreviated F), body centered (I), or side centered (A, B, or C depending on which side).

Those unit cells that are not centered are said to form a primitive lattice (denoted P).

Space group symbols are written to identify one of the 230 different space groups. For example, Pnma means that the space group would have a **p**rimitive lattice, an **n**-glide perpendicular to the unit cell a-axis, a **m**irror plane perpendicular to the b-axis, and an **a**-glide perpendicular to the c-axis.

Descriptions of the symmetry elements for the various space groups are to be found in x-ray diffraction texts.

Further Reading

Drago, R.S. *Physical Methods in Chemistry.* W. B. Saunders Co., Philadelphia, 1977.

Holmes, K.C. and D.M. Blow. In *Methods of Biochemical Analysis,* D. Glick, ed. Interscience, New York, Vol. 13, p. 113, 1965.

Vincent, A. *Molecular Symmetry and Group Theory.* Wiley, New York, 1977.

Illustration and Table Credits

Figure 1.8 from page 438 in H. D. Law, *Amino Acids, Peptides, and Proteins*, 1974. Reproduced with permission of the Royal Society of Chemistry.

Figures 2.6, 2.8, 2.16, 2.17, and **2.18** from pages 13, 13, 29, 34, and 35 in R. E. Dickerson and I. Geis, *The Structure and Action of Proteins*, 1969. Reproduced with permission of Harper and Row.

Figure 2.30 from page 46 in B. Lotz et al., *First Cleveland Symposium on Macromolecules* (A. G. Walton, ed.), 1977. Reproduced with permission of Elsevier North-Holland.

Table 2.1 from page 77 in G. N. Ramachandran, *Structural Chemistry and Molecular Biology* (A. Rich and N. Davidson, eds.), 1968. Reproduced with permission of W. H. Freeman and Company.

Table 2.2 from page 351 in A. G. Walton and J. Blackwell, *Biopolymers*, 1973. Reproduced with permission of Academic Press, Inc.

Figures 3.4, 3.5, 3.6, and **3.9** from pages 68 and 96 in M. Levitt, *Journal of Molecular Biology* 104, 1976. Reproduced with permission of Academic Press, Inc. (London), Ltd.

Table 3.2 from page 1202 in G. D. Fasman and P. Y. Chou, *Biophysical Journal* 16, 1976. Reproduced with permission of Biophysical Society.

Table 3.3 from page 71 in P. Y. Chou and G. D. Fasman, *Advances in Enzymology* 27, 1978. Reproduced with permission of John Wiley & Sons, Inc.

Table 3.7 from page 140 in E. A. Kabat et al., *Nature* 250, 1974. Reproduced with permission of Macmillan Journals Ltd.

Table 3.8 from 570 in W. A. Hiltner and A. G. Walton, *Journal of Molecular Biology* 92, 1975. Reproduced with permission of Academic Press Inc. (London), Ltd.

Figure 4.10 from page 56 in L. Alexander, *X-ray Diffraction Methods in Polymer Science*, 1969. Reproduced with permission of John Wiley & Sons, Inc.

Figure 4.15 from page 538 in L. Brown and I. F. Trotter, *Transactions of the Faraday Society* 52: 537, 1956. Reproduced with permission of authors.

Figure 4.16 from page 519 in A. G. Walton and J. Blackwell, *Biopolymers*, 1973. Reproduced with permission of Academic Press, Inc.

Figure 4.20 from page 443 in M. S. Weininger and L. J. Banaszak, *Journal of Molecular Biology* 119, 1978. Reproduced with permission of Academic Press, Inc. (London) Ltd.

Figures 4.21, 4.22, 4.23, 4.24, 4.25, 4.26, 4.27, 4.28, and **4.29** from various pages in J. Richardson, *The Protein Structure Coloring Book,* 1979. Reproduced with permission of author.

Figure 5.1 from pages 137, 144, and 148 in A. G. Walton and J. Blackwell, *Biopolymers*, 1973. Reproduced with permission of Academic Press, Inc.

Figure 5.2 from page 42 in B. Lotz, in A. G. Walton (ed.), *First Cleveland Symposium of Macromolecules,* 1977. Reproduced with permission of Elsevier/North-Holland, Inc.

Figure 5.3 from page 172 in B. Lotz, *Journal of Molecular Biology* 87, 1974. Reproduced with permission of Academic Press Inc. (London), Ltd.

Figure 5.4 (**a** and **b**), **5.5,** and **5.7** from pages 62, 307, and 444 in R. D. B. Fraser and T. P. MacRae, *Conformation in Fibrous Proteins,* 1973. Reproduced with permission from Academic Press, Inc.

Figure 5.6 from plate 3, page 235 in P. N. T. Urwin, *Journal of Molecular Biology* 98, 1975. Reproduced with permission from Academic Press Inc. (London), Ltd.

Figure 5.10 courtesy Dr. James Lake, New York University Medical School.

Figures 6.14, 6.15, 6.16, and **6.17** from pages 474, 319, and 474 in R. D. B. Fraser and T. P. MacRae, *Conformation of Fibrous Proteins,* 1973. Reproduced with permission of Academic Press, Inc.

Figure 6.17 courtesy of Prof. J. L. Koenig, Department of Macromolecular Science, Case Western Reserve University.

Figure 7.3 from page 2163 in T. J. Yu et al., *Biopolymers* 12, 1973. Reproduced with permission of John Wiley & Sons, Inc.

Figure 7.4 from page 7076 in J. L. Lippert et al., *Journal of the American Chemical Society* 98, 1976. Reproduced with permission of the American Chemical Society.

Figure 7.5 from page 192 in A. C. Walton and J. L. Blackwell, *Biopolymers*, 1974. Reproduced with permission of Academic Press, Inc.

Figure 7.6 courtesy J. L. Koenig, Department of Macromolecular Science, Case Western Reserve University.

Figure 7.7 from page 2390 in P. R. Carey et al., *Biochemistry* 15, 1976. Reproduced with permission of the American Chemical Society.

Tables 7.3–7.6 from pages 7078–7079 in J. L. Lippert et al., *Journal of the American Chemical Society* 98, 1976. Reproduced with permission of the American Chemical Society.

Figure 8.20 from page 1133 in J. Brahms et al., *Proceedings of the National Academy of Sciences* (USA) 60, 1968. Reproduced with permission of the authors.

Illustrations and Table Credits

Figures 9.1 and **9.4** from pages 206 and 211 in K. E. Van Holde, *Physical Biochemistry*, 1971. Reprinted with permission of Prentice-Hall, Inc.

Figures 9.8 and **9.9** from pages 4110 and 4111 in N. J. Greenfield and G. D. Fasman, *Biochemistry* 8, 1969. Reproduced with permission of the American Chemical Society.

Figure 9.10 from page 971 in V. P. Saxena and D. B. Wetlaufer, *Proceedings of the National Academy of Sciences* (USA) 68, 1971. Reproduced with permission of authors.

Table 9.2 and **9.3** from pages 4109 and 4113 in N. J. Greenfield and G. D. Fasman, *Biochemistry* 8, 1969. Reproduced with permission of the American Chemical Society.

Figures 10.1, 10.2 and **10.7** from pages 148, 152, 158 and **Tables 10.2** and **10.3** from pages 67 and 74 in S. Answorth, *Introduction to the Spectroscopy of Biological Polymers* (D. W. Jones, ed.), 1976. Reproduced with permission of Academic Press, Inc.

Figures 10.3, 10.4, 10.5, and **10.6** from pages 10, 52, 54, 57 and **Table 10.1** from page 10 in S. V. Konev, *Fluorescence and Phosphorescence of Proteins and Nucleic Acids*, 1967. Reproduced with permission of Plenum Publishing Corp.

Figure 10.8 from page 173 in K. E. Van Holde, *Physical Biochemistry*, 1971. Reproduced with permission of Prentice-Hall, Inc.

Figure 10.9 from page 584 in J. Yguerabide et al., *Journal of Molecular Biology* 51 1970. Reproduced with permission of Academic Press Inc. (London), Ltd.

Figure 10.11 from page 977, and **Table 10.4** from page 978 in P. W. Schiller, *Proceedings of the National Academy of Sciences* (USA) 69, 1972. Reproduced with permission of author.

Figure 10.15 from page 5815 in R. F. Chen and J. C. Kernohan, *Journal of Biological Chemistry* 242, 1967. Reproduced with permission from the American Society of Biological Chemists, Inc.

Figures 11.2 and **11.3** from page 112 in R. M. Silverstein and J. C. Brassler, *Spectrometric Identification of Organic Compounds*, 1974. Reproduced with permission from John Wiley & Sons, Inc.

Figure 11.4 courtesy of Varvian Associates, Palo Alto, California.

Figure 11.5 and **11.8** from pages 892 and 894 in F. R. N. Gurd and P. Klein, *Methods in Enzymology XXVII*, 1973. Reproduced courtesy of Academic Press, Inc.

Figure 11.9 from page 70 in P. F. Knowles et al., *Magnetic Resonance of Biomolecules*, 1976. Reproduced courtesy of John Wiley & Sons Ltd.

Figure 11.10 from page 467 in J. J. Villafrance and R. E. Viola, *Archives of Biochemistry and Biophysics* 160, 1974. Reproduced with permission of author.

Figure 11.11b from page 204 in M. Ptak, *Jerusalem Symposium on Quantum Chemistry and Biochemistry*, Vol. 5, 1973. Reproduced with permission of the Israel Academy of Sciences and Humanities.

Figure 11.12 from page 527 in G. E. K. Roberts and O. Jardetzky, *Advances in Protein Chemistry* 24, 1970. Reproduced with permission of the author.

Figure 11.13 from page 260 in O. Jardetzky et al., *Cold Spring Harbor Symposium on Quantitative Biology*, 36, 1971. Reproduced with permission of Cold Spring Harbor Laboratory.

Figures 11.14b and **11.15** from pages 142 and 143 in G. Warner et al., *Biophysical Structural Mechanisms* 2, 1976. Reproduced with permission of Springer-Verlag.

Figures 12.1, 12.4a, 12.6, 12.9, 12.32, 12.36, and **Table 12.1** from pages 303, 311, 320, 323, 329, 332, and 324 in A. G. Walton and G. L. Blackwell, *Biopolymers,* 1973. Reproduced with permission of Academic Press, Inc.

Figure 12.2 from page 2210 in F. J. Castellino and R. Barker, *Biochemistry* 7, 1968. Reproduced with permission of the American Chemical Society.

Figure 12.3 from page 3689 in G. Guidotti, *Journal of Biological Chemistry* 242, 1967. Reproduced with permission of the American Society of Biological Chemists, Inc.

Figure 12.5 from page 255 in E. P. Pittz and S. N. Timashiff, *Advances in Enzymology XXVII,* 1973. Reproduced with permission of Academic Press, Inc.

Figure 12.7 from page 483 in E. L. Uhlenhapp and B. H. Zinn, *Advances in Enzymology XXVII,* 1973. Reproduced with permission of Academic Press, Inc.

Figure 12.11 from page 406 in R. T. Holzbach et al., *Micellization, Solubilization and Microemulsions* (K. L. Mittal, ed.), 1977. Reproduced with permission of Plenum Publishing Corp.

Figure 12.15 from page 353 in G. K. Akers, *Advances in Protein Chemistry* 24, 1970. Reprinted with permission of Academic Press, Inc.

Figure 12.16 from page 4992 in W. W. Fish et al., *Journal of Biological Chemistry* 244, 1969. Reprinted with permission of the American Society of Biological Chemists, Inc.

Figure 12.17 from page 597 in P. Andrews, *Biochemical Journal* 96, 1965. Reprinted with permission of the Biochemical Society.

Figure 12.18 and **Table 12.2** from pages 4410 and 4411 in K. Weber and M. Osborn, *Journal of Biological Chemistry* 244, 1969. Reproduced with permission of the American Society of Biological Chemists.

Figure 12.19 from page 393 in D. C. Teller, *Methods in Enzymology XXVII,* 1973. Reproduced with permission of Academic Press, Inc.

Figure 12.20 from page 1612 in K. C. Aune and S. N. Timashiff, *Biochemistry* 10, 1971. Reproduced with permission of American Chemical Society.

Figure 12.21 from page 312 in A. G. Walton and G. L. Blackwell, *Biopolymers,* 1973. Reproduced with permission of Academic Press, Inc.

Figure 12.22 from page 1376 in A. M. Holtzer and S. Lowey, *Journal of the American Chemical Society* 81, 1959. Reproduced with permission of the American Chemical Society.

Figure 12.23 from page 443 in E. P. Geiduschek and A. Holtzer, *Advances in Biology, Medicine and Physiology* 6, 1958. Reproduced with permission of the author.

Figure 12.24 from page 2553 in H. Boedtker and N. S. Simmons, *Journal of the American Chemical Society* 80, 1958. Reproduced with permission of the American Chemical Society.

Figures 12.25 and **12.26** from pages 776 and 777 in H. Fujita et al., *Biopolymers* 4, 1966. Reproduced with permission of John Wiley and Sons, Inc.

Illustrations and Table Credits

Figure 12.27 from page 759 in W. Burchard and J. M. G. Cowie, *Light Scattering from Polymer Solutions* (M. B. Hugbin, ed.), 1972. Reproduced with permission of Academic Press, Inc.

Figure 12.28 and **Table 12.4** from pages 158 and 200 in H. Pessen et al., *Advances in Enzymology XXVII*, 1978. Reproduced with permission of Academic Press, Inc.

Figure 12.29 from page 170 in J. Witz et al., *Journal of the American Chemical Society* 86, 1964. Reproduced with permission of the American Chemical Society.

Figures 12.30, 12.31, 12.33, 12.35, and **12.38,** and **Table 12.5** from pages 335, 326, 441, 112, 383, and 395 in C. Tanford, *Physical Chemistry of Macromolecules*, 1961. Reproduced with permission of John Wiley & Sons, Inc.

Figure 12.34 from page 746 in P. Moser et al., *Journal of Physical Chemistry* 70, 1966. Reproduced with permission of the American Chemical Society.

Figure 12.37 from page 476 in G. Sprach et al., *Journal of Molecular Biology* 7, 1963. Reproduced with permission of Academic Press Inc. (London), Ltd.

Figure 12.39 from page 284 in L. M. Gilbert and G. A. Gilbert, *Methods in Enzymology XXVII*, 1973. Reproduced courtesy of Academic Press, Inc.

Figure 12.40 from page 2242 in T. J. Herbert and F. D. Carlson, *Biopolymers* 10, 1971. Reproduced with permission of John Wiley & Sons, Inc.

Figures 13.3a and **13.8** from pages 352 and 398 in R. D. B. Fraser and T. P. MacRae, *Fibrous Proteins*, 1973. Reproduced with permission of Academic Press, Inc.

Figure 13.4 from page 107 in G. N. Romachandran, *Collagen*, 1967. Reproduced with permission of Academic Press, Inc. (London), Ltd.

Figure 13.9 and **Table 13.1** from pages 139 and 143 in D. J. S. Hulmes, *Journal of Molecular Biology* 79, 1973. Reproduced with permission of Academic Press Inc. (London), Ltd.

Figure 13.10 from page 144 in A. G. Walton, *Structure of Fibrous Biopolymers*, Colston Papers No. 26, 1975. Reproduced with permission of Butterworths, Ltd.

Figures 14.2, 14.4, 14.13 from pages 38, 38, and 39 in C. Cohen, *Scientific American* 233, 1975. Reproduced with permission of W. H. Freeman and Co.

Figure 14.3 from page 304 in H. E. Huxley, *The Structure and Function of Muscle* (G. H. Bourne, ed.), 1972. Reproduced with permission of Academic Press, Inc.

Figures 14.5, 14.9 (a, b), 14.11 (b) from pages 428, 442, and 437 in R. D. B. Fraser and T. P. MacRae, *Conformation of Fibrous Proteins*, 1973. Reproduced with permission of Academic Press, Inc.

Figure 14.6 from plate IIa in J. Kendrick Jones et al., *Journal of Molecular Biology* 59, 1971. Reproduced with permission of Academic Press Inc. (London), Ltd.

Figure 14.7 from plate IIIb in S. Lowey et al., *Journal of Molecular Biology* 23, 1967. Reproduced with permission of Academic Press Inc. (London), Ltd.

Figure 14.8 from plate X in H. E. Huxley, *Journal of Molecular Biology* 7, 1963. Reproduced with permission of Academic Press Inc. (London), Ltd.

Figure 14.9 (c) from page 383 in H. E. Huxley and W. Brown, *Journal of Molecular Biology* 30, 1967. Reproduced with permission of Academic Press Inc. (London), Ltd.

Figures 14.10 and **14.11a** from pages 53 and plate IIIb in J. Hanson and J. Lowy, *Journal of Molecular Biology* 6, 1963. Reproduced with permission of Academic Press Inc. (London), Ltd.

Figure 14.12 from plate IIa in P. B. Moore et al., *Journal of Molecular Biology* 50, 1970. Reproduced with permission of Academic Press Inc. (London), Ltd.

Figure 14.16 from page 1204 in J. R. Pearlstone et al., *Proceedings of the National Academy of Sciences* 73, 1976. Reproduced with permission of the authors.

Figure 14.17 and **Table 14.5** from pages 142 and 138 in J. M. Wilkinson and R. J. A. Grand, *FEBS Proceedings* 31, 1975. Reproduced with permission of the Federation of European Biochemical Societies.

Figure 14.18 from page 19 in H. E. Huxley, *Scientific American* 213, 1965. Reproduced with permission of W. H. Freeman and Co.

Table 14.1 from page 15 in S. Lowey et al., *Journal of Molecular Biology* 42, 1969. Reproduced with permission of Academic Press Inc. (London) Ltd.

Table 14.2 from page 2689 in M. Elzinga et al., *Proceedings of the National Academy of Sciences* (USA) 70, 1973. Reprinted with permission of the authors.

Table 14.3 from page 127 in D. Stone et al., *FEBS Proceedings* 31, 1974. Reproduced with permission of the Federation of European Biochemical Societies.

Table 14.4 from page 1904 in J. R. Pearlstone et al., *Proceedings of the National Academy of Sciences* (USA) 73, 1976. Reproduced with permission of the authors.

Table 14.6 from page 6357 in J. H. Collins et al., *Journal of Biological Chemistry* 252, 1977. Reproduced with permission of the American Society of Biological Chemists, Inc.

Figure 15.3 and **Table 15.3** from pages 211 and 210 in T. L. Blundell, *Amino-acids, Peptides and Proteins*, Vol. 4, 1972. Reproduced with permission of The Chemical Society.

Figure 15.7 from page 4628 in M. Schiffer et al., *Biochemistry* 12, 1973. Reproduced with permission of the American Chemical Society.

Figure 15.8 from page 4300 in D. M. Segal et al., *Proceedings of the National Academy of Sciences* 71, 1974. Reproduced with permission of the authors.

Figure 15.9 from page 22 in J. L. Fox, *Chemical and Engineering News* 57, 1979. Reproduced with permission of The American Chemical Society.

Figure 15.10 from page 591 in P. A. Behrens et al., *Federation Proceedings* 34, 1975. Reproduced with permission of the Federated American Society of Experimental Biology.

Figure 15.12 from page 76 in A. G. Walton et al., *Proceedings of the International Symposium on Biomolecular Structure* (R. Srinivasan, ed.), 1980. Reproduced with permission of Pergamon Press Ltd.

Figure 15.14 from page 543 in R. D. B. Fraser and T. P. MacRae, *Conformation of Fibrous Proteins*, 1973. Reproduced with permission of Academic Press, Inc.

Illustrations and Table Credits   **385**

Figure 15.15 from page 559 in H. Bouma et al., *Thrombosis Research* 13, 1978. Reproduced with permission of Pergamon Press.

Figure 15.17 from page 24 in N. M. Tooney and C. Cohen, *Nature* 237, 1972. Reproduced with permission of Macmillan Journals Ltd.

Figure 15.18 from cover page in E. Bernstein and E. Kairinen, *Science* 173, 1971. Reproduced with permission of the American Association for the Advancement of Science.

Index of Peptides and Proteins

A
Actin, 324
Adenylkinase, 60, 99
ACTH, 203
Alcohol dehydrogenase, 242, 264
Aldolase, 241, 242, 261, 262, 264
Amino acid oxidase, 264
Angiotension, 14, 15
Antihemophilic factor, 357
Apoferritin, 262
Aspartate transcarbamylase, 264

C
Carbonic anhydrase, 208, 264
Carboxypeptidase A, 100, 151, 184, 264
Casein, 138
Christmas factor, 357
Chymotrypsin, 148, 149, 198, 264
Chymotrypsinogen, 148, 149, 198, 256, 262, 264, 269
Collagen, 115, 116, 186, 291, 299
Concanavalin A, 148, 149
Creatine kinase, 264
Crotonase, 261
Cytochrome c, 262, 264

E
Elastin, 232
Enolase, 241, 242, 264

F
Ferredoxin, 101
Ferritin, 262
Fibrin, 360
Fibrinogen, 198, 291, 357, 360
Fibrin stabilizing factor, 357
Fibroin, 110, 135
Fumarase, 262, 264

G
β-Galactosidase, 262, 264
γ-Globulin, 202, 262, 264, 347
Glucagon, 262
Glutamate dehydrogenase, 264
Glyceraldehyde phosphate dehydrogenase, 262, 264
Gramicidin S, 222

H
Hageman factor, 211, 357
Hemoglobin, 97, 136, 137, 138, 264, 291, 341

I
Insulin, 148, 149, 197

K
Keratin, 132

L
Lactate dehydrogenase, 242, 264
β-Lactoglobulin, 138, 261, 264, 291
Lactoperoxidase, 262
Leucine amino peptidase, 264
Lysozyme, 57–59, 102, 103, 150, 184, 185, 198, 264

M
Malate dehydrogenase, 94, 95, 262
Malformin, 226, 227
Methemoglobin, 241, 242
Myoglobin, 184, 185, 262, 264, 345
Myohemerythrin, 97
Myosin, 264, 275, 291, 319

N
Nuclease, 228, 229

O
Ovalbumin, 198, 241, 242, 261, 262, 264
Ovomucoid, 262

P
Pancreatic trypsin inhibitor, 65–67, 230, 231
Papain, 152, 264
Paramyosin, 113, 114, 264
Pepsin, 148, 149
Pepsinogen, 148, 149
Phosphorylase, 264
Plasma thromboplastin antecedent, 357
Proaccelerin, 357
Proconvertin, 357
Prothrombin, 187, 357, 358
Pyruvate kinase, 264

R
Ribonuclease, 138, 184, 185, 197, 247, 264, 291
Rubredoxin, 101

S
Serum albumin, 138, 148, 149, 197, 198, 200, 209, 210, 241, 242, 250, 259, 261, 262, 264, 291, 356
Stuart factor, 357
Subtilisin, 264

T
Thrombin, 357, 359
Transferrin, 261, 262
Triose-P-isomerase, 98
Troponin, 329, 333
Tropomyosin, 111, 259, 264, 291, 329
Trypsin, 100, 198, 264
Trypsinogen, 198

U
Urease, 262

Z
Zein, 197

Index

A
Absorbance, definition of, 166–168
Absorption spectroscopy, 155
  and conformation, 169
  of peptide groups, 158–159
  shape of bands, 167
ACTH, conformation and fluorescence, 203–204
Actin
  chain structure of, 326
  helical arrangement of, 327
  electron imaging of, 328
  primary structure of, 325
Active site, 102, 103
Allowed transitions, 377
α-Helix, absorption spectrum of
  antiparallel, 96, 98
  circular dichroism of, 180
  conformational parameters and, 169
  content of proteins, by O.R.D., 178, 179
  description of, 33
  fiber diffraction from, 90, 91
  illustration of, 34
  infrared spectrum of, 131, 132
Amide bands, 121
Amide bond
  dipole moment and, 123
  double bond character, 20
Amide vibrations, 122
Amino acids
  conformation directing properties of, 55
  frequency in β-turns, 56
  optical configuration of, 4
  structures of, 4–8
Anisotropy of fluorescence, 202
Antigen/antibody, 347, 353, 354
Antiparallel β-sheet
  circular dichroism and, 180, 182
  infrared spectroscopy and, 129
  Raman spectroscopy and, 144
  structural properties, 34–36
Atomic interaction potentials, 25
Asymmetry (axial), 281
Axial ratio, 281

B
Beer-Lambert Law, 165
Bence-Jones protein, 352
Bessel function, 86
β-Helices (π_{LD} helices), 46
β-Sheet
  antiparallel, 34–35
  C.D. spectra of, 180, 182
  cross-β, 38
  infrared spectra and, 129
  parallel, 36
  pleated, 35, 37
  prediction of, 50
  vibrations in, 127–129
Blood clotting pathway, 357
Blood clotting proteins, 356, 357
  primary structure of, 358
  tertiary structure of, 359
Bond rotation, 17, 18
Bravais lattices, 73
Bragg equation, 71

C
^{13}C NMR spectra, 221
Cabannes correction factor, 245
Capillary viscometer, 248
Character tables, 370, 371
Chemical potential of protein solutions, 239
Chemical shift, 219, 220
Chondroitin sulfate, 315

Circular dichroism spectroscopy, 172–179
Circularly polarized light, 156
Coiled coils, diffraction from, 90–92, 93
Coil structure, definition of, 46
Collagen, 299
   biopolymer models, 301
   circular dichroism of, 307
   conformation of, 307
   electron microscopy of, 115, 116
   fibrillar structure of, 309–311
   forces directing structure, 311, 312
   infrared spectra of, 308
   primary structure, 302, 303
   x-ray diffraction of, 305, 306
Concentration of proteins, by fluorescence, 195, 196
Conformational analysis
   Chou-Fasman method, 54, 56–60
   circular dichroism and, 180, 183–185
   energy calculations and, 23
   equivalence condition in, 22
   Raman spectroscopy and, 145–147
Conformation, definition of, 5
Conformation, energy maps, 29
Conformation map
   alanine, 29
   glycine, 29
Copolypeptides, 9
Crystal imperfections, 80
   systems, 73–75
Cyanogen bromide cleavage, 11

D
Del Re equations, 26, 27
Denaturation of proteins, 238
   of collagen, 186
Density gradient ultracentrifugation, 254
Depolarization
   of fluorescence, 199
   of Raman lines, 188
Dichroism. *See* C.D. spectroscopy
   circular, 172–179
   infrared, 131
   ultraviolet (linear), 171
Dielectric relaxation, 288
Differential refractometer, 244
Diffraction from helices, 84–88
Diffusion coefficients (and molecular weight), 252
Dipole moment, 123
Dispersion forces, 25, 26
Dispersion phenomena, 155
Disulfide bonds, 10
Domain structure, 96–101
Donnan effect, 240
Drude equation, 177
Drug interactions, 205–211
Dynamic light scattering, 254–256

E
Edman reagent, 10
Einstein-Simha equation, 288
Electric birefringence, 288
Electron diffraction, 105
   from silk fibroin, 110
Electron microscopy, of fibrous proteins, 113
Electronic absorbance, lifetime of states, 191
Electronic emission, 189
Electronic spectroscopy, 154
   symmetry and, 376
Electronic transitions, 157–160
   allowed, 377
Electrophoretic mobility, and molecular weight, 263
Electrostatic interactions, 26
Elliptically polarized light, 157
Ellipticity, 179
Emission anisotropy, 202
Energetics of chain folding, 61
Energy levels
   electronic, 160, 163
   vibrational, 164
Energy transfer (fluorescence), 202, 203
Enzymes. *See* specific proteins
Enzyme degradation, 12
Enzyme/substrate binding, by resonance Raman spectroscopy, 152
Equivalence condition, 22
Extinction coefficient, 165, 166

F
Fiber diffraction, 87
Fibrin, electron microscopy of, 363
Fibrinogen
   conformational analysis of, 360
   primary structure, 360
   structure of, 361, 362
Fibroin, silk, 52
Fibrous proteins
   infrared spectroscopy of, 134
   microscopy of, 113
Floating rotor viscometer, 249
Flow birefringence, 287, 291
Flow dichroism, 286, 287
Fluorescence, 189
   assay of proteins, 195, 196
   and complexes, 205
   and conformation, 196
   depolarization, 198, 199
   quenching by hydrogen bonds, 197
   quenching by peptide bonds, 196
   titration, 208
Fluorescent lifetime, 201
Folding of polypeptide chains, 61
Forbidden transitions, 164, 377
Fourier transform analysis, 365
   IR spectroscopy, 136
Free energy. *See* thermodynamics
Free induction decay, 221

G
Gel permeation chromatography, 259
Gelatin, 307
Gels, polyacrylamide, 265
Globular proteins
   circular dichroism spectra of, 183
   conformational analysis of, 50
   domain structure of, 96

Index 391

    infrared spectra of, 136–139
    Raman spectra of, 145
    x-ray diffraction and, 94–96
Guanidine hydrochloride, denaturation by, 238
Guinier plots, 273, 280

H

Halobacterium halobium, electron imaging of, 116, 117
Helical vibrations, 125
Helices
    α-helices, 33, 34
    β-helices, 46
    definition of, 30
    and line symmetry, 375
    ω-helices, 40, 41
    polyglycine II (pGII), 38, 39
    polyproline II, 38–41
    and rotation parameters, 32
Helix (α) content
    from C.D. spectroscopy, 184
    from O.R.D. spectroscopy, 179
    from Raman spectroscopy, 146
    circular dichroism of, 180, 182
    infrared spectrum of, 126
    structure of, 33, 34
Hemoglobin
    abnormal, 344, 345
    primary structure, 340, 341
    secondary structure, 342
    tertiary structure, 343, 344
Hierarchy
    of collagen, 300
    of muscle, 317
Homology, 15
Homoserine, 13
Hormones, 13
Hydrogen bonds, 28
    length from I.R. spectroscopy, 133
Hydrodynamic volume, 282

I

Immunoglobulins, 374
    conformational analysis, 351
    computer simulation, 354
    primary structure, 349, 350
    schematic diagram, 349
    subclasses, 348
    tertiary structure of, 352, 353
Indexing of diffraction patterns, 77
Inductive coefficient, 26
Infrared spectroscopy (I.R.), 120
    dichroism, 123, 131
    vibrational analysis and, 124, 125
Interatomic potential, 25
Intersystem crossing, in fluorescence, 190
Intrinsic fluorescence, 193
Intrinsic viscosity, 247, 248
Irregular structure, definition of, 45

J

Jablonski diagram, 190

K

Keratin, infrared dichroism, 132
Kinetics of chain folding, 62
Kronig-Kramers transform, 179

L

Laser light scattering. See dynamic light scattering or light scattering
Laser Raman spectroscopy. See Raman spectroscopy
Lattice planes, 72–74
Lennard-Jones potential, 25
Levitt, approach to structure prediction, 64
Lifetime of fluorescence, 201
Light scattering
    determination of molecular weight, 243
    methodology, 244
    virial coefficients and, 246
Linear dichroism, ultraviolet, 170, 171
London dispersion forces, 25
Lysozyme. See Index of Peptides and Proteins
    active site of, 103
    binding to substrate, 102
    prediction of conformation, 57

M

Magnetic dipole, 176
Mandelkern-Flory equation, 258
Membrane osmometry, 240
Meridional reflections, 88
Miller indices, 72–74
Miyazawa equation, 124
Moffitt-Yang equation, 178
Molar ellipticity, 179. See ellipticity
Molar rotation, 173
Molecular dynamics, 62–68
Molecular orbitals, aromatic, 161
Molecular weight
    number average, 239
    wave functions for, 160
    weight average, 245
Mulliken equation, 27
Multiple resonance, 223
    hierarchy, 317
Muscle structure, 316
    assembly of, 321, 322
Myosin
    conformation, 320
    radius of gyration, 275
    reciprocal scattering, 275
    structure, 317, 319

N

Near infrared, amide bands, 122
Negative staining, in electron microscopy, 113
Newtonian viscosity, 248
Nuclear Overhauser effect, 224
Nuclear magnetic resonance (N.M.R.), 213
    of β-turns, 232
    of nuclease, 238, 239
    of pancreatic trypsin inhibitor, 230, 231
    relaxation in pentapeptides, 219
    spectrometer, 216

Nucleation, of protein structure, 62
Number average molecular weight, 239

O

Oblate ellipsoid, 285
Optical activity, 175
Optical rotation, 177
Optical rotatory dispersion (O.R.D.), 172
Orbital representations, 161, 162
Oscillator strength, 168
Osmometer, automated, 240
Osmotic pressure, 238
    virial coefficients and, 239

P

Parallel β-sheet, domains, 98, 99
Paramagnetic complexes, 224
Paramyosin, electron microscopy, 113, 114
Partial molal volume, 239
Partial specific volume, 257
Partition coefficients, 261
Patterson function, 95
Pentapeptides, nmr relaxation and, 219
Peptide bond, 18
    bond angles in, 22
    electron delocalization in, 19
Peptide group
    absorption spectrum of, 160
    infrared spectrum of, 122
    planar structure, 22
    vibrations in, 122
Perrin equation, 200
Perrin factor, 285
Phenylalanine, fluorescence of, 194
Phenylisothiocyanate, 10
Phosphorescence, 191
Phosphotungstic acid staining, 111
π Orbitals, 159, 161, 162
π-π* Transitions, 160
Pitch of helix, 85
Pleated sheet, 35, 37
Polar interactions, 26
Polarizability, 142, 143
Polarization of fluorescence, 199, 200
Polarized light, 154–156
Poly-O-acetyl-L-hydroxyproline, crystals of, 106
Polyacrylamide gels, 265
Poly-L-alanine
    crystals of, 106
    infrared vibrational modes, 130
    x-ray diffraction from, 90, 91
Poly(amino acids), summary of conformation, 51
Poly-γ-benzyl-L-glutamate
    electron diffraction from, 107
    fibers of, 106
    helix/random transition, 279
    light scattering from, 278
    peptide repeat in solution, 277
Poly(L-glutamic acid), 47
    Raman spectrum of, 144
Polyglycine I
    electron diffraction of, 108
    morphology, 108

Polyglycine II helix (pGII), 38
Poly-L-lysine
    circular dichroism spectrum of, 180–183
    Raman spectra of, 144, 146
Polypeptides, amide bands, 126
Poly(L,D-peptides), 47
Polyproline II helix (PpII), 38
    circular dichroism spectrum of, 185, 186
    light scattering, 271
Porcupine quill, infrared dichroic spectrum, 135
Potential energy calculations, 25–30
Protein domains, 96
Protein structure
    prediction of structure, 50
    secondary structure, 149
    solid and solution comparison, 150
    tertiary structure, 96. *See also* specific protein and specific technique
Protein/drug complexes, 205–211
Proteins, molecular dimensions, 291
Proteoglycans, 314, 315
Proton magnetic resonance (P.M.R.). *See* nuclear magnetic resonance (N.M.R.)

Q

Quantum yield, 192
Quenching, fluorescence, 196

R

Radius of gyration, 272, 274
Ramachandran diagrams, 25
Raman scattering, 140
    and conformation, 144
    polylysine spectra, 144, 146
Random structure, 45
Rayleigh ratio, 246
Rayleigh scattering, 141
Relaxation (N.M.R.), 218
Resonance Raman spectroscopy, 151
Rotational diffusion coefficient, 200
Rotational strength, 176

S

Scatchard plot, and fluorescence, 206, 207
Secondary structure, definition, 17
Sedimentation coefficient, 250, 251
Sedimentation equilibrium, 265, 266
Selection rules
    for electronic transitions, 164, 377
    for vibrational transitions, 374, 375
Sequential analysis, 9
Silk fibroin, electron diffraction from, 110
    infrared dichroic spectrum, 135
Simha factor, 283
Singlet transition, 190
Small angle x-ray scattering, 268
Soulé-Porod plot, 280
Spin angular momentum, 214
Spin decoupling of lysine, 223
Spin-lattice relaxation, 217
Spin quantum number, 213, 214
Spin-spin coupling, and conformation, 225, 226

Spin-spin relaxation, 216, 217
Spin-spin splitting, 220
Staining procedures in electron microscopy, 111, 113
Stokes and anti-Stokes lines, 141
Stokes radius, 284
Superhelices, 92, 93
Symmetry
  application to vibrational spectroscopy, 373–375
  in crystallography, 377, 378
  and electronic transitions, 163–165
  elements of, 369
  operations and matrices, 371, 372
  and structure, 369

T

Tertiary structure, prediction of, 61
Thermodynamics, solution, 237–241
Thick filaments of muscle, 317, 318
Thin filaments of muscle, 324
Topographic mapping, 171
Transition dipole
  electronic, 168
  vibrational, 123
Triple helix (of collagen), 307, 309
Tropomyosin
  circular dichroism of, 331
  electron microscopy, 111
  interchain forces, 332
  predicted conformation, 53
  primary structure, 330
Tryptophan, fluorescence of, 193
Tyrosine, fluorescence of, 194

U

Ubbelohde viscometer, 249

Ultracentrifuge, 253, 254
Ultraviolet absorbance, and conformation, 168–170
Unit cell, 72
Uranyl acetate staining, 111

V

Vacuum ultraviolet absorbance, 159
Van der Waals radii, 23
Vector decomposition, of infrared vibrations, 127
Vibrational analysis, 124–130
Vibronic transition, 164
Virial coefficient, 239, 245, 246
Viscometer (Ubbelohde), 249
Viscosity, 247

W

Wave function, 158, 160
Weight, average molecular weight, 245

X

X-ray diffraction, 71
  of coiled coils, 90–93
  intensity of, 82
  from oriented samples (fiber diffraction), 81, 82
  of polyalanine, 90
  by powders, 75–78
  from silk fibroin, 110

Z

Zimm plot, 273, 274